Aircraft Landing Gear Systems

PT-37

Edited by
John A. Tanner

(Selected papers through 1989)
Prepared under the auspices of the SAE A-5 Aerospace Landing Gear Systems Committee

Published by:
Society of Automotive Engineers, Inc.
400 Commonwealth Drive
Warrendale, PA 15096

ISBN 1-56091-074-7
SAE/PT-90/37

Library of Congress Catalog Card Number: 90-83427

PREFACE

Landing gear systems continue to be one of the most maintenance intensive systems for aircraft operators. Environmental and performance demands impose tremendous pressures on today's landing gear designs and the future looks even more challenging. Modern landing gear systems are required to operate for years in environments which can be very corrosive and to provide adequate braking and steering forces for safe aircraft ground handling operations under all-weather conditions. These landing gear systems may also be designed to accommodate special mission requirements such as soft field operations or takeoff and landing operations from the deck of an aircraft carrier. The Space Shuttle orbiter routinely lands at speeds greater than 200 knots and exposes its main-gear tires to loads in excess of 100,000 lbs. Designers of the National Aerospace Plane are talking about ground operations at speeds approaching 400 knots.

The challenges associated with aircraft landing gear systems has stimulated considerable research activity for many years. The SAE Committee A-5 for Aerospace Landing Gear Systems has been and continues to be an advocate of this much needed research. The thirty (30) papers selected for this book have been presented at previous SAE Aerotech meetings dating from 1982. These papers cover a variety of research activities in the industry, academic, and government communities and are divided into the following categories:

- Test Facilities and Runway Effects
- Modeling and Analyses
- Braking and Steering Systems
- Struts, Couplings, and Actuators
- Ground Operational Problems
- Tire Friction, Wear, and Mechanical Properties

For each topic at least four papers have been selected to provide some idea of the recent landing gear systems research activities. A bibliography of research papers is also included which represents important contributions to the landing gear literature.

John A. Tanner
SAE Committee A-5 Aerospace Landing Gear Systems

TABLE OF CONTENTS

Test Facilities and Runway Effects

Modeling and Analyses

Braking and Steering Systems

Struts, Couplings and Actuators

Ground Operational Problems

Tire Friction, Wear, and Mechanical Properties

Test Facilities and Runway Effects

Laboratory Simulation of Landing Gear Pitch-Plane Dynamics

John J. Enright
Aerospace and Defense Division
BFGoodrich Co.

** Paper 851937 presented at the Aerospace Technology Conference and Exposition, Long Beach, California, October, 1985.

ABSTRACT

A technique for laboratory dynamometer simulation of landing gear-brake dynamics is discussed. The method was developed as a means of improving certain limitations of conventional dynamometer testing with and without actual strut hardware. The test fixture, the basis for its similitude, assumptions, and design criteria are described. Background includes descriptions of the form and significance of brake-induced vibration and the concept of a critical torque-speed slope as it relates to system stability criteria. Use of this fixture enables the duplication of brake squeal modes otherwise masked by standard dynamometer fixturing. The large linear range of amplitude and damping permit operation in a deliberately unstable condition to verify stability margins. Examples of the application and verification of the technique are included.

THE AIRCRAFT LANDING GEAR is by nature a complex multi-degree-of-freedom dynamic system. As such, it may encounter various vibratory modes which can be induced by brake frictional characteristics and design features. Aircraft design specifications and industry practices require that these modes be assessed during the design concept stage and verified during hardware development.

Dynamometer simulation of the landing gear is a desirable way to accomplish the necessary brake/landing gear dynamic assessment since brake testing is routinely accomplished by landing a complete wheel, brake, and tire unit against an inertial wheel. The requirements for a sufficient gear simulation should include the major system degrees-of-freedom that affect the transient brake/landing-gear load dynamics.

Historically, landing gear simulation has been achieved by mounting a two-wheel gear (or two-wheel simulator with equivalent dynamic response of a four-wheel gear) in an overhead position on a roadwheel dynamometer. This paper discusses a simplified technique for laboratory dynamometer simulation of landing gear-brake dynamics providing a means of improving certain limitations of conventional dynamometer testing with and without actual strut hardware.

FRICTION-INDUCED VIBRATION

Friction-induced vibration has been described and analyzed in a number of published studies of various degrees of sophistication. (1, 2, 3)* The potential for dynamic instability exists in dynamic friction systems such as brakes and landing gears because the rate of change in friction with respect to rubbing speed is typically negative and has a negative damping effect on the motion of such systems. This negative damping effect always decreases the total system damping, making it more responsive to forced vibration. If the friction-speed sensitivity is too great, the system will be unstable, with the result that any disturbance will cause exponentially increasing oscillation of the gear, brake, or both.

Brake friction acts in the pitch-plane of the landing gear system, and so affects the stability of three pitch-plane modes of vibration, illustrated in Figure 1.

Brake Squeal is defined as "A self-induced brake vibration mode with a frequency greater than 100 Hz." (4). The fundamental squeal mode is the torsional motion of the non-rotating brake parts about the axle and against the elastic restraint of the brake-gear attachment hardware.

* Numbers in parentheses designate references at end of paper.

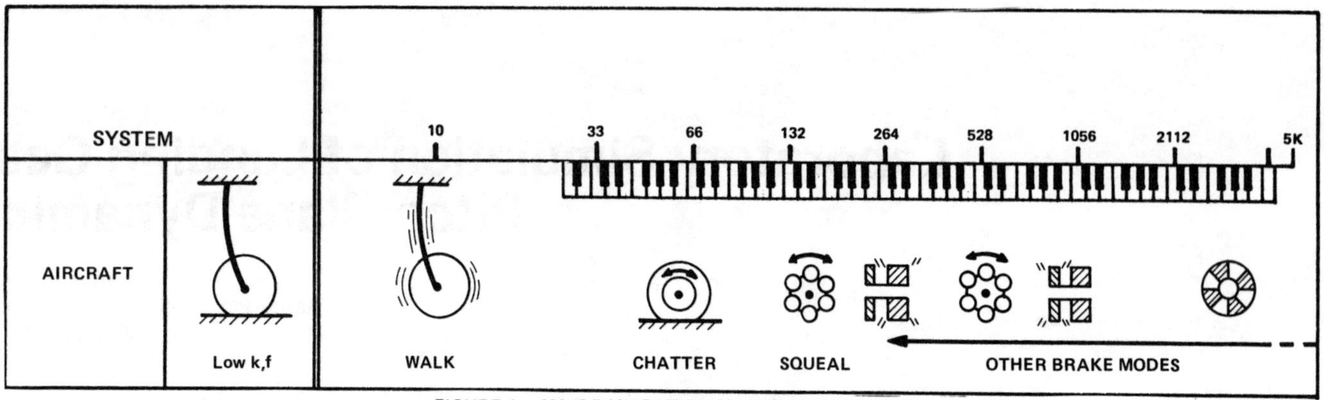

FIGURE 1. MAJOR VIBRATION MODES AND FREQUENCIES

Brake Chatter is defined as "A self-induced brake vibration of less than 100 Hz, excited by the friction characteristics of the rubbing surfaces." (4). The fundamental chatter mode is the torsional motion of the rotating parts of the brake-wheel-tire assembly about the axle and against the elastic restraint of the tire. It is typically above 50 Hz and coupled with the squeal mode.

Gear Walk is defined as "Cyclic fore and aft motion of the landing gear strut assembly about a normally static vertical strut centerline. Caused by drag loads applied at the tire/ground interface and the natural spring rate of the gear structure. Sometimes aggravated by anti-skid braking action (cycling) and resonant frequency of vibration of the strut." (4). The fundamental walk frequency is typically below 20 Hz and is relatively uncoupled from squeal and chatter.

Gear walk may be brake induced and stabilized (5) or destabilized by the brake control system. Unstable gear walk is illustrated by the time histories of gear deflection, brake torque, and dynamometer speed shown in Figure 2. As speed decreases under the action of increasing brake torque, gear oscillations diverge in an exponential manner. At Point A in the motion, the growing wheel speed oscillation accompanying the gear walk becomes equal to the decreasing rolling speed and the divergence is limited as the brake torque instantaneously reverses. Between A and B, the gear oscillation limit cycle amplitude is proportional to decreasing speed and torque reversal occurs once per oscillation. The gear is seen to act effectively as a single-degree-of-freedom system.

PITCH-PLANE GEAR SIMULATION

A valid gear simulation is one having the same dynamic response to brake torque as the actual gear. Basically, this means that the simulated gear must be designed to have the same equation of motion in its walk mode under the action of speed-dependent braking friction.

The squeal and walk frequencies previously described are typically well separated and are not strongly coupled natural modes except in the sense that both are excited by brake torque, which may or may not be affected by either vibration. The squeal mode is coupled to other brake component and assembly modes as well as wheel, tire, axle, and torque equalizer rod modes, all of which are duplicated exactly by using actual hardware suitably isolated from fixture effects.

The walk mode is coupled to other strut and aircraft modes, but may be regarded as an equivalent one-degree-of-freedom system, in the form of a cantilever beam with root fixity and inertia. The characteristics of this equivalent system are, of course, affected by the strut extension, aircraft weight, and weight distribution. Consequently, the range of these variables defines a family of fundamental gear walk modes.

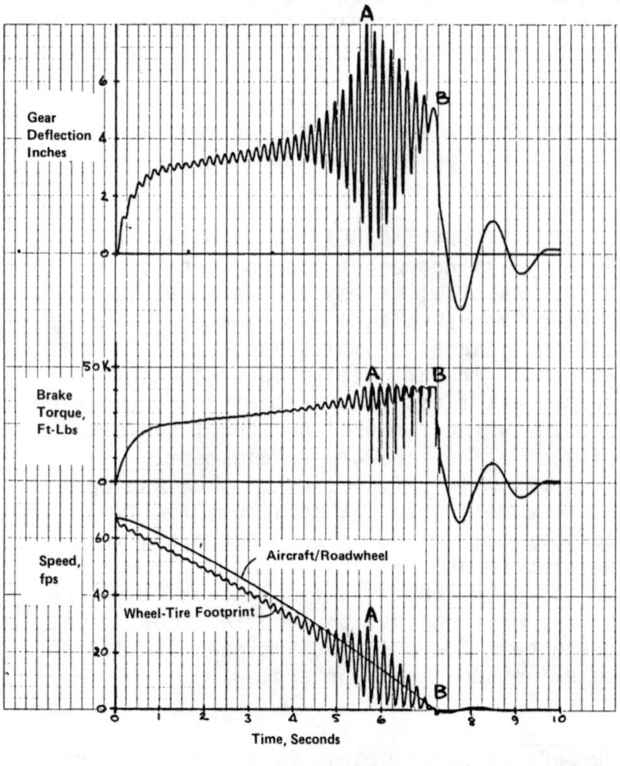

FIGURE 2. TIME HISTORY OF GEAR WALK INSTABILITY

The one member of this family that is most easily destabilized by brake torque is the gear walk mode of interest. Any structure which duplicates the characteristics of this mode is a valid gear simulation.

The simulation can be achieved, of course, by the use of the actual landing gear structure with the appropriate attachment compliance and compensation for roadwheel curvature. But an alternate simulation structure is necessary if the system compatibility is to be evaluated in the design stage of the aircraft before actual landing gear structure is available.

An alternate structure may be preferred in some cases because there are limitations and complexities in using the actual gear, some of which are discussed in SAE AIR 1064A on "Brake Dynamics." (6) One such alternate which is the subject of this paper, is a modification of the dynamometer fixture such that one of its fundamental modes duplicates the dynamic characteristics of the gear walk mode of interest.

This fixture modification may be called a "Pitch-Plane Simulator" because it specifically simulates the three fundamental degrees-of-freedom of the landing gear brake system in the aircraft's pitch-plane which are pertinent to the brake design objectives of ensuring that the brake has no destabilizing effect on the gear.

All other gear modes, such as shimmy, bounce, bogie pitch, or roll-plane gear modes, are present on the actual system, but unrelated to this objective and therefore can be regarded as unnecessary complexities unrelated to this aspect of brake dynamics.

With the pitch-plane simulator, dynamic equivalence to the actual system is achieved by duplicating the following pertinent system characteristics as compared to the actual landing gear system:

- System of equations of pitch-plane motion.
- Fundamental natural frequencies.
- Angular deflection of the stator and rotor parts of the brake per unit torque.
- Relative angular velocity between brake rotors and stators.
- Instability per unit of negative brake torque-speed slope, i.e., the same amount of effective negative system damping due to the brake.
- Dynamic response, i.e., the same kinetic energy of oscillation per unit vibration amplitude.
- Stop conditions of energy, speed, pressure, and tire load.
- Wheel, brake, tire, and hydraulic system hardware and controls.
- Axle and equalizer rod stiffness and interface geometry.

All dynamometer testing is, or course, simulation to a certain extent. Good simulation lies in compromising the superficial aspects and eliminating unnecessary complexities while duplicating the pertinent characteristics of the system and its environment.

The following compromises are inherent to the pitch-plane technique:

One Wheel-Tire-Brake Assembly - One wheel is sufficient and can be run on the dynamometer without excessive inertia dictated by a roadwheel width sufficient to land two wheels on a full axle with a full landing gear.

Oscillating Inertia Instead of Landing Gear Strut - The fore-aft motion of the gear may be simulated by the angular motion of an inertia pivoted about a non-translating axle and elastically restrained by a torsion spring. Torque is transmitted through the axle flange or a torque-equalizer rod into this inertia. The tire does not move back and forth over the roadwheel, but rotates only. Thus, nonlinear, tire off-loading effects and ground load direction variations which do not exist on the aircraft are avoided. The equivalence and validity of this mode inversion is analogous to representing a moving aircraft on a stationary runway by a moving "roadwheel" and a stationary axle in typical dynamometer tests.

Absence of Strut Angle - Since there is no strut in the single-degree-of-freedom system, there will be no strut angle. The effect of this is that a small component of the gear static deflection (due to the fact that the ground load resultant is aft of the strut centerline) is absent in the simulator. This has no bearing on dynamic compatibility.

Omission of Extraneous Gear Modes - The dynamic interactions of importance in brake design are those involving the gear walk mode. Few non-walk gear modes are affected by brake dynamics. The shimmy mode of the gear can be excited by differences in brake torque on multi-brake gears, but is normally not a major concern to aircraft manufacturers.

EQUATIONS OF PITCH-PLANE MOTION OF GEAR AND SINGLE-DEGREE-OF-FREEDOM SIMULATOR

Consider the brake friction to be a function of relative rubbing speed between rotors and stators. (7) Then, at any general operating point speed during the deceleration of the aircraft or dynamometer roadwheel, the brake torque during perturbations about the operating point can be described as a linearized function:

$$T = \bar{T} + (\partial T/\partial \Omega)(\Omega - \bar{\Omega})$$

$$= \bar{T} - S(\Omega - \bar{\Omega})$$

where:

Ω = Relative Rotor-Stator Angular Speed

$\bar{\Omega}$ = General Operating Point Value of Ω

$\bar{T}$ = Brake Torque at Speed $\bar{\Omega}$

S = Negative Torque-Speed slope at Speed $\bar{\Omega}$

The average torque per brake of n brakes at the same speed is:

$$T_n = \bar{T}_n - S_n(\Omega - \bar{\Omega})$$

Assume tires are rolling without slip. Then the relative angular speed between rotors and stators is:

$$\Omega = V/r - (\dot{X}/\lambda + \dot{\theta}_B + \dot{\theta}_C)$$

where,

V = ground speed of aircraft or roadwheel

r = tire rolling radius

X = fore-aft motion of gear at the axle relative to aircraft

λ = rL/(r + L)

L = gear effective rigid-strut length

θ_B = angular motion of brake relative to strut

θ_C = angular motion of wheel relative to tire footprint

and the torque per brake can be expressed as:

$$T_n = T_n + S_n(\dot{X}/\lambda + \dot{\theta}_B + \dot{\theta}_C)$$

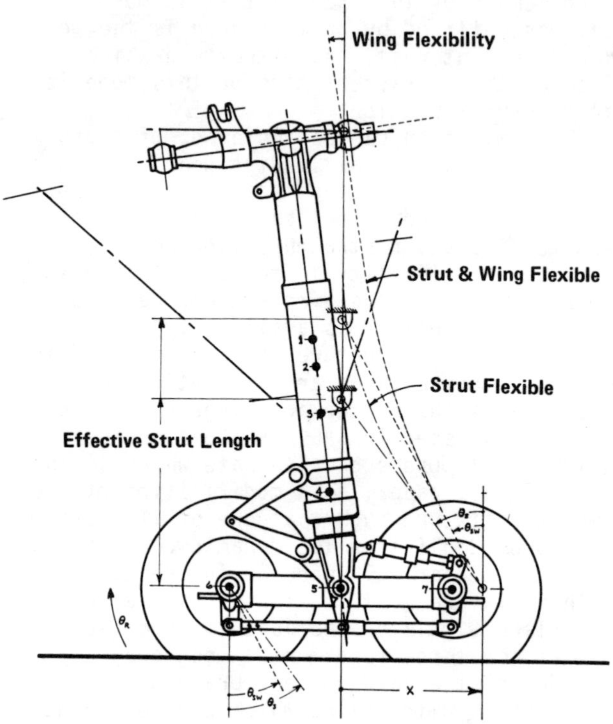

FIGURE 3. MODE SHAPE OF TYPICAL BOGIE GEAR

The concept of the effective rigid strut length (which is the ratio between the linear and angular deflection of the strut at the axle) is illustrated in the mode shape of a bogie gear shown in Figure 3. The contribution of wing twist to effective fore-aft stiffness is also illustrated.

Consider the gear to be a three-degree-of-freedom linear system in its pitch-plane modes of vibration. Then the gear's unforced squeal, chatter, and walk motions during landed braking may be described by three simultaneous differential equations, which, when second-order effects are eliminated, reduce to:

BRAKE SQUEAL $\quad I_B\ddot{\theta}_B + (C_B - S)\dot{\theta}_B + K_B\theta_B = \bar{T} + S(\dot{\theta}_C + \dot{\theta}_G)$

WHEEL CHATTER $\quad I_W\ddot{\theta}_C + (C_W - S)\dot{\theta}_C + K_W\theta_C = \bar{T} + S(\dot{\theta}_G + \dot{\theta}_B)$

GEAR WALK $\quad \Lambda M\ddot{\theta}_G + (\Lambda C - S_n)\dot{\theta}_G + \Lambda K\theta_G = \bar{T}_n + S_n(\dot{\theta}_B + \dot{\theta}_C)$

where:

θ_G = X/λ

Λ = $\ell\lambda/n$

ℓ = $RL/(R+L)$

and R = tire deflected radius

I_B = inertia of nonrotating parts of brake

I_W = inertia of rotors, wheel, and part of tire

M = effective mass of landed gear at axle

K_B = torsional stiffness of brake relative to strut

K_W = torsional stiffness of wheel relative to landed tire

K = effective fore-aft dynamic stiffness of landed gear at axle

The viscous damping coefficients of the unbraked system can be expressed in terms of their stiffnesses and uncoupled natural frequencies:

$$C_B = \zeta_B K_B / \pi f_B$$

$$C_W = \zeta_W K_W / \pi f_C$$

$$C = \zeta_G K / \pi f_G$$

where: $\zeta_B, \zeta_W, \zeta_G$ = equivalent viscous damping ratios of the unbraked system's uncoupled modes

f_B, f_C, f_G = natural frequencies

A single brake mounted on a dynamometer axle attached to a large inertia mandrel which is elastically restrained in torsion is a dynamic system similar to a landing gear. This concept is illustrated in Figure 4. "Dynamometer walk" may be defined to be the angular motion of the mandrel and attached inertia. Then the equations of squeal, chatter, and walk motions during landed braking are:

BRAKE SQUEAL $\quad I_B\ddot{\theta}_B + (C_B - S)\dot{\theta}_B + K_B\theta_B = \overline{T} + S(\dot{\theta}_C + \dot{\theta}_D)$

WHEEL CHATTER $\quad I_W\ddot{\theta}_C + (C_W - S)\dot{\theta}_C + K_W\theta_C = \overline{T} + S(\dot{\theta}_D + \dot{\theta}_B)$

DYNO WALK $\quad I_M\ddot{\theta}_D + (C_M - S)\dot{\theta}_D + K_M\theta_D = \overline{T} + S(\dot{\theta}_B + \dot{\theta}_C)$

where: I_M = inertia of dynamometer mandrel

K_M = torsional stiffness of mandrel's elastic restraint

The equations of motion of this system are the same as those for a landing gear system, subject only to the assumption that:

$$\overline{T}_n = \overline{T}$$

$$S_n = S$$

This system, therefore, is the dynamic equivalent of a landing gear system in which the gear walk motion is proportional to the dynamometer walk motion:

$$X = \lambda\theta_G = \lambda\theta_D$$

From the coefficients of their respective walk equations, the terms of equivalence of the gear and gear simulator are:

$$\Lambda K = K_M$$

$$\Lambda M = I_M$$

$$\Lambda C = C_M$$

The assumption of rolling without slip does not affect the equivalence of the two systems described since the tire footprint experiences slip on the aircraft or dynamometer in the same manner. Slip does, however, affect the stability to a small degree. Critical torque-speed slopes calculated on the basis of a stability analysis of the equations of motion have been shown to be conservative. (7)

LABORATORY SIMULATION HARDWARE

The preceding equations have shown that the angular windup motion of a single-degree-of-freedom simulator fixture, in response to one brake, is directly proportional to the fore-aft motion of a T-gear in response to two brakes, or a bogie-gear in response to four brakes. The dynamic equivalence of the gear and this simulator is based on the fact that the squeal, chatter, and walk equations of braked motion are the same.

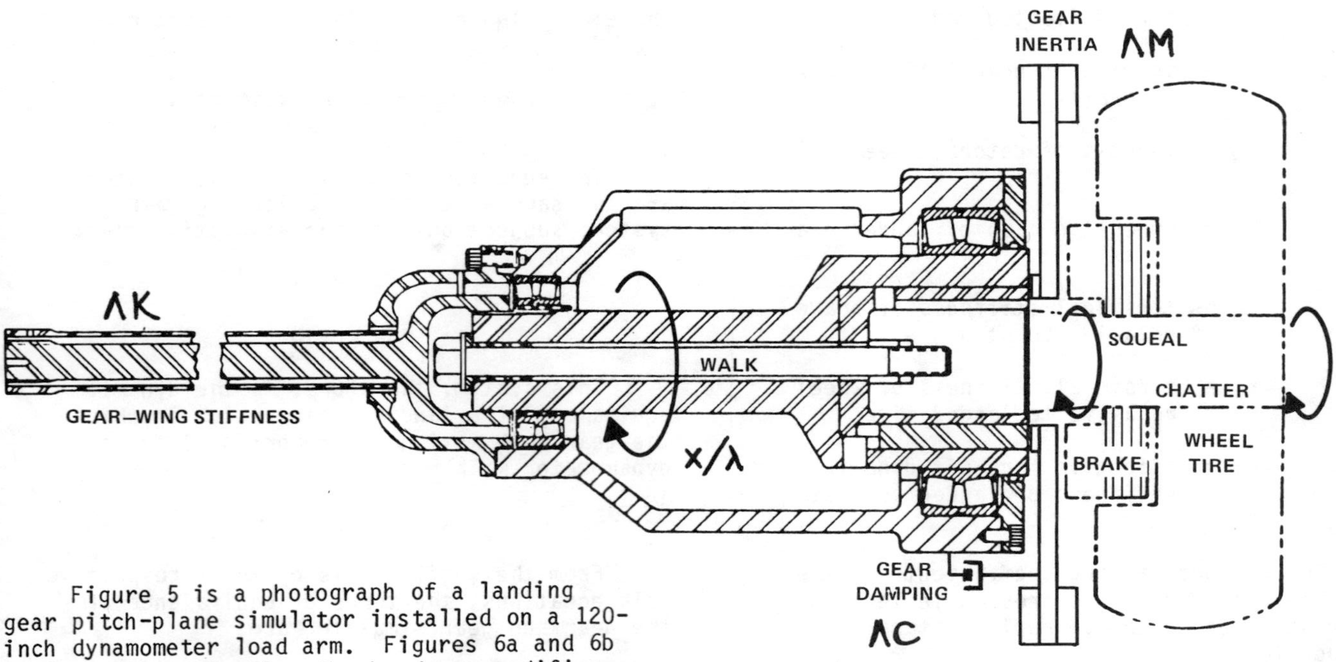

FIGURE 4. SINGLE-DEGREE-OF-FREEDOM GEAR MODE SIMULATOR

Figure 5 is a photograph of a landing gear pitch-plane simulator installed on a 120-inch dynamometer load arm. Figures 6a and 6b depict, schematically, the hardware modifications necessary to convert the standard dynamometer arm into a single-degree-of-freedom pitch-plane gear simulator. Figure 6a is a top-view cross section of the loading arm and mandrel of the right arm of BFGoodrich's 120-inch dynamometer before modification. Part Number 2 is a large mandrel which accepts a test axle and brake assembly in the keyed socket on the right and transfers the torque directly to the fixed housing (Part 3) through a splined flange (Part 1) bolted to the housing. The mandrel barrel section is strain gaged to serve as a torque cell. It has a torsional stiffness of 97.2×10^6 in lb/rad and a natural frequency of 140 to 170 Hz. A brake, axle, wheel, and tire are shown outlined in phantom. The fixture is used for both link and flange-type torque-takeout brakes.

Figure 6b depicts the simulator. It is the same as Figure 6a, except for two modifications that enable it to represent a landing gear in the manner previously described. The rigid splined flange torque anchor is replaced by a torsion spring (Part 1 of Figure 6b), and a spherical roller bearing (Part 4) is added to support the left end of the mandrel. The torsion spring consists of a shaft inside of and in series with a tube. The spring shown in Figure 5 has stiffness of 0.688×10^6 in lb/rad. The mandrel, oscillating in the roller bearings against this spring has a natural frequency of 13.3 Hz or lower with an axle and brake installed. The natural frequency can be lowered by bolting inertia plates to the exposed right face of the mandrel.

FIGURE 5. PITCH-PLANE SIMULATOR INSTALLATION

The damping associated with a gear mode of interest is provided by a linear, viscous damper consisting of a double-rod-end hydraulic cylinder with recirculating silicone fluid of the appropriate viscosity. For the simulation of bogie gears, coulomb damping of the bogie pivot friction is provided by a small caliper brake acting on the inertia plate bolted to the mandrel.

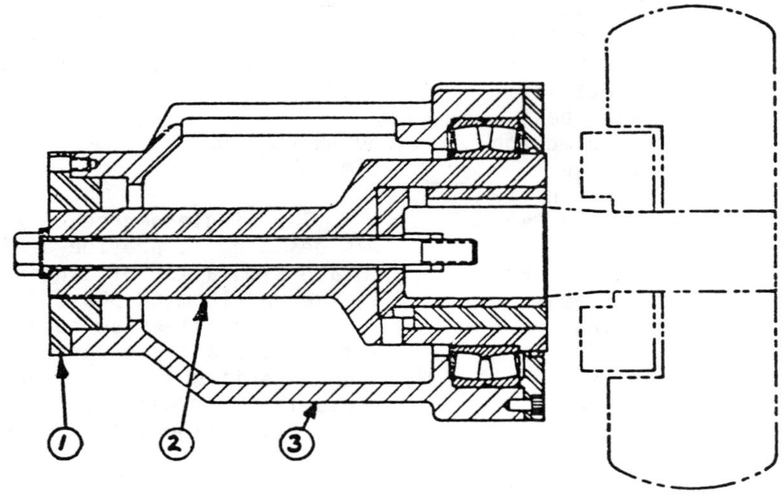

FIGURE 6a. 120-INCH DYNAMOMETER ARM BEFORE MODIFICATION

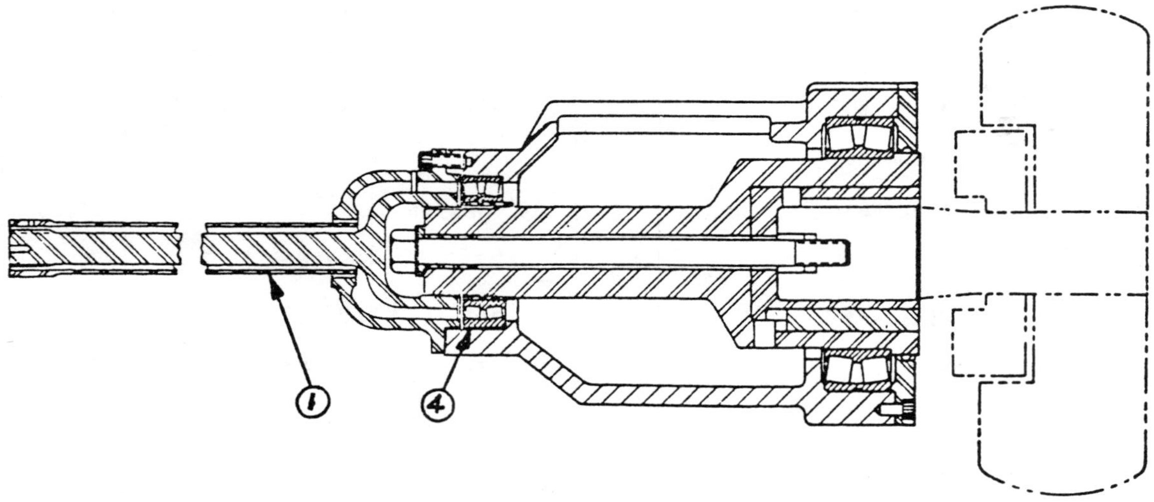

FIGURE 6b. TEST FIXTURE EQUIVALENT OF GEAR

The procedure involved in providing any damping ratio from 1.6% to 15% is straightforward, but subleties are involved in determining the appropriate value of damping for the simulation. Natural frequency alone is insufficient to define the damping that will deliver a damping ratio equivalent to the actual system. Mode shape and stiffness information are also necessary. If exact information is unavailable or uncertain, best estimates may be used and tests conducted for both high and low values of damping. The stability of the gear mode of interest in response to torque is directly affected by the gear mode damping.

A range of landing gear simulations have been built and tested including three gear stiffnesses, selectable viscous damping ratios, and adjustable coulomb damping. Different configurations of the single-degree-of-freedom gear simulator have been used for development and qualification tests on the Boeing 737, 757, and 767 aircraft. Table 1 summarizes the characteristics of the three configurations for these aircraft. The listed characteristics were derived from basic gear information supplied by the customer, and the symbols refer to system parameters described in the differential equations previously discussed. A fourth torsion spring fixture has been designed to simulate 17 Hz and 20 Hz gear modes of the Space Shuttle in tests scheduled to improve stability of its gear-brake-control system.

ADVANTAGES OF THE SINGLE-DEGREE-OF-FREEDOM SIMULATOR

The single-degree-of-freedom gear simulator was designed to provide a simple system for accurate simulation of landing gear-brake dynamics in the development phase.

Its relative low cost, linear characteristics over a large range, and ease of installation make it both an attractive enhancement to conventional dynamometer testing, and an attractive alternative to full landing gear simulation.

TABLE 1

DYNAMIC EQUIVALENCE OF TEST CONFIGURATIONS

SYSTEM PARAMETER		UNITS	AIRCRAFT SIMULATED		
			737	767	757
Equivalent Rigid Strut Length	L	in	47.8	69.7	82.9
Effective Gear Weight	gM	lbs	1,029	2,099	3,438
Landed Walk Frequency	f_G	Hz	9.6	10.5	6.52
Effective Landed Gear Stiffness	K	10^3 lbs/in	9.70	23.7	14.95
Damping Ratio	ζ_G	%	7.6	7.2	6.7
Equivalence Constant	Λ	in^2	93.67	57.03	60.77
Deflection Ratio	λ	in	12.54	14.24	17.23
Critical Slope	S	ft lbs/RPM	19.2	32.8	26
Angular Stiffness	K_M	ft lbs/deg	1,321	1,960	1,321

As an enhancement to conventional dynamometer testing, the simulator fixture acts to isolate the brake and decouple it from fixture modes that can mask the vibration characteristics of the brake itself. For this application, the exact gear mode frequency is not important, and a high value of gear mode damping is deliberately used to focus on the accurate study of brake dynamics alone. Using this technique, a 737 brake vibration problem not anticipated with conventional testing was duplicated and solution alternatives were substantiated. The only restriction to the full-time use of this fixture in conventional dynamometer brake tests to date has been the torque capacity limits of current fixtures.

The pitch-plane simulator also avoids some of the problems and limitations associated with full-gear testing and enables the simulation to include the verification of the stability margin of the system.

Because the desired simulation is achieved by the use of only one wheel-brake-tire assembly, which does not translate fore and aft, installation on a wider range of dynamometer test sites is possible. Dynamometer face width and curvature are not factors. With full-gear simulation, the face width necessary to accommodate a two-wheel axle set not only limits the available test sites, but may dictate an inertia too high to represent the aircraft.

Simulation with excessive inertia extends the duration of any instability that may occur, and therefore prolongs divergence to unrealistic amplitudes. This was a necessary compromise on half-scale simulation testing of the L-1011 at WPAFB, requiring extensive post-processing of the results for correct interpretations.

A unique advantage of the pitch-plane fixture stems from the fact that its inherent damping (1.6%) is very low compared to the actual landing gear. The simulated damping used can therefore be reduced to deliberately destabilize the gear mode and verify the stability margin of the gear. This technique was used to produce the test results shown in Figure 7, which graphically presents "dynamometer walk." The figure further illustrates the concepts of instability and limit cycle previously discussed.

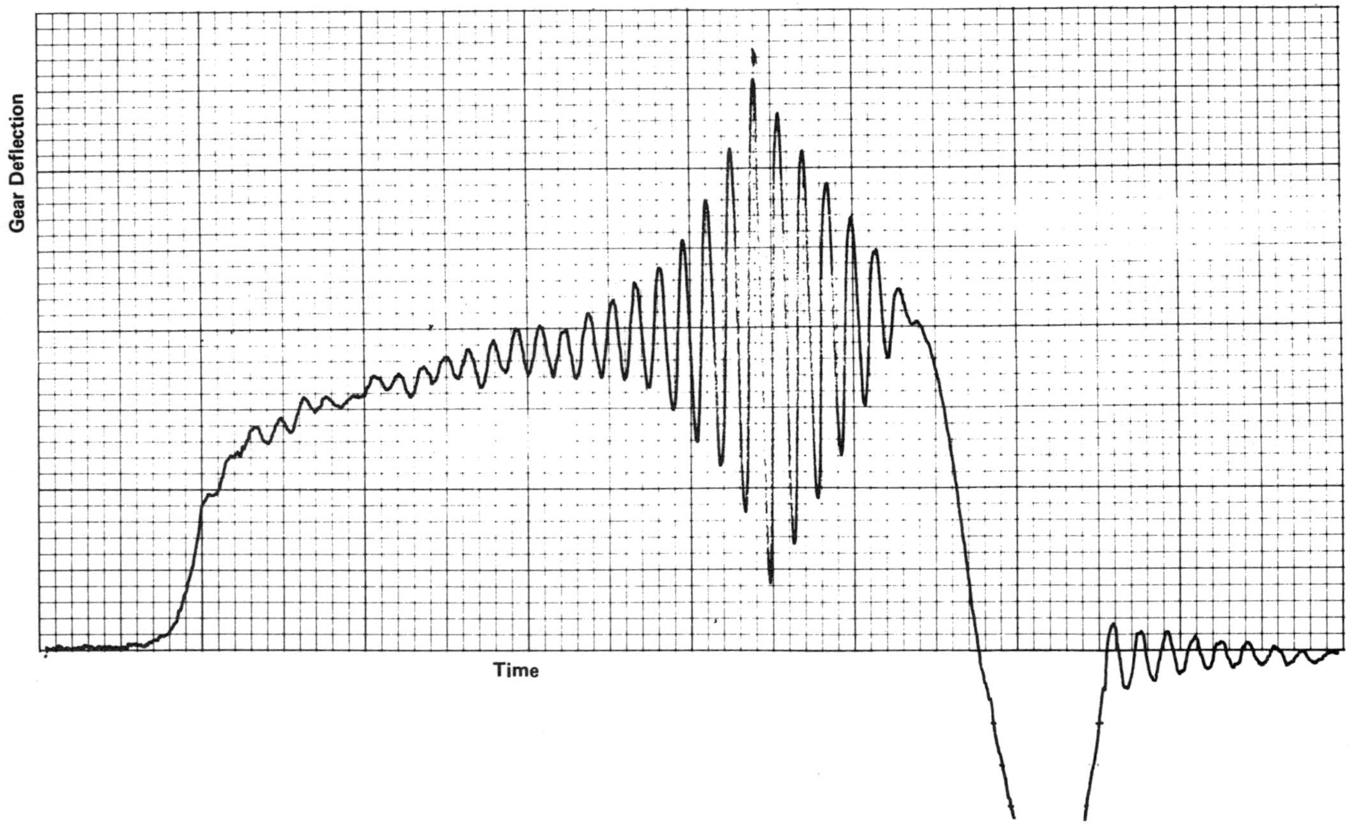

FIGURE 7. DYNAMOMETER-SIMULATED GEAR WALK

CONCLUSIONS

This paper has discussed a simplified technique for laboratory dynamometer simulation of landing gear-brake dynamics. The technique has been successfully and effectively applied to a number of aircraft and subsequent results obtained during on-aircraft testing have verified the adequacy of the simulation.

Several advantages derive from the fact that the technique provides very clearly a low frequency single-degree-of-freedom gear walk mode with a large range of linear viscous damping. Design of the system is straightforward and fabrication of the fixturing can commence far in advance of the availability of actual landing gear hardware components. The inherent low-damping of the fixturing involved in the gear mode simulation enables it to be used uniquely as a means to substantiate the stability margin of a brake's friction characteristics. Furthermore, simplicity and low cost enable it to be used as a matter of routine to study brake dynamics accurately by isolating the brake from conventional dynamometer fixturing which can couple with significant assembly modes of the brake in conventional dynamometer testing.

REFERENCES

1. Wignot and Hoblit, "Landing Gear Oscillations due to Unstable Skidding Friction," Journal of the Aeronautical Science, V16, #8, August 1949.
2. Zimmerman, N.H., "Landing Gear Vibrations Induced by Skidding Tires - Theoretical and Experimental Study," McDonnell Aircraft Corporation, 1958.
3. Singh and Busby, "Aircraft Multidisk Brake Squeal, Phase I, Preliminary Analytical and Experimental Studies," Ohio State University, 1984.
4. Anon., "Aircraft Terminology," SAE Document AIR 1489, April 1977.
5. Kaiser, W.D., "Vibration Suppressor for Braked Wheels," Goodyear Tire and Rubber Company, United States Patent No. 3,630,578, December 28, 1971.
6. Anon., "Brake Dynamics," SAE Document AIR 1064A, July 1979.
7. Enright, J.J., "Dynamic Compatibility Analysis of Landing Gears," BFGoodrich Engineering Report 4774, January 1979.

Aircraft Landing Dynamics Facility, A Unique Facility with New Capabilities

Pamela A. Davis, Sandy M. Stubbs, and John A. Tanner
NASA Langley Research Center

* Paper 851938 presented at the Aerospace Technology Conference and Exposition, Long Beach, California, October, 1985.

ABSTRACT

The Aircraft Landing Dynamics Facility (ALDF), formerly called the Landing Loads Track, is described. The paper gives a historical overview of the original NASA Langley Research Center Landing Loads Track and discusses the unique features of this national test facility. Comparisions are made between the original track characteristics and the new capabilities of the Aircraft Landing Dynamics Facility following the recently completed facility update. Details of the new propulsion and arresting gear systems are presented along with the novel features of the new high-speed carriage. The data acquisition system is described and the paper concludes with a review of future test programs.

THE LANGLEY RESEARCH CENTER Aircraft Landing Dynamics Facility, formerly known as the Landing Loads Track, is the only facility in the world capable of testing full size aircraft landing gear systems under closely controlled conditions on actual runway surfaces to simulate landing and take off operations of various aircraft. Testing at this facility is advantageous over flight testing for several reasons including safety, economy, parameter control, and versatility. Essentially any landing gear can be accommodated in the test carriage including those exhibiting new concepts and any runway surface and weather condition can be duplicated on the track. Research on slush drag, hydroplaning, tire braking, steering performance and runway grooving was accomplished at this facility. This paper presents a description of the Aircraft Landing Dynamics Facility and indicates how this facility has been upgraded to a higher speed capability. The main features of the facility update-the high pressure propulsion system, the arresting gear system, the high speed carriage and the track extension are described. The upgraded facility is scheduled to become operational during the summer of 1985.

DESCRIPTION OF OLD FACILITY, 1956-1982

A photograph of the old Landing Loads Track facility taken from the propulsion end looking towards the arresting gear is shown in figure 1. The compressor building houses a control room, a high pressure water pump, and an air compressor used to pressurize the three large air storage tanks to a pressure of approximately 21.7 MPa (3150 psi). The air flows from the three storage tanks, through a manifold, and a large goose neck shaped pipe up to the top of the "L" shaped vessel. The "L" vessel is filled with water using the high pressure water pump located in the compressor building.

Figure 1 - Landing Loads Track.

In front of the "L" vessel in figure 1 is the old test carriage. The test carriage is supported by steel rails that are 15.2 cm x 15.2 cm (6 in. x 6 in.) in cross section and extend the full length of the track. The test surface, shown in the center of the track, can be made of concrete, asphalt or other road bed type materials. Tests can be conducted on dry, damp or flooded runway surfaces, and a small section of runway can be covered with ice or ice slush. At the far end of the track is an arresting gear system that brings the carriage to a stop after the run is completed and the carriage building used for set up and calibration of various test articles.

A side view of the old test carriage in front of the old "L" vessel is shown in figure 2. The main features of the test carriage are the turning bucket, the drop carriage and the nose block. The turning bucket is located in the rear of the carriage and accepts the high pressure water jet from the quick opening valve located on the end of the "L" vessel. The turning bucket turns the jet 180° which produces the force necessary to accelerate the carriage to the desired speed. The vertical rails located in the center of the test carriage are the drop rails on which rides the drop carriage. The test article is attached to the drop carriage to accomplish vertical impact or loading of the test specimen. On the front end of the carriage is a nose block which consists of five grooves to intercept five arresting gear cables which are stretched across the track at the arresting gear end of the track.

Figure 2 - Old test carriage.

The valve on the end of the old "L" vessel is a ten inch plug valve and external to that valve is a 17.8 cm (7 in.) diameter nozzle which directs the water flow into the turning bucket. At the maximum air pressure of 21.7 MPa (3150 psi) the speed of the water jet is 207 m/sec (680 ft/sec). A photograph of a typical catapult is shown in figure 3. The carriage will accelerate to its maximum speed before reaching the raised runway shown in the right foreground of the photograph. The raised runway in this figure is concrete onto which the test article will be lowered by the drop carriage. A typical test article is shown in the inset of figure 3. The inset shows the test wheel and tire mounted on a dynamometer used to accurately measure the forces exerted on the tire when the tire strikes the runway.

Figure 3 - Typical catapult.

Touchdown speeds for a wide range of commercial aircraft as a function of the year those aircraft were introduced into service are shown in figure 4. The Landing Loads Track facility became operational in 1956 and at that time, the track capability of 110 kts (1 knot equals 0.5 m/sec) was adequate to cover the landing speeds of commercial propeller driven aircraft. With the advent of the commercial jet aircraft, however, landing speed climbed to a different plateau. The plateau was even higher for military aircraft, not shown on figure 4. The

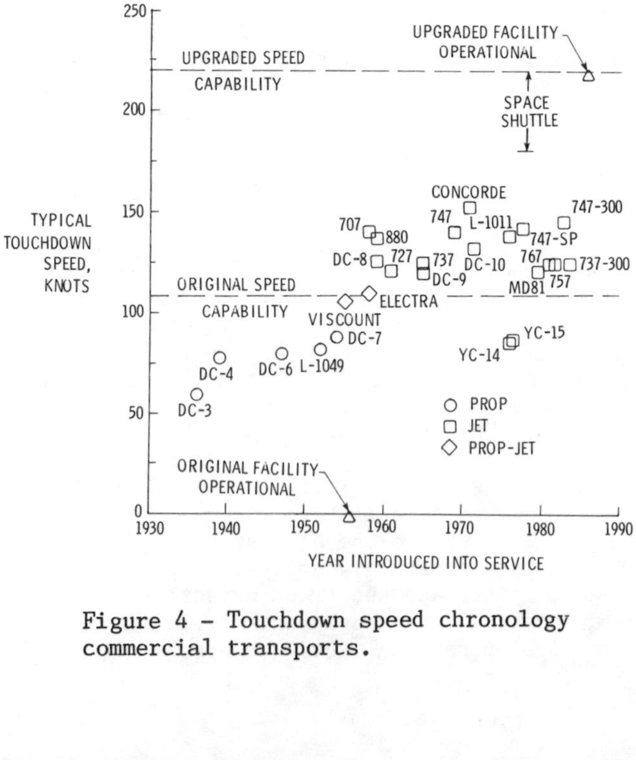

Figure 4 - Touchdown speed chronology commercial transports.

Space Shuttle lands at speeds between 175 and 220 kts. The updated facility is designed to enable testing at landing speeds up to 220 kts. The reason for upgrading the Aircraft Landing Dynamics Facility was to obtain the testing capability at landing speeds covering all current commercial and future aircraft.

RESEARCH REQUIREMENTS FOR UPDATED FACILITY

There were four basic requirements for the updated facility. The first was to increase the maximum test speed capability from 110 kts to 220 kts. A second requirement was to extend the track 183 m (600 ft) to obtain meaningful test times at these higher speeds. The third requirement was to provide a larger open bay test carriage to be able to accommodate large landing gear test articles for test loads up to 222 kN (50,000 lbs). The new carriage was designed to withstand much higher acceleration forces than the old carriage to facilitate the higher speed capability. The fourth requirement was to design-in the flexibility of utilizing the existing test carriage of the old facility. By meeting these requirements, NASA will have a unique national facility with increased speed capability to study current military and commercial aircraft landing problems, and to investigate landing systems of future aircraft including the Space Shuttle Orbiter.

DESCRIPTION OF MAJOR HARDWARE FOR UPDATED FACILITY, 1985. . .

A sketch of the new facility is shown in figure 5. On the left side of the figure is the "L" vessel and air piping system. The three existing air storage bottles were used and a new 122 cm (48 in.) diameter air pipe was fabricated to carry air from the air bottles to the top of the new "L" vessel. The new "L" vessel holds 98.4 kℓ (26,000 gals) of water in contrast to the 37.8 kℓ (10,000 gals) of the old "L" vessel. A quick opening shutter valve is mounted on the end of the new "L" vessel to control the water jet which catapults the carriage to the desired speed. The new carriage, with the large open bay, is a major addition. Another major addition is the new arresting gear system. It consists of five arresting gear units on each side of the track connected with five arresting gear cables that intercept the nose block of the carriage and bring it to a stop. The carriage building (Building 1261) for experiment preparation and calibration has been relocated at the end of the 183 m (600 ft) track extension shown in cross section. A new transfer system, shown in an auxiliary view, and an addition to the shop area behind Building 1262 has been built to facilitate a two carriage operation.

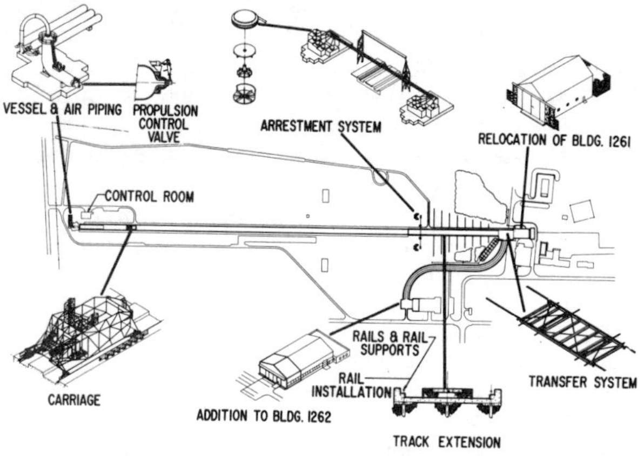

Figure 5 - Aircraft Landing Dynamics Facility.

PROPULSION SYSTEM - A photograph of the upgraded facility from the propulsion end is shown in figure 6. On the end of the "L" vessel is a fast acting shutter valve which controls the water jet. In front of the shutter valve is a flow straightener that takes the water during initial valve opening that would be deflected downward by the shutter and redirects it into the turning bucket at the rear of the carriage. The foundation for the "L" vessel is a slab of concrete that is 3.7 m (12 ft) thick and weighs approximately 31.7 MN (7 million lbs). A sketch

Figure 6 - Aircraft Landing Dynamics Facility.

of the high speed shutter valve that is mounted on the end of the "L" vessel is shown in figure 7. A spherical valve body was chosen to contain the high pressure water. A safety shutter, internal to the valve body, is used to obtain a water tight seal between runs. During a typical catapult operation, the safety shutter opens toward the top of the valve body and then the high speed shutter, which is connected by linkages to the hydraulic actuator, is moved in 0.4 seconds to an open position, held open for the dwell time necessary to obtain the desired carriage speed, and then returned to the closed position in 0.3 seconds. The internal nozzle shown in figure 7 provides a smooth contour from the end of the "L" vessel and forms a water jet 45.8 cm (18 in.) in diameter. The nozzle is centrally positioned within the 50.8 cm (20 in.) diameter valve body opening. At a maximum pressure of 21.7 MPa (3150 psi), the 45.8 cm (18 in.) jet will produce a thrust of approximately 8 MN (1.8 million lbs) on the new carriage. With a carriage weight of approximately 480 kN (108,000 lbs) this thrust creates a peak acceleration of approximately 17 g's.

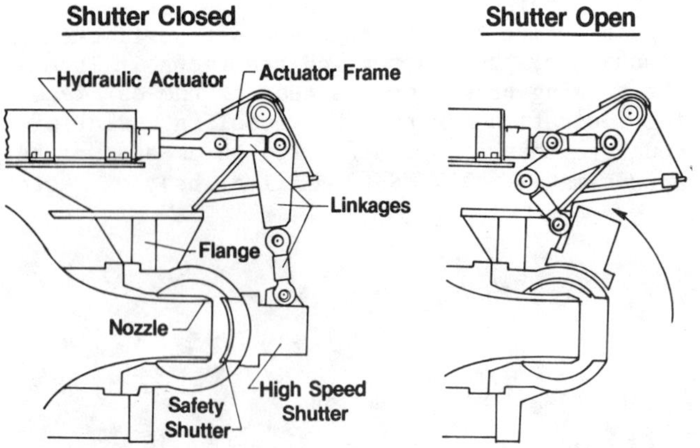

Figure 7 - Propulsion control valve.

Figure 8 is a photograph of the high speed shutter valve during final assembly. Noted in the figure is the high speed shutter and the linkage mechanism for opening and closing the shutter. The hydraulics and nitrogen supply system controls are shown near the top of the valve. These systems control the pressure and flow of oil in the actuator which opens and shuts the high speed shutter. These systems also control the flow of oil to the safety shutter and several safety pins around the

Figure 8 - High speed shutter valve.

valve. Figure 9 shows the "L" vessel, the goose neck pipe and the new valve sitting beside the "L" vessel before installation. The internal stainless steel nozzle can be seen protruding from the end of the "L" vessel. The valve slips over the nozzle and is bolted to the end of the

Figure 9 - Propulsion system.

"L" vessel. Figure 10 is a photograph of the flow straightener at the end of the high speed shutter valve.

Figure 10 - Flow straightener.

CARRIAGE - A photograph of the new carriage is shown in figure 11. The carriage is constructed of tube memebers with a central open bay 12.2 m (40 ft) long and 6.1 m (20 ft) wide to enable mounting a wide variety of test article shapes and sizes. The aft end of the carriage contains the large turning bucket which is approximately 3 m (10 ft) high. At the front end of the carriage is the nose block that intercepts the five arresting cables that stretch across the track. In the center of the carriage is the drop carriage to which the test article is attached. The drop carriage rides on four vertical rails and two hydraulic lift cylinders are used to raise and lower the test

Figure 11 - New test carriage.

article and to apply vertical loads to the test article as needed. Also noted in figure 11, on the near side of the carriage, is the hydraulic system for positioning the drop carriage, applying loads, obtaining free fall at speeds up to 6.1 m/sec (20 ft/sec), and applying wing lift to simulate an aircraft touchdown. Data from the test article are routed through an instrumentation box on the left side of the carriage and transmitted to a recording station at the propulsion end of the facility. Also shown in the photograph are outriggers on each corner of the carriage with rollers that run under the hold down rails shown in figure 12. The hold down rollers and rails are designed to hold the carriage on the main rails during the catapult stroke when upward force vectors might cause the carriage to be lifted off the rails.

Figure 12 - Propulsion end of ALDF track.

ARRESTMENT SYSTEM - A sketch of the arrestment system is shown in figure 13. A massive concrete foundation is located on either side of the track with five arresting engines mounted on each foundation. A close up view of an arrest-

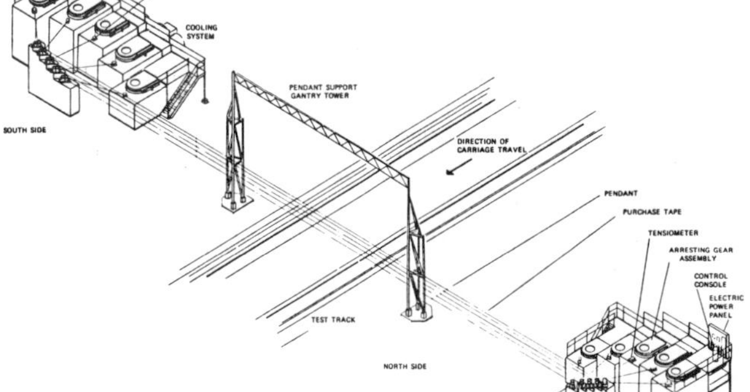

Figure 13 - Arresting gear system.

ing gear engine is shown in figure 14. Noted in the figure is a tub which holds a mixture of water and antifreeze. The tubs (5 on each side of the track) have stator vanes inside (not shown) on the bottom and top surfaces. Between the top and bottom stator vanes are rotor vanes attached to the rotating shaft that protrude through the top of the tub and to which is attached a spool which contains nylon tape that is approximately 20.3 cm (8 in.) wide and about 1 cm (3/8 in.) thick. The tape is connected to a cable that goes across the track and picks up the nose block of the moving carriage during arrestment. As the nose block pulls the tape from the spool, the rotor vanes churn the water in the tub. This churning action dissipates the kinetic energy of the carriage. The new arresting system was designed to stop the carriage in approximately 152 m (500 ft) even if only three of the five sets of engines are operating.

Figure 14 - Arresting gear engine.

A photograph of the complete arresting gear system is shown in figure 15. The new carriage is shown approximately half way down the track and the propulsion system is shown at the far end of the track. The new 183 m (600 ft) track extension, shown in the foreground, was not completed when this picture was taken.

Figure 15 - Arresting gear end of ALDF track.

Figure 16 is a photograph of a maximum speed shot (220 kts) of the new high speed carriage. Figure 17 is a carpet plot of the carriage speed as a function of propulsion valve dwell time, "L" vessel pressure, and water

Figure 16 - New carriage during maximum speed catapult, July 3, 1985.

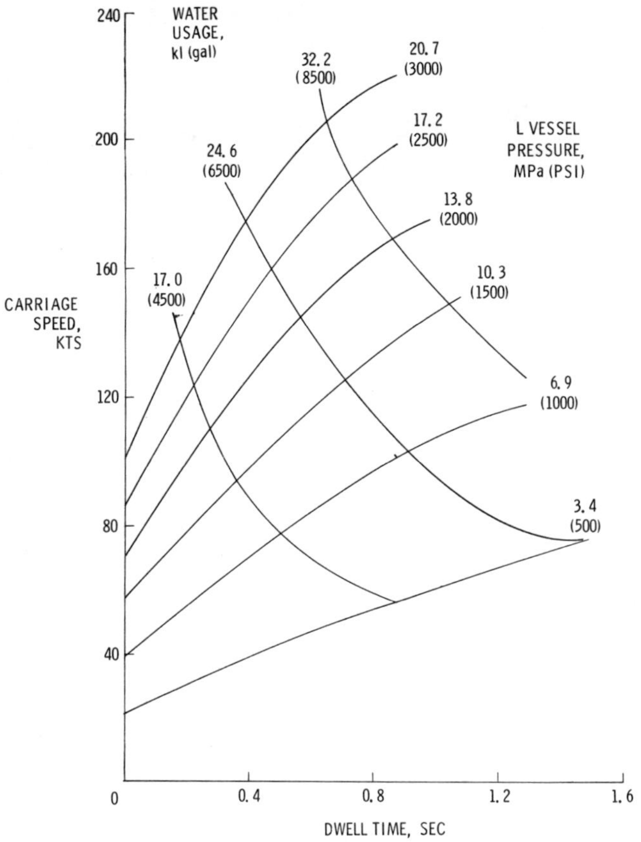

Figure 17 - Carriage speed as a function of dwell time.

usage. Table 1 summarizes the old and updated capabilities of the Aircraft Landing Dynamics Facility.

	Old Capability	Updated Capability
Max. Test Speed, kts	110	220
Lengths, m (ft)		
Overall Track	670 (2200)	853 (2800)
Test Section	366 (1200)	549 (1800)
Test Duration, sec	7@100 kts	11@100 kts 5@220 kts
Max. Vertical Loading, kN (lbs)	222 (50k)	222 (50k) @200 kts >222 (50k) @ lower V
Catapult		
Max. Accel, g units	3.3	17-18
Max. Force, MN (lbs)	1.6 (350k)	8 (1.8k)
Stroke, m (ft)	122 (400)	122 (400)
Pressure, kPa (PSI)	21 (3000)	21.7 (3200)
Nozzle Diameter, cm (in.)	18 (7.16)	46 (18)
H_2O Consumption, kℓ (gal) (Max. Speed Test)	11 (3000)	38 (10k)
Carriage		
Open Bay Size, m (ft)	3x5 (10x15)	6x12 (20x40)
Vertical Speed on Test Article, m/sec (ft/sec)	0-6 (0-20)	0-6 (0-20)
Vertical Load on Test Article, kN (lbs)	0-200 (0-45k)	0-222* (0-50k)*

*With growth capability of 445 (100k)

Table 1 - Aircraft Landing Dynamics Facility Capabilities Comparison

RESEARCH PROGRAMS

The first research programs performed at the updated facility will investigate Shuttle Orbiter main and nose gear tire spin up wear characteristics and cornering force characteristics at high speeds. Another test program will collect data on the frictional characteristics of radial and H-type aircraft tires for comparison with conventional bias ply tires. This radial and H-type tire program will be a joint effort between NASA, the FAA, the Air Force, the Society of Automotive Engineers, and the U.S. tire industry. A third program will be a joint NASA-FAA runway surface traction program studying the effects of different runway surface textures and various runway grooving patterns on the stopping and steering characteristics of aircraft tires. A fourth program will support a National Tire Modeling Program, that is a joint effort structured between NASA, and the U.S. tire industry to generate analytical models for the design of new types of tires. Research from the ALDF facility will be used to verify analytical work currently underway.

CONCLUDING REMARKS

A description of the original NASA Langley Research Center Landing Loads Track facility and its upgraded facility referred to as the Aircraft Landing Dynamics Facility (ALDF) is presented. Operational characteristics of the ALDF propulsion system, high speed test carriage, and arresting gear system are reviewed. The upgraded facility is scheduled to become operational during the summer of 1985.

As a result of the facility upgrade, the maximum speed capability of the Aircraft Landing Dynamics Facility has been doubled. The facility can also handle heavier and more bulky test articles than it could in the past. With this upgraded capability, the Aircraft Landing Dynamics Facility is now equipped to conduct research on present and future landing gear systems under realistic operational conditions.

Shuttle Landing Runway Modification to Improve Tire Spin-Up Wear Performance

Robert H. Daugherty, Thomas J. Yager, and Sandy M. Stubbs
NASA Langley Research Center

* Paper 881402 presented at the Aerospace Technology Conference and Exposition, Anaheim, California, October, 1988.

ABSTRACT

Landings of the Space Shuttle Orbiter at 200 knot speeds on the rough, grooved Kennedy Space Center runway have encountered greater than anticipated tire wear, which resulted in limiting landings on that runway to crosswinds of 10 knots or less. The excessive wear stems from wear caused during the initial tire touchdown spin-up. Tire spin-up wear tests have been conducted on a simulated KSC runway surface modified by several different techniques in an effort to reduce spin-up wear while retaining adequate wet cornering coefficients for directional control. The runway surface produced by a concrete smoothing machine using cutters spaced 1 3/4 blades per centimeter was found to give adequate wet cornering while limiting spin-up wear to that experienced in spinups on smooth concrete. As a result of these tests, the KSC runway has been smoothed for approximately 1066 m at each end leaving the original high friction surface for better wet steering and braking in the 2438 m central section of the runway.

EARLY LANDINGS OF THE SPACE SHUTTLE ORBITER were made on a lakebed and smooth concrete runway at Edwards Air Force Base to allow the greatest margin possible for errors in setup for landing or anomalies during landing rollout. It is desirable, however, to land on the Kennedy Space Center runway to minimize the cost, time, and hazards associated with ferry flights of the orbiter from Edwards AFB to KSC. Also, the orbiter must always have the capability to land safely at KSC in the event of an abort during ascent or poor weather conditions at other landing sites. Thus far, five landings have been made at KSC and for some of these landings greater than expected tire wear has occurred.

Tests were started in 1985 at NASA Langley Research Center's Aircraft Landing and Dynamics Facility (ALDF) to investigate the tire wear problem for landings at KSC with emphasis on tire rubber compounds and modifications to the runway surface that might give greater safety margins and higher crosswind landing capability. Results from early tests of cornering and wear characteristics of the orbiter main gear tire have been presented in references 1 and 2.

The purpose of this paper is to present results of tests to alleviate tire spin-up wear by modifying the KSC runway surface. Tests at touchdown speeds of 220 kts. will be presented as well as high and low speed friction tests on the KSC runway investigating various smoothing techniques.

APPARATUS

This investigation was conducted at the NASA Langley Research Center's Aircraft Landing Dynamics Facility (ALDF). The facility consists of a set of rails 850 m long on which a 49,000 kg carriage travels. The facility is shown in figure 1. The carriage is propelled at speeds up to 220 kts. using a high pressure water jet and is arrested using a set of water turbines connected by nylon tapes. A more detailed description of the facility can be found in reference 3.

The tires used in this study were 44.5 x 16.0 - 21 bias-ply aircraft tires with a 34-ply rating. The tires have a 5-groove tread pattern made of natural rubber with the grooves 2.54 mm deep. The tires have 16 actual carcass plies and their rated load and pressure are 271 kN and 2.17 MPa respectively. The tires were mounted on Orbiter main gear wheels with mass added to simulated brake rotor inertia and installed in a force measurement dynamometer on the test carriage. Data generated at the dynamometer were digitally telemetered to a receiving station where they were converted into engineering units by a desktop computer and stored.

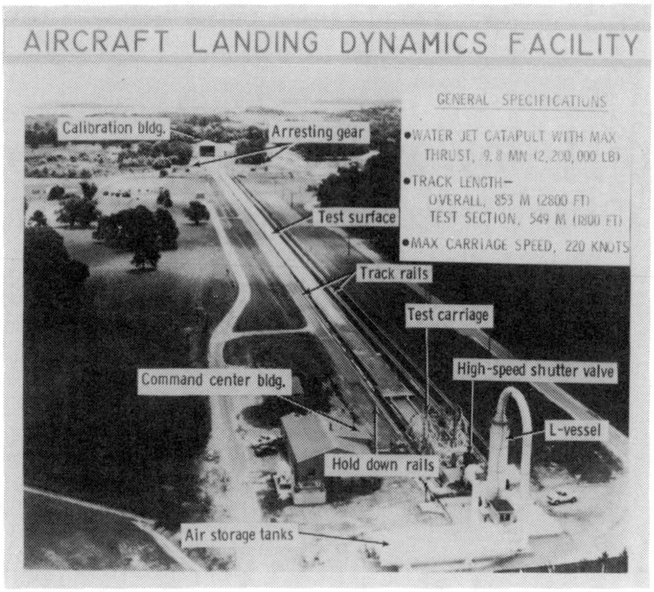

Figure 1.- The Langley Research Center Aircraft Landing Dynamics Facility.

A 550 m long simulated Kennedy Space Center (KSC) runway was installed at the facility to conduct these tests. The KSC runway shown in figure 2 has an extremely rough longitudinally-brushed texture combined with transverse grooves 6.4 mm wide by 6.4 mm deep with 29 mm spacing. The runway was designed to retain good friction characteristics when wet, which it does exceptionally well, but associated with that comes increased tire spin-up wear.

Figure 2.- Kennedy Space Center runway surface.

RESULTS AND DISCUSSION

Most airplanes land at speeds significantly lower than the orbiter landing speeds of 200-220 kts. and they land on relatively smooth asphalt or concrete surfaces. Even at the high shuttle landing speeds, little wear was experienced for lakebed and smooth concrete runway landings in low crosswinds. Figure 3 shows typical spin-up wear for a touchdown on smooth concrete. The wear spot is only to a depth of 1.5 mm and is not to the bottom of the tire tread. By contrast, the spin-up wear on a simulated KSC runway surface at 222 kts., shown in figure 4, was 5.5 mm and into the second cord layer of the tire. This spin-up wear alone is not a great cause for concern but it is the catalyst for increased wear during subsequent landing rollout if there is a crosswind or if there is a need for numerous steering inputs by the pilot. Figure 5 shows the wear progression of the spin-up spot during a rollout in a 15 kt. crosswind with minimal pilot steering inputs. The wear in the spin-up spot has progressed into the 4th cord layer and landing simulator studies have indicated that with substantial steering inputs in a 15 kt crosswind, the spin-up wear spot could progress to the 7th cord layer.

Figure 3.- Spinup wear for touchdown on smooth concrete at 220 knots.

Figure 4.- Spinup wear for touchdown on a simulated Kennedy Space Center runway at 222 knots.

Figure 5.- Spinup wear spot after rollout at simulated minimum wear 15 knot crosswind conditions.

One method of alleviating the wear problem is to reduce the spin-up wear depth by modifying the runway surface in the touchdown zones at either end of the runway. Several techniques were tested in an attempt to minimize wear and at the same time preserve as much side force friction coefficient capability as possible for directional control of the vehicle. All dry landing surfaces give adequate friction for control of the orbiter, but vehicle control during wet runway landing simulation studies have indicated that control problems can start when the side force friction coefficient drops below about 70% of the dry value. Although here is no intention to fly the orbiter in the rain, there is still a possibility that the runway could be damp from dew or from an unpredicted shower.

Figure 6 is a bar chart of the spin-up wear for the shuttle main gear tire touching down at ground speeds of 210 to 220 kts. and at sink speeds of 1.2 - 2.4 kts. on various dry surfaces and figure 7 is a bar chart of side force friction coefficient for the same surfaces wet. The maximum acceptable spin-up wear for crosswind landings at KSC is 1.5 mm as denoted in figure 6, and the minimum wet cornering friction value is 70% of the dry KSC value as denoted in figure 7.

Spin-up wear and friction data are shown for the original rough, grooved KSC surface and the smooth, ungrooved Edwards surface for comparison with the modified surfaces. The first modification

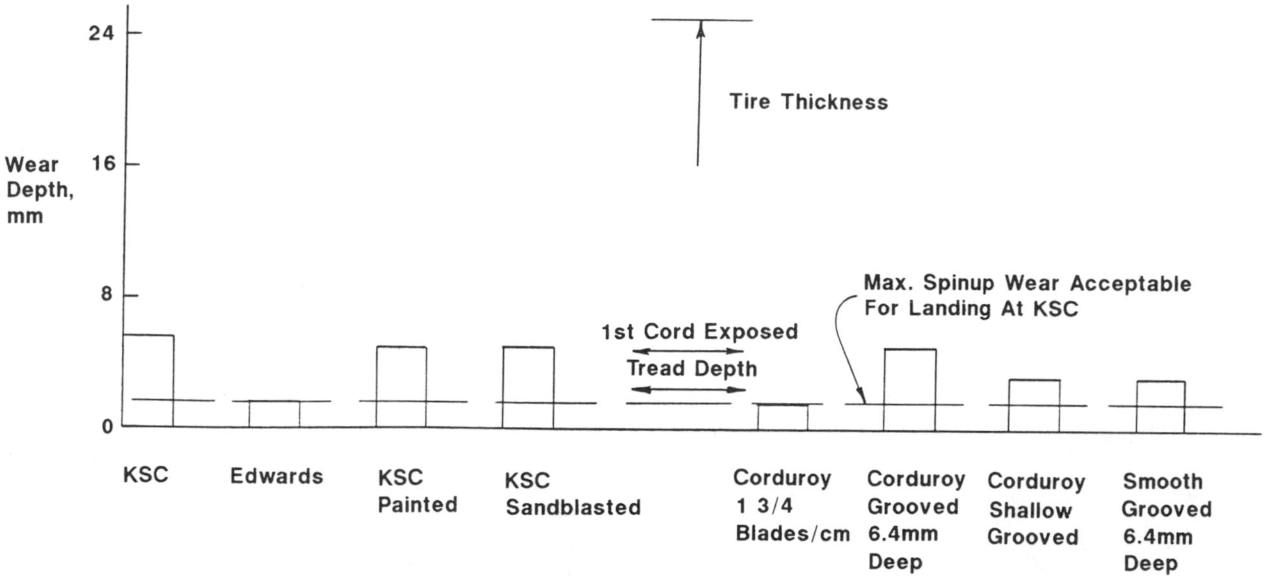

Figure 6.- Shuttle main tire spin-up wear for several dry runway surface treatments. Touchdown speeds 210-220 knots, 0° Yaw, and 1.2-2.4 knot sink speed.

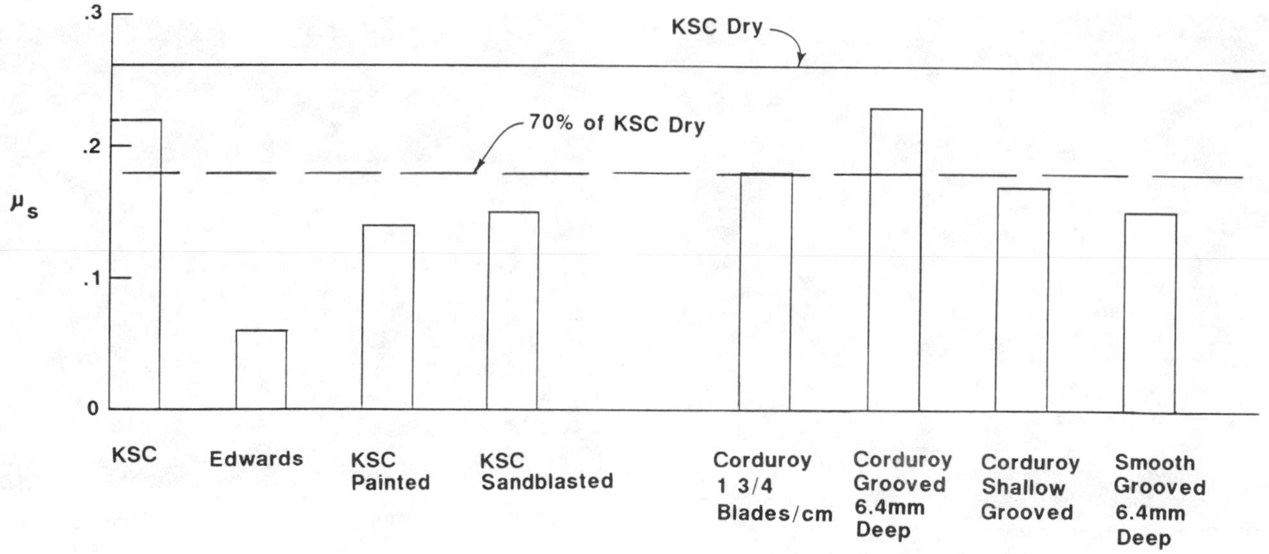

Figure 7.- Shuttle main gear tire side force friction coefficients for wet surfaces. Test speeds 200 knots, approximately 267 kN vertical load, 4° Yaw.

tested was painting the surface with runway marking paint because it was perhaps the easiest and cheapest modification if it would solve the wear problem. The painted KSC surface gave unacceptable spin-up wear of 5 mm, almost as bad as the original surface. Sandblasting the surface gave the same spin-up wear as painting and for both of these modifications, the friction levels dropped to unacceptable levels.

A highway smoothing technique was then examined which utilized diamond tipped blades spaced close together to grind the concrete surface. This technique was well established for use on highways but had only been used once on an airport runway and there was no wear or cornering data for tire pressures of 2.17 MPa and touchdown speeds in the 210 kt. range. Several blade spacings and grind depths were used to find the ideal setup to give acceptable wear and friction characteristics.

A photograph of the grinding head with 1 3/4 blades per cm. is shown in figure 8. The 1 3/4 blades/cm. grinder, when used to cut to the bottom of the runway grooves produced a corduroy surface texture shown in figure 9. Also shown and labeled in figure 9 is the KSC simulated surface, the sandblasted surface, and traces of the painted surface. The corduroy finish cut to a depth to remove the grooves gave spin-up wear of only 1.5 mm (see figure 6) and a cornering coefficient of .18 at a yaw angle of 4° (see figure 7). Figure 10 is a photograph of the spin-up wear on the corduroy surface. Although the wear is not to the bottom of the tire tread, the corduroy ridges sliced the tire circumferentally, as can be seen in the figure. There was concern that these cuts in the tire surface might precipitate increased wear rate during subsequent crosswind simulated rollout on the unmodified rough KSC surface but this concern proved to be unfounded.

Figure 8.- Grinding head cutters for smoothing the runway surface. .

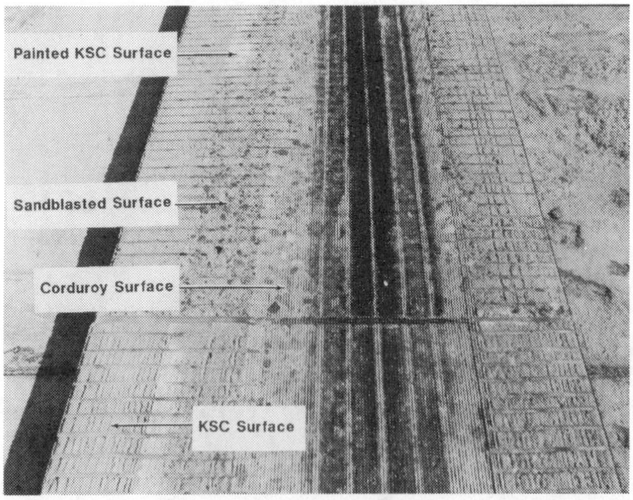

Figure 9.- Simulated Kennedy Space Center runway with modifications to reduce spin-up wear.

Figure 10.- Spin-up wear for touchdown on corduroy surface (1 3/4 blades/cm) at 210 knots.

A plot of tire wear as a function of side energy from ref. 1 is shown in figure 11. The solid line in the figure shows the progression of wear in the spin-up patch for spin-up and rollout on the KSC surface. The dashed curve is a fairing of the wear progression for two tires: one spun up on smooth concrete (circular symbols) and the other spun up on the corduroy surface smoothed by the 1 3/4 blades/cm cutter head (square symbols). The wear progression of the spinup spots for both tires occurred during rollout on the unmodified KSC surface. The effect of both the smooth and corduroy surfaces was to reduce tire spin-up wear as denoted by the lower initial point on the dashed curve in the figure. The narrow scatter band for the smooth concrete spin up and the corduroy spin up suggests that the circumferential cuts observed following the corduroy spin up did not adversely affect the tire wear characteristics during subsequent rollout on the unmodified KSC surface. The dashed and solid curves in figure 11 are parallel because each rollout was conducted on the unmodified KSC surface.

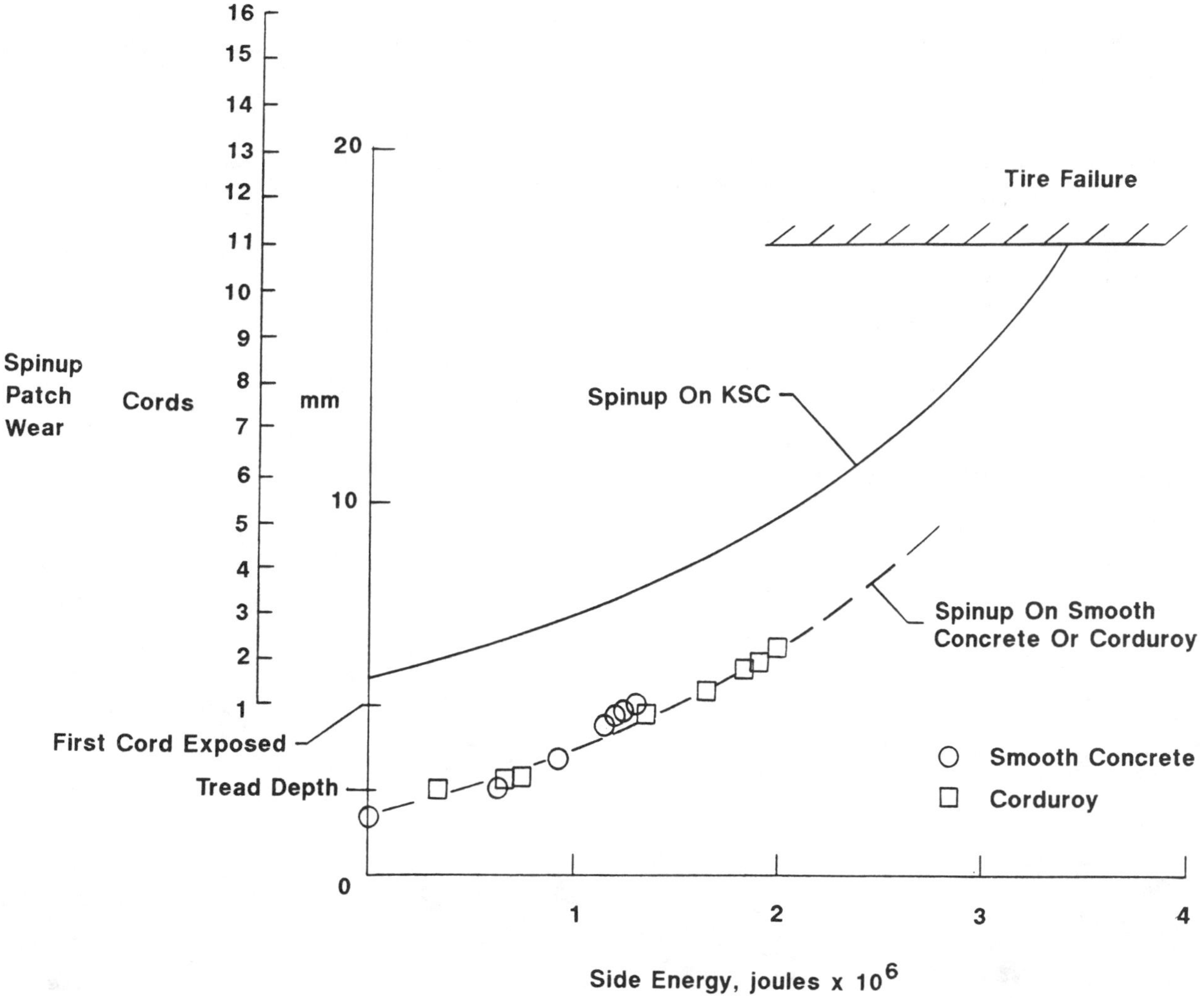

Figure 11.- Wear progression of spin-up patch for spinups on Kennedy Space Center surface and smooth concrete or corduroy surfaces.

Limited titled slab tests at the ALDF indicated that removing all of the transverse grooves in the runway would be detrimental to its water drainage capability, therefore additional tests were conducted on a simulated KSC surface that had been smoothed with the 1 3/4 blades/cm. corduroy cutter head set to shallower depths thus leaving some depth of transverse grooves in place. This smoothing operation was done both in an area that had been previously sandblasted and on the original KSC surface. In the first case, the grinding produced a corduroy surface superimposed on shallow grooves (Figure 12) and in the second case, the grinding produced a corduroy surface on 6.4 mm deep grooves (figure 13). The corduroy surface on deep grooves gave acceptable friction coefficient (figure 7) while the shallow grooved corduroy coefficient was slightly below acceptable levels, but both of these surfaces resulted in excessive spin-up wear (figure 6). An interlocking blade cutter head shown in figure 8, was set at a shallow depth and used to eliminate the corduroy effect producing a very smooth finished surface superimposed on sharp edged 6.4 mm deep grooves. Smoothing with the interlocking blades still produced excessive spin-up wear and insufficient cornering coefficient. The only smoothing configuration that resulted in adequate cornering coefficient and acceptable spin-up wear was the 1 3/4 blades per cm. corduroy cutting head grinding the KSC surface to a depth to eliminate the grooves. Since drainage of the ungrooved corduroy surface was slower than drainage of the original KSC grooved surface, it was decided that only the touchdown areas of the KSC runway should be smoothed with the corduroy cutter, leaving the center of the runway with the original high friction surface.

The aggregate used in the concrete runway at KSC was soft limestone whereas the aggregate in

Figure 12.- Shallow grooved corduroy surface.

Figure 13.- Corduroy surface with grooves 6.4 mm deep.

the simulated test runway at LaRC was a combination of much harder granite and river rock. Several different cutter blade spacings were used to smooth a sample area of the actual KSC runway to determine which blade configuration would produce a corduroy surface finish that best matched the surface at LaRC that gave good spin-up wear and acceptable wet friction. Blade spacings of 1.5, 1 3/4, 2, 2 3/4 blades/cm and interlocking were used and produced results shown in figures 14 and 15. Cornering friction and wear tests were conducted on these surfaces under both dry and wet conditions. Although there is a difference in aggregate in the KSC runway compared with the simulated runway at LaRC, the 1 3/4 blades/cm cutter produced almost identical corduroy texture on both surfaces, thus, it was chosen as the configuration to be used in smoothing the KSC runway.

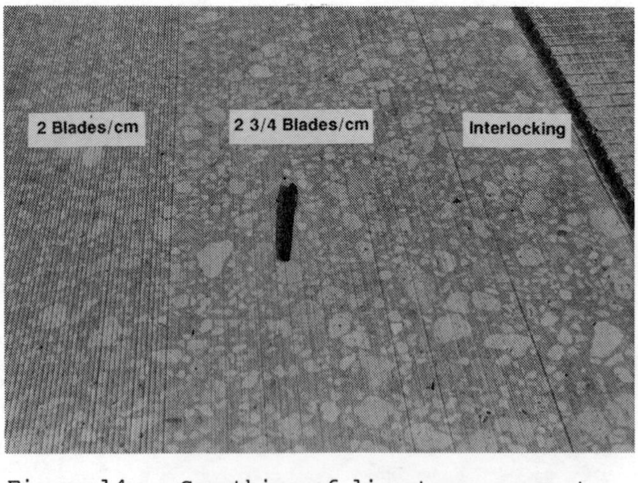

Figure 14.- Smoothing of limestone aggregate Kennedy Space Center runway surface with varying blade spacings.

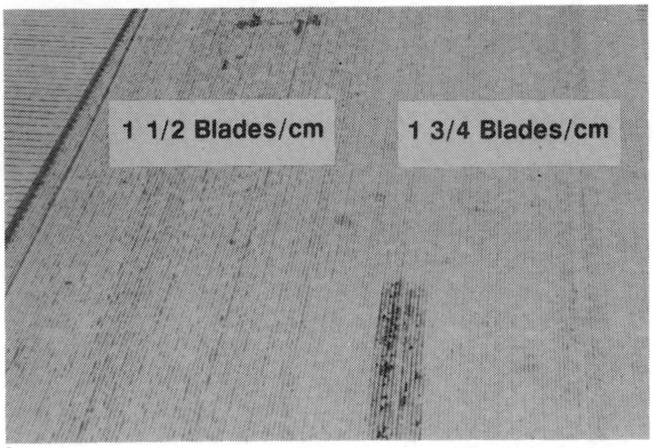

Figure 15.- Smoothing of Kennedy Space Center surface with 1 1/2 and 1 3/4 blades/cm spacings.

A photograph of the modified KSC runway is shown in figure 16. The runway was smoothed for a length of 1066 m at both ends using the 1 3/4 blades per cm. corduroy cutters. The middle 2438 m of the runway was not altered, thus retaining maximum wet cornering coefficient for the major portion of the runout. Smoothing the runway touchdown zones will minimize tire spin-up wear and has resulted in a three knot increase in the crosswind limit for shuttle landing operations.

Figure 16.- Kennedy Space Center runway with landing touchdown zones smoothed with 1 3/4 blades/cm cutter heads.

CONCLUDING REMARKS

Landings of the Space Shuttle Orbiter at ground speeds of 200 kt. speeds on the rough-grooved Kennedy Space Center runway have resulted in greater tire wear than anticipated. Experimental tests have been conducted at the Langley Research Center on a simulated KSC runway surface to investigate the causes of this tire wear and to examine several treatments to the runway surface to minimize tire wear while retaining sufficient wet cornering characteristics of that runway.
The excessive tire wear stems from wear in the touchdown spin-up spot that, in the presence of a crosswind or pilot steering inputs, can grow to unacceptable levels and threaten the integrity of the tire during subsequent rollout.

Modifications to the simulated KSC runway surface aimed at reducing spin-up wear included painting, sandblasting, and smoothing with highway grinding equipment. Painting and sandblasting the surface resulted in reductions of wet side force coefficient to unacceptable levels and did not decrease spin-up wear significantly. Smoothing the runway with a 1 3/4 blades per cm. cutter head gave a corduroy texture to the surface which reduced spin-up wear depth to that of smooth concrete and maintained wet side force friction coefficient at 70 percent of the dry KSC value. The KSC runway has been smoothed to the corduroy texture for 1066 m at each end to reduce spin-up wear while leaving the middle 2438 m rough-grooved surface for maximum wet braking and cornering friction coefficient.

REFERENCES

1. Daugherty, Robert H.; Stubbs, Sandy M.: Cornering and Wear Behavior of the Space Shuttle Orbiter Main Gear Tire. SAE 871867. Presented at the 1987 Aerospace Technology Conference and Exposition, Long Beach, CA.

2. Daugherty, Robert H.; Stubbs, Sandy M.; Robinson, Martha P.: Cornering Characteristics of the Main Gear Tire of the Space Shuttle Orbiter. NASA TP 2790, March 1988.

3. Davis, Pamela A.; Stubbs, Sandy M.; and Tanner, John A.: Langley's Aircraft Landing Dynamics Facility. NASA RP 1189, October 1987.

A Summary of Recent Aircraft/Ground Vehicle Friction Measurement Tests

Thomas J. Yager
NASA Langley Research Center

* Paper 881403 presented at the Aerospace Technology Conference and Exposition, Anaheim, California, October, 1988.

ABSTRACT

Tests with specially instrumented NASA B-737 and B-727 aircraft together with several different ground friction measuring devices have been conducted for a variety of runway surface types and wetness conditions. This effort is part of the Joint FAA/NASA Aircraft/Ground Vehicle Runway Friction Program aimed at obtaining a better understanding of aircraft ground handling performance under adverse weather conditions and defining relationships between aircraft and ground vehicle tire friction measurements. Aircraft braking performance on dry, wet, snow-, and ice-covered runway conditions is discussed together with ground vehicle friction data obtained under similar runway conditions. For a given contaminated runway surface condition, the relationship between ground vehicles and aircraft friction data is identified. The influence of major test parameters on friction measurements such as speed, test tire characteristics, and surface contaminant type are discussed. The test results indicate that use of properly maintained and calibrated ground vehicles for monitoring runway friction conditions should be encouraged particularly under adverse weather conditions.

THERE IS AN IMPERATIVE OPERATIONAL NEED for information on runway which may become slippery due to various forms and types of contaminants. Experience has shown that since the beginning of "all weather" aircraft operations, there have been landing and aborted takeoff incidents and/or accidents each year where aircraft have either run off the end or veered off the shoulder of low friction runways. From January 1981 to January 1988, more than 400 traction-related incident/accidents have occurred according to Federal Aviation Administration (FAA) and National Transportation Safety Board (NTSB) records. These cases have provided the motivation for various government agencies and aviation industries to conduct extensive tests and research programs to identify the factors which cause the runway friction to be less than acceptable [1-15]. Aircraft takeoff and landing accidents continue to occur as depicted in figure 1 for a variety of tire/pavement related causes.

Figure 1.- Consequences of aircraft takeoff and landing accidents.

The fuel-fed fire which destroyed a NASA Convair 990 aircraft at March AFB, CA in July 1985 happened following right gear tire failure(s) just prior to reaching takeoff decision speed. The pilot aborted the takeoff but during the subsequent rollout the right wing was punctured and fuel leakage was ignited prior to the aircraft coming to a stop near the runway end. None of the nineteen crew members and scientists on board the aircraft was injured but the aircraft was destroyed. Failure of one tire initially led to overload failure of the remaining three tires on the right main gear. This accident happened on a long, dry, clean, high friction runway surface whereas the DC-9 aircraft shown straddling the roadway in the lower photograph of figure 1 was exposed to more severe runway conditions. With a tailwind component during the approach and landing on a snow-covered runway at Erie, PA in February 1986, the aircraft touchdown was delayed and occurred at a higher than normal speed. The relatively low braking friction developed on the snow-covered runway was insufficient to stop the aircraft prior to going off the end of the runway and the aircraft finally stopped at a roadside embankment. Only one person was injured and the damage to the aircraft has since been repaired. These accidents emphasize the need for improved measurement techniques and inspection procedures related to tire and runway conditions. NASA Langley's Landing and Impact Dynamics Branch is involved in several research programs directed towards obtaining a better understanding of how different tire properties interact with varying pavement surface characteristics to produce acceptable performance for aircraft ground handling requirements. The following sections of this paper describe one such effort which was jointly supported by NASA and the FAA.

SCOPE OF PROGRAM

The Joint FAA/NASA Aircraft/Ground Vehicle Runway Friction Program is aimed at obtaining a better understanding of aircraft ground handling performance under a variety of adverse weather conditions and to define relationships between aircraft and ground vehicle tire friction measurements. These tests involved a specially instrumented NASA B-737 aircraft and several different ground friction measuring vehicles shown in figure 2. The diagonal-braked vehicle developed by NASA measured locked wheel sliding friction values. The FAA mu-meter trailer monitored side force variation on two tires yawed to an included angle of 15° degrees. Both the surface friction tester automobile and the Swedish BV-11 skiddometer trailer measured tire braking friction near the peak of the tire friction/slip ratio curve. The FAA B-727 instrumented test aircraft is shown in figure 3 during rain wet runway tests. A new runway friction tester van which measured peak tire braking friction was used during the B-727 aircraft tests, and is shown in figure 4 during testing under winter runway conditions at Brunswick Naval Air Station in Maine. A Navy runway condition reading vehicle (RCR), shown in figure 5, was also used to collect vehicle deceleration values. Both a Tapley meter and a Bowmonk brakemeter were installed in this RCR vehicle to indicate vehicle braking deceleration

Figure 2.- Instrumented B-737 test aircraft and several ground friction measuring vehicles.

Figure 3.- Instrumented B-727 test aircraft.

Figure 4.- Runway friction tester during test run.

Figure 5.- Navy runway condition reading (RCR) test vehicle.

levels under snow and ice conditions. Additional information describing these different ground friction measuring vehicles is contained in references 16 to 19. Between June 1983 and March 1986, tests were performed on 12 different concrete and asphalt runways, grooved and ungrooved, including porous friction coarse, under dry, truck wet, rain wet, snow-, slush-, and ice-covered surface conditions. A limited assessment of some runway chemical de-icing treatments was also obtained. Over 200 test runs were made with the two transport aircraft and over 1100 runs were made with the different ground test vehicles. Most of the dry and wet runway surface test runs were performed at NASA Wallops Flight Facility and the FAA Technical Center airport. All the winter runway test conditions were evaluated at Brunswick NAS [20]. The test procedure for wet runway conditions was to make ground vehicle runs before and after each aircraft braking run. For the winter runway conditions of compacted snow and solid ice, a series of ground vehicle runs were made immediately following the aircraft test runs on each surface contamination condition. The principal objective of this paper is to define the relationship between the friction measurements for the test aircraft and the friction measurements for the ground vehicles under typical wet runway conditions as well as compacted snow- and ice-covered runway conditions. The influence of major test parameters on friction measurements such as speed and the tire characteristics are discussed.

TEST RESULTS AND DISCUSSION

A substantial tire friction database has been collected during this Joint FAA/NASA Runway Friction Program and extensive data reduction and analysis have been accomplished at NASA Langley. However, only a very limited amount of aircraft and ground vehicle friction data are presented and discussed herein to indicate some of the major test findings and data trends.

WET RUNWAYS - An example of some of the B-737 aircraft braking friction data is plotted as a function of speed in figure 6 for different runway

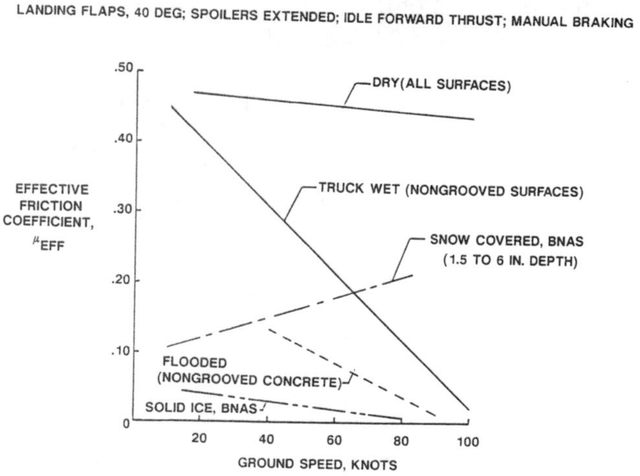

Figure 6.- Comparison of B-737 aircraft braking performance.

conditions. The range of effective friction coefficients is from nearly 0.5 on dry runways to 0.05 on glare ice surface at Brusnwick Naval Air Station (BNAS). The range of B-737 aircraft and ground vehicle friction measurements obtained on nongrooved and grooved surfaces under truck wet conditions is shown in figure 7. The grooved runway surface

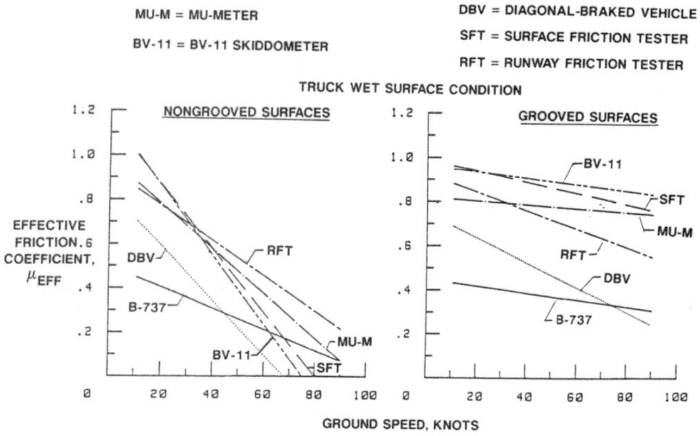

Figure 7.- Range of B-737 aircraft and ground vehicle friction measurements.

friction data is significantly greater than the nongrooved data, particularly at the higher speeds. Most of the ground vehicle friction values were higher than those developed by the B-737 aircraft because of differences in braking test mode, tire tread design, and tire inflation pressure. When these major factors are considered in estimating aircraft wet runway braking performance from ground vehicle friction measurements using the techniques and calculations described in reference 21, the relationship between actual braking friction coefficient for the B-737 and estimated braking friction coefficients of the airplane obtained from the ground vehicles is shown in figure 8. For most of the

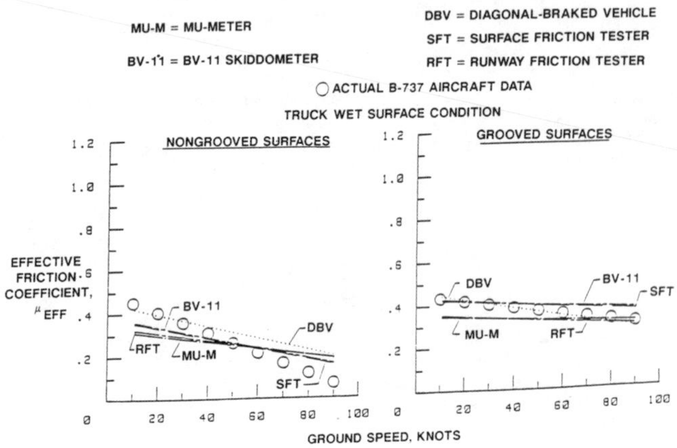

Figure 8.- Relationship between actual and estimated B-737 aircraft braking performance.

ground vehicle friction measurements, the estimated aircraft performance is in good agreement with the actual measured aircraft braking friction level. The available data suggest that the ground vehicle friction data for wet runway conditions can estimate aircraft tire friction performance to within about 15 percent of the actual measured aircraft friction values and in some cases, within 5 percent. An example of this predictive capability is shown for the runway friction tester on truck wet nongrooved surfaces in figure 9 and on truck wet grooved surfaces in figure 10. Four different speed increments were selected

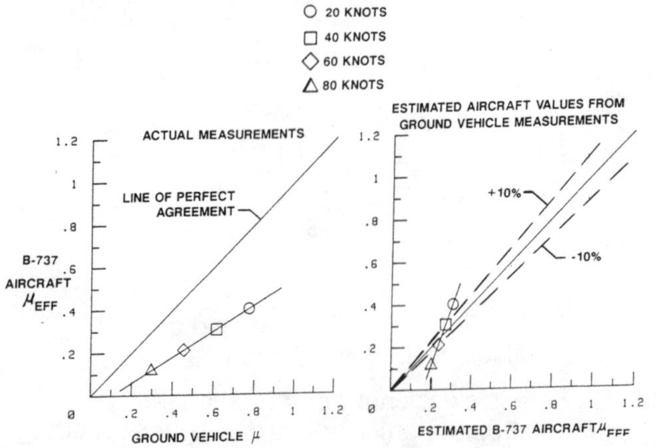

Figure 9.- Relationship between B-737 aircraft and runway friction tester friction data on truck wet nongrooved surfaces.

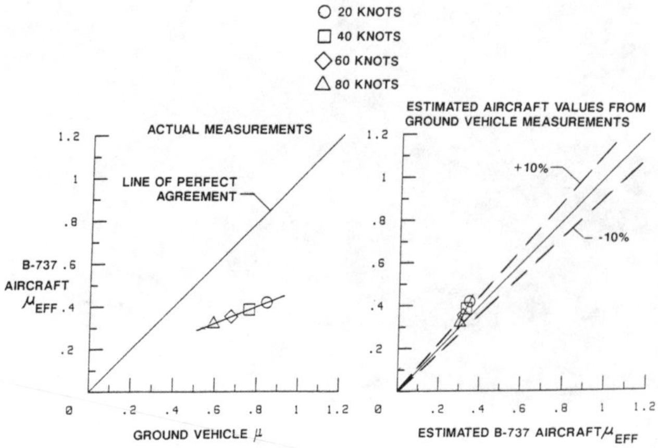

Figure 10.- Relationship between B-737 aircraft and runway friction tester friction data on truck wet grooved surfaces.

for this analysis using the B-737 aircraft data. The left hand plots in figures 9 and 10 show the relationship of actual ground vehicle friction values with the aircraft friction values and the right hand plots show the correlation between actual and estimated aircraft performance based on runway friction tester data. For both types of runway surfaces under truck wet conditions, the estimates are accurate to within about 15 percent. The relationship between ground vehicle estimates and actual aircraft tire friction values will vary with changes in wetness conditions. Hence, ground vehicle measurements should be taken on any given runway for a range of wetness conditions.

SNOW- AND ICE-COVERED RUNWAYS - The friction measurements obtained with the various ground test devices indicated that forward speed had little effect on the magnitude of the friction values for compacted snow-covered and ice-covered conditions. Furthermore, the friction values obtained from each vehicle were in close agreement for either surface condition. The Tapley and Bowmonk meters were installed in the Navy runway condition reading vehicle and the manually recorded friction values for each instrument were in close agreement for each test run. Reference 22 describes the RCR test procedure used at U.S. military bases for monitoring runway friction conditions. Figure 11 provides a listing of friction values obtained at Brunswick NAS. Tire conditions, ambient temperatures, and test speeds are indicated in the notes accompanying the figure. Qualitative verbal braking action terms namely, excellent, good, marginal, and poor, were used to identify four distinct levels or ranges in friction readings for each device. The relationship determined between each of the ground vehicle friction measurements and the values measured by the Tapley meter under similar winter runway conditions is identified below along with standard

Runway Surface Conditions: Compacted Snow and Ice

VERBAL BRAKING ACTION	GROUND VEHICLE FRICTION READINGS						
	MU-METER	TAPLEY METER	RUNWAY CONDITION READINGS (RCR)	BOWMONK METER	SAAB FRICTION TESTER	RUNWAY FRICTION TESTER	BV-II SKIDDOMETER
EXCELLENT	0.50 and above	0.48 and above	16 and above	0.46 and above	0.58 and above	0.50 and above	0.58 and above
GOOD	0.49 to 0.36	0.46 to 0.35	15 to 12	0.44 to 0.34	0.56 to 0.42	0.48 to 0.35	0.56 to 0.42
MARGINAL	0.35 to 0.26	0.33 to 0.25	11 to 9	0.32 to 0.24	0.39 to 0.29	0.33 to 0.24	0.39 to 0.29
POOR	0.25 and below	0.24 and below	8 and below	0.23 and below	0.27 and below	0.23 and below	0.27 and below

NOTES:

(1) Mu-meter equipped with smooth RL-2 tires inflated to 69 kPa (10 lb/in.^{2})

(2) Runway friction tester equipped with smooth RL-2 tire inflated to 207 kPa (30 lb/in.^{2})

(3) Saab friction tester and BV-II skiddometer equipped with grooved aero tire inflated to 690 kPa (100 lb/in.^{2})

(4) Ambient air temperature range, -15 to +5^0 C (5 to 41^0 F)

(5) Test speed range, 32 to 97 km/h (20 to 60 mph)

Figure 11.- Ground vehicle friction reading correlation table.

deviation values:

Mu-meter = -0.08 + 1.20 Tapley;
Standard Deviation = 0.00016
Bowmonk = -0.01 + 0.96 Tapley;
Standard Deviation = 0.0032
BV-11 & SFT = -0.05 + 1.30 Tapley;
Standard Deviation = 0.0011
RFT = -0.05 + 1.13 Tapley;
Standard Deviation = 0.0012

In general, the excellent friction readings were close to some wet surface values, e.g. 0.5 and above, whereas, the poor friction readings were normally below a friction coefficient levels of 0.25. The BV-11 skiddometer and Saab friction tester measured similar friction values as expected since the test tire and braking slip operation were identical. The range of friction values at each of the four qualitative levels is nearly the same for the mu-meter, Tapley meter, runway friction tester, and the Bowmonk meter. Slightly higher friction values were obtained with the Saab friction tester and the BV-11 skiddometer for each qualitative value probably due to the use of a higher test tire inflation pressure and the use of a grooved tread pattern on the tire instead of a smooth tread.

The range of aircraft effective braking friction coefficient values with ground speed for compacted snow- and ice-covered runway conditions is shown in figure 12. The data symbols and line codes denote the different test conditions and aircraft. The best fit least squares linear curve for the compacted snow-covered surface friction data, denoted by the solid line, is nearly four times greater than the data from the glare ice-covered surface denoted by the dashed line. The difference in braking performance between the two test aircraft under these winter runway conditions was found to be insignificant. The aircraft braking performance on the snow-covered and ice-covered surfaces was

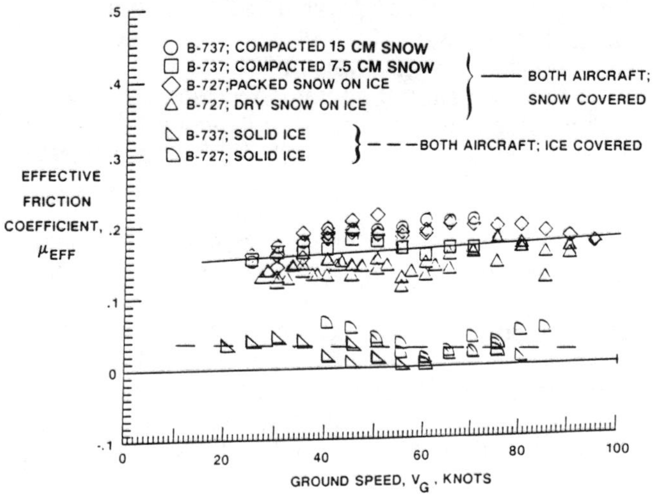

Figure 12.- Aircraft braking friction performance on compacted snow- and ice-covered runways.

relatively insensitive to ground speed variations.

Since each test aircraft indicated a significant difference between the compacted snow-covered and the ice-covered surface conditions, two ranges or means of aircraft braking friction data were selected to define the relationship with the ground vehicle friction measurements. The resulting aircraft and ground vehicle friction correlation chart is shown in figure 13 where the compacted snow-covered and

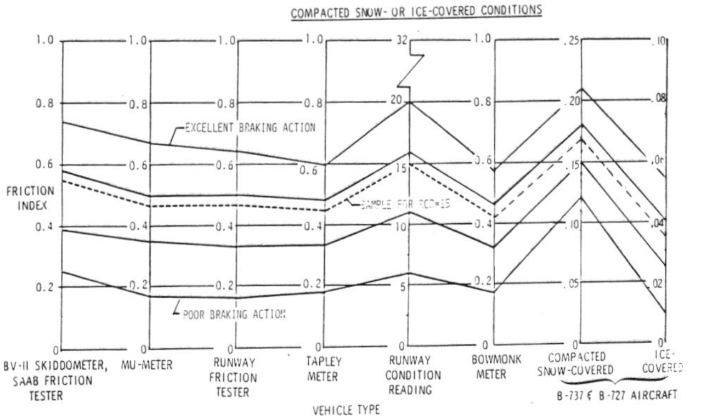

Figure 13.- Aircraft and ground vehicle friction correlation chart.

ice-covered surface condition is delineated for the two aircraft. For the compacted snow-covered surface condition, an aircraft effective braking friction coefficient value of 0.21 was selected for excellent braking action level and 0.12 was used for the poor braking action level. An effective braking friction coefficient range from 0.055 to 0.01 was selected for comparable aircraft braking action levels on the ice-covered surface condition. The dashed line in the figure depicts comparable values for other ground vehicles and the two aircraft/surface conditions for an RCR value of 15.

From an aircraft operator's viewpoint, these values of friction for a snow- or ice-covered runway must be considered in respect to the actual runway geometry and such environmental conditions as pressure/altitude, winds, and ambient temperature at the time of a particular aircraft operation. It should also be recognized that aircraft operations can occur on runways which have a nonuniform mixture of compacted snow-covered area and exposed solid ice-covered surfaces. In such circumstances, additional ground vehicle friction measurements need to be taken to adequately determine average friction numbers for each portion of the runway. How well this established relationship between aircraft and ground vehicle friction values remains for other aircraft types is somewhat questionable although the available data tends to suggest a similar relationship. The use of actual friction numbers in place of qualitative braking action terms is strongly recommended because with experience, these runway friction values measured by a ground vehicle will provide the pilot a more precise and accurate gage on the safety margins available for landing on a given runway. Proper and timely use of snow removal equipment and runway chemical treatments to minimize and/or remove snow and ice contaminants is still recognized as a necessity to return to dry runway friction levels as soon as possible.

CONCLUDING REMARKS

An overview of the Joint FAA/NASA Aircraft/Ground Vehicle Runway Friction Program has been given. A substantial friction database has been collected from tests with two instrumented transport aircraft and several different ground test vehicles on a variety of runway surfaces and wetness conditions. Major factors influencing tire friction performance have been identified and a promising relationship between ground vehicle and aircraft tire friction performance has been defined. Greater usage of ground vehicle friction measurements at airports can be expected to define runway surface maintenance requirements and to monitor current runway friction-levels under adverse weather conditions.

REFERENCES

1. Horne, Walter B.; Yager, Thomas J.; and Taylor, Glenn R.: Review of Causes and Alleviation of Low Tire Traction on Wet Runways, NASA TN D-4406, 1968.
2. Anon.: Pavement Grooving and Traction Studies, NASA SP-5073, 1969.
3. Yager, Thomas J.; Phillips, W. Pelham; Horne, Walter B.; and Sparks, Howard C. (Aeronautical Systems Division, WPAFB): Comparison of Aircraft and Ground Vehicle Stopping Performance on Dry, Wet, Flooded, Slush- and Ice-Covered Runways. NASA TN D-6098, November 1970.
4. Merritt, Leslie R.: Impact of Runway Traction on Possible Approaches to Certification and Operation of Jet Transport Aircraft. Paper 740497, Society of Automotive Engineers, April-May 1974.
5. Horne, Walter B.: Status of Runway Slipperiness Research. NASA SP-416, October 1976, pp. 191-245.
6. Horne, Walter B.; Yager, Thomas J.; Sleeper, Robert K.; and Merritt, Leslie R. (Federal Aviation Administration): Preliminary Test Results of the Joint FAA-USAF-NASA Runway Research Program, Part I - Traction Measurements of Several Runways Under Wet and Dry Conditions with a Boeing 727, a Diagonal-Braked Vehicle, and a Mu-Meter. NASA TM X-73909, 1977.
7. Horne, Walter B.; Yager, Thomas J.; Sleeper, Robert K.; Smith, Eunice, G.; and Merritt, Leslie R. (Federal Aviation Administration): Preliminary Test Results of the Joint FAA-USAF-NASA Runway Research Program, Part II - Traction Measurements of Several Runways Under Wet, Snow-Covered, and Dry Conditions with a Douglas, DC-9, a Diagonal-Braked Vehicle, and a Mu-Meter. NASA TM X-73910, 1977.
8. Yager, Thomas J.; and White, Ellis J.: Recent Progress Towards Predicting Aircraft Ground Handling Performance. NASA TM 81952, March 1981.
9. Yager, Thomas J.: Review of Factors Affecting Aircraft Wet Runway Performance. Paper 83-0274, American Institute of Aeronautics and Astronautics 21st Aerospace Sciences Meeting, January 1981.
10. Model L-1011 (Base Aircraft) Landing Performance Report for FAA Evaluation of Concorde SST Special Condition 25-43-EU-12. Report No. LR 26227 Lockheed Aircraft Corp., January 14, 1974.
11. Yager, Thomas J.: Factors Influencing Aircraft Ground Handling Performance. NASA TM 85652, June 1983.
12. Fristedt, Knut; and Norrbom, Bo: Studies Cornering Snow, Ice, and Slush on Runways. FFA Memo 106, Aeronautical Research Institute of Sweden, 1975.
13. Anon.: Bowmonk Brakemeter - Dynometer Technical Evaluation AK-71-09-211 Airports and Construction, Transport Canada, May 1985.

14. Fristedt, Knut; and Norrbom, Bo: Studies of Contaminated Runway, FFA Memo 121, Aeronautical Research Institute of Sweden, 1980.
15. Herb, H. R.: Problems Associated with the Presence of Water, Slush, Snow, and Ice on Runways. AGARD Report 500, 1965.
16. Yager, Thomas J.; Fowler, Ray; and Daiutolo, Hector: FAA/NASA Runway Operational Experiments in Ice and Snow. Presented at 65th Annual Transportation Research Board Meeting, Washington, DC, January 13-17, 1986.
17. Yager, Thomas J.; and Daiutolo, Hector: Recent Winter Runway Test Findings from Joint FAA/NASA Program. Presented at the 20 Annual International Aviation Snow Symposium, American Association of Airport Executives, Allentown, PA, April 28-May 1, 1986.
18. Yager, Thomas J.: Tire and Runway Surface Research. Paper 861618, Society of Automotive Engineers, October 1986.
19. Yager, Thomas J.: Winter Runway Testing in Joint FAA/NASA Aircraft/Ground Vehicle Runway Friction Program. Presented at the Annual International Aviation Snow Symposium, American Association of Airport Executives, Buffalo, NY, April 26-30, 1987.
20. Yager, Thomas J.; Vogler, William A.; and Baldasare, Paul: Summary Report on Aircraft and Ground Vehicle Friction Correlation Test Results Obtained Under Winter Runway Conditions During Joint FAA/NASA Runway Friction Program. NASA TM 100506, March 1988.
21. Horne, Walter B.; and Buhlmann, F.: A Method for Rating the Skid Resistance and Micro/Macrotexture Characteristics of Wet Pavements, Frictional Interaction of Tire and Pavement, ASTM STP 793, W. E. Meyer and J. D. Walter, Eds. pp 191-218, 1983.
22. Anon.: Airport Services Manual, Doc. 9137-An/898 Part 2 - Pavement Surface Conditions. International Civil Aviation Organization, second edition, 1984.

Aircraft and Ground Vehicle Friction Measurements Obtained Under Winter Runway Conditions

Thomas J. Yager
NASA Langley Research Center

* Paper 891070 presented at the General Aviation Aircraft Meeting and Exposition, Wichita, Kansas, April, 1989.

ABSTRACT

Tests with specially instrumented NASA B-737 and B-727 aircraft together with several different ground friction measuring devices have been conducted for a variety of runway surface types and wetness conditions. This effort is part of the Joint FAA/NASA Aircraft/Ground Vehicle Runway Friction Program aimed at obtaining a better understanding of aircraft ground handling performance under adverse weather conditions and defining relationships between aircraft and ground vehicle tire friction measurements. Aircraft braking performance on dry, wet, snow-, and ice-covered runway conditions is discussed together with ground vehicle friction data obtained under similar runway conditions. For the wet, compacted snow- and ice-covered runway conditions, the relationship between ground vehicles and aircraft friction data is identified. The influence of major test parameters on friction measurements such as speed, test tire characteristics, and surface contaminant type are discussed. The test results indicate that use of properly maintained and calibrated ground vehicles for monitoring runway friction conditions should be encouraged particularly under adverse weather conditions.

THERE IS AN IMPERATIVE NEED for information on runways which may become slippery due to various forms and types of contaminants. Experience has shown that since the beginning of "all weather" aircraft operations, there have been landing and aborted takeoff incidents and/or accidents each year where aircraft have either run off the end or veered off the shoulder of low friction runways. From January 1981 to January 1988, more than 400 traction-related incident/accidents have occurred according to Federal Aviation Administration (FAA) and National Transportation Safety Board (NTSB) records. These cases have provided the motivation for various government agencies and aviation industries to conduct extensive tests and research programs to identify the factors which cause the runway friction to be less than acceptable [1-15]. The continued occurrence of aircraft takeoff and landing accidents emphasize the need for improved measurement techniques and inspection procedures related to tire and runway conditions. NASA Langley's Landing and Impact Dynamics Branch is involved in several research programs directed towards obtaining a better understanding of how different tire properties interact with varying pavement surface characteristics to produce acceptable performance for aircraft ground handling requirements. The following sections of this paper describe one such effort which was jointly supported by not only NASA and the FAA but by several aviation industry groups.

The Joint FAA/NASA Aircraft/Ground Vehicle Runway Friction Program is aimed at obtaining a better understanding of aircraft ground handling performance under a variety of adverse weather conditions and to define relationships between aircraft and ground vehicle tire friction measurements. Major parameters influencing tire friction performance such as speed, contaminant type and amount, test tire inflation pressure, and runway surface texture were evaluated during the test program. These tests involved a specially instrumented NASA B-737 aircraft and an FAA B-727 aircraft shown during test runs in figure 1. Several

Figure 1.- Specially instrumented test aircraft.

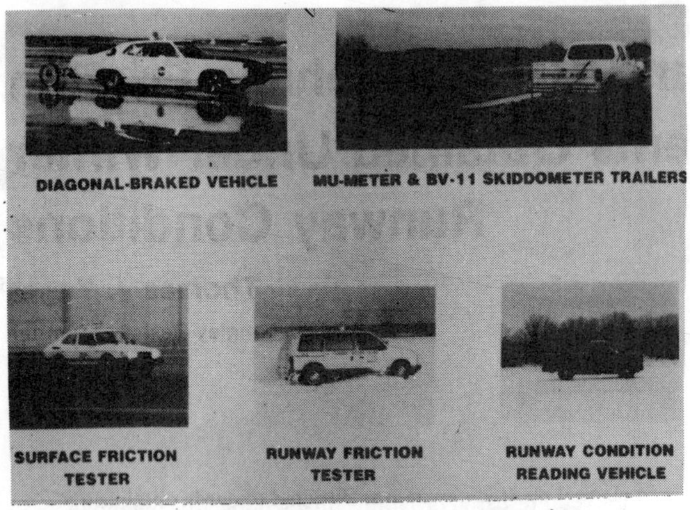

Figure 2.- Ground friction measuring devices.

different ground friction measuring vehicles used during the program are shown in figure 2. The diagonal-braked vehicle developed by NASA measures locked wheel sliding friction values. The FAA mu-meter trailer monitors side force variation on two tires yawed to an included angle of 15 degrees. Both the surface friction tester automobile and Swedish BV-11 skiddometer trailer measure tire braking friction near the peak of the tire friction/slip ratio curve. A relatively new runway friction tester van also measures peak tire braking friction. Both a Tapley meter and a Bowmonk brakemeter were installed in the runway condition reading (RCR) vehicle to indicate vehicle braking deceleration levels under snow and ice conditions. Additional information describing these different ground friction measuring vehicles is contained in references 16 to 19. With these known differences in ground vehicle test tire operational modes, different levels of tire friction measurements were expected, and obtained, for the same runway surface condition. Between June 1983 and March 1986, tests were performed on 12 different concrete and asphalt runways, grooved and nongrooved, including porous friction coarse, under dry, truck wet, rain wet, snow-, slush-, and ice-covered surface conditions. A limited assessment of some runway chemical de-icing treatments was also obtained. Over 200 test runs were made with the two transport aircraft and over 1100 runs were made with the different ground test vehicles. Most of the dry and the truck wet runway surface test runs were performed at NASA Wallops Flight Facility in Virginia and the FAA Technical Center airport in New Jersey. A limited number of rain wet tests were performed at Langley Air Force Base, Virginia, Pease Air Force Base, New Hampshire, and Portland International Jetport, Maine [20]. All the winter runway test conditions were evaluated at Brunswick Naval Air Station in Maine. The test procedure for wet runway conditions was to make ground vehicle runs before and after each aircraft braking run. For the winter runway conditions of compacted snow and solid ice, a series of ground vehicle runs were made immediately following the aircraft test runs on each surface contamination condition. At loose snow depths equal to or greater than 5 cm (2 in.), test runs with the two trailer devices were suspended because constant speed could not be maintained.

TEST RESULTS AND DISCUSSION

A substantial tire friction database has been collected during this Joint FAA/NASA Runway Friction Program with extensive data reduction and analysis being accomplished at NASA Langley. All of the runway friction data will be discussed and analyzed in a soon-to-be-published NASA technical report that has undergone both FAA and NASA technical reviews. Only a very limited amount of aircraft and ground vehicle friction data are presented herein to indicate some of the major test findings and data trends.

Wet runways - The range of B-737 aircraft and ground vehicle friction measurements obtained on nongrooved and grooved surfaces under truck wet conditions is shown in figure 3. As expected, the grooved

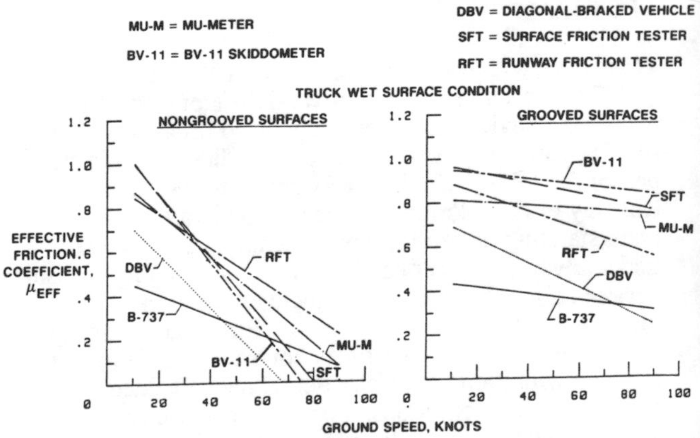

Figure 3.- Range of B-737 aircraft and ground vehicle friction measurements.

runway surface friction data is significantly greater than the nongrooved data, particularly at the higher speeds. Most of the ground vehicle friction values were higher than those developed by the B-737 aircraft because of differences in braking test mode, tire tread design, and tire inflation pressure. When these major factors are properly considered using techniques and methodologies being developed at NASA Langley [21], aircraft wet runway braking performance can be estimated from ground vehicle friction measurements. The relationship between actual braking

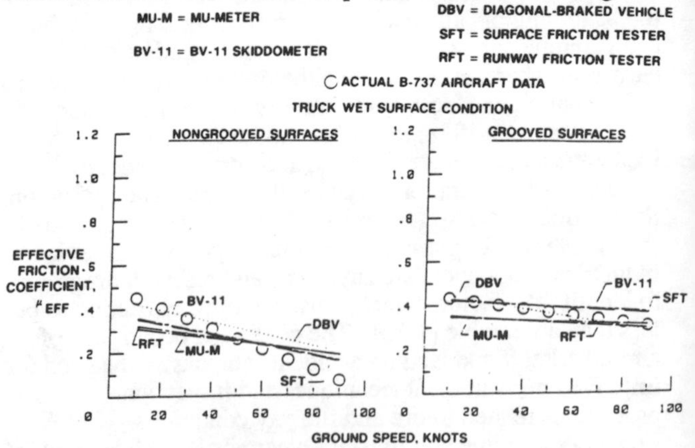

Figure 4.- Relationship between actual and estimated B-737 aircraft braking performance.

friction coefficient for the B-737 and estimated braking friction coefficients of the airplane obtained from the ground vehicle measurements is shown in figure 4. For most of the ground vehicle friction measurements, the estimated aircraft performance is in good agreement with the actual measured aircraft braking friction level. The available data suggest that the ground vehicle friction data for wet runway conditions can estimate aircraft tire friction performance to within about 15 percent of the actual measured aircraft friction values and in some cases, within 5 percent. The relationship between ground vehicle estimated and actual aircraft tire friction values will vary with changes in wetness conditions. Hence, ground vehicle friction measurements should be taken on a runway for a range of wetness conditions related to different precipitation rates and surface winds.

Snow- and ice-covered runways - A comparison of B-737 aircraft braking performance for snow- and ice-covered runways as well as dry, truck wet, and flooded conditions is given in figure 5. The range of aircraft

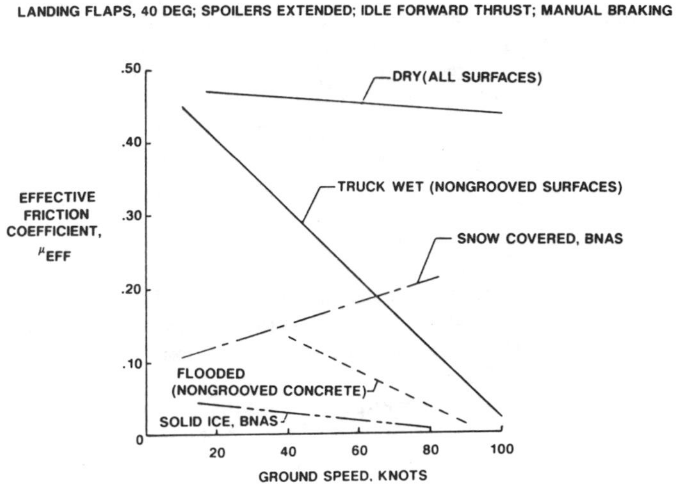

Figure 5.- Comparison of B-737 aircraft braking performance.

effective friction coefficients is from nearly 0.5 on dry runways to 0.05 on the solid ice surface at Brunswick Naval Air Station (BNAS). Similar results were obtained during the B-727 aircraft tests. Several nonbraking test runs were conducted with the B-737 aircraft rolling through 15 cm (6 in.) of loose snow on the runway. The variation in aircraft deceleration due to impingement drag with ground speed is shown in figure 6. The magnitude of this deceleration on the aircraft is sufficient to prevent the B-737 aircraft from reaching rotational speed for takeoffs. For compacted snow- and ice-covered conditions, the friction measurements obtained with the various ground test devices indicated that forward speed had little effect on the magnitude of the friction values. Furthermore, the friction values obtained from each vehicle showed no significant difference between compacted snow- and ice-covered conditions. The Tapley and Bowmonk meters were both installed in the Navy runway condition reading vehicle and the manually recorded friction values for each instrument were in close agreement for a given test run. Reference 22 describes the RCR test procedure used at U. S. military bases for monitoring runway friction conditions.

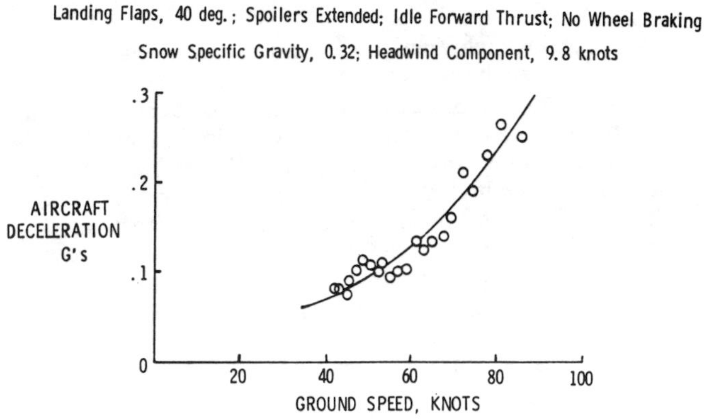

Figure 6.- Deceleration of B-737 aircraft in 15 cm (6 in.) loose snow.

The range of aircraft effective braking friction coefficient values with ground speed for compacted snow- and ice-covered runway conditions is shown in figure 7. The data symbols and line codes denote the different test conditions and aircraft. The best fit, least squares, linear curve for the compacted snow-covered surface friction data, denoted by the solid line, is nearly four times greater than the data from the glare ice-covered surface denoted by the dashed line. These aircraft results differ from the ground vehicle measurements which indicated no significant difference between compacted snow-covered runway condition and the solid ice-covered condition. The difference in braking performance shown in figure 7 between the two test aircraft under these winter runway conditions was considered insignificant. The aircraft braking performance on the snow-covered and ice-covered surfaces was relatively insensitive to ground speed variations which was also found for the ground vehicle measurements.

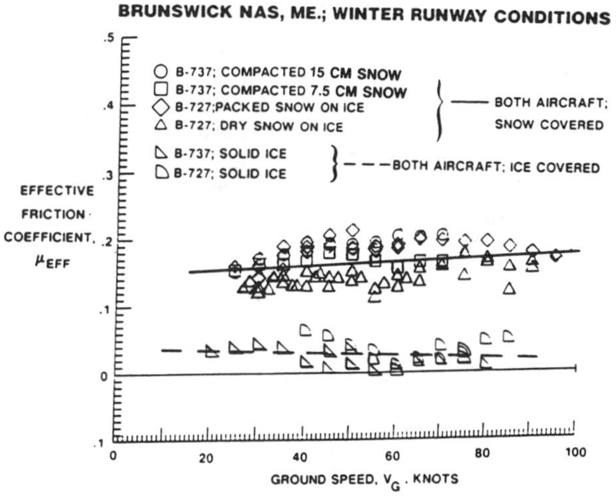

Figure 7.- Aircraft braking friction performance on compacted snow- and ice-covered runways.

Since each test aircraft indicated a significant difference between the compacted snow-covered and ice-covered surface conditions, two ranges or means of aircraft braking friction data were selected to define the relationship with the ground vehicle friction measurements. The resulting aircraft and ground vehicle friction correlation chart is shown in figure 8 where the compacted snow-covered and ice-covered surface condition is delineated for the two aircraft. For the compacted snow-covered surface condition, an aircraft effective braking friction coefficient value of 0.21 was selected for the highest braking action level and 0.12 was used for the lowest braking action level. An effective braking friction coefficient range from 0.055 to 0.01 was selected for comparable aircraft braking action levels on the ice-covered surface condition. The dashed line in figure 8 depicts comparable values for other ground vehicles and the two aircraft/surface conditions for an RCR value of 15.

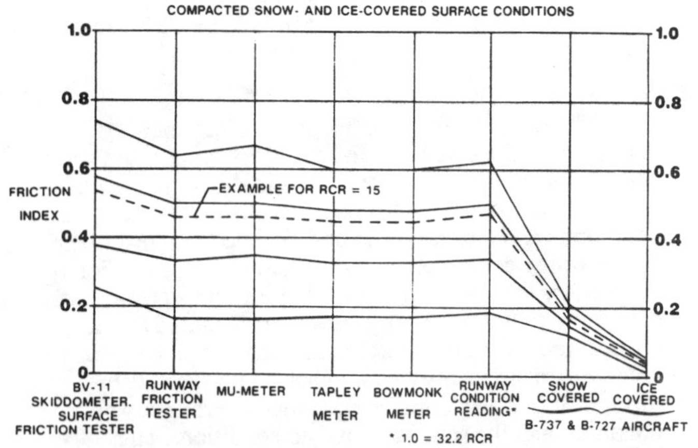

Figure 8.- Aircraft and ground vehicle friction correlation chart.

From an aircraft operator's viewpoint, these values of friction for a snow- or ice-covered runway must be considered in respect to the actual runway geometry and such environmental conditions as pressure/altitude, winds, and ambient temperature at the time of a particular aircraft operation. It should also be recognized that aircraft operations can occur on runways which have a nonuniform mixture of compacted snow-covered area and exposed solid ice-covered surfaces. In such circumstances, additional ground vehicle friction measurements need to be taken to adequately determine average friction numbers for each portion of the runway. How well this established relationship between aircraft and ground vehicle friction values remains for other aircraft types is somewhat questionable although the available data tends to suggest a similar relationship. The use of actual friction numbers in place of qualitative braking action terms is strongly recommended because with experience, these runway friction values measured by a ground vehicle will provide the pilot a more precise and accurate gage on the safety margins available for landing on a given runway. Proper and timely use of snow removal equipment and runway chemical treatments to minimize and/or remove snow and ice contaminants is still recognized as a necessity to return to dry runway friction levels as soon as possible.

CONCLUDING REMARKS

An overview of the Joint FAA/NASA Aircraft/Ground Vehicle Runway Friction Program has been given. A substantial tire friction database has been collected from tests with two instrumented transport aircraft and several different ground test vehicles on a variety of runway surfaces and wetness conditions. A better understanding of the major factors influencing tire friction performance has been achieved. The relationships defined between the different ground vehicles and between ground vehicle and aircraft tire friction performance are very encouraging. Greater usage of ground vehicle friction measurements at airports is strongly encouraged to define runway surface maintenance requirements and to monitor current runway friction levels under adverse weather conditions.

In October 1988, a Runway Friction Workshop was held at NASA Langley to discuss with the aviation community the preliminary test results from the joint program and to obtain their comments and recommendations. Eighteen formal presentations were made to approximately 80 attendees representing U. S., Canadian, and Swedish government agencies, airframe manufacturers, airlines and pilots, airport managers, ground test vehicle manufacturers/suppliers, and aircraft tire and brake companies. Separate presentations were given concerning runway friction work being conducted in Sweden, England, France, Japan, and Canada. Based upon workshop discussion, the Joint Runway Friction Program draft report has been modified and improved. Future plans include a Joint NASA/FAA Surface Traction Program using the Aircraft Landing Dynamics Facility at Langley to evaluate radial-constructed transport aircraft tires. Work in designing a new standardized form for use at all U. S. airports for reporting and documenting ground vehicle/aircraft friction data will be initiated. Additional meetings with aviation industry representations are planned at FAA Headquarters to discuss how the joint program test findings impact existing advisory circulars, standards, and regulations. With new improved test tires, brake systems, and other equipment becoming available for airport operations in future years, the need is recognized for continued testing of aircraft/ground vehicle runway friction performance.

REFERENCES

1. Horne, Walter B.; Yager, Thomas J.; and Taylor, Glenn R.: Review of Causes and Alleviation of Low Tire Traction on Wet Runways, NASA TN D-4406, 1968.
2. Anon.: Pavement Grooving and Traction Studies, NASA SP-5073, 1969.
3. Yager, Thomas J.; Phillips, W. Pelham; Horne, Walter B.; and Sparks, Howard C. (Aeronautical Systems Division, WPAFB): Comparison of Aircraft and Ground Vehicle Stopping Performance on Dry, Wet, Flooded, Slush- and Ice-Covered Runways. NASA TN D-6098, November 1970.
4. Merritt, Leslie, R.: Impact of Runway Traction on Possible Approaches to Certification and Operation of Jet Transport Aircraft. Paper 740497, Society of Automotive Engineers, April-May 1974.
5. Horne, Walter B.: Status of Runway Slipperiness Research. NASA SP-416, October 1976, pp. 191-245.
6. Horne, Walter B.; Yager, Thomas J.; Sleeper, Robert K.; and Merritt, Leslie R. (Federal Aviation Administration): Preliminary Test Results of the Joint FAA-USAF-NASA Runway Research Program, Part I - Traction Measurements of Several Runways Under Wet and Dry Conditions with a Boeing 727, a Diagonal-Braked Vehicle, and a Mu-Meter. NASA TM X-73909, 1977.

7. Horne, Walter B.; Yager, Thomas J.; Sleeper, Robert K.; Smith, Eunice, G.; and Merritt, Leslie R. (Federal Aviation Administration): Preliminary Test Results of the Joint FAA-USAF-NASA Runway Research Program, Part II - Traction Measurements of Several Runways Under Wet, Snow-Covered, and Dry Conditions with a Douglas, DC-9, a Diagonal-Braked Vehicle, and a Mu-Meter. NASA TM X-73910, 1977.
8. Yager, Thomas J.; and White, Ellis J.: Recent Progress Towards Predicting Aircraft Ground Handling Performance. NASA TM 81952, March 1981.
9. Yager, Thomas J.: Review of Factors Affecting Aircraft Wet Runway Performance. Paper 83-0274, American Institute of Aeronautics and Astronautics 21st Aerospace Sciences Meeting, January 1981.
10. Model L-1011 (Base Aircraft) Landing Performance Report for FAA Evaluation of Concorde SST Special Condition 25-43-EU-12. Report No. LR 26227 Lockheed Aircraft Corp., January 14, 1974.
11. Yager, Thomas J.: Factors Influencing Aircraft Ground Handling Performance. NASA TM 85652, June 1983.
12. Fristedt, Knut; and Norrbom, Bo: Studies Cornering Snow, Ice, and Slush on Runways. FFA Memo 106, Aeronautical Research Institute of Sweden, 1975.
13. Anon.: Bowmonk Brakemeter - Dynometer Technical Evaluation AK-71-09-211 Airports and Construction, Transport Canada, May 1985.
14. Fristedt, Knut; and Norrbom, Bo: Studies of Contaminated Runway, FFA Memo 121, Aeronautical Research Institute of Sweden, 1980.
15. Herb, H. R.: Problems Associated with the Presence of Water, Slush, Snow, and Ice on Runways. AGARD Report 500, 1965.
16. Yager, Thomas J.; Fowler, Ray; and Daiutolo, Hector: FAA/NASA Runway Operational Experiments in Ice and Snow. Presented at 65th Annual Transportation Research Board Meeting, Washington, DC, January 13-17, 1986.
17. Yager, Thomas J.; and Daiutolo, Hector: Recent Winter Runway Test Findings from Joint FAA/NASA Program. Presented at the 20th Annual International Aviation Snow Symposium, American Association of Airport Executives, Allentown, PA, April 28-May 1, 1986.
18. Yager, Thomas J.: Tire and Runway Surface Research. Paper 861618, Society of Automotive Engineers, October 1986.
19. Yager, Thomas J.: Winter Runway Testing in Joint FAA/NASA Aircraft/Ground Vehicle Runway Friction Program. Presented at the Annual International Aviation Snow Symposium, American Association of Airport Executives, Buffalo, NY, April 26-30, 1987.
20. Yager, Thomas J.; Vogler, William A.; and Baldasare, Paul: Summary Report on Aircraft and Ground Vehicle Friction Correlation Test Results Obtained Under Winter Runway Conditions During Joint FAA/NASA Runway Friction Program. NASA TM 100506, March 1988.
21. Horne, Walter B.; and Buhlmann, F.: A Method for Rating the Skid Resistance and Micro/Macrotexture Characteristics of Wet Pavements, Frictional Interaction of Tire and Pavement, ASTM STP 793, W. E. Meyer and J. D. Walter, Eds. pp 191-218, 1983.
22. Anon.: Airport Services Manual, Doc. 9137- An/898 Part 2 - Pavement Surface Conditions. International Civil Aviation Organization, second edition, 1984.

Current Status of Joint FAA/NASA Runway Friction Program

Thomas J. Yager
NASA Langley Research Center
William A. Vogler
Planning Research Corp.

* Paper 892340 presented at the Aerospace Technology Conference and Exposition, Anaheim, California, September, 1989.

ABSTRACT

Tests with specially instrumented NASA B-737 and FAA B-727 aircraft together with several different ground friction measuring devices have been conducted for a variety of runway surface types and wetness conditions. This effort is part of the Joint FAA/NASA Aircraft/Ground Vehicle Runway Friction Program aimed at obtaining a better understanding of aircraft ground handling performance under adverse weather conditions and defining relationships between aircraft and ground vehicle tire friction measurements. Aircraft braking performance on dry, wet, snow-, and ice-covered runway conditions is discussed together with ground vehicle friction data obtained under similar runway conditions. For the wet, compacted snow- and ice-covered runway conditions, the relationship between ground vehicles and aircraft friction data is identified. The influence of major test parameters on friction measurements such as speed, test tire characteristics, and surface contaminant type are discussed. The test results indicate that use of properly maintained and calibrated ground vehicles for monitoring runway friction conditions should be encouraged particularly under adverse weather conditions. The current status of the runway friction program is summarized and future test plans are identified.

THERE IS AN IMPERATIVE NEED for information on runways which may become slippery due to various forms and types of contaminants. Experience has shown that since the beginning of "all weather" aircraft operations, there have been landing and aborted takeoff incidents and/or accidents each year where aircraft have either run off the end or veered off the shoulder of low friction runways. From January 1981 to January 1988, more than 400 traction-related incident/accidents have occurred according to Federal Aviation Administration (FAA) and National Transportation Safety Board (NTSB) records. These cases have provided the motivation for various government agencies and aviation industries to conduct extensive tests and research programs to identify the factors which cause the runway friction to be less than acceptable [1-15]. The continued occurrence of aircraft takeoff and landing accidents emphasize the need for improved measurement techniques and inspection procedures related to tire and runway conditions. NASA Langley's Landing and Impact Dynamics Branch is involved in several research programs directed towards obtaining a better understanding of how different tire properties interact with varying pavement surface characteristics to produce acceptable performance for aircraft ground handling requirements. The following sections of this paper describe one such effort which was jointly supported by not only NASA and the FAA but by several aviation industry groups.

SCOPE OF PROGRAM

The Joint FAA/NASA Aircraft/Ground Vehicle Runway Friction Program is aimed at obtaining a better understanding of aircraft ground handling performance under a variety of adverse weather conditions and to define relationships between aircraft and ground vehicle tire friction measurements. Major parameters influencing tire friction performance such as speed, contaminant type and amount, test tire inflation pressure, and runway surface texture were evaluated during the test program. These tests involved a specially instrumented NASA B-737 aircraft and an FAA B-727 aircraft shown during test runs in figure 1. Several

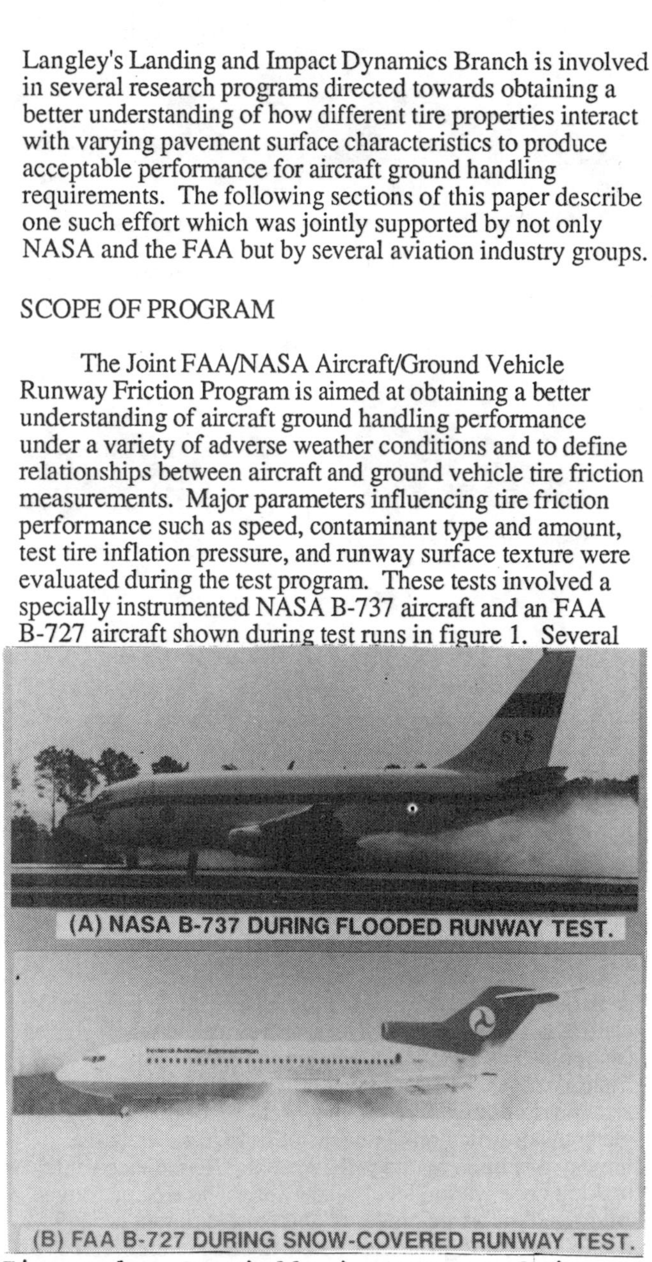
(A) NASA B-737 DURING FLOODED RUNWAY TEST.
(B) FAA B-727 DURING SNOW-COVERED RUNWAY TEST.
Figure 1.- Specially instrumented aircraft.

different ground friction measuring vehicles used during the program are shown in figures 2 and 3. The FAA mu-meter trailer monitors side force variation on two tires yawed to an included angle of 15 degrees. The Swedish BV-11 skiddometer trailer measures tire braking friction near the

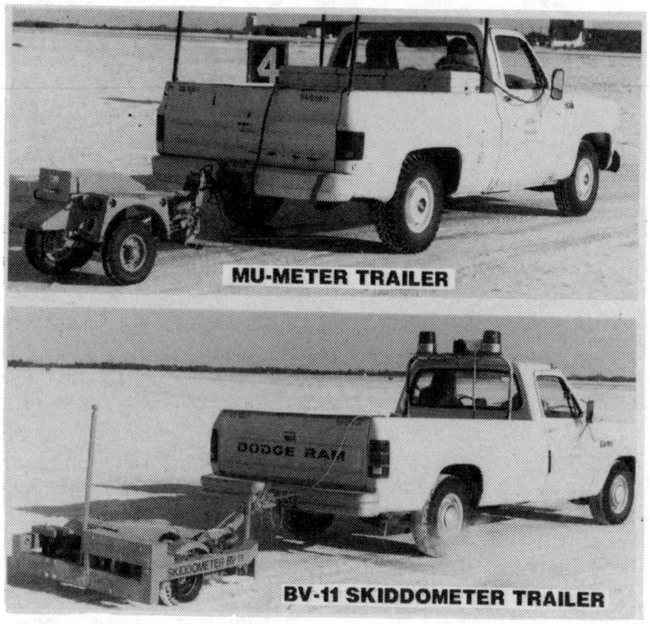

Figure 2.- Ground friction measuring trailer devices.

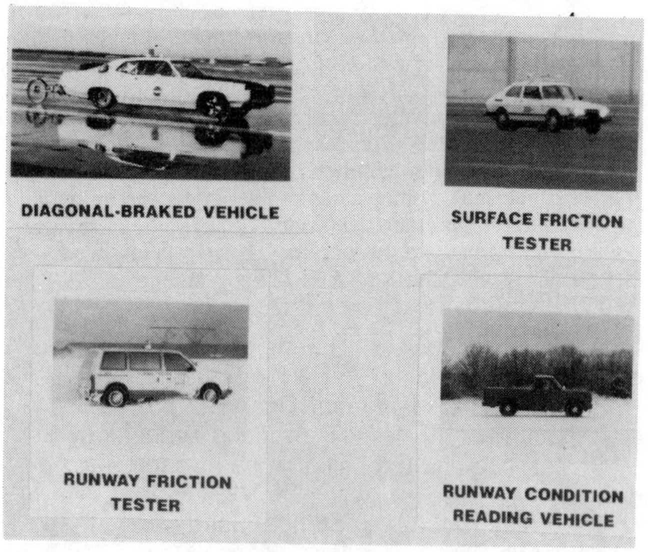

Figure 3.- Ground friction measuring vehicles.

peak of the tire friction/slip ratio curve. The diagonal-braked vehicle developed by NASA measures locked wheel sliding friction values and the surface friction tester measures peak friction values. A relatively new runway friction tester van also measures peak tire braking friction. Both a Tapley meter and a Bowmonk brakemeter were installed in the runway condition reading (RCR) vehicle to indicate vehicle braking deceleration levels under snow and ice conditions. Additional information describing these different ground friction measuring vehicles is contained in references 16 to 19. With these known differences in ground vehicle test tire operational modes, different levels of tire friction measurements were expected, and obtained, for the same runway surface condition. Between June 1983 and March 1986, tests were performed on 12 different concrete and asphalt runways, grooved and nongrooved, including porous friction coarse, under dry, truck wet, rain wet, snow-, slush-, and ice-covered surface conditions. A limited assessment of some runway chemical de-icing treatments was also obtained. Over 200 test runs were made with the two transport aircraft and over 1100 runs were made with the different ground test vehicles. Most of the dry and the truck wet runway surface test runs were performed at NASA Wallops Flight Facility in Virginia and the FAA Technical Center airport in New Jersey. A limited number of rain wet tests were performed at Langley Air Force Base, Virginia, Pease Air Force Base, New Hampshire, and Portland International Jetport, Maine [20]. All the winter runway test conditions were evaluated at Brunswick Naval Air Station in Maine. The test procedure for wet runway conditions was to make ground vehicle runs before and after each aircraft braking run. For the winter runway conditions of compacted snow and solid ice, a series of ground vehicle runs were made immediately following the aircraft test runs on each surface contamination condition. At loose snow depths equal to or greater than 5 cm (2 in.), test runs with the two trailer devices were suspended because constant speed could not be maintained.

TEST RESULTS AND DISCUSSION

A substantial tire friction database has been collected during this Joint FAA/NASA Runway Friction Program with extensive data reduction and analysis being accomplished at NASA Langley. All of the runway friction data will be discussed and analyzed in a soon-to-be-published NASA technical report that has undergone both FAA and NASA technical reviews. Only a very limited amount of aircraft and ground vehicle friction data are presented herein to indicate some of the major test findings and data trends.

Wet runways - The range of B-737 aircraft and ground vehicle friction measurements obtained on nongrooved and grooved surfaces under truck wet conditions is shown in figure 4. As expected, the grooved runway surface friction data is significantly greater than the nongrooved data, particularly at the higher speeds. Most of the ground vehicle friction values were higher than those developed by the B-737 aircraft because of differences in braking test mode, tire tread design, and tire inflation

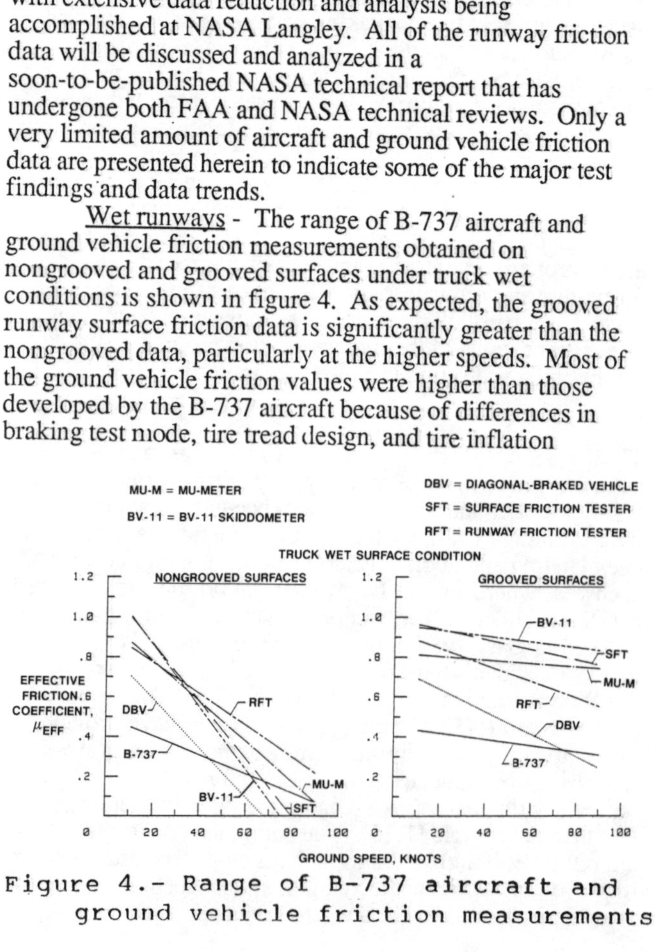

Figure 4.- Range of B-737 aircraft and ground vehicle friction measurements.

pressure. When these major factors are properly considered using techniques and methodologies being developed at NASA Langley [21], aircraft wet runway braking performance can be estimated from ground vehicle friction measurements. The relationship between actual braking friction coefficient for the B-737 and estimated braking friction coefficients of the airplane obtained from the ground vehicle measurements is shown in figure 5. For most of the

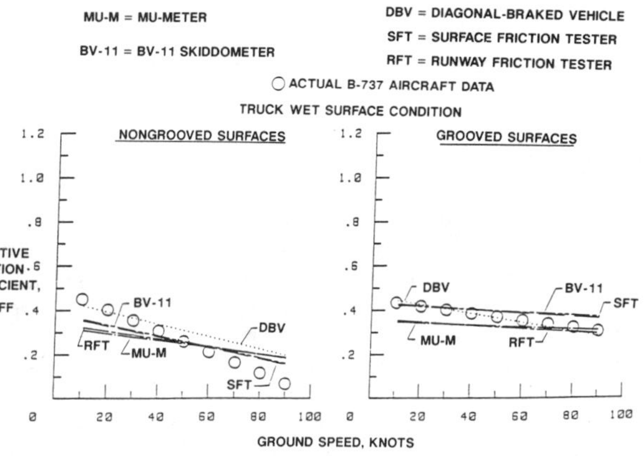

Figure 5.- Relationship between actual and estimated B-737 aircraft braking performance.

ground vehicle friction measurements, the estimated aircraft performance is in good agreement with the actual measured aircraft braking friction level. The available data suggest that the ground vehicle friction data for wet runway conditions can estimate aircraft tire friction performance to within about 15 percent of the actual measured aircraft friction values and in some cases, within 5 percent. The relationship between ground vehicle estimated and actual aircraft tire friction values will vary with changes in wetness conditions. Hence, ground vehicle friction measurements should be taken on a runway for a range of wetness conditions related to different precipitation rates and surface winds.

Snow- and ice-covered runways - A comparison of B-737 aircraft braking performance for snow- and ice-covered runways as well as dry, truck wet, and flooded conditions is given in figure 6. The range of aircraft

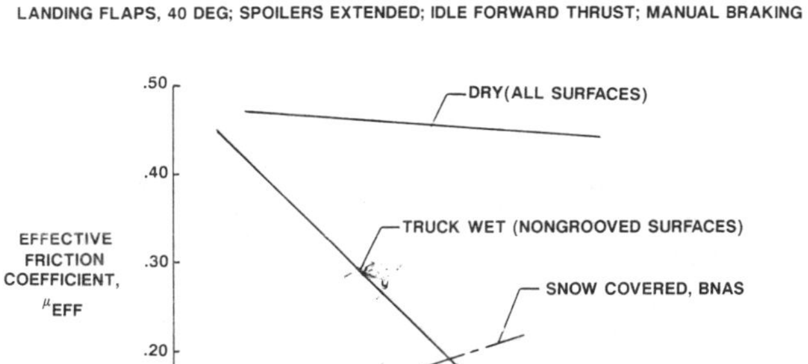

Figure 6.- Comparison of B-737 aircraft braking performance.

effective friction coefficients is from nearly 0.5 on dry runways to 0.05 on the solid ice surface at Brunswick Naval Air Station (BNAS). Similar results were obtained during the B-727 aircraft tests.

For winter runway conditions, the friction measurements obtained with the various ground test vehicles indicated that forward speed had little effect on the magnitude of the friction values. Furthermore, the friction values obtained from each vehicle showed no significant difference between compacted snow- and ice-covered conditions. The Tapley and Bowmonk meters were both installed in the Navy runway condition reading vehicle and the manually recorded friction values for each instrument were in close agreement for a given test run. Reference 22 describes the RCR test procedure used at U. S. military bases for monitoring runway friction conditions.

Figure 7 provides a listing of the range of friction readings for four braking action classifications derived from the test results at Brunswick NAS as well as from other similar winter runway test results obtained at other locations.

Runway Surface Conditions: Compacted Snow and Ice

BRAKING ACTION LEVEL	GROUND VEHICLE FRICTION READINGS						
	MU-METER	TAPLEY METER	RUNWAY CONDITION READINGS (RCR)	BOWMONK METER	SURFACE FRICTION TESTER	RUNWAY FRICTION TESTER	BV-11 SKIDDOMETER
EXCELLENT	0.50 and above	0.53 and above	17 and above	0.51 and above	0.53 and above	0.50 and above	0.58 and above
GOOD	0.47 to 0.35	0.50 to 0.38	16 to 12	0.48 to 0.37	0.50 to 0.37	0.47 to 0.35	0.54 to 0.41
MARGINAL	0.33 t0 0.26	0.35 to 0.28	11 to 9	0.34 to 0.27	0.34 to 0.28	0.33 to 0.26	0.37 to 0.31
POOR	0.24 and below	0.26 and below	8 and below	0.25 and below	0.25 and below	0.24 and below	0.27 and below

NOTES:

(1) Mu-meter equipped with smooth RL-2 tires inflated to 69 kPA (10 lb/in.^{2})
(2) Runway friction tester equipped with smooth RL-2 tire inflated to 207 kPA (30 lb/in.^{2})
(3) Surface friction tester and BV-11 skiddometer equipped with grooved aero tire inflated to 690 kPA (100 lb/in.^{2})
(4) Ambient air temperature range, -15 to +5 °C (5 to 41 °F)
(5) Test speed range, 32 to 97 km/h (20 to 60 mph)
(6) RCR values equal Tapley meter reading x 32.

Figure 7.- Ground vehicle friction reading correlation table.

The vehicle test tire conditions, range of ambient temperatures, and test speeds are indicated in the notes accompanying figure 7. Qualitative verbal braking action terms, namely, excellent, good, marginal, and poor, were used to identify four distinct levels or ranges in friction readings for each device.

In general, the excellent friction readings were close to some wet surface values, e.g. 0.5 and above, whereas, the poor friction readings were normally below 0.25 and found on the solid glare ice-covered surface. The data obtained in figure 7 are plotted in the chart given in figure 8 to illustrate the friction relationship between the different ground vehicle devices as well as the two test aircraft. The format for this chart was derived from a chart contained in reference 22 used by European countries. The mu-meter and the runway friction tester which measured similar friction values, are plotted together. The compared snow- and ice-covered surface condition is delineated for the two aircraft since a significant difference in aircraft friction performance was found for these two surface conditions. The four lines represent a sample derivation of the vehicle friction measurements comparable or equivalent to RCR values of 5, 10, 15, and 20. The range of friction values at each of these four levels is nearly the same for the mu-meter. Runway

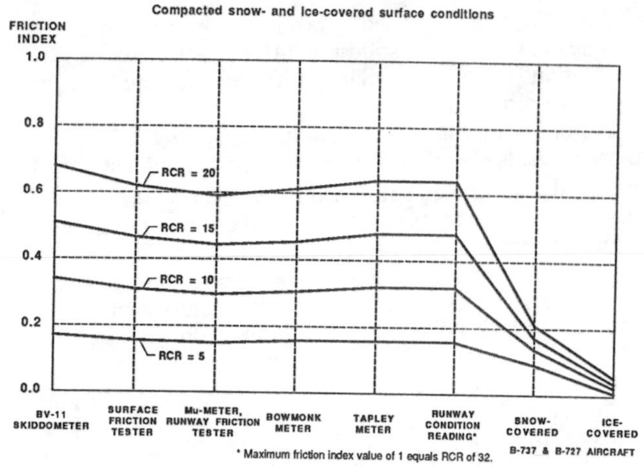

Figure 8.- Aircraft and ground vehicle friction correlation chart.

friction tester, tapley meter, and the Bowmonk meter. Slightly higher values of friction for each level were obtained with the surface friction tester and the BV-11 skiddometer mainly because of using a higher test tire inflation pressure (100 vs. 30 psi or less) combined with a grooved tread pattern on the tire instead of a smooth (black) tread.

Some limited ground vehicle friction data collected using the Tapley meter has been evaluated in an effort to better define the effects of ambient temperature and solar heating on tire friction performance. These data are given in figure 9 with the upper solid line indicating the variation in friction readings with temperature during overcast or at night (minimum solar heating). The lower, dashed curve indicates tire friction variation with temperature measured during daylight hours with bright sun (maximum solar heating).

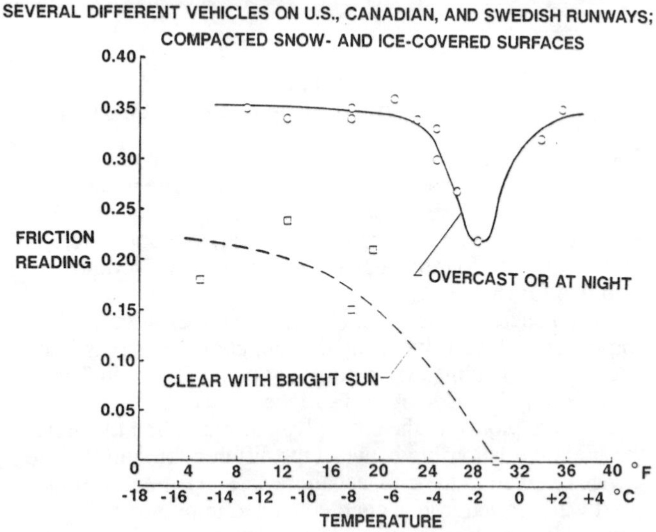

Figure 9.- Temperature effect on Tapley meter friction measurements.

These comparable data indicate that solar heating has a significant effect on tire friction performance and that temperature only is significant near ($\pm 5^{o}$ F) the freezing point.

From an aircraft operator's viewpoint, these values of friction for a snow- or ice-covered runway must be considered in respect to the actual runway geometry and such environmental conditions as pressure/altitude, winds, and ambient temperature at the time of a particular aircraft operation. It should also be recognized that aircraft operations can occur on runways which have a nonuniform mixture of compacted snow-covered area and exposed solid ice-covered surfaces. In such circumstances, additional ground vehicle friction measurements need to be taken to adequately determine average friction numbers for each portion of the runway. How well this established relationship between aircraft and ground vehicle friction values remains for other aircraft types is somewhat questionable although the available data tends to suggest a similar relationship. The use of actual friction numbers in place of qualitative braking action terms is strongly recommended because with experience, these runway friction values measured by a ground vehicle will provide the pilot a more precise and accurate gage on the safety margins available for landing on a given runway. Proper and timely use of snow removal equipment and runway chemical treatments to minimize and/or remove snow and ice contaminants is still recognized as a necessity to return to dry runway friction levels as soon as possible.

CONCLUDING REMARKS

An overview of the Joint FAA/NASA Aircraft/Ground Vehicle Runway Friction Program has been given. A substantial tire friction database has been collected from tests with two instrumented transport aircraft and several different ground test vehicles on a variety of runway surfaces and wetness conditions. A better understanding of the major factors influencing tire friction performance has been achieved. The relationships defined between the different ground vehicles and between ground vehicle and aircraft tire friction performance are very encouraging. Greater usage of ground vehicle friction measurements at airports is strongly encouraged to define runway surface maintenance requirements and to monitor current runway friction levels under adverse weather conditions.

In October 1988, a Runway Friction Workshop was held at NASA Langley to discuss with the aviation community the preliminary test results from the joint program and to obtain their comments and recommendations. Eighteen formal presentations were made to approximately 80 attendees representing U. S., Canadian, and Swedish government agencies, airframe manufacturers, airlines and pilots, airport managers, ground test vehicle manufacturers/suppliers, and aircraft tire and brake companies. Separate presentations were given concerning runway friction work being conducted in Sweden, England, France, Japan, and Canada. Based upon workshop discussion, the Joint Runway Friction Program draft report has been modified and improved. Future plans include a Joint NASA/FAA Surface Traction Program using the Aircraft Landing Dynamics Facility at Langley to evaluate radial-constructed transport aircraft tires. Work in designing a new standardized form for use at all U. S. airports for reporting and documenting ground vehicle/aircraft friction data will be initiated. Additional meetings with aviation industry representations are planned at FAA Headquarters to discuss how the joint program test findings impact existing advisory circulars, standards, and regulations. With new improved test tires, brake systems, and other equipment becoming available for airport operations in future years, the need is recognized for continued testing of aircraft/ground vehicle runway friction performance.

REFERENCES

1. Horne, Walter B.; Yager, Thomas J.; and Taylor, Glenn R.: Review of Causes and Alleviation of Low Tire Traction on Wet Runways, NASA TN D-4406, 1968.
2. Anon.: Pavement Grooving and Traction Studies, NASA SP-5073, 1969.
3. Yager, Thomas J.; Phillips, W. Pelham; Horne, Walter B.; and Sparks, Howard C. (Aeronautical Systems Division, WPAFB): Comparison of Aircraft and Ground Vehicle Stopping Performance on Dry, Wet, Flooded, Slush- and Ice-Covered Runways. NASA TN D-6098, November 1970.
4. Merritt, Leslie, R.: Impact of Runway Traction on Possible Approaches to Certification and Operation of Jet Transport Aircraft. Paper 740497, Society of Automotive Engineers, April-May 1974.
5. Horne, Walter B.: Status of Runway Slipperiness Research. NASA SP-416, October 1976, pp. 191-245.
6. Horne, Walter B.; Yager, Thomas J.; Sleeper, Robert K.; and Merritt, Leslie R. (Federal Aviation Administration): Preliminary Test Results of the Joint FAA-USAF-NASA Runway Research Program, Part I - Traction Measurements of Several Runways Under Wet and Dry Conditions with a Boeing 727, a Diagonal-Braked Vehicle, and a Mu-Meter. NASA TM X-73909, 1977.
7. Horne, Walter B.; Yager, Thomas J.; Sleeper, Robert K.; Smith, Eunice, G.; and Merritt, Leslie R. (Federal Aviation Administration): Preliminary Test Results of the Joint FAA-USAF-NASA Runway Research Program, Part II - Traction Measurements of Several Runways Under Wet, Snow-Covered, and Dry Conditions with a Douglas, DC-9, a Diagonal-Braked Vehicle, and a Mu-Meter. NASA TM X-73910, 1977.
8. Yager, Thomas J.; and White, Ellis J.: Recent Progress Towards Predicting Aircraft Ground Handling Performance. NASA TM 81952, March 1981.
9. Yager, Thomas J.: Review of Factors Affecting Aircraft Wet Runway Performance. Paper 83-0274, American Institute of Aeronautics and Astronautics 21st Aerospace Sciences Meeting, January 1981.
10. Model L-1011 (Base Aircraft) Landing Performance Report for FAA Evaluation of Concorde SST Special Condition 25-43-EU-12. Report No. LR 26227 Lockheed Aircraft Corp., January 14, 1974.
11. Yager, Thomas J.: Factors Influencing Aircraft Ground Handling Performance. NASA TM 85652, June 1983.
12. Fristedt, Knut; and Norrbom, Bo: Studies Cornering Snow, Ice, and Slush on Runways. FFA Memo 106, Aeronautical Research Institute of Sweden, 1975.
13. Anon.: Bowmonk Brakemeter - Dynometer Technical Evaluation AK-71-09-211 Airports and Construction, Transport Canada, May 1985.
14. Fristedt, Knut; and Norrbom, Bo: Studies of Contaminated Runway, FFA Memo 121, Aeronautical Research Institute of Sweden, 1980.
15. Herb, H. R.: Problems Associated with the Presence of Water, Slush, Snow, and Ice on Runways. AGARD Report 500, 1965.
16. Yager, Thomas J.; Fowler, Ray; and Daiutolo, Hector: FAA/NASA Runway Operational Experiments in Ice and Snow. Presented at 65th Annual Transportation Research Board Meeting, Washington, DC, January 13-17, 1986.
17. Yager, Thomas J.; and Daiutolo, Hector: Recent Winter Runway Test Findings from Joint FAA/NASA Program. Presented at the 20th Annual International Aviation Snow Symposium, American Association of Airport Executives, Allentown, PA, April 28-May 1, 1986.
18. Yager, Thomas J.: Tire and Runway Surface Research. Paper 861618, Society of Automotive Engineers, October 1986.
19. Yager, Thomas J.: Winter Runway Testing in Joint FAA/NASA Aircraft/Ground Vehicle Runway Friction Program. Presented at the Annual International Aviation Snow Symposium, American Association of Airport Executives, Buffalo, NY, April 26-30, 1987.
20. Yager, Thomas J.; Vogler, William A.; and Baldasare, Paul: Summary Report on Aircraft and Ground Vehicle Friction Correlation Test Results Obtained Under Winter Runway Conditions During Joint FAA/NASA Runway Friction Program. NASA TM 100506, March 1988.
21. Horne, Walter B.; and Buhlmann, F.: A Method for Rating the Skid Resistance and Micro/Macrotexture Characteristics of Wet Pavements, Frictional Interaction of Tire and Pavement, ASTM STP 793, W. E. Meyer and J. D. Walter, Eds. pp 191-218, 1983.
22. Anon.: Airport Services Manual, Doc. 9137- An/898 Part 2 - Pavement Surface Conditions. International Civil Aviation Organization, second edition, 1984.

Modeling and Analyses

Recent Aircraft Tire Thermal Studies

Richard N. Dodge and Samuel K. Clark
University of Michigan

* Paper 821392 presented at the Aerospace Congress and Exposition, Anaheim, California, October, 1982.

ABSTRACT

A method has been developed for calculating the internal temperature distribution in an aircraft tire while free rolling under load. The method uses an approximate stress analysis of each point in the tire as it rolls through the contact patch, and from this stress change the mechanical work done on each volume element may be obtained and converted into a heat release rate through a knowledge of material characteristics. The tire cross-section is then considered as a body with internal heat generation, and the diffusion equation is solved numerically with appropriate boundary conditions at the wheel and runway surface. Comparisons with buried thermocouples in actual aircraft tires shows good agreement.

THE EVERYDAY USE of aircraft tires can produce interactions of temperature and stress that can cause tires to fail even if the stress levels are not above design values. These failures are due to reduced carcass structural integrity brought about by adverse interactions of stress and temperature during aircraft ground operations. One particular condition which precipitates such an occurrence is the long taxi required at some airports because of the relatively long distance from the terminal to the runway. During these long taxi runs the tire can generate a great deal of heat which interacts with the stresses which are already high because the aircraft is operating under its heaviest load. In most cases the tires have been adequately designed for the loading stresses but little or no allowance has been made for the reduced structural integrity due to the interaction of stress and temperature. Thus there is a need for methods for understanding the complex nature of the interaction of stress, temperature and time during ground operations of aircraft tires.

The purpose of this paper is to review the results of recent efforts to analytically predict and experimentally measure the temperature distribution in the carcass of an aircraft tire under free rolling conditions. These efforts have been carried out jointly by the Impact Dynamics Branch, Langley Research Center, NASA, Hampton, Virginia and by the University of Michigan, Ann Arbor, Michigan.

ANALYTICAL DEVELOPMENT

A measure of the heat generation in aircraft tires is necessary to define the strength and fatigue limitations of the tire carcass structure. The basic material property used in the analytical model for predicting this heat generation is the hysteretic loss characteristics of the tire materials during cyclic stressing. Because of these characteristics of the carcass materials there is a loss of mechanical work each time the tire passes through a stress cycle. This lost work is converted into heat and is either diffused through the carcass and eventually lost from its surfaces or must be accounted for in a heat balance.

It has been demonstrated that the energy loss during cyclic stressing depends on the change in strain energy between two non zero stress points. (see ref. 1) The methods for calculating this change in strain energy will be discussed in a general manner.

MAJOR ASSUMPTIONS - The fundamental method used for calculating the change in strain energy is based on two major assumptions which are explained with the aid of figure 1. The first is that the tire is made up of a series of material points, each of which undergoes a change in stress state corresponding to points A and B in figure 1(a). Point A represents the inflated, non-

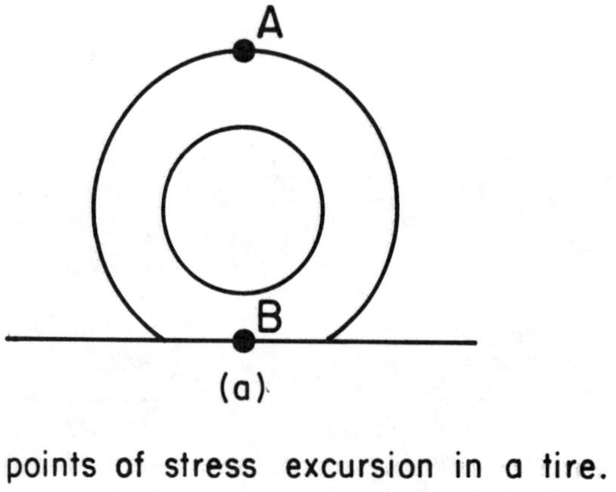

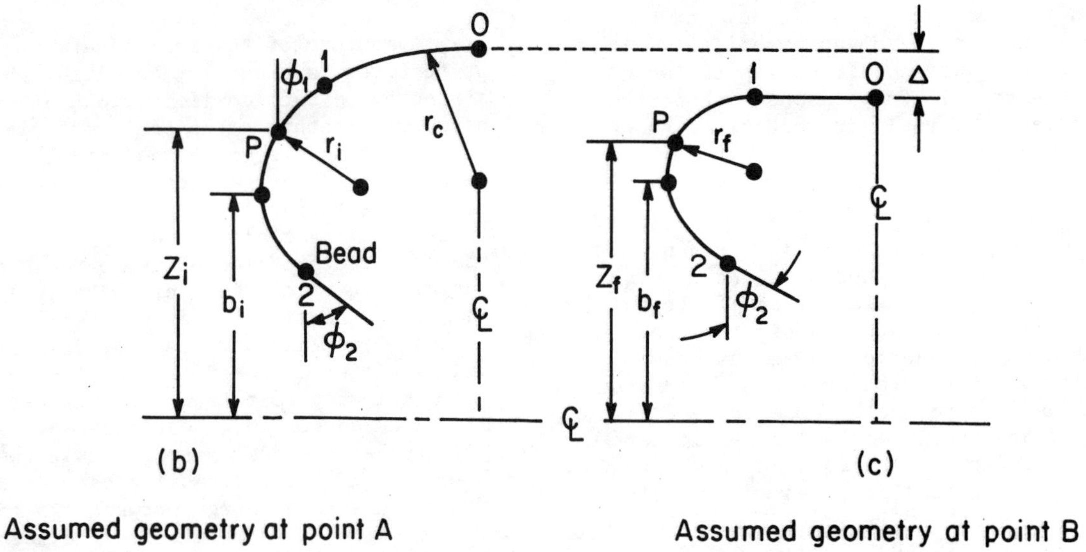

Figure 1. (a) Extreme points of stress excursion in a tire.
(b) Assumed geometry at point A. (c) Assumed geometry at point B.

deflected state of stress and point B represents the inflated, deflected state of stress. It is assumed that the general wave form of the stress cycle from A to B and back to A is unimportant, and higher harmonics of cyclic stress are neglected in favor of a single fundamental cyclic stress change, which can be described in terms of assumed geometry of the tire at points A and B.

The second major assumption made in calculating the strain energy change of the tire as it rolls through the contact region is that at both points A and B the cross-section of the tire may be represented geometrically by its netural axis, around which acts both membrane and bending strains.

TIRE DEFORMATION - The undeformed and deformed radii of curvature of the neutral axis are necessary to calculate the stresses and strains required to determine the strain energy change as the tire rolls through contact. For the inflated, undeflected state the cross-section and its neutral axis may be approximated by several methods. However, the numerical data generated for this paper were obtained by fitting two intersecting arcs of radii r_c and r_1 (see figure 1(b)) through an approximate inflated tire cross-sectional drawing.

It is more difficult to approximate the cross-sectional shape of the deflected tire. For the computations in this paper, the portion of the cross-section in contact with the flat surface was represented by an infinite radius of curvature while the sidewall was approximated by a segment of a circular arc of radius r_f. (see figure 1(c)). This shape is surprisingly realistic in spite of its approximate character.

The results of this simplified geometry are illustrated in figure 1(b) and 1(c) where the neutral axis of the structural part of the carcass is represented by the lines 0-1-2. The inflated, undeflected tire geometry is presumed to be represented by figure 1(b), and the deformed tire by figure 1(c). The bead, point 2, is assumed fixed. The computational details for the geometric representation of the deformed tire once the tire deflection and initial geometry are known under inflated conditions are given in reference 1.

TIRE STRESSES, STRAINS AND STORED ENERGY - The strain energy density of a tire as it rolls through contact depends on both the strain state and the stress state. This can be represented generally by

$$U = \tfrac{1}{2}(\Delta\sigma_\phi \cdot \Delta\varepsilon_\phi + \Delta\sigma_\theta \cdot \Delta\varepsilon_\theta \qquad (1)$$

where: U = elastic energy stored due to change in stress and strain from A to B in figure 1(a)

$\Delta\sigma_\phi$ = change in meridional stress

$\Delta\varepsilon_\phi$ = change in meridianal strain

$\Delta\sigma_\theta$ = change in circumferential stress

$\Delta\varepsilon_\theta$ = change in circumferential strain

The stress and strain state within the tire relies heavily on the assumed geometry in figure 1.

For purposes of computing changes in strain energy states during the cyclic loading process the tire is divided into five regions, denoted I-V in figure 2.

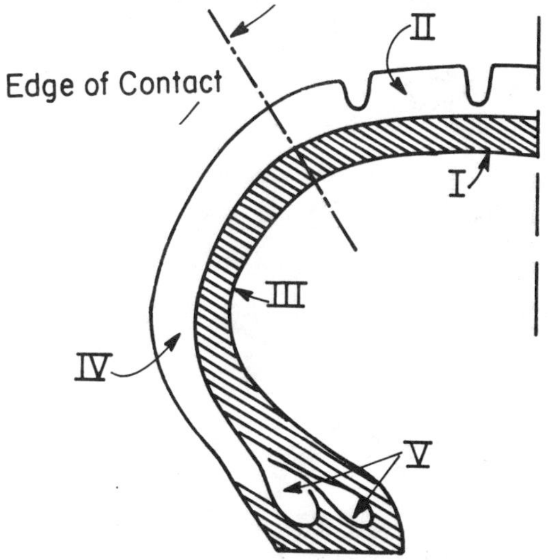

Figure 2 - Tire cross-section showing regions of different characteristics and loading patterns

The rationale behind the division of the tire into these parts is that each section exhibits a different load-deformation characteristic and thus requires a separate analysis to determine the stresses and strains required to calculate the change in strain energy. For instance, region II represents the tread region of the tire, which is considered to be in direct contact with the runway surface, and to consist of isotropic rubber-like material.

Region I, on the other hand, is immediately beneath the tread region of the tire and is firmly affixed to it, so that its membrane strain state is the same as region II. However, region I is assumed to be orthotropic in its material characteristics and thus the relationship between stress and strain is not the same as it is in region II.

Regions III and IV represent the sidewall areas of the tire which are subjected also to changes of curvature as the tire deforms. These areas do not contact the surfaces so that their stresses and strains are not determined as in regions I and II.

Finally, the beads of the tire undergo large force fluctuations, and these must be accounted for separately since they are caused by changing forces in the sidewall which are reacted entirely by the bead ring. For that reason the bead area is denoted by region V and is handled as a separate analysis.

The actual calculation of the strain energy stored in the tire during the deformation process is accomplished by dividing each region of the tire into a number of cells, each cell being denoted by a pair of numbers in matrix form. This is illustrated in figure 3.

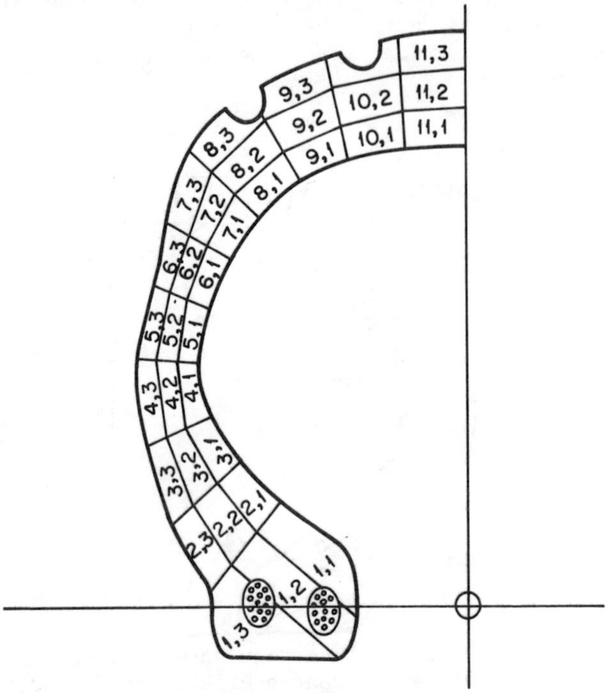

Figure 3 - Element sectioning for thermal analysis

The details for calculating the stress and strain changes and the resulting strain energy change for each cell are given in reference 1.

HEAT GENERATION RATES - The conventional Kelvin-Voight mechanical model of an elastic solid which exhibits loss characteristics under cyclic stress is used to determine the heat generation rates from strain energy stored during the elastic cycling of the tire. This model predicts that the energy loss is directly proportional to the elastic strain energy and the loss tangent of the material. For a tire under free rolling conditions the heat generation rate can be described by

$$\dot{q} = C \cdot U \cdot v_o \cdot \tan\delta / r_r \qquad (2)$$

where: $\dot{q}$ = heat generation rate, heat energy/vol/time
U = elastic energy, energy/vol
C = numerical constant to unify units
v_o = aircraft ground speed
$\tan\delta$ = loss tangent modulus of material
r_r = rolling radius of tire

It is emphasized that the energy lost in going from one stress state to another is represented by using the change in stress or strain when computing the strain energy term in equation (2).

TEMPERATURE DISTRIBUTION MODEL - Referring to figure 3, the aircraft tire is divided into segments or cells in each of which the rate of heat generation is considered constant, and in which temperatures and stresses are calculated on the basis of being averaged across the cell area. In the analysis discussed here quadrilateral elements were consistently utilized in constructing the numerical analog. Thus each element either loses or gains heat from four faces and this may be through conduction when the element in question is bounded by a similar element or to the runway surface or to the rim flange, depending on the location of the element.

Using the well known expression for the diffusion of heat in a solid body under conditions of internal heat generation, such as given by Carslaw and Jaeger, reference 2, the equation for the conduction of heat for a typical element, expressed for finite difference purposes, takes the form

$$\begin{aligned}\frac{\Delta\theta}{\Delta t} = {} & \dot{q} + \bar{\alpha}_{K-1,k,\ell}(\theta_{K-1,\ell} - \theta_{k,\ell}) \\ & + \bar{\alpha}_{K,K+1,\ell}(\theta_{K+1,\ell} - \theta_{K,\ell}) \\ & + \bar{\alpha}_{K,\ell-1,\ell}(\theta_{K,\ell-1} - \theta_{K,\ell}) \\ & + \bar{\alpha}_{K,\ell,\ell+1}(\theta_{K,\ell+1} - \theta_{K,\ell}) \qquad (3)\end{aligned}$$

where: $\bar{\alpha}_{m,n}$ = heat conduction coefficients
$\Delta\theta$ = temperature differences

The basic method of computation is to use small time increments and to calculate the temperature rise in each cell utilizing the temperatures of adjoining elements from the previous time increment computation. It has been demonstrated that such a time base forward computation is relatively convergent as it is applied in this analysis. This computational procedure is sequential, thus requiring minimum memory storage in a computer, and may conveniently be carried out on relatively unsophisticated equipment such as a microcomputer.

It has also been shown that for the type of computation used here the dominant term is the rate of heat generation, $\dot{q}$, and for purposes of aircraft taxi and take-off the heat build up occurs over a relatively short time period, so that the heat generation term almost completely controls the temperature rise in the tire. Heat diffusion is a minimal effect. This means that heat conduction using adjoining cell temperatures from the previous time interval incurs only small errors in the final temperature distribution in the tire.

EXPERIMENTAL PROGRAM

The problem of developing an analytical model for predicting the temperature profiles of free rolling tires, as described above, is an extremely complex one and could not be expected to be verified without considerable experimental data. A number of assumptions are used in the computations of the rate of internal heat generation in the tire, and many of these assumptions can only be verified by direct measurement of temperature fields in the actual tire. For that reason a comprehensive experimental program was carried out jointly by the Impact Dynamics Branch, NASA and the University of Michigan.

TIRE PREPARATION - Several tires of each of four different tire configurations were used in this experimental study. These included 22 x 5.5, 8-ply rated and 12-ply rated aircraft tires as well as 40 x 14, 22-ply rated and 28-ply rated aircraft tires. Each tire was instrumented with an extensive set of thermocouples.

The installation of the thermocouples began by clearly laying out the desired locations on the surface of each tire. A typical set of such locations is illustrated in figure 4. It can be

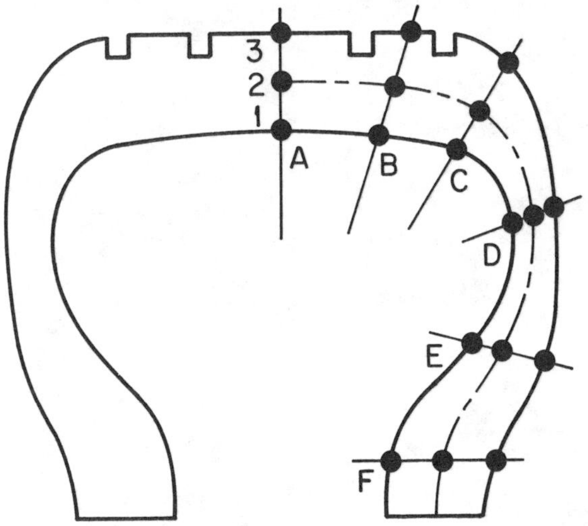

Figure 4 - Location of thermocouples in tire thermal study

seen in this figure that a thorough study of the temperature distribution can be obtained with these locations, both through the thickness of the carcass as well as around the meridian of the cross-section.

Following the layout, small holes were drilled in each tire for implanting all interior thermocouples. The tires were then transferred to a commercial retreader where the treads were buffed off and the thermocouples in the tread region installed just prior to the retreading. After the retreading process the tires were returned to the laboratory where all remaining thermocouples were inserted in their proper locations and sealed with a hot patch process. A 22 x 5.5, 12-ply rated tire instrumented with a complete set of thermocouples, mounted on an appropriate wheel, and wired for recording all thermocouples simultaneously, is illustrated in figure 5.

TEMPERATURE PROFILE TESTS - All of the instrumented 22 x 5.5 tires were tested on the Impact Dynamics Branch Ground Test Vehicle shown in figure 6. This vehicle is a production-model truck which has been modified to accommodate a rear-mounted tire test fixture. The fixture consists of the tire and pneumatic loading cylinders. The tire vertical load is measured by strain gage beams in the test fixture. The output from the thermocouples, as collected through slip rings, are routed to an automatic data logger in the vehicle driving compartment. An instrumented trailing wheel is used to provide an accurate measurement of vehicle speed and distance traveled.

The NASA tests consisted of temperature profile measurements for many different operating conditions, including a variety of pressures, deflections and speeds.

All of the 40 x 14 tires and some of the 22 x 5.5 tires were tested by the University of Michigan on the automated 120-inch diameter dynamometer at the Air Force Flight Dynamics Laboratory, Wright-Patterson Air Force Base. A fully instrumented 40 x 14 tire is shown loaded against the dynamometer wheel in figure 7. This dynamometer features a cantilevered wheel mounting fixture which is hydraulically loaded against the driven dynamometer wheel. The radial load on the tire is measured by strain gage beams in the mounting fixture. The thermocouple signals are again collected through slip rings and recorded in an exterior data logger.

The University of Michigan tests were also conducted under a variety of pressures, deflections and speeds.

A great deal of temperature data were obtained from these test programs of which a small sampling is shown in figures 8 through 10.

Figure 8 illustrates a composite plot of all of the temperature measurements in one of the test tires as a function of time for a speed of twenty miles per hour at rated operating conditions. The ordinate of these plots is the temperature rise in each of the measurements above the initial temperature value. The letters, A,B,C,D,E,F refer to the thermocouple locations noted in figure 4. In general it can be seen that the inside temperatures are highest when the measurements are being made in a region outside of the contact region. However, in the contact region, the mid-line and inside measurements are nearly the same. The next two figures will present data at a fixed time to show the effect of two operating parameters, namely tire deflection and tire inflation pressure.

The effect of deflection on temperature distribution in the tire carcass is illustrated in figure 9. The ordinate of these plots also represents the change in temperature above the initial temperature. It is also noted that these plots indicate the temperature rise when time equals 150 seconds at 20 miles per hour. As expected, the highest temperatures occur with higher deflections. The inside measurements indicate that temperatures increase with increasing deflection from the crown to the bead. The midline measurements indicate a similar pattern, especially in the shoulder region where flexing is greatest. However, the inside measurements indicate very little change with deflection. The

Figure 5 - 22 x 5.5, 12 ply rated tire instrumented with thermocouples

Figure 6 - Impact Dynamics Branch Ground Test Vehicle

Figure 7 - Instrumented 40 x 14, 28 ply-rated tire loaded against the Air Force Flight Dynamics Laboratory 120-inch diameter dynamometer

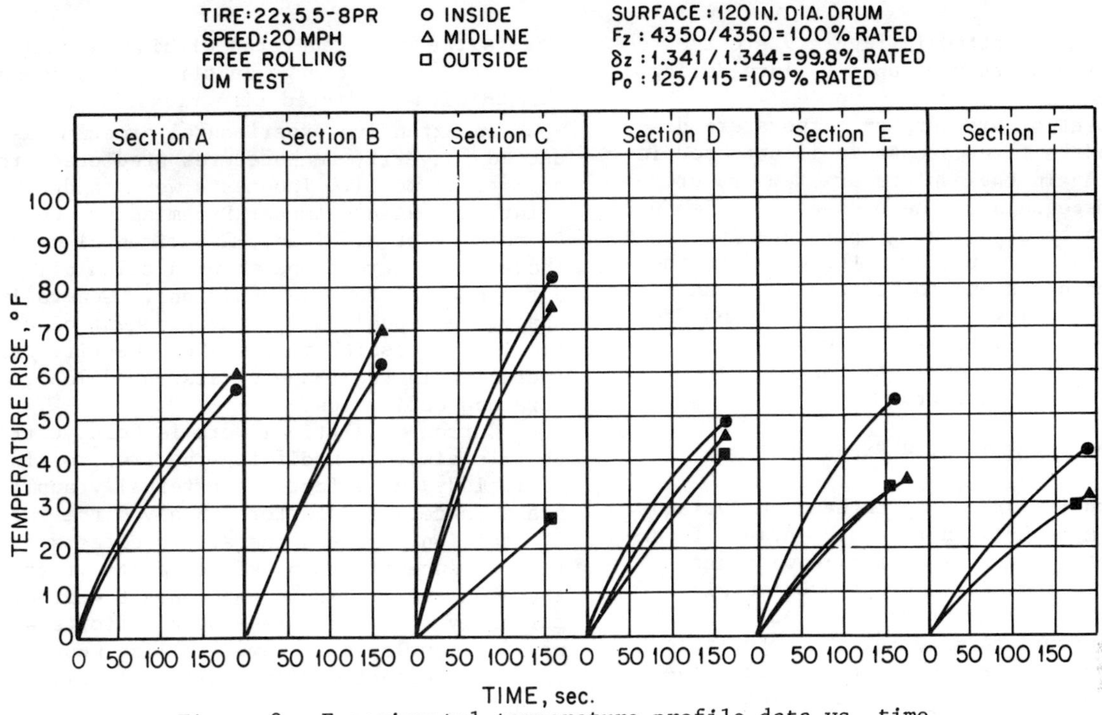

Figure 8 - Experimental temperature profile data vs. time

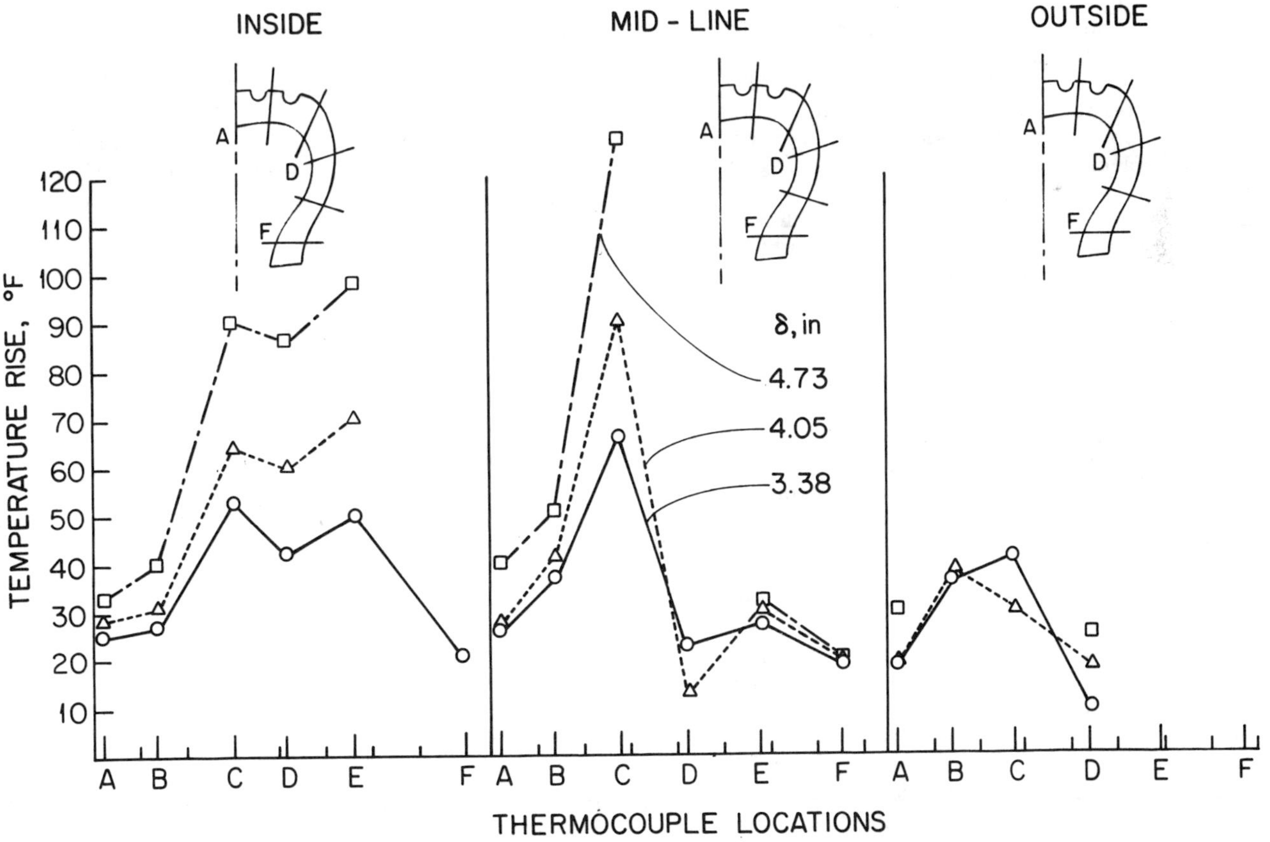

Figure 9 - Effect of deflection on temperature distribution in tire carcass 40 x 14, 22 ply-rated, P_o = 160 psi, time = 150 secs. at 20 mph.

gaps in the data plotted in figure 9 are the result of individual thermocouple failure during the test program.

The effect of pressure on temperature distribution in the tire carcass is illustrated in figure 10. Again these plots are temperature rise for time equals 150 seconds at 20 miles per hour. As can be seen, the temperature increases with increasing pressure from the crown to the bead. The change seems to be relatively uniform throughout the thickness as well as around the cross-section. Again there are gaps in the plotted data due to individual thermocouple failure during the test program.

COMPARISON OR ANALYSIS WITH EXPERIMENT

With experimental data in hand comparisons can then be made between the calculation of temperature distributions in the tire with data measured both from NASA tests and University of Michigan tests. In order to carry out the computations appropriate material characteristics had to be assigned to each of the tires. These data were provided from actual measurements as well as from various literature sources.

Using a time increment of t = 10 seconds, and using the appropriate material properties and geometries, computations were carried out for the temperature rise in the tires under operating conditions experienced by both the NASA tests and the tests conducted by the University of Michigan. Some typical comparisons between calculation and experiment are shown in figures 11 through 13.

A typical comparison of experimental and calculated temperature changes in the contact region of the tire during free roll is illustrated in figure 11. As is observed from this figure, the temperature change predicted from the analytical model is in good agreement with the actual measurements. This is true for both the inside and outside portions of the tire and holds well for different loading conditions.

A similar comparison for the sidewall portion of the tire out of the contact region is illustrated in figure 12. Again there is relatively good agreement between the analytical and experimetnal results.

Figures 11 and 12 illustrate typical comparisons of individual locations in the carcass. However, many more comparisons were made of the complete temperature profiles, as illustrated in figure 13, where all of the meridianal and thickness locations are compared simultaneously. From a study of these comparisons, and many more like them, it is concluded that the major trends in the temperature rise problem can be calculated adequately for short term taxi-takeoff conditions with the analytical model presented here.

CONCLUDING REMARKS - An analytical model has been developed for approximating the internal temperature distribution in an aircraft tire operating under conditions of unyawed free rolling. The model employs an assembly of finite elements to represent the tire cross-section and treats the heat generated within the tire as a function of the change in strain energy associated with predicted tire flexure.

An extensive experimental program was conducted to verify temperatures predicted from the analytical model. The tests were conducted jointly by NASA's Impact Dynamics Branch and the University of Michigan. Experimental data from these tests were compared with calculation over a range of operating conditions. Generally the analytical model gave good agreement of calculation with measured values when the tire was operating in the range of its rated load, pressure and deflection.

Current analytical work is being carried out that is aimed at modifying the free rolling model to include the effects of externally applied brake force. This effort involves the inclusion of additional stress and strain effects due to the presence of the brake force itself as well as the effects of slip in the contact region. Concurrently with this brake work, efforts are under way for developing a yawed-rolling model to include the effects on temperature profiles due to cornering.

REFERENCES

1. Clark, Samuel K.; Dodge, Richard N.: Heat Generation in Aircraft Tires Under Free Rolling Conditions. NASA TN

2. Carslaw, H.S.; Jaeger, J.C.: Conduction of Heat in Solids, Second ed., Oxford University Press Inc., 1959.

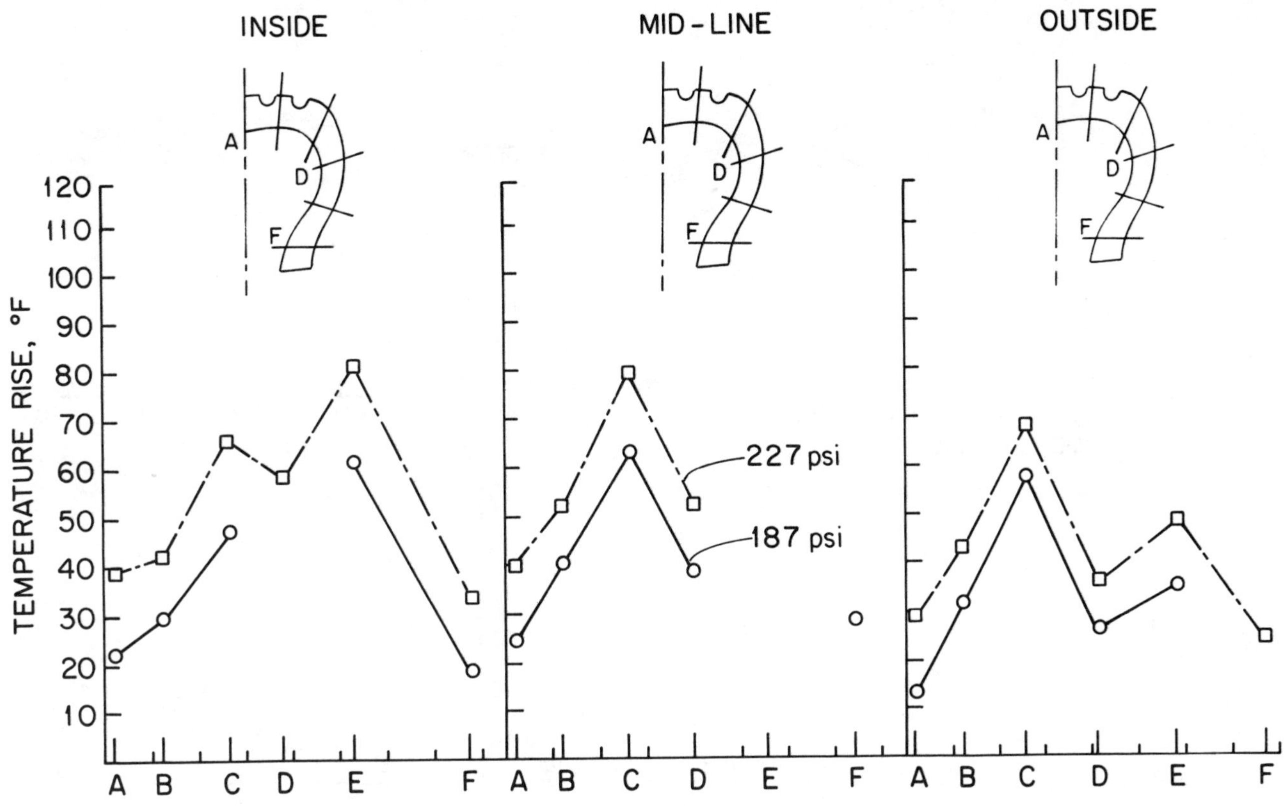

Figure 10 - Effect of pressure on temperature distribution in tire carcass 40 x 14, 28 ply-rated, vertical deflection = 3.48 inches, t = 150 sec. at 20 mph

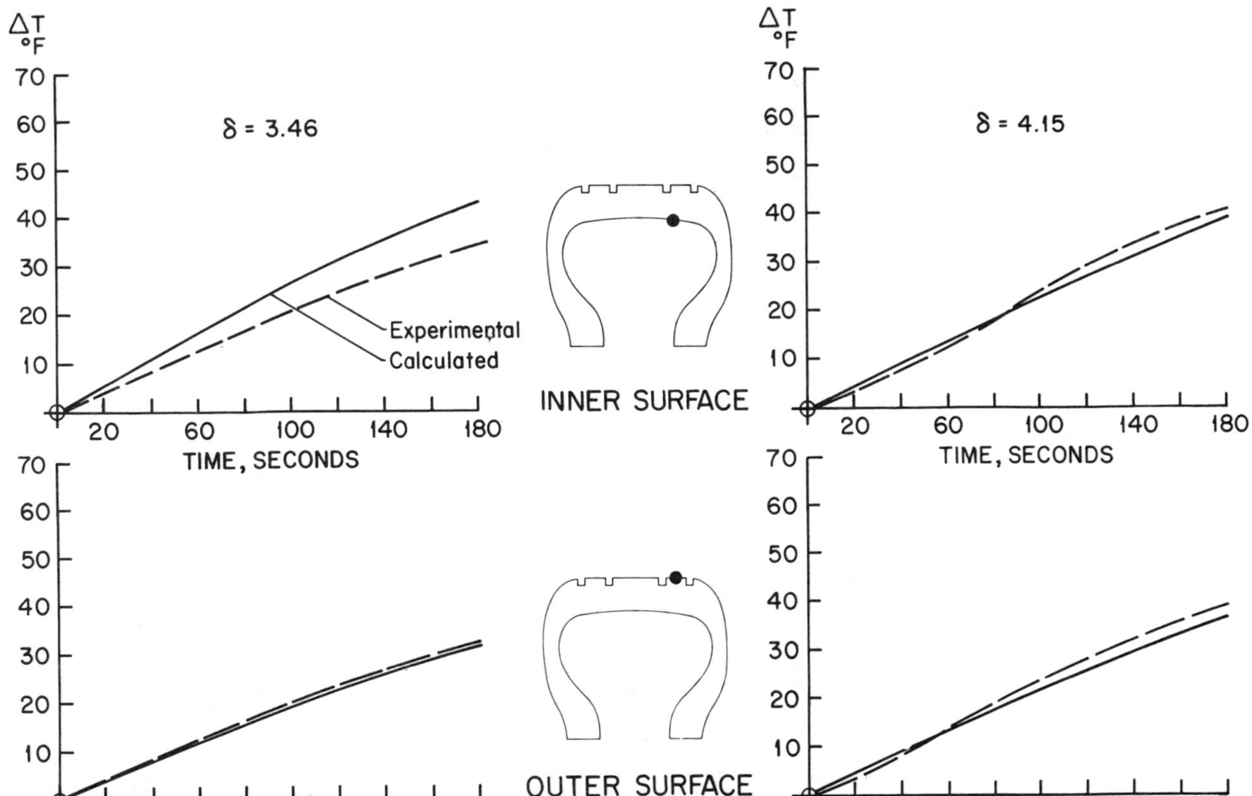

Figure 11 - Comparison of experimental and calculated temperature in the contact region during free roll. 40 x 14, 28-ply rated, p_o = 207 psi, V = 20 mph

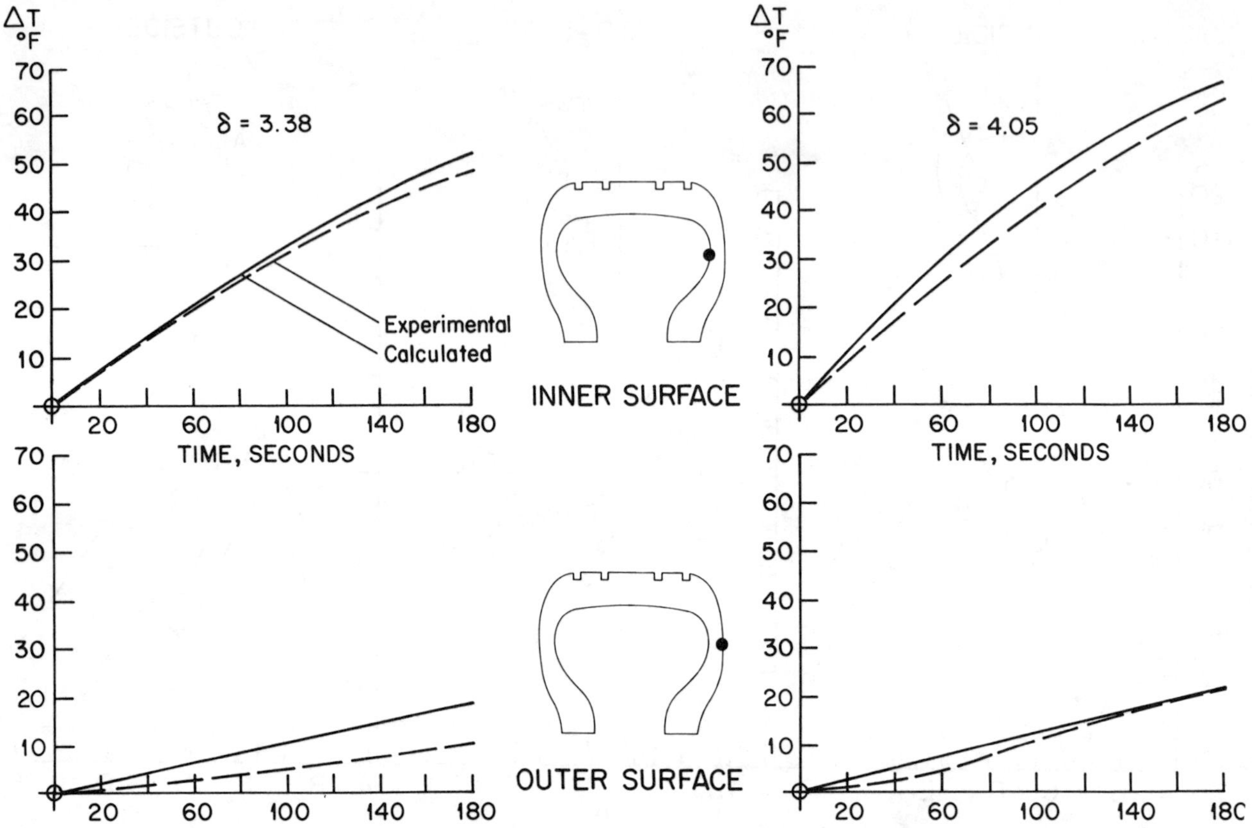

Figure 12 - Comparison of experimental and calculated temperatures in the sidewall region out of contact during free roll. 40 x 14, 22 ply-rated, P_o = 160 psi, V = 20 mph

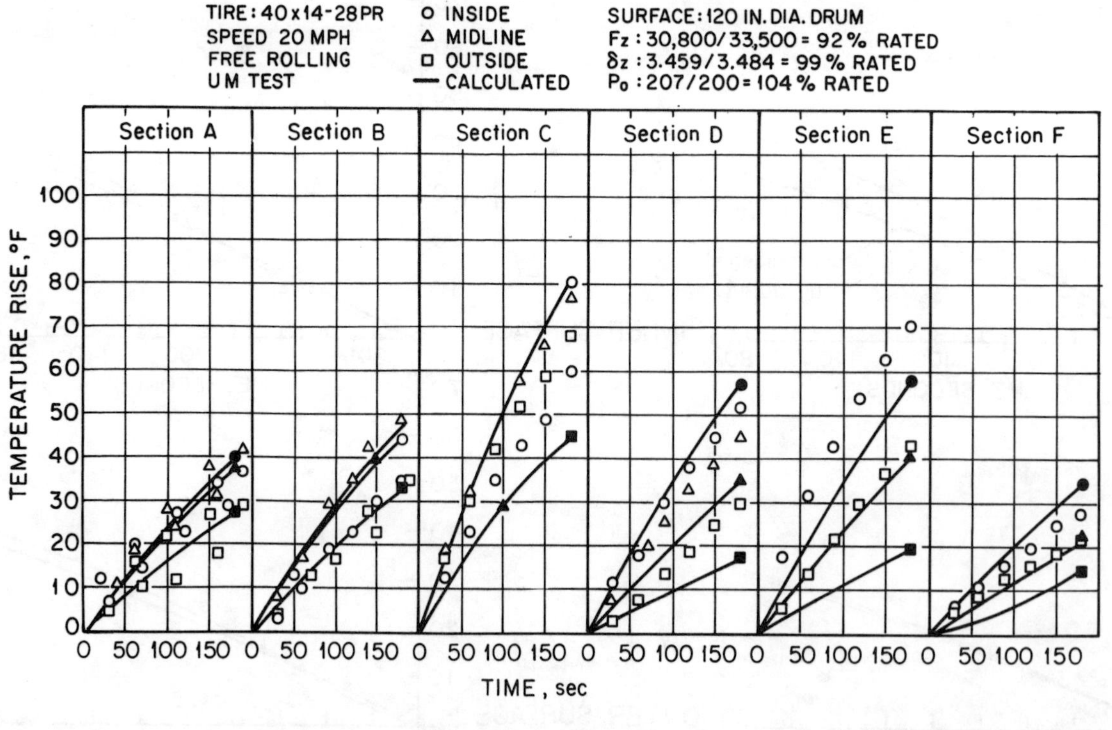

Figure 13 - Measured and calculated tire temperature profiles

Non-Linear Cord-Rubber Composites

S.K. Clark and R.N. Dodge
Department of Mechanical Engineering and Applied Mechanics
University of Michigan

** Paper 892339 presented at the
Aerospace Technology and Exposition,
Anaheim, California, September, 1989.

ABSTRACT

A method is presented for calculating the stress-strain relations in a multi-layer composite made up of materials whose individual stress-strain characteristics are non-linear and possibly different. The method is applied to the case of asymmetric tubes in tension, and comparisons with experimentally measured data are given.

THE RAPID DEVELOPMENT OF DIGITAL capabilities over the last decade has caused considerable interest in the tire industry in finite element models for a pneumatic tire. The material characteristics of the tire constitute a vital part of those models, both in the case of the essentially isotropic rubber compounds as well as the highly anisotropic textile cord construction.

For a number of reasons sheets of textile cord are calendered in rubber to make laminae or plies which are highly anisotropic, and laminates formed from these plies make up the carcass of the tire. Calculation of the stiffness properties of this structure is important to any numerical model since it has a much higher stiffness than the rubber compounds. Conventional laminate theory is described in numerous textbooks, and in much of the literature relies heavily on linear relationships between stress and strain in order to predict strain states under known loads. Unfortunately, the textile materials commonly used in tire construction are not linear, and so application of conventional laminate theory to these materials is not immediately obvious and can lead to serious errors.

The problem of non-linear reinforcement in a composite matrix is not a new one. Generally in the past it has been recognized in two forms. The first is the large difference observed in stiffness of composites between compression and tension in the reinforcement direction. This phenomenon is mostly caused by local buckling of small filaments. It has been reviewed in the work of Tabaddor [1] * and Bert [2], while measurements of this phenomenon have been carried out on commercially useful materials by Bert and Kumar [3]. Problems of some structural complexity have been solved using such materials in the work of Kincannon, Bert and Reddy [4]. Most of this literature has been based on the concept of bi-modular materials, that is, materials exhibiting one set of linear modulus properties in tension and a different set in compression.

A second source of non-linearity occurs in the region of pure cord tension, or compression, where even under monotonically increasing loads the stress-strain response is not linear. This has often been attributed to curvilinear textile filaments associated with the twist usually imparted to textile

*Numbers in parentheses designate references at end of paper.

cord during processing. As these filaments straighten under extension they usually cause an increasing modulus to be observed. This phenomenon has been studied theoretically by Tabaddor and Chen [5], while correlartion of experiment with theory has been published recently by Kuo, Takahashi and Chou [6].

The present paper has as its objective the formation of a procedure for incorporating the particular and special non-linearities observed in cord-rubber composites into plane laminate theory. The non-linearities may be completely general, of either type just described, since the procedure is numerical.

REVIEW OF TEXTILE CORD MATERIAL PROPERTIES

Any general review of textile cord-rubber composite material properties must begin with a single laminae. These properties are reasonably well known for a number of specific materials, and the following general description will apply to those which are commonly used in tire construction, namely nylon, polyester, glass, steel and aramid.

The most elementary form of the description is obtained by reference to Fig. 1, where the notation L and T are used to describe the cord and transverse directions of a simple element. The general characteristics of these materials may be described in Figures 2 through 4. The properties are the conventional ones associated with Hooke's Law in the form shown in Eqs. (1). Our observations over the range of materials indicated shows that most nonlinear effects are seen in the relationship between σ_L and ε_L, with specific reference to the regions of tension and compression. Other nonlinearities appear in the Poisson's ratio as shown, while the transverse modulus and shear modulus are essentially constant.

$$\varepsilon_L = \frac{\sigma_L}{E_L} - \upsilon_{TL}\frac{\sigma_T}{E_T} \qquad e_T = -\upsilon_{LT}\frac{\sigma_L}{E_L} + \frac{\sigma_T}{E_T}$$

$$\gamma_{LT} = \frac{\tau_{LT}}{G_{LT}} \tag{1}$$

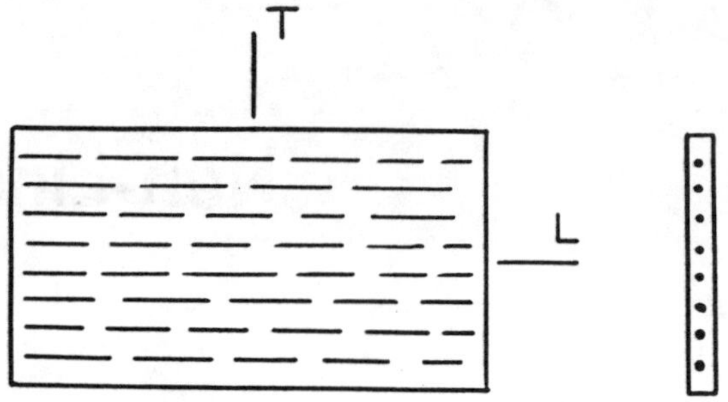

FIG. 1 Notation for a lamina

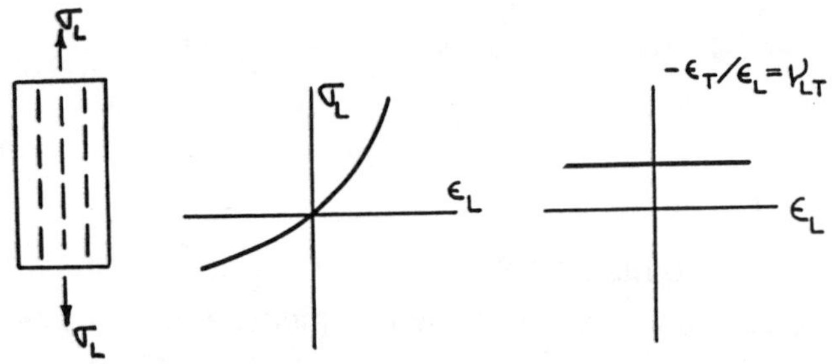

FIG. 2 Typical longitudinal properties for cord-rubber composites

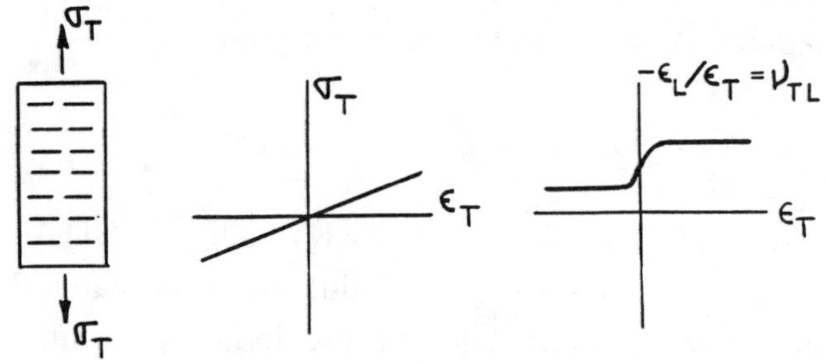

FIG. 3 Typical transverse properties for cord-rubber composites

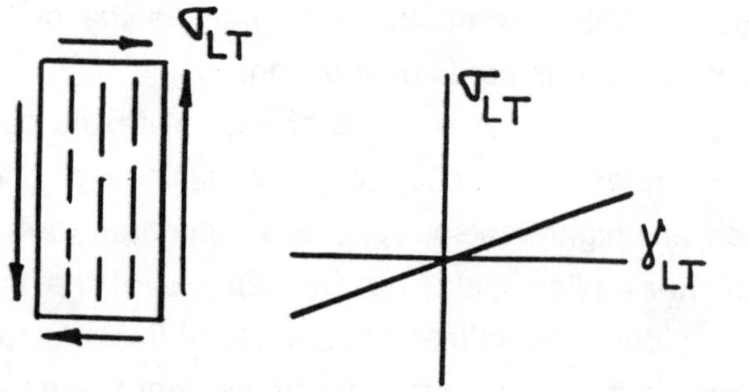

FIG. 4 Typical shear properties for cord-rubber composites

This type of description arises naturally from the mechanics of isotropic materials, where loads are specified and at least average stresses can be obtained by analysis. The stresses are the independent variables. It has become more or less standard practice to think in these terms. However, this is not a convenient or useful format for laminates, since the stress levels in individual lamina may vary widely due to anisotropic effects.

It is common in laminate theory to invert Eqs. (1) in the well known form given in Eqs. (2):

$$\sigma_L = \frac{E_L}{1-\upsilon_{LT}\upsilon_{TL}}(\varepsilon_L + \upsilon_{TL}\varepsilon_T);$$

$$\sigma_T = \frac{E_T}{1-\upsilon_{LT}\upsilon_{TL}}(\varepsilon_T + \upsilon_{LT}\,\varepsilon_L); \quad \tau_{LT} = \gamma_{LT}\,G_{LT} \qquad (2)$$

Here the <u>strains</u> are the independent variables, and this is important since only the strains are well defined in laminates. However, these equations depend on the linear definitions based on Eq. (1).

It is useful to consider how to convert Eqs. (2) to accept non-linear materials. In doing this one may draw on the physical observation that the primary source of non-linearity lies in the widely different behavior of textile cords in tension and in compression. This implies that most, if not all, non-linearities will be functionally dependent on ε_L.

Based on that concept one may rewrite Eqs. (2) in the form

$$\sigma_L = f(\varepsilon_L) + g(\varepsilon_L).\varepsilon_T;$$

$$\sigma_T = h(\varepsilon_L)\,.\,\varepsilon_T + j(\varepsilon_L)\,.\,\varepsilon_L;$$

$$\tau_{LT} = G_{LT}(\varepsilon_L)\,.\,\gamma_{LT} \qquad (3)$$

It is useful to consider Eqs. (3) as the basic equations defining a material, and to directly measure the five elastic functions f, g, h, j, and G_{LT}.

Due to the fact that in cord-rubber composites the Poisson ratio υ_{TL} is quite small, and due to the fact that in approximate terms E_T and G_{LT} are nearly independent of ε_L, the following observations have been made based on experimental evidence:

(1) G_{LT} is a constant independent of ε_L

(2) $h(\varepsilon_L)$ is a constant independent of ε_L

(3) $j(\varepsilon_L)$ is small.

The direct way of obtaining these functions is to impose a given strain and to measure the corresponding stresses or forces. This is most easily done in a multi-axial specimen, and due to the low modulus of these materials it has been found that tubular specimens are most convenient here, since both axial loads and internal pressure/vacuum can be used to control the two strains.

The function f can be obtained by using a 0° tubular specimen to impose a strain ε_L while measuring σ_L, at the same time using internal pressure to maintain ε_T zero. Alternately, pressure can be used to vary ε_T and σ_L while simultaneously holding ε_L zero. This is shown in Fig. 5.

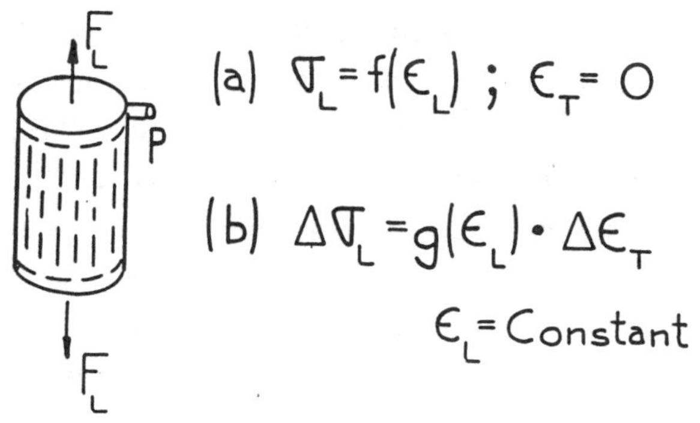

FIG. 5 Experimental determination of non-linear functions f and g

A 90° tubular specimen can be used in a similar way. By using direct extension while controlling internal pressure, $h(\varepsilon_L)$ can be found by maintaining $\varepsilon_L = 0$ as shown in Fig. 6. Similarly by varying internal pressure while holding length constant, the relationship $j(\varepsilon_L)$ can be determined by measuring diameter change.

Finally, conventional torque tests can be used on either a 0° or 90° tubular specimen in order to obtain the relationship between τ_{LT} and γ_{LT}, a property which appears from our experimental data to be linear.

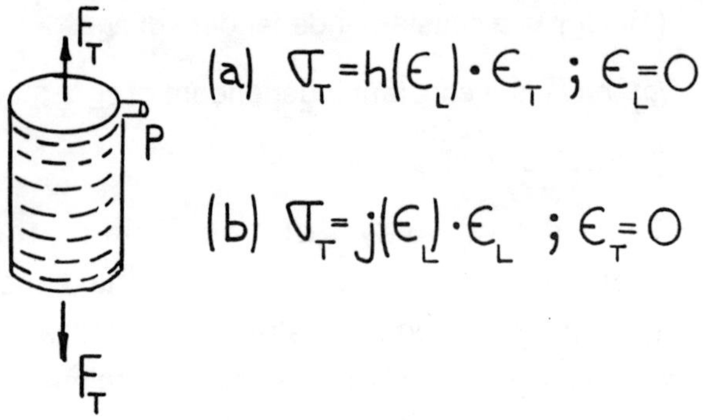

FIG. 6 Experimental determination of non-linear functions h and j

Treating these quantities as indicated in Eqs. 3 now allows one to carry out computations on laminate structures even though the characteristics are nonlinear. Here the key parameter is the strain ε_L, since all nonlinearities depend on it. Each of these may be fit with a polynominal or other functional form to whatever degree of accuracy is required, so that the quantities f and g may be considered as functions of ε_L, while the other properties may be considered as constants. There may very well be different functional forms for different regions of ε_L. For example, one can often take the $f(\varepsilon_L)$ to be linearily proportional to ε_L for $\varepsilon_L \leq 0$.

A general laminate may have many plies, but within membrane and bending deformation there exists just two sets of possible strain states, that of plane extension and Kirchhoff bending with the strain variables shown:

$$\text{Membrane} = \begin{bmatrix} \varepsilon x \\ \varepsilon y \\ \gamma xy \end{bmatrix} \qquad \text{Bending} = \begin{bmatrix} kx \\ ky \\ kxy \end{bmatrix}$$

A proposed method of solution for a non-linear n ply laminate will be proposed for membrane extension, but it could equally well be written for bending. Let the principal material directions with respect to an arbitrary x-y axis be θ_j for the jth ply, and the ply thicknesses t_j. The computational steps are as follows, given a known set of imposed overall average membrane forces N_{x_o}, N_{y_o}, N_{xy_o}.

1. Assume values εx, εy, εxy

2. Compute $\begin{bmatrix} \varepsilon_L \\ \varepsilon_T \\ \gamma_{LT} \end{bmatrix}_j = \left(T_\varepsilon(\theta_j)\right) \begin{bmatrix} \varepsilon x \\ \varepsilon y \\ \varepsilon xy \end{bmatrix}$

3. From Eqs. (3) obtain $\begin{bmatrix} \sigma_L \\ \sigma_T \\ \sigma_{LT} \end{bmatrix}_j$

4. Compute $\begin{bmatrix} \sigma x \\ \sigma y \\ \sigma xy \end{bmatrix}_j = \left(T_\sigma(\theta_j)\right) \begin{bmatrix} \sigma_L \\ \sigma_T \\ \sigma_{LT} \end{bmatrix}$

5. $\begin{bmatrix} N_x \\ N_y \\ N_{xy} \end{bmatrix}_1 = \sum_{j=1}^{n} \begin{bmatrix} \sigma x \\ \sigma y \\ \sigma xy \end{bmatrix}_j \cdot t_j$

6. These membrane forces are compared to the known imposed membrane forces N_{x_o}, N_{y_o}, N_{xy_o}, and corrections to the assumed strain state can then be made and the process repeated until correspondence is obtained.

This process is admittedly iterative and must be implemented by Newton-Raphson or other computational techniques in most cases. For simple cases of uniaxial loading it can be done by trial and error methods.

In the case of bending moments a similar procedure can be used, depending on the Kirchhoff hypothesis for the linearity between curvature and strain. In this case the quantities which must be assumed at the beginning of the computational process are the curvature terms, as opposed to the membrane strains in the illustration. The remainder of the process would be identical.

The results of steps (1) through (6) just described can be visualized as a set of equations relating membrane forces $\{N\}$ to membrane strains $\{\varepsilon\}$.

Since the relations are general, these six steps take the place of what in linear theory is called [A], a matrix connecting the vectors {N} and {ε}.

It should be emphasized that the most important concept here is that the elastic constants now should be considered as the functions appearing in Eqs. (3), and that these can be obtained directly from experiments on 0° and 90° tube specimens made from the materials in question. Since cord-rubber composite specimens are easy to fabricate this should be little burden on experimental work.,

EXPERIMENTAL EXAMPLE

In order to demonstrate the techniques proposed here a series of four asymmetric tubes were constructed of material which is known to exhibit substantial nonlinear behavior, namely 840/2 nylon 66 calendered in a natural rubber compound. Two-ply tubes were constructed with the angular combinations 0°/10°, 0°/30°, 0°/50° and 0°/70°. The material properties defined in Eqs. (3) were obtained on 0° and 90° tubular specimens using a combination of axial extension and inflation. They are illustrated Figs. 7 and 8. The major nonlinearitiy occurs in the function $f(\varepsilon_L)$. The remaining relationships are essentially linear.

For the particular experiments in question the nonlinear functions f and g could be approximated by a polynominal expression using first and third order terms in ε_L.

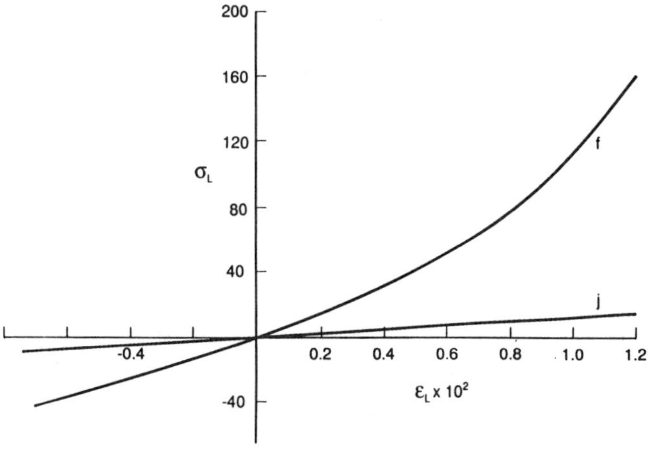

FIG. 7 Functions f and j for 840/2 nylon 66 in NR

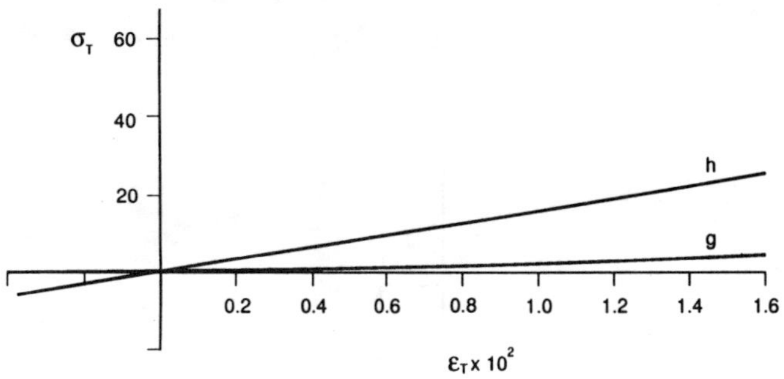

FIG. 8 Functions h and g for 840/2 nylon 66 in NR

The actual experiments involved loading of the asymmetric nylon tubes in pure tension by use of dead weights and observing the rotation of each tube under various load conditions. The data is compared with calculations in Fig. 9. The calculations were carried out on an iterative basis using the principle that under this pure axial load the transverse stress N_y and the shear force N_{xy} must both vanish for the laminate.

The tubes chosen for these experiments were fabricated with l/d ratios greater than 30 so that aside from small end effects near the grips no curvatures {k} were exhibited under tensile loading. The resulting deformation was purely plane, involving {ε} caused by a non-zero N_x.

While the agreement between calculation and experiment is only general, it does illustrate the trends of the data quite well. Further refinement is to be expected as automated convergence techniques are developed for digital computation.

ACKNOWLEDGEMENTS

The authors would like to thank the National Aeronautics and Space Administration for support of this work, and in particular the cooperation of the technical monitor Mr. John Tanner.

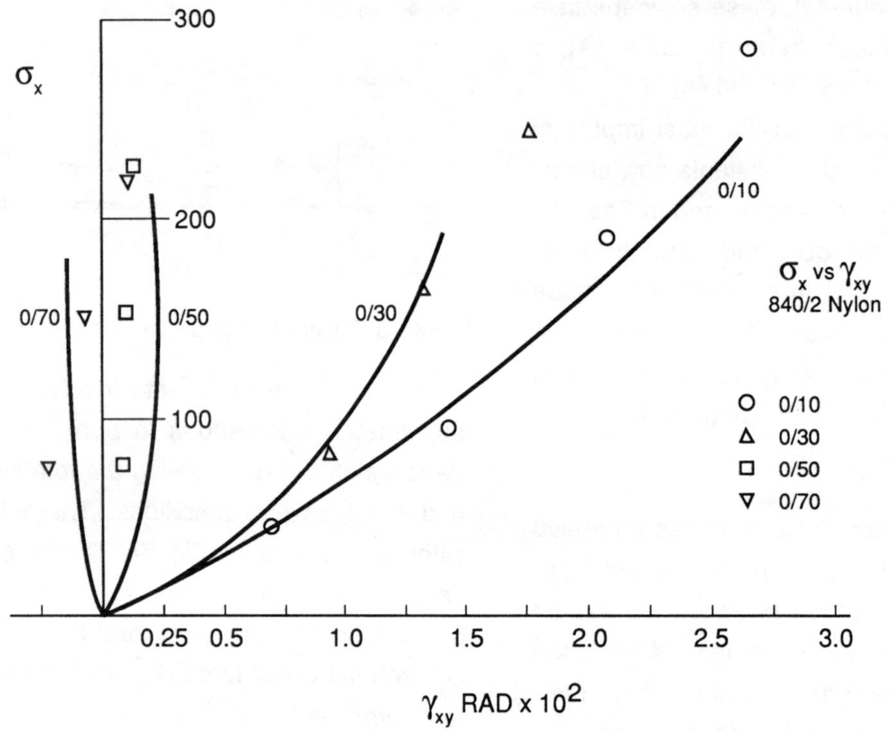

FIG. 9 Results of calculated vs predicted shear in axial loading of asymmetric tubes

REFERENCES

1. Tabaddor, F., "A Survey of Constitutive Equations of Bimodulus Elastic Materials", Mechanics of Bimodulus Materials, ASME AMD, v. 33, Dec. 1979.

2. Bert, C. W., "Micromechanics of the Different Elastic Behavior of Filamentary Composites in Tension and Compression" Ibid.

3. Kumar, M. and C. W. Bert, "Experimental Characterization of Mechanical Behavior of Cord-Rubber Composites," Tire Sci. Tech., v. 10, Jan. 1982.

4. Kincannon, S. K., C. W. Bert and V. S. Reddy, "Cross-Ply Elliptic Plates of Bimodulus Material," J. Struct. Div., ASCE, v. 106, p. 1437, 1980.

5. Tabaddor, F. and C. H. Chen, "Reinforced Composite Materials with Curved Fibers," J. Spacecraft and Rockets, v. 8, n. 2, 1971.

6. Kuo, C.-M., Takahashi, K. and Chuo, T.-W., "Effect of Fiber Waviness on the Nonlinear Elastic Behavior of Flexible Composites," Jour. Composite Mat., v. 22, 1988.

Evaluation of Critical Speeds in High Speed Aircraft Tires

Joe Padovan and Amir Kazempour
Department of Mechanical Engineering
University of Akron
Farhad Tabaddor
Uniroyal Goodrich Corp.
Bob Brockman
University of Dayton

** Paper 892349 presented at the Aerospace Technology Conference and Exposition, Anaheim, California, September, 1989.

ABSTRACT

By recasting the rolling tire system as a large deformation nonlinear eigenvalue problem, this paper develops a methodology capable of handling the critical speed/standing wave response. Based on a moderate/small deformation superposed on large finite element formulation, tangent properties are developed to enable the establishment of the appropriate critical speed eigenvalue problem. The methodology enables the handling of large deformation kinematics material nonlinearity, rotational inertial fields, as well as contact boundary conditions. Employing the model, the results of a benchmark example involving the NASA space shuttle main tires are presented. This includes evaluating the effects of rotation, material stiffness, pressurization and large deformation contact on the critical speed properties.

IN RECENT YEARS, THE increased size and operating speeds of aircraft has required higher takeoff (T/O) and landing speeds. The trend in the design of future aircraft and aerospace (space plane/shuttle) vehicles is toward systems with even higher T/O and landing speed requirements. This is exemplified by the mission limits desired by the Air Force and NASA for the National Aerospace Plane (NASP). Some of these requirements are a final T/O and landing speed of 450 mph and 315 mph respectively. To maintain runway integrity, internal pressures are limited to approximate 300 psia. Additionally, to provide reasonable landing gear loads, rolling contact deformations of between 20% - 40% are typically sought. The foregoing requirements are significantly above existing technology.

The upper bound delimiting behavior of high speed tires is the so-called standing wave problem. The critical speed which generates this wave yield sinusoidal like deformations of the tread and sidewall regions. Generally such waves tend to lead to catastrophic failure. This is due to a combination of factors. These include i) overstressing, ii) severe heat generation, as well as iii) a loss of cohesive and adhesive properties.

Generally the critical speed is dependent on many factors. These include such effects as: i) internal pressure, ii) construction/geometry, and iii) loading/contact deformation. Most of the analyses presently employed are usually on very simplistic models. Normally these involve a closed form type analytical solution. Such simulations treat tires as rings or beams on elastic foundations (1-3).* Most of these formulations do not account for such important factors as centripetal loading, the ground contact interface, direct pressure effects, large strain/rotation etc. In this context, a costly amount of experimental results is required to compensate for such deficiencies. Currently, the design of tires relies heavily on such approaches. With the operating conditions defined for NASP, experimental results cannot be easily obtained.

At this juncture, we recall the work of Padovan et al. (4) wherein the standing wave problem is simulated directly through finite element (FE) analysis involving the use of so-called moving element strategies (4). This procedure enables the modelling of all the various features noted earlier, i.e. large deformation kinematics, contact (slip, stick, friction) rolling inertial fields, thermal effects, material nonlinearity and anisotropy. Overall, the moving FE analysis (4) provides the ultimate modelling scheme for rolling tires.

The main emphasis of this paper will be to establish a methodology which provides a cost effective "shortcut" CADCAM type procedure to establish the bounds on critical speeds as controlled by the various features of tire design and loading environment. Once the "shortcut" procedure establishes the range of critical speeds, the moving FE strategy would then be used to refine the answer. Here the main thrust will be to develop a

*Numbers in parentheses designate references at end of paper.

reliable methodology which being based on first principles will not rely on the extensive experimental supporting data required by the classical lumped models (1,2.3).

Recalling the early work of Padovan, it follows that the frequencies of the tire are related to the critical speed problem. Specifically, it was shown that the critical speed is directly dependent on the circumferential frequency branch. More recently, the formal relationship between the spectral characteristics and the traveling load speed were worked out for general structural configurations (5).

In the context of the spectral-critical speed connection (5), the "shortcut" methodology will be based on evaluating the nonlinear eigenvalue properties of the dynamic response of the tire. Here special attention will be given to accounting for such factors as a) pressurization (follower forces (6), b) material properties (cord stiffness, end count, ply angles, etc.), c) contact (friction/slip/stick), d) rotational speed.

ANALYTICAL METHODOLOGY

It is well known that tires exhibit a variety of nonlinear response characteristics. Such behavior occurs during pressurization, contact and rolling/maneuvering (cornering, braking, skidding etc.). We need to establish the influence of such effects on the frequencies/critical speed of the tire. Recalling the recent theoretical work of (5), we shall consider the critical speed problem as moderate deformations superposed on an initial large deformation state. For example, noting Fig. 1, if a pressurized tire were bumped to yield a spectral analysis of the dynamic response, the resulting vibrations would be about the prestress state. In the case of a small/moderate bumping process, the resulting vibrations are essentially controlled by the tangent stiffness characteristics of the prestate.

As was stated earlier, (5), has formly shown that the critical speed of the tire is related to its natural frequencies. Specifically, it has been shown that each frequency eigenvalue has its own critical speed. In particular, as the contact force field moves circumferentially around the tire, various circumferential frequency modes may be excited. The first such excited initiates the critical speed, i.e. standing wave process in the tire.

Noting Fig. 2, consider the $(m)^{th}$ circumferential vibration mode at the start of its period T_m and half way through at $T_m/2$. Assume that a circumferentially moving force covers the distance π/m during the half period. In this context, in terms of Fig. 2, we see that the force is tuned into the vibrations process and will continuously pump energy into the given mode. Such a situation is identical to a classical resonance point. Based on the foregoing, we see that each circumferential mode of vibration will be resonated when the travelling force velocity is given by the expression

$$V_{cm} = \frac{\text{circumferential distance for half period}}{\text{Time of half period}}$$

$$= \frac{\pi/m}{T_m/2} = \frac{2\pi}{m}\frac{1}{T_m} \tag{1}$$

where since

$$\omega_m \text{ (natural frequency in radians)} = \frac{2\pi}{T_m} \tag{2}$$

we yield that

$$V_{cm} = \frac{\omega_m}{m} \tag{3}$$

Next, to establish the appropriate FE analysis, we recall that the governing relation is typically written in the form (4)

$$\int_R [B^*]^T \underset{\sim}{S}\, dv = F + [M_1]\, \ddot{\underset{\sim}{x}} + \ldots \tag{4}$$

where here S is the 2nd Piola-Kirchhoff stress, $\underset{\sim}{F}$ is the nodal force, $\underset{\sim}{x}$ the nodal deflection, $(\ddot{\ })$ denotes acceleration and the coefficient $[B^*]$ rises out of the use of the Green-Lagrange (4) strain measure. Considering that the vibratory response can be defined as moderate/small dynamic excursions about an initial static prestress field, we yield that

$$\underset{\sim}{X} = (\underset{\sim}{X})_o + \Delta \underset{\sim}{X};\ \underset{\sim}{S} = (\underset{\sim}{S})_o + \Delta \underset{\sim}{S};\ \underset{\sim}{F} = (\underset{\sim}{F})_o + \Delta\underset{\sim}{F} \tag{5}$$

where here it follows that (4)

$$\Delta \underset{\sim}{S} \sim [D_T((\underset{\sim}{S})_o)][B^*((\underset{\sim}{X})_o)]\Delta \underset{\sim}{X} \tag{6}$$

Note, $[D_T((S)_o)]$ denotes the so-called tangent material stiffness. It represents the "slope" of the stress-strain relation. Based on Eqs 5 and 6, Eq 4 reduces to the following more tractable form namely

$$[M]\,\Delta\ddot{\underset{\sim}{X}} + [K_T]\,\Delta\underset{\sim}{X} = \Delta\underset{\sim}{F} \tag{7}$$

where here (4), the so-called tangent stiffness matrix takes the form

$$[K_T] \sim \Omega^2[M_1] + \int_R \{[G]^T[(\underset{\sim}{S})_o][G] + [B^*]^T[D_T((S)_o)][B^*]\}dv \tag{8}$$

such that $\Omega^2[M_1]$ denotes the centripetal stiffening. The matrix $[K_T]$ represents the slope of the force deflection surface of the overall tire response.

To determine the frequency eigenvalues, we assume that the small/moderate dynamic excursions about the initial stress state are also harmonic. In this context

$$\Delta\underset{\sim}{X} \cong \underset{\sim}{Z}\cos(\omega t) \tag{9}$$

where here ω represents the frequency and $\tilde{Z}$ its associated mode shape. Employing Eq 9, we yield the following expression for the frequency of the preloaded tire, namely

$$[\omega^2 [M] + [K_T]] \tilde{Z} = 0 \qquad (10)$$

The solution of Eq 10 requires the usual eigenvalue analysis, i.e. the power method, determinant search, subspace iteration etc.

Based on the foregoing, to determine the critical speed of the tire, we must solve Eq 10 for the eigenvalue ω which yields the lowest critical speed. At this juncture, it must be noted that great care must be taken to sort out the appropriate $(\omega, \tilde{Z})$ eigenvalue/vector pair. Overall, due to its complex geometry, the tire possesses a wide variety of vibrational modes. These include the following types namely: meridional bending, meridional inplane, circumferential bending, circumferential inplane, torsional, thickness, and combination type modes.

Since the rolling process of the tire is circumferential in nature, the type of critical speed excited involves predominantly circumferential behavior which is a combination of circumferential-meridional modes of bending. In this context, once the various $(\omega, \tilde{Z})$ pairs of Eq 10 are calculated, one must sort out the appropriate circumferential type as well as its mode number. For problems which involve significant displacements, the manner of calculating the critical speed must be modified. In particular, the circumferential mode number must be reinterpreted. For the noncontacted tire, the circumferential waves are equally spaced. Hence, the m appearing in Eq 3 is a direct measure of the arc length traversed by the moving load. In situations wherein the contact involves a minimum portion of the circumference, Eq 3 can continue to be used to yield a good estimate of the relationship between frequency and critical speed. This applies to many automotive and truck applications. In aircraft tires, since a 20-40% deformations of the section height is possible, Eq 3 needs to be modified to account for the variation in spacing of the mode shape.

Figure 3 illustrates a prototypical circumferential mode shape of a contacted tire. Since the spacing between waves is not uniform, we must consider the critical speed in a range sense, i.e. an upper and lower bound speed. This is a direct outgrowth of the nonlinear effects induced by the contact process. In terms of Fig. 3, we see that

$$V_{cm} \sim \frac{\alpha_m}{Tm/2} \sim \frac{\alpha_m}{\pi} \omega_m \qquad (11)$$

where here α_m is the half arc wave length (in radians) of a typical vibration wave. Based on this relation, we see that V_{cm} is bound by the relation

$$\left(\frac{\alpha_m \omega_m}{\pi}\right) \Big|_{min} \leq V_{cm} \leq \left(\frac{\alpha_m \omega_m}{\pi}\right) \Big|_{max} \qquad (12)$$

MODEL DEVELOPMENT

Overall the model development process consists of several different stages. Depending on which parametric sensitivity is sought, the preloading history must be varied to account for the correct sequence of events. For instance, if we seek the noncontacted critical speed, the modelling process consists of the following phases of loading, i.e., 1) initial mounting, 2) pressurization, 3) spin up and, 4) the development of tangent properties of mounted-pressurized-spinning tire.

In this case, the $(\tilde{S})_o$ and $(\tilde{X})_o$ fields defining the [M] and $[K_T]$ matrices appearing in Eq 10 must reflect the stresses and deformations which characterize the mounted-pressurized-spinning tire.

To determine the influence of pressure and rotation effects, $[K_T]$ and [M] must be evaluated for the appropriate range under consideration. Once these matrices are available, the eigenvalue extraction process associated with Eq 10 can be undertaken. As noted earlier, for the noncontacted case, the circumferential wave form has uniform spacing. In this context Eq 3 applies.

As will be seen later, prototypically an FE model requires a large number of degrees of freedom to define the problem $[K_T]$ behavior. For the noncontacted tire, typically the lowest critical speed occurs for m (mode numbers) below say 20. Hence, we need not evaluate all the model frequencies. To determine the correct value, one must realize that for the actual tire, the various modes associated with the meridional, circumferential radial, torsional and combined behavior are all interspersed. In this context, the first 20 circumferential modes may be embedded in the initial 150 frequencies of the tire. Hence, care must be taken to sort out the various modes. What we seek is a circumferential mode which yields a tire crown motion similar to that described in Fig. 2. The one that initiates the lowest critical speed must satisfy the relation

$$V_{cm} = \text{Min} \left(\frac{\omega_m}{m}\right) \qquad (13)$$

One interesting point must be raised at this juncture. While ω_m is typically monotone increasing namely $\omega_1 < \omega_2 < \omega_3 < \ldots\ldots$, the critical speed can occur for any ω_m value. This depends on its rate of increase with m. Figure 4 illustrates typical (ω, m) and (V_{cm}, m) dependencies. As can be seen, quite unexpectedly, the circumferential critical speed does not grow monotonically with mode number. See Padovan (5) for a more formal discussion of this characteristic.

Note, the foregoing approach also applies for contacted tires wherein the contact process does not yield large deformations of the circumferential cross section of the tire (10% or less of the circumferential arc length). For large deformation situations, i.e. 30-40% of section height, the contacted modelling approach must be employed. Such a model requires the following phases of loading, i.e., 1) initial mounting, 2) pressurization, 3) static contact (4,5), 4) spin up-contact redefinition, 5) the development of tangent properties of tire subjected to items (1-4). For this case, the contact and spin up phases must be properly modified. Since such a model cannot properly represent true rolling friction, two modelling alternatives are possible:

i) Disregard friction: simply apply appropriate normal stress/displacement constraints at interface as rotational speed is incrementally increased.

ii) Apply torque through axle which will generate tangential contact force whose magnitude is the same as rolling resistance load.

While ii) is more realistic, for the sake of simplicity, i) is more easily applied. Regardless, the friction forces have little effect on the frequency/critical speed.

Once contact is established, the spin up process consists of incrementally increasing the rotational speed. This yields a modified contact region. In auto and truck applications, prototypically at highway type rolling speeds, the centripetal force is small relative to the internal pressure. At aircraft T/O and landing speeds, the centripetal loads are essentially on the order of the pressure load. Figure 5 illustrates potential centripetal loads for a typical nylon type carcass construction where $\rho_{carcass}$ = .0001 lbm/3. As can be seen from Fig. 6, whereas the pressure load is always perpendicular to the walls, the centripetal load is always outward radial facing. In this context, while pressure tends to force the tire profile into a circular toroidal configuration, the centripetal load tends to ovalize the tire cross section. As will be seen during the benchmarking activity, both pressurization and rotation effects yield a hardening effect on the frequency eigenvalues.

As a final point, care must be taken to establish models with the appropriate element densification. Since 3-d dynamic behavior is sought, the requisite refinement is required in the radial, meridional as well as the circumferential directions. Since the critical speed response is strongly influenced by the circumferential vibrational characteristics, care must be taken to properly model such behavior. As noted earlier, prototypically, critical speeds occur in the range $0 < m < 20$. In this context, at least 60 3-D 20 node brick elements are required in the circumferential direction. For the half model, 30 elements are required along with the appropriate boundary conditions. To retain a reasonable overall problem bandwidth, care must be taken to establish a cost effective meridional thickness simulation. This can be achieved by using the axisymmetric pressurization model to pare back the meridional thickness densification. Note, in such a process, care must be taken to guarantee the representation of meridional effects which are in the range of the required critical speed behavior.

BENCHMARKING

For benchmarking purposes, the shuttle tire was used to access the potential of the scheme. In this context, Fig. 7 illustrates the basic cross-section of configuration. The plies noted feature biased construction with some 16 plies in the crown zone and 32 in the bead area. The turn ups end at various points along the sidewall of the tire.

To illustrate the paring of the meridional model densification, Figs. 8 and 9 depict a moderately refined and pared simulation. Note the paring process was cut off when the stress and deformations were within preset tolerances, i.e. 2%. The mesh was also required to provide a preset accuracy for the meridional axisymmetric frequencies in the anticipated range of the critical speed (300-450 mph). As can be seen from Fig. 9, a total of 14 elements was found adequate in the meridional direction. Since 30 were used circumferentially, a total of 420,20 noded 3-D brick elements were employed in the model. These lead to 3163 nodes and after applying the boundary conditions a total of 8731 degrees of freedom. By optimally numbering the nodes, the mean half bandwidth was restricted to 260 terms while the maximum was limited to 313. Overall, the resulting tangent stiffness possessed 2.27 mega stiffness terms. This exceeded available memory in the AU vectorized IBM-3090-200. Hence, an out of core solver was employed. Due to the blocking methodology used, some 4 blocks were required by the solver.

As a final check, the axisymmetric pressurization results were compared with those of the 3-D simulation. This also included a comparison of the appropriate meridional frequencies. Overall, the results remained within 2% accuracy. Note, the 3-D pressurization run required 13 minutes on the IBM 3090. This covered the $0 < P < 400$ psia range of pressures.

As noted earlier, the 2-D and 3-D models were used to establish a data base defining the effects of the different parametric variations. Where possible, these were compared with experimental data. Note, once a basic design is defined, it is often useful to parametrically vary the different design parameters to define the potential response envelop. For the current purposes, this will include the following factors namely: i) Internal pressure, ii) Material stiffness, iii) rotational speed, iv) contact.

To start the parametric considerations, Figs. 10 and 11 illustrate the effects of pressure, rotational speed and material stiffness on the crown deflection. As can be seen, while pressure effects are of a softening type, the rotationally generated load yields hardening response behavior. With regard to material properties, noting Fig. 11, it follows that increases in stiffness yield basically softening characteristics.

Next, we shall consider the frequency behavior. Figure 12 illustrates the influence of rotation. As can be seen, various modes exhibit essentially a hardening relation to rotational effects. In contrast, the pressure and stiffness tend to exhibit a softening interaction. These are depicted in Figs. 13 and 14.

Continuing, Figs. 15-18 define the parametric influences on the modal critical speeds. The nonmonotonicity of the modal critical speed behavior is clearly illustrated. For instance, Fig. 15 illustrates the pressure effects. As can be seen, the lowest critical speed tends to shift to higher mode numbers. Noting Fig. 16, this trend is reaffirmed for higher pressures and stiffness characteristics.

At normal highway speeds, 55-65 mph, centripetal loads tend to have a minor influence on the dynamic response behavior. This was clearly shown in Fig. 5. In aircraft applications with high T/O and landing speeds, the Ω^2 effect of centripetal inertia is more significant especially for low inflation pressures, i.e. Fig. 17. Again, as with pressure effects, the noncontacted lowest critical speed tends to shift to higher mode numbers, i.e. shorter wave lengths. Similar trends are also evidenced as material stiffness is increased, i.e. Fig. 18.

The foregoing results point to the fact that increases in pressure and stiffness (ply angles) have a limited effect on raising the potentially lowest modal critical speed. To finish the discussion, we shall consider the effects of contact on the dynamic response. As will be seen, unlike normal auto and trucking applications, the high contact deformation associated with aircraft tires has a very significant effect on the frequency spectrum. Figure 19 illustrates the variation of the lowest critical velocity as a function of percent contact, i.e. percentage of section height. As can be seen, a very significant reduction is possible.

Beyond the overall frequency reduction, large deformation contact tends to initiate a significant amount of modal shifting. Such behavior is a direct outgrowth of the changing boundary conditions induced by contact. In particular, in the noncontacted situation, the tire is a periodic structure wherein the bending and inplane waves are free to traverse the entire 360° and full meridional trajectory, i.e. from bead to bead. For the heavily contacted case, the various meridional, circumferential and torsional bending waves tend to terminate and reflect off the edges of the contact zone. As contact increases, the associated edge account for increasing portions of the tire circumference and meridional topology. Overall, contact induces an asymmetry in the mode shape.

The modal shifting is a result of the stiffening of some modes and softening of others. For instance, the circumferential frequencies tend to increase with growing contact deformation. Concomitant with the frequency increase, a given mode undergoes a wave length reduction induced by the reduction of free circumference. Note, the combined increase in circumferential bending wave frequencies is counterbalanced by the reduction in wave length. Overall, this leads to a drop in critical velocity. Such behavior is proportional to the increase in contact.

As can be seen from Fig. 19, excellent agreement is obtained with experimental data. In this context, the procedure outlined herein not only provides useful design insights, if judicially applied, it can yield excellent results. As a consequence of the foregoing parametric studies, we see that for biased constructions:

i) Material stiffness and pressure increases raise the critical speed limit for a given geometric design configuration;

ii) Note, such improvements asymptotically diminish to a negligible level for the upper end range of economically viable material properties and environmentally acceptable (i.e. runway limits);

iii) At low inflation pressures, high speed centripetal loads yield a significant effect on the critical speed range;

iv) Increasing contact deformation can significantly reduce critical speeds, i.e. 0-40% drops forrealistic vehicle load ranges: Since the onset of critical speed behavior severly hampers the life, aircraft loading limits need to be observed.

Note, the methodology developed herein can easily be applied to both radial and biased type construction. In the case of the radial tire, the meridional representation may need a more complex model to represent the belt edge zone. Regardless, since the evaluation of 50 frequencies and associated mode shapes required under 70.0 minutes of CPU time, more complex models don't represent a major burden. Overall, the procedure fits in the current CAD-CAM type environment.

ACKNOWLEDGMENT

The author wish's to thank, J. champion and P. Wagner of U.S. Airforce Wright Patterson for their helpful suggestions and support.

REFERENCES

1) Clark, S.K., Mechanics of Pneumatic Tire, US DOT Printing Office, Washington, DC, 1979.
2) Champion, U.H., Wagner, P.M., A Critical Speed Study for Aircraft Bias Ply Tires, U.S. Airforce, AFWAL-TR-88-3006, 1988.
3) Padovan, J., On Standing Waves in Tires, The Science and Tech., Vol. 5, pp 83, 1977.
4) Padovan, J., Kennedy, R., and Nakajima, T., Finite Element Analysis of Steady and Transiently Moving/Rolling Nonlinear Viscoelastic Structure, Parts I, II, II, Comp. Struct. 27, 249, 1988.
5) Padovan, J., Spectral/Critical Speed Characteristics of Structure Subject to Moving Loads, Int. Jr. Engrg. Sci. 20, 77, 1982.
6) Stafford, J., Tabaddor, F., ADINA Load Updating for Pressurized Structures, Proc. of ADINA Confer., MIT 82448-9, pp 537, 1979.

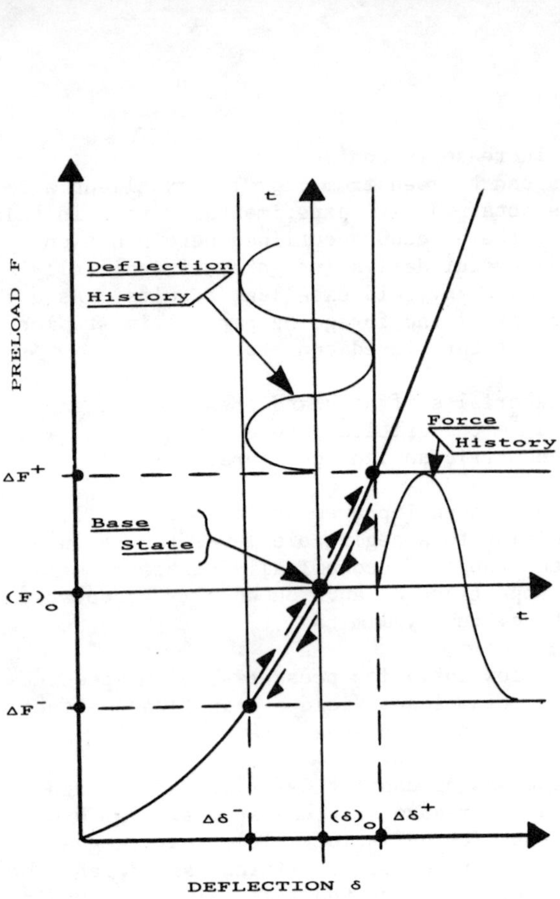

Fig. 1 Vibrations about a force-deflection base state.

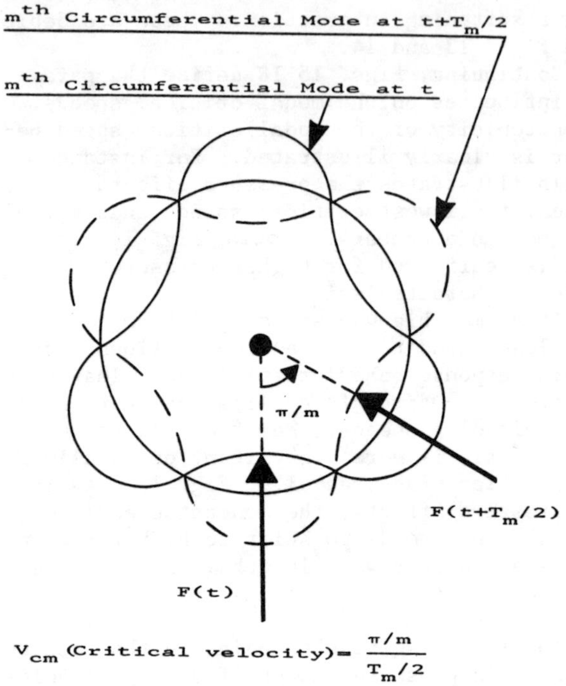

Fig. 2 Relationship between critical speed and frequency: uncontacted.

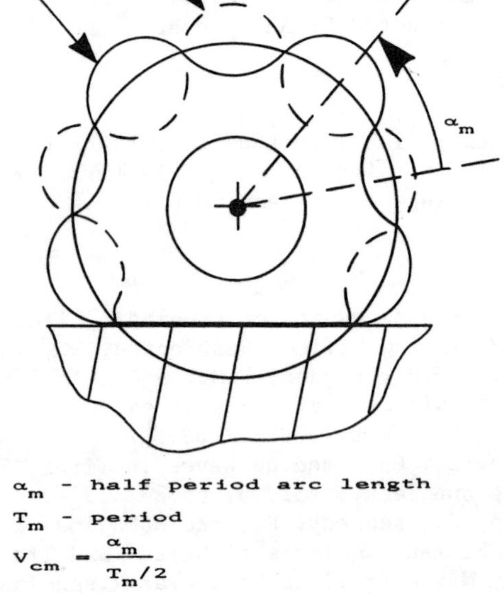

Fig. 3 Relationship between critical speed and frequency: contacted.

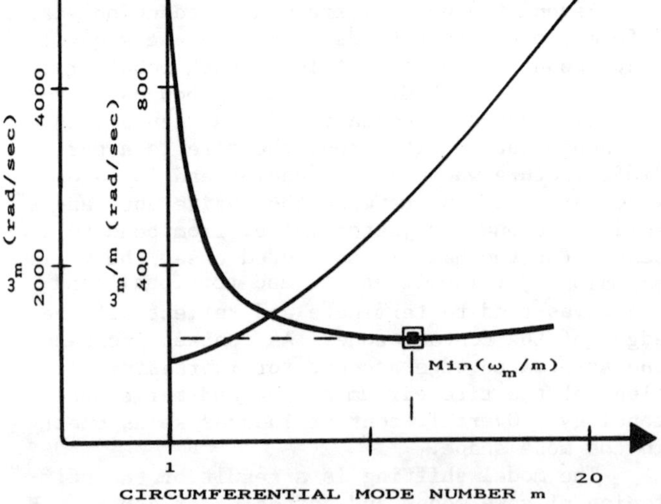

Fig. 4 Mode number dependency of circumferential critical speed: noncontacted.

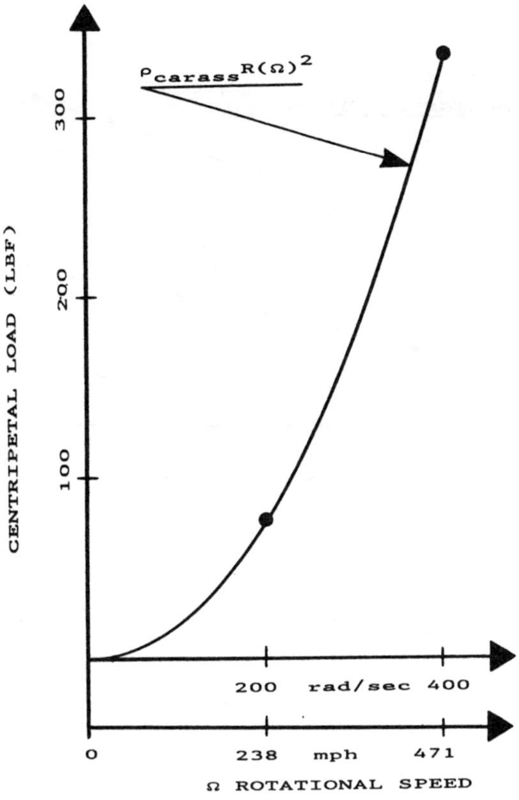

Fig. 5 Rotation effects on centripetal load.

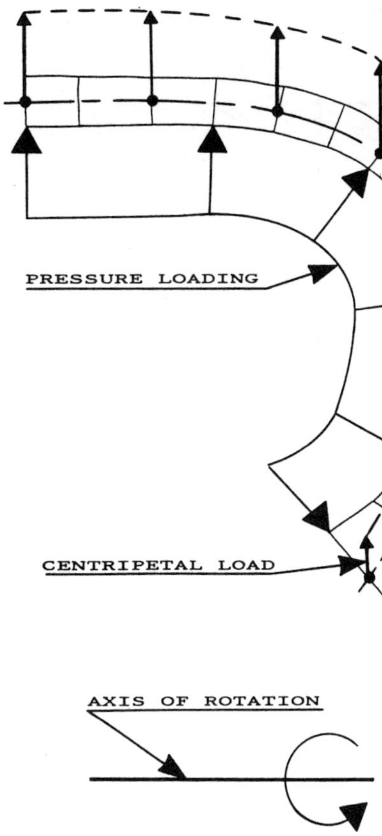

Fig. 6 Pressure and load on tire cross-section.

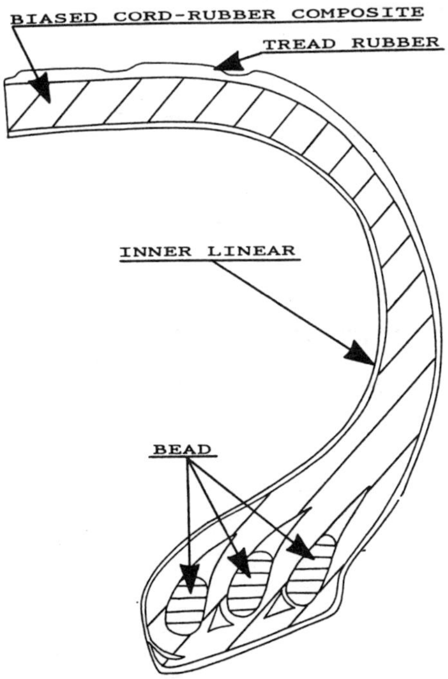

Fig. 7 Basic cross-section of shuttle tire.

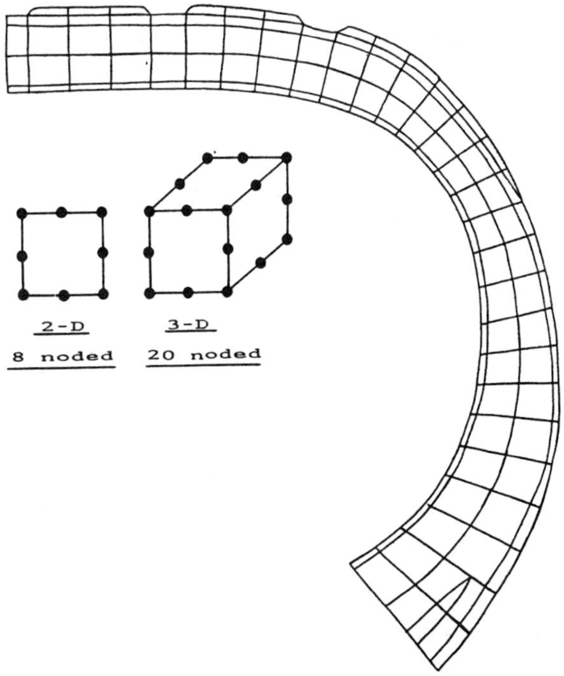

Fig. 8 Moderately refined FE model.

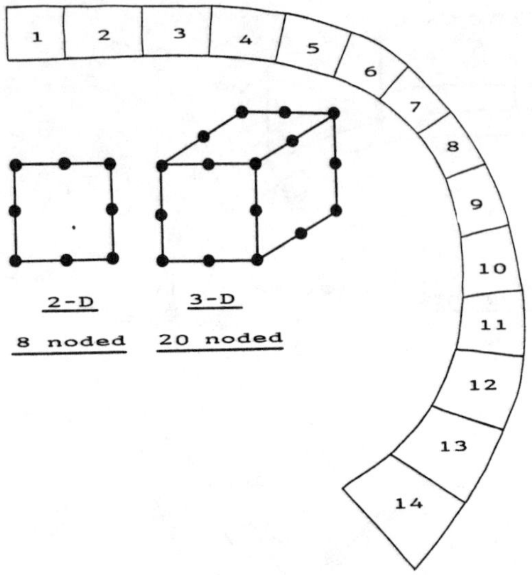

Fig. 9 Pared model.

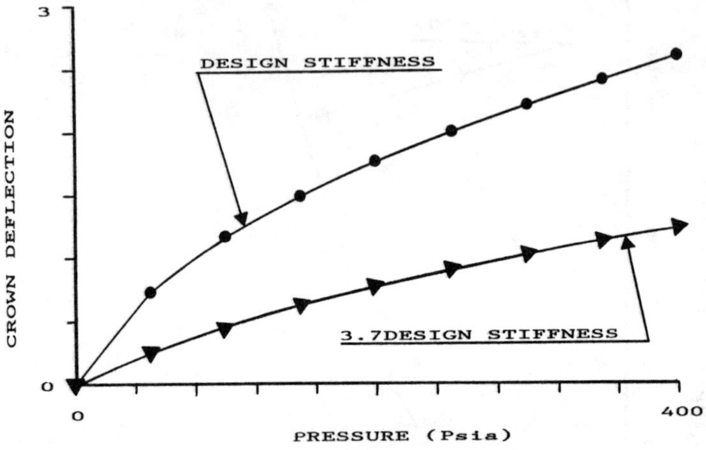

Fig. 11 Material stiffness and pressure effects.

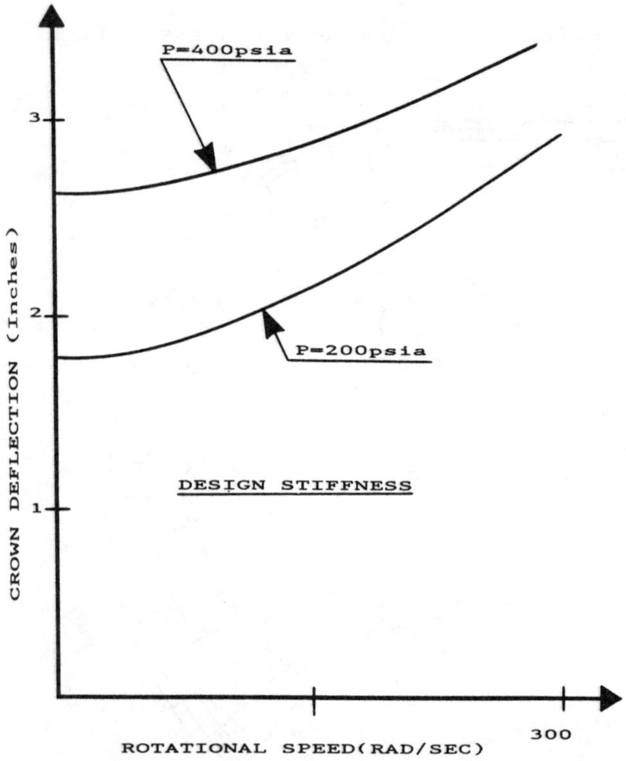

Fig. 10 Pressure and rotation effects on crown deflection.

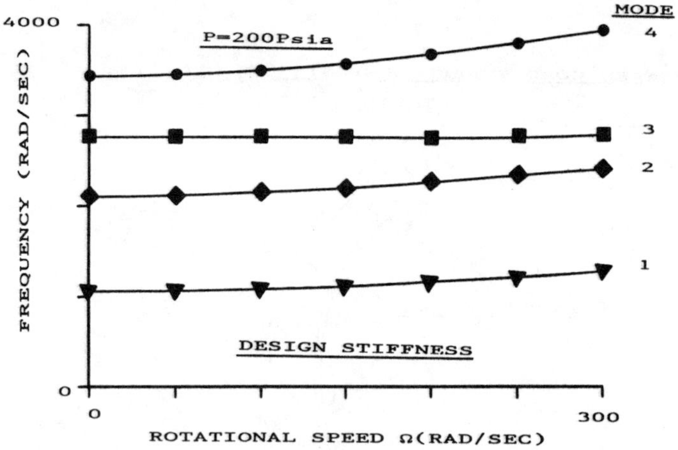

Fig. 12 Rotational effects on axisymmetric frequencies.

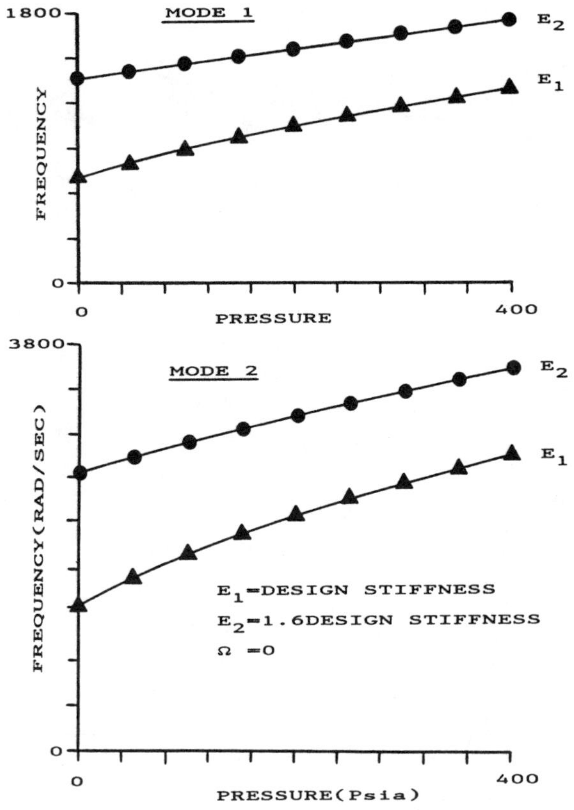

Fig. 13 Material and pressure effects on frequencies.

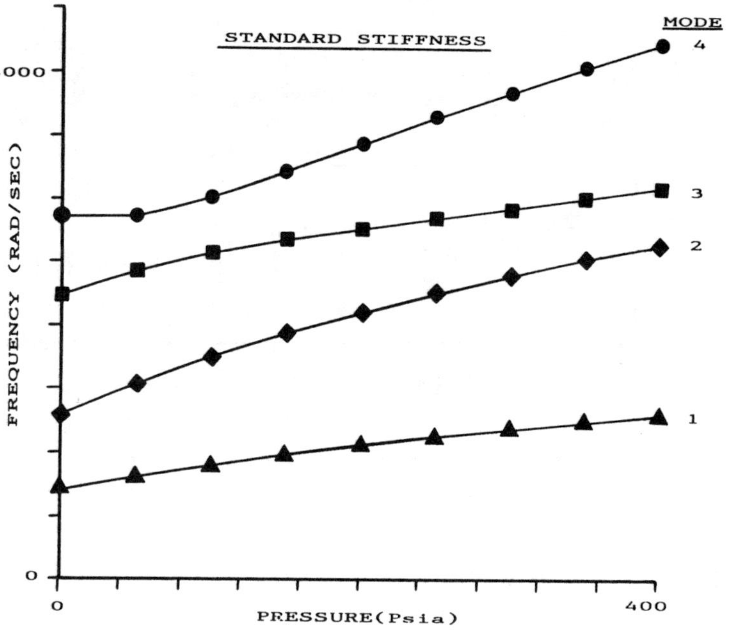

Fig. 14 Pressure effects: frequencies.

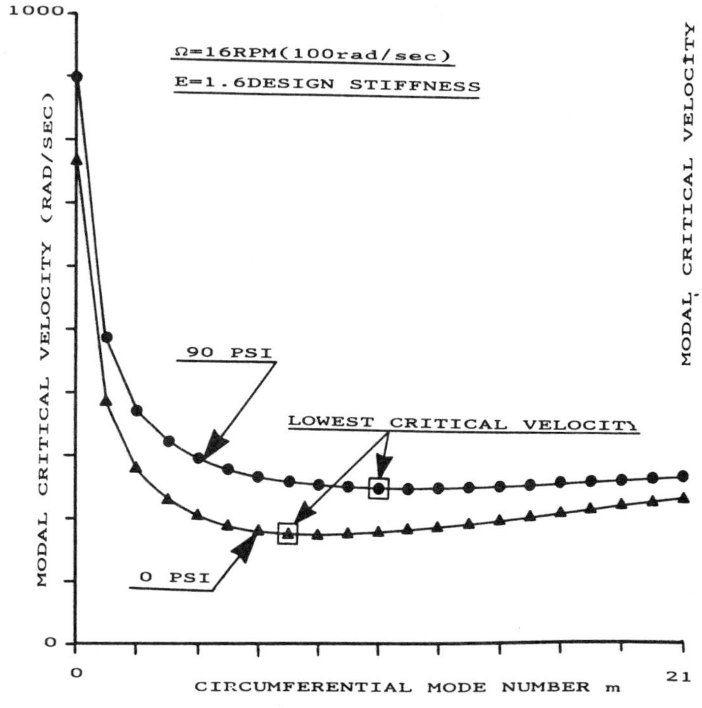

Fig. 15 Pressure effects on modal critical velocity.

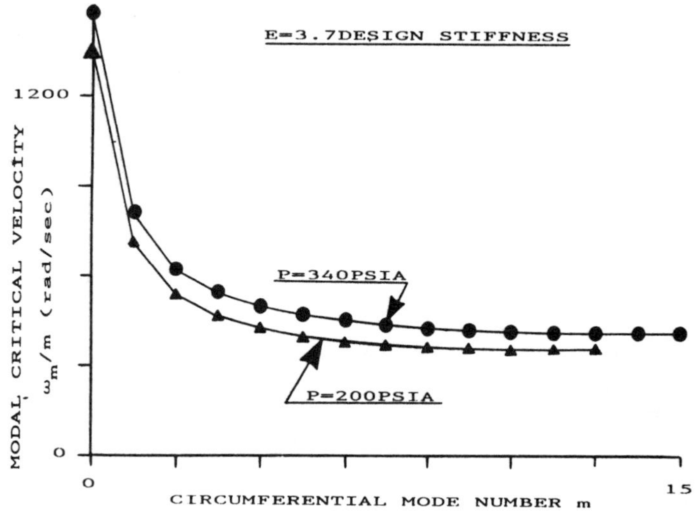

Fig. 16 Pressure effects on modal velocity.

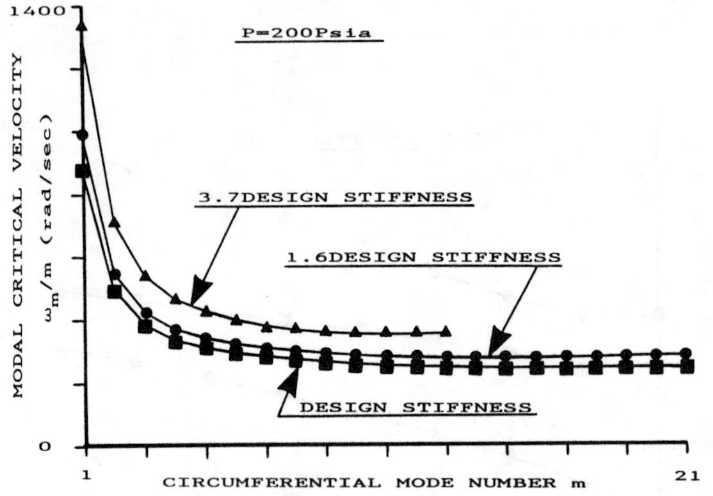

Fig. 17 Material stiffness effects on critical velocities.

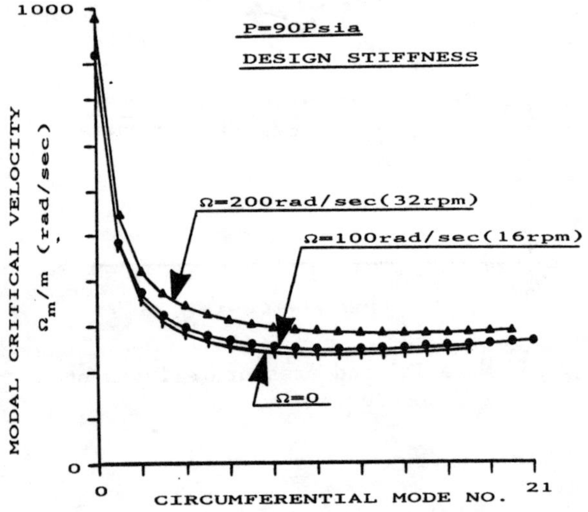

Fig. 18 Rotational effects on critical velocity.

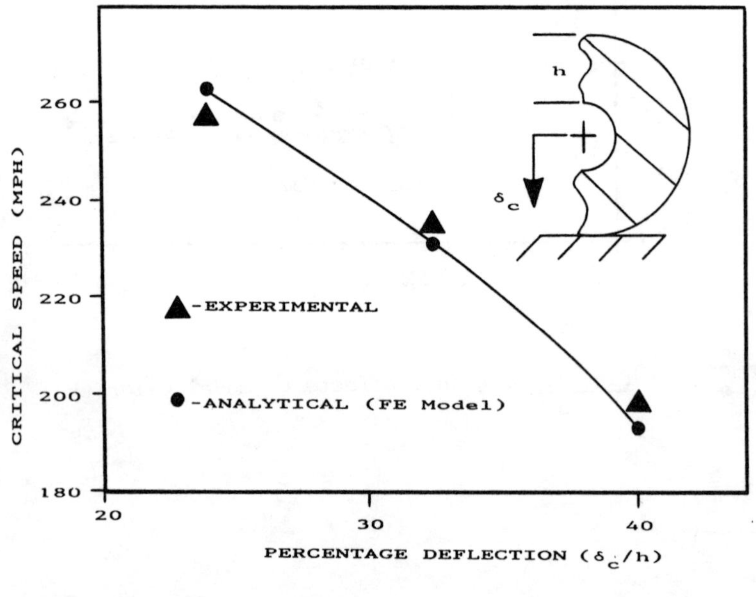

Fig. 19 Deformation effects on critical speed: contacted case.

Frictionless Contact of Aircraft Tires

** Paper 892350 presented at the Aerospace Technology Conference and Exposition, Anaheim, California, September, 1989.

Kyun O. Kim, John A. Tanner, and Ahmed K. Noor
George Washington University
NASA Langley Research Center

SUMMARY

A computational procedure is presented for the solution of frictionless contact problems of aircraft tires. The space shuttle nose-gear tire is modeled using a two-dimensional laminated anisotropic shell theory with the effects of variation in material and geometric parameters, transverse shear deformation, and geometric nonlinearities included. The contact conditions are incorporated into the formulation by using a perturbed Lagrangian approach with the fundamental unknowns consisting of the stress resultants, the generalized displacements, and the Lagrange multipliers associated with the contact conditions. The elemental arrays are obtained by using a modified two-field mixed variational principle.

Numerical results are presented for the shuttle nose-gear tire when subjected to inflation pressure and pressed against a rigid pavement. Comparison is made with the experiments conducted at NASA Langley. The detailed information presented herein is considerably more extensive than previously reported and helps in gaining physical insight about the response of the tire. The numerical studies have demonstrated the high accuracy of the mixed models and the effectiveness of the computational procedure which combines both the geometrically nonlinear terms and the contact conditions in one iteration loop.

INTRODUCTION

Numerical modeling of the response characteristics of aircraft tires remains one of the most challenging applications of computational structural mechanics. There are several aspects of this problem that can lead to numerical difficulties and/or excessive computational expense. During normal aircraft ground operations tires are subjected to large displacements and temperature gradients. The tire is a composite structure composed of rubber, textile, and steel constituents which exhibits anisotropic and nonhomogeneous material properties. Furthermore, all the forces developed by the tire associated with takeoff and landing operations are developed through the tire-pavement interface; thus, any practical modeling tool must include a good contact algorithm. These facts and attending difficulties emphasize the need to develop modeling strategies and analysis methodologies including efficient contact algorithms which are both powerful and economical. In recent years nonlinear analyses of static and dynamic problems involving contact have been the focus of intense research activities. Novel techniques which have emerged from these efforts include semianalytic finite element models for nonlinear analysis of shells of revolution (1 and 2)*, reduction methods (3 and 4), and operator splitting techniques (5-7). Application of these new techniques to tire modeling is summarized in (4, 7 and 8).

SCOPE OF INVESTIGATION

Current research on tire modeling and analysis at NASA Langley Research Center is aimed at developing an accurate and efficient strategy for predicting aircraft tire response. The focus of this paper is directed toward the developments in tire contact techniques. These contact algorithms are incorporated into a mixed formulation, two-field, two-dimensional finite element code based on the moderate-rotation Sanders-Budiansky shell theory with the effects of transverse-shear deformation and laminated anisotropic material response included (9 and 10). The contact algorithm is based on the perturbed Lagrangian formulation (11 and 12) and utilizes the preconditioned conjugate gradient (PCG) iteration procedure (13-15) to determine contact area and pressure distribution. To demonstrate the capabilities of the analysis techniques, numerical studies were conducted with a space shuttle orbiter nose gear tire model. Analysis results are presented for an inflated shuttle nose gear tire statically loaded on a flat surface. The analysis assumes frictionless contact. The analytical results are compared with experimental measurements.

NOTATION

c	number of nodal points in contact in the element
c_{ij}, d_{ij}, f_{ij}	tire stiffness coefficients (i,j = 1,2,6)
c_{44}, c_{45}, c_{55}	transverse-shear stiffness coefficients of the tire
$\vec{e}_s, \vec{e}_\theta$	tangential unit vectors in the meridional and circumferential directions
$[F]$	flexibility matrix for an individual element
$\{\tilde{f}(Z,p)\}$	vector defined in Eqs. (7)
$\{G(X)\}$, $\{\dot{M}(H,X)\}$	vectors of nonlinear terms in the element equations
$\{\overline{G}(X)\}$	vector of nonlinear terms in Eqs. (4) and (5)
$\{\tilde{G}(Z)\}$	vector of nonlinear contributions to the global equations
$\overline{g}$	current gap (measured in the direction of the normal to the contact surface)
$\overline{g}_o$	initial gap
$\{g_o\}$	vector of initial gaps for the contact element
$\{H\}$	vector of stress resultant parameters
h	total thickness of the tire

*Designates references at end of paper.

$\bar{h}$	nondimensional thickness of the tire (see figure 3)
$[\tilde{K}]$	global linear matrix of the tire
$M_s, M_\theta, M_{s\theta}$	bending and twisting stress resultants (see figure 1)
m	number of displacement nodes in the element
N	shape functions used for approximating the generalized displacements and the Lagrange multipliers
$N_s, N_\theta, N_{s\theta}$	extensional stress resultants (see figure 1)
n	total number of degrees of freedom
$\{P\}$	normalized external load vector
$\{\tilde{P}\}$	global vector of normalized external loads and initial gaps
P_n	nodal (contact) force normal to the contact surface
p_o	intensity of inflation pressure
p_s, p_θ, p	intensity of external loading in the coordinate directions (see figure 1)
$[Q]$, $[R]$	elemental matrices associated with the contact condition and the regularization term in the functional
Q_s, Q_θ	transverse shear stress resultants (see figure 1)
q	load parameter
R_1, R_2	principal radii of curvature in the meridional and circumferential directions
r	normal distance from the tire axis to the reference surface
$[S]$	strain-displacement matrix for an individual element
s	meridional coordinate of the tire (see figure 1)
$[T]$	transformation matrix
T_n	intensity of contact pressure (acting normal to the contact surface)
U	strain energy density (strain energy per unit area)
u,v,w	displacement components of the reference surface of the tire in the meridional, circumferential, and normal directions (see figure 1)
$\bar{w}$	normal displacement at $\xi=\theta=0$ (see figure 4)
$\{X\}$	vector of nodal displacements in the shell coordinate system
$\{\bar{X}\}$	vector of nodal displacements in the Cartesian coordinate system
x,y,z	Cartesian coordinate system
x_3	coordinate normal to the tire reference surface (see figure 1)
$\{Z\}$	global response vector
ε	penalty parameter
$\varepsilon_s, \varepsilon_\theta, 2\varepsilon_{s\theta}$	extensional strains of the reference surface of the tire
$2\varepsilon_{s3}, 2\varepsilon_{\theta 3}$	transverse shear strains of the tire
θ	circumferential (hoop) coordinate of the tire (see figure 1)
$\kappa_s, \kappa_\theta, 2\kappa_{s\theta}$	bending strains of the tire
$\bar{\lambda}$	Lagrange multiplier, representing the intensity of the contact pressure (acting normal to the contact surface)
$\{\lambda\}$	vector of nodal values of the Lagrange multiplier
ξ	dimensionless coordinate along the meridian (see figure 3)
π	functional
ϕ	rotation around the normal to the tire reference surface
ϕ_s, ϕ_θ	rotation components of the reference surface of the tire (see figure 1)
$\Omega^{(e)}$	element domain
Ω_c	contact surface
∂_s .	$\equiv \partial/\partial s$
∂_θ	$\equiv \partial/\partial\theta$

<u>Superscripts:</u>

(e)	individual elements
i,j	indices of shape functions for approximating Lagrange multipliers
i'	index of shape function for approximating generalized displacements; ranges from 1 to m
r	number of iteration cycles
t	matrix transposition.

Finite elements are designated M9-4 as shown in Table 1.

MATHEMATICAL FORMULATION

The analytical formulation for frictionless contact of aircraft tires is based on a form of moderate rotation, Sanders-Budiansky shell theory with the effects of large displacements and transverse shear deformation included. A mixed formulation is used in which the fundamental unknowns consist of the five generalized displacements and the eight stress resultants. The sign convention for the generalized displacements and the stress resultants is given in figure 1. The fundamental equations of the shell theory used herein are given in (9 and 10) and are summarized in Appendix A.

CONTACT CONDITION - Figure 2 shows the characteristics of the frictionless contact of a shell pressed against a rigid plate: Ω_c refers to the contact region; $\bar{g}_o$ is the initial gap between the shell and the plate; $\bar{g}$ is the current gap (both $\bar{g}_o$ and $\bar{g}$ are measured in the direction of the normal to Ω_c); and T_n is the normal traction on Ω_c. The contact condition can be expressed by the following inequalities which must be satisfied at each point on the contact surface, Ω_c:

$$\bar{g} \geq 0\,, T_n \leq 0 \text{ and } T_n \bar{g} = 0 \tag{1}$$

The first inequality in Eqs. (1) represents the kinematic condition of no penetration of the rigid plate (zero gap for the contact points); the second inequality is the static condition of compressive (or zero) normal tractions; and the third condition is that of zero work done by the contact stresses (i.e., the contact stresses exist only at the points where the structure is in contact with the rigid plate). The inequalities $\bar{g} > 0$, $T_n > 0$ are henceforth referred to as the *inactive contact conditions.*

GOVERNING FINITE ELEMENT EQUATIONS - The discrete equations governing the response of the structure are obtained by applying a modified form of the two-field, Hellinger-Reissner mixed variational principle. The modification consists of augmenting the functional of that principle by two terms: the Lagrange multiplier associated with the nodal contact pressures and a regularization term which is quadratic in the Lagrange multipliers. For a detailed discussion of the perturbed and the augmented Lagrangian formulations, see (11, 12 and 16).

The modified functional has the following form:

$$\pi = \pi_{HR} + \int_{\Omega_c} \left[\bar{\lambda} \bar{g} - \frac{1}{2\varepsilon} (\bar{\lambda})^2 \right] d\Omega \tag{2}$$

where π_{HR} is the functional of the Hellinger-Reissner variational principle; $\bar{\lambda}$ is the Lagrange multiplier; and ε is a penalty parameter associated with the regularization term. The explicit forms of π_{HR} for axisymmetric shells is given in (4). Note that the addition of the regularization term amounts to approximating the rigid plate by continuously distributed springs with stiffness of ε, for sufficiently large ε. As $1/\varepsilon$ approaches zero, the continuous springs become the rigid plate.

The shape functions used in approximating the generalized displacements and the Lagrange multipliers are selected to be the same and differ from those used in approximating the stress resultants. Moreover, because of the nature of the functional π in Eq. (2), the continuity of both the stress resultants and the Lagrange multiplier is not imposed at interelement boundaries.

The finite element equations for each individual element can be cast in the following compact form:

$$\begin{bmatrix} -F & S & \cdot \\ S^t & \cdot & Q \\ \cdot & Q^t & \frac{R}{\varepsilon} \end{bmatrix}^{(e)} \begin{Bmatrix} H \\ X \\ \lambda \end{Bmatrix}^{(e)} + \begin{Bmatrix} G(X) \\ M(H,X) \\ \cdot \end{Bmatrix}^{(e)} - \begin{Bmatrix} \cdot \\ pP \\ g_o \end{Bmatrix}^{(e)} = 0 \tag{3}$$

where $\{H\}$, $\{X\}$ and $\{\lambda\}$ are the vectors of the stress resultant parameters, nodal values of the generalized displacements and nodal values of the Lagrange multipliers; $[F]$ is the matrix of linear flexibility coefficients; $[S]$ is the strain-displacement matrix; $[Q]$ and $[R]$ are matrices associated with the contact condition and the regularization term in the functional (see Appendix B); $\{G(X)\}$ and $\{M(H,X)\}$ are vectors of nonlinear terms; $\{g_o\}$ is the vector of initial gaps in the contact region; a dot refers to a zero submatrix or a zero subvector; superscript (e) refers to individual elements; $\{P\}$ is the normalized external load vector; and p is a load parameter. As the load is incremented, only the value of the load parameter p changes and the normalized vector $\{P\}$ is constant. The formulas for the elemental arrays $[F]$, $[S]$, $\{G(X)\}$, $\{M(H,X)\}$ and $\{P\}$ are given in (7). The formulas for the elemental arrays $[Q]$ and $[R]$ are given in Appendix B.

Note that the size of the coefficient matrices $[R]$, $[Q]$ and $\{g_o\}$ varies with the number of active contact conditions. The difficulty associated with an equation system whose size varies during the solution process can be alleviated by allowing the Lagrange multipliers to be discontinuous at interelement boundaries and eliminating them on the element level. This was done in the present study. If the stress resultant parameters and the Lagrange multiplier parameters are eliminated from Eqs. (3), one obtains the following equations in the nodal displacements $\{X\}$:

$$\begin{aligned} &[[S]^t [F]^{-1} [S] - \varepsilon [Q] [R]^{-1} [Q]^t]^{(e)} \{X\}^{(e)} \\ &+ \{\bar{G}(X)\}^{(e)} + \varepsilon [Q] [R]^{-1} \{g_o\}^{(e)} - p \{P\}^{(e)} = 0 \end{aligned} \tag{4}$$

where

$$\{\bar{G}(X)\}^{(e)} = [S]^t [F]^{-1} \{G(X)\}^{(e)} + \{M(H,X)\} \tag{5}$$

and the vector $\{H\}$ in $\{M(H,X)\}$ is replaced by its expression in terms of $\{X\}$.

To simplify the treatment of the contact conditions, the displacement components are transformed from the shell coordinates $(s,\ \theta,\ x_3)$ to the global Cartesian coordinates (x,y,z) before assembly. The relations between the displacement vector in the shell coordinates, $\{X\}^{(e)}$, and the corresponding vector in the Cartesian coordinates, $\{\bar{X}\}^{(e)}$, can be written in the following compact form:

$$\{X\}^{(e)} = [T] \{\bar{X}\}^{(e)} \tag{6}$$

where $[T]$ is a transformation matrix. The different arrays in the finite element equations are transformed accordingly. The explicit form of the transformation relations is given in Appendix C.

SOLUTION OF NONLINEAR ALGEBRAIC EQUATIONS - The discrete equations governing the response of the tire are obtained by assembling the elemental contributions in Eqs. (3) or (4), and can be written in the following compact form:

$$\{\tilde{f}(Z,p)\} = [\tilde{K}] \{Z\} + \{\tilde{G}(Z)\} - p \{\tilde{P}\} = 0 \tag{7}$$

where $[\tilde{K}]$ is the global linear matrix of the structure; $\{\tilde{G}(Z)\}$ is the vector of nonlinear contributions; $\{\tilde{P}\}$ is the global vector of normalized external loads and initial gaps; and $\{Z\}$ is the global response vector of the structure (obtained by assembling the contributions from the subvectors $\{H\}$, $\{X\}$ and $\{\lambda\}$).

The nonlinear algebraic equations (Eqs. (7)), are solved and the contact region and the contact pressures are determined by using an incremental/iterative technique (i.e., a predictor-corrector continuation method) in which the response vector $\{Z\}$, corresponding to a particular value of the load parameter, p, is used to calculate suitable approximation (predictor) for $\{Z\}$ at a different value of p. This approximation is then chosen as an initial estimate for $\{Z\}$ in a corrective iterative scheme such as the Newton-Raphson technique. In each Newton-Raphson iteration the contact conditions are checked and updated.

COMPUTATIONAL PROCEDURE AND DETERMINATION OF CONTACT PRESSURES

The computational procedure used in determining the contact region and the contact pressures is outlined in Appendix D. The nonlinearities due to the large displacements (moderate rotations) and the contact condition are combined into a single iteration loop. Ref. (15) advocated the use of two-level (nested) iteration scheme. The inner iteration loop accounts for the contact conditions associated with the contact pressures, and the outer iteration loop uses Newton-Raphson iteration scheme. Numerical experiments have demonstrated that for frictionless contact problems the two-level iterative scheme requires more iterations than the single-level scheme (see 17).

The solution of the governing discrete equations of the entire structure generates the nodal displacements, the internal force parameters, and the values of the Lagrange multipliers at the contact nodes. For each individual element in contact, the intensity of the contact pressure at a node T_n is equal to the value of the Lagrange multiplier $\bar{\lambda}$ at the same node. The contact pressures are also related to the nodal forces normal to the contact surface P^i_n as follows:

$$P^i_n = \int_{\Omega^{(e)}} N^i N^j \, d\Omega \, T^j_n \tag{8}$$

where N^i are the shape functions used in approximating the Lagrange multiplier, and the generalized displacements; and $\Omega^{(e)}$ is the domain of the contact element. The range of both i and j in Eqs. (8) is from 1 to the number of displacement nodes. Other approaches for determining the contact pressures are discussed in (18).

COMMENTS ON THE MIXED MODELS, PERTURBED LAGRANGIAN FORMULATION, AND COMPUTATIONAL PROCEDURE

The following comments regarding the mixed models, the perturbed Lagrangian formulation and the computational procedure used herein are in order:

1. The nonlinear terms in the finite element equations of the mixed model (Eqs. (3)) have a considerably simpler form than those of the corresponding displacement model (Eqs. (5)).

2. Equations (3) include both those of the Lagrange multiplier approach and the penalty method as special cases, as follows:

a) By letting the penalty parameter ε go to infinity, Eqs. (3) reduce to those of the Lagrange multiplier approach.

b) By eliminating the Lagrange multiplier terms from Eqs. (3), the resulting equations are identical to those of the penalty method.

3. The perturbed Lagrangian formulation alleviates two of the drawbacks of the Lagrange multiplier approach and the penalty method, namely:

a) The regularization term in the functional results in replacing one of the zero diagonal blocks in the discrete equations of the Lagrange multiplier approach by the diagonal matrix $[R]/\varepsilon$ in Eqs. (3).

b) The contact condition is satisfied exactly by transforming the constrained problem to an unconstrained one through the introduction of Lagrange multipliers (the term $\int_{\Omega_c} \bar{\lambda}\, \bar{g}\, d\Omega$ in Eq. (2)) rather than approximately as in the penalty method. However, the presence of the regularization term (the term $-\int_{\Omega_c} \frac{1}{2\varepsilon} (\bar{\lambda})^2 \, d\Omega$ in Eq. (2)) results in replacing the contact condition by the perturbed condition:

$$\frac{1}{\varepsilon}[R]\{\lambda\} + [Q]^t\{X\} - \{g_o\} = 0 \tag{9}$$

4. An important consideration in the perturbed Lagrangian formulation (and in any penalty formulation) is the proper selection of the penalty parameter ε. With the foregoing mixed models the penalty parameter can be chosen independently of the element size, without adversely affecting the performance of the model. The accuracy of the solution increases with increasing value of the penalty parameter. However, for very large values of ε, the equations become ill-conditioned and thus round-off errors increase.

5. The elemental arrays $[F]$, $[S]$, $\{G(X)\}$, $\{M(H,X)\}$ and $\{P\}$ are evaluated numerically using a Gauss-Legendre formula. The arrays $[Q]$, $[R]$ and $\{g_o\}$ are evaluated using a Newton-Cotes formula. In both cases the same number of quadrature points used is the same as the number of displacement nodes in the element. This results in underintegrating the arrays $[Q]$ and $[R]$ and avoids the oscillatory behavior of

the contact pressures which has been observed when these arrays are fully integrated. Note that the use of Newton-Cotes formula allows the contact pressures to be evaluated at the displacement nodes.

NUMERICAL STUDIES

Numerical studies were performed to assess the accuracy of the two-dimensional shell code by modeling the space shuttle nose-gear tire. The cord-rubber composite was treated as a laminated anisotropic material. For the purpose of computing stiffness variations in the meridional direction, half the tire cross-section was divided into seven regions as shown in figure 3. The geometric and resulting material characteristics of the tire can be found in (19). The outer surface of the tire was chosen to be the reference surface for the two-dimensional shell model.

A two-field mixed finite element model was used for discretization of the tire in both the meridional and circumferential directions. Due to the tire being angle-ply laminated anisotropic material, the whole tire was discretized using 1080 M9-4 elements (see Table 1), 30 elements in the meridional direction, and 36 elements in the circumferential direction as shown in figure 4, where tire was shown as four-node elements (i.e., total 4320 elements). Single-level iteration scheme which combines Newton-Raphson iteration and contact iteration was used. This combined single-iteration scheme was proven to be more efficient than the two-level iteration scheme (17).

Uniform inflation pressure of $\overline{p}_o$ = M Pa, acting normal to the inner surface, were applied to obtain the inflated deformed shape of the tire. The deformed shape of the tire subjected to the inflation pressure was compared well with experiments (19). The vertical load was then applied against a rigid pavement by controlling the displacements. The load-deflection curve for the frictionless shell model is compared in figure 5 with the experimental data obtained on the shuttle nose-gear tire. The close agreement reveals that the friction at the interfaces of the tire and the pavement has little effect on the global response of the tire. The deformed configurations and footprint areas of the tire for the inflation pressure and the applied displacement are shown in figures 6 and 7, respectively. In figures 6 and 7, tire mesh was shown as discretized with four-node elements.

Figure 8 shows the variation of the contact normal pressure distribution with applied displacements. Considering the fact that grooves of the tire tread were not modeled in the present study, it is expected that computed contact pressures will be slightly less than measured ones. The distribution of total and shear strain energy densities are shown in figure 9. The results in figures 7, 8 and 9 show that the response characteristics of the tire exhibits the symmetry with respect to the coordinate center (inversion or polar symmetry). This symmetry condition can be exploited for reduction of computational efforts. It is noted in figure 9 that the shear energy density increases at the footprint area (up to a maximum 25% of the total strain energy) and it shows the importance of the shear deformation).

The effect of the magnitude of the penalty parameter on the accuracy of the total strain energy and the total contact force obtained at $\overline{p}_o = 2.2$ M Pa and $\overline{w} = 0.9$ inch, is depicted in figure 10. As can be seen, the accuracy of the total strain energy and the total contact force is fairly insensitive to the choice of the penalty parameter ε in the range of 10^6 to 10^{15}.

The convergence study was carried out on the contact pressure distribution with mesh refinements and is shown in figure 11. The results justify the validity of the finite element mesh and refinement at the contact area chosen in the present study.

CONCLUDING REMARKS

A computational procedure is presented for the solution of frictionless contact problems of aircraft tires. The space shuttle nose-gear tire is modeled using a two-dimensional laminated anisotropic shell theory with the effects of variation in material and geometric parameters, transverse shear deformation, and geometric nonlinearities included. The contact conditions are incorporated into the formulation by using a perturbed Lagrangian approach with the fundamental unknowns consisting of the stress resultants, the generalized displacements, and the Lagrange multipliers associated with the contact conditions. The elemental arrays are obtained by using a modified two-field mixed variational principle. The modification consists of augmenting the functional of that principle by two terms: the Lagrange multiplier vector associated with the nodal contact pressures and a regularization term which is quadratic in the Lagrange multiplier vector.

The shape functions used in approximating the generalized displacements and the Lagrange multipliers are selected to be the same and differ from those used in approximating the stress resultants. The stress resultants and the Lagrange multipliers are allowed to be discontinuous at interelement boundaries. The nonlinearities due to both the large displacements, moderate rotations, and the contact conditions are combined into the same iteration loop and are handled by using the Newton-Raphson iterative scheme.

Numerical results are presented for the shuttle nose-gear tire when subjected to inflation pressure and pressed against a rigid pavement. Comparison is made with the experiments conducted at NASA Langley. The detailed information presented herein is considerably more extensive than previously reported and helps in gaining physical insight about the response of the tire. The numerical studies have demonstrated the high accuracy of the mixed models and the effectiveness of the computational procedure which combines both the geometrically nonlinear terms and the contact conditions in one iteration loop.

REFERENCES

1. Schaeffer, H. G. and Ball, R. E., "Nonlinear Deflections of Asymmetrically Loaded Shells of Revolution," AIAA Paper No. 68-292, April 1968.
2. Wunderlich, W., Cramer, H. and Obrecht, H., "Application of Ring Elements in the Nonlinear Analysis of Shells of Revolution Under Nonaxisymmetric Loading," Computer Methods in Applied Mechanics and Engineering, Vol. 51, No. 1-3, Sept. 1985, pp. 259-275.
3. Noor, A. K., "On Making Large Nonlinear Problems Small," Computer Methods in Applied Mechanics and Engineering, Vol. 34, Nos. 1-3 1982, pp. 955-985.
4. Noor, A. K., Andersen, C. M. and Tanner, J. A., "Mixed Models and Reduction Techniques for Large-Rotation Nonlinear Analysis of Shells of Revolution with Application to Tires," NASA TP-2343, 1984.
5. Noor, A. K., "Reduction Method for the Nonlinear Analysis of Symmetric Anisotropic Panels," International Journal for Numerical Methods in Engineering, Vol. 23, 1986, pp. 1329-1341.
6. Noor, A. K. and Peters, J. M., "Nonlinear Analysis of Anisotropic Panels," AIAA Journal, Vol. 24, No. 9, Sept. 1986, pp. 1545-1553.
7. Noor, A. K., Andersen, C. M. and Tanner, J. A., "Exploiting Symmetries in the Modeling and Analysis of Tires," NASA TP-2649, 1987.
8. Noor, A. K. and Tanner, J. A., "Advances in Contact Algorithms and Their Application to Tires," NASA TP-2781, 1988.
9. Sanders, J. L., Jr., "Nonlinear Theories for Thin Shells," Quarterly of Applied Mathematics, Vol. XXI, No. 1, April 1963, pp. 21-36.
10. Budiansky, B., "Notes on Nonlinear Shell Theory," Journal of Applied Mechanics, Transactions of ASME, Series E, Vol. 35, No. 2, June 1968, pp. 393-401.
11. Simo, J. C., Wriggers, P. and Taylor, R. L., "A Perturbed Lagrangian Formulation for the Finite Element Solution of Contact Problems," Computer Methods in Applied Mechanics and Engineering, Vol. 50, No. 2, Aug. 1985, pp. 163-180.
12. Stein, E., Wagner, W. and Wriggers, P., "Finite Element Postbuckling Analysis of Shells with Nonlinear Contact Constraints," in Finite Element Methods for Nonlinear Problems, edited by P. G. Bergan, K. J. Bathe and W. Wunderlich, Springer-Verlag, 1986, pp. 719-744.
13. Concus, P., Golub, G. H. and O'Leary, D. P., "A Generalized Conjugate Gradient Method for the Numerical Solution of Elliptic Partial Differential Equations," in Sparse Matrix Computations, edited by James R. Bunch and Donald J. Rose, Academic Press, 1976, pp. 309-332.
14. Adams, L., "m-Step Preconditional Conjugate Gradient Methods," SIAM Journal of Science and Statistical Computing, Vol. 6, No. 2, April 1985, pp. 452-463.
15. Wriggers, P. and Nour-Omid, B., "Solution Methods for Contact Problems," Rep. No. UCB/SESM-84/09 (Contact N00014-76-C-0013), Dept. of Civil Engineering, University of California, July 1984.
16. Wriggers, P., Wagner, W. and Stein, E., "Algorithms for Nonlinear Contact Constraints with Applications to Stability Problems of Rods and Shells," Computational Mechanics, Vol. 2, 1987, pp. 215-230.
17. Noor, A. K. and Kim, K. O., "Mixed Formulation for Frictionless Contact Problems," Finite Elements in Analysis and Design, Vol. 4, 1989, pp. 315-332; also published as NASA TP-2897, 1989.
18. Torstenfelt, B., "Finite Elements in Contact and Friction Applications," Linkoping Studies in Science and Technology, Dissertation No. 103, Division of Solid Mechanics and Strength of Materials, Dept. of Mechanical Engineering, Linkoping University, Linkoping, Sweden, 1983.
19. Kim, K. O., Noor, A. K. and Tanner, J. A., "Modeling and Analysis of Space Shuttle Nose-Gear Tire Using Semianalytic Finite Elements," NASA TP, 1989 (to appear).

APPENDIX A - FUNDAMENTAL EQUATIONS OF SHELL THEORY USED IN THE PRESENT STUDY

The fundamental equations of the Sanders-Budiansky type shell of revolution used in the present study are summarized herein. The effects of laminated, anisotropic material response and transverse-shear deformation are included.

STRAIN-DISPLACEMENT RELATIONSHIPS - The relationships between the strains and the generalized displacements of the middle surface are given by

$$\varepsilon_s = \partial_s u + \frac{w}{R_1} + \underline{\frac{1}{2}\left(\frac{u}{R_1} - \partial_s w\right)^2 + \frac{1}{2}\phi^2} \tag{A1}$$

$$\varepsilon_\theta = \frac{\partial_s r}{r} u + \frac{1}{r}\partial_\theta v + \frac{w}{R_2} + \underline{\frac{1}{2}\left(\frac{v}{R_2} - \frac{1}{r}\partial_\theta w\right)^2 + \frac{1}{2}\phi^2} \tag{A2}$$

$$2\varepsilon_{s\theta} = \frac{1}{r}\partial_\theta u + \left(\partial_s - \frac{\partial_s r}{r}\right) v + \underline{\left(\frac{u}{R_1} - \partial_s w\right)\left(\frac{v}{R_2} - \frac{1}{r}\partial_\theta w\right)} \tag{A3}$$

$$\kappa_s = \partial_s \phi_s \tag{A4}$$

$$\kappa_\theta = \frac{\partial_s r}{r}\phi_s + \frac{1}{r}\partial_\theta \phi_\theta \tag{A5}$$

$$2\kappa_{s\theta} = \frac{1}{r}\,\partial_\theta \phi_s + \left(\partial_s - \frac{\partial_s r}{r}\right)\phi_\theta + \left(\frac{1}{R_2} - \frac{1}{R_1}\right)\phi \qquad \text{(A6)}$$

$$2\varepsilon_{s3} = -\frac{u}{R_1} + \partial_s w + \phi_s \qquad \text{(A7)}$$

$$2\varepsilon_{\theta 3} = -\frac{v}{R_2} + \frac{1}{r}\,\partial_\theta w + \phi_\theta \qquad \text{(A8)}$$

where ε_s and ε_θ are the extensional strains in the meridional and circumferential directions, $2\varepsilon_{s\theta}$ is the in-plane shearing strain, κ_s and κ_θ are the bending strains in the meridional and circumferential directions, $2\kappa_{s\theta}$ is the twisting strain, $2\varepsilon_{s3}$ and $2\varepsilon_{\theta 3}$ are the transverse-shear strains, $\partial_s \equiv \partial/\partial s$, $\partial_\theta \equiv \partial/\partial\theta$, and ϕ is the rotation around the normal to the shell, which is given by

$$\phi = \frac{1}{2}\left[-\frac{1}{r}\,\partial_\theta u + \left(\partial_s + \frac{\partial_s r}{r}\right)v\right] \qquad \text{(A9)}$$

The nonlinear terms which account for moderate rotations are underlined with dashes in Eqs. (A1) to (A3).

CONSTITUTIVE RELATIONS - The shell is assumed to be made of a laminated, anisotropic, linearly elastic material. Every point of the shell is assumed to possess a single plane of elastic symmetry parallel to the middle surface. The relationships between the stress resultants and the strain measures of the shell are given by

$$\begin{Bmatrix} N_s \\ N_\theta \\ N_{s\theta} \\ M_s \\ M_\theta \\ M_{s\theta} \\ Q_s \\ Q_\theta \end{Bmatrix} = \begin{bmatrix} c_{11} & c_{12} & c_{16} & f_{11} & f_{12} & f_{16} & \cdot & \cdot \\ & c_{22} & c_{26} & f_{12} & f_{22} & f_{26} & \cdot & \cdot \\ & & c_{66} & f_{16} & f_{26} & f_{66} & \cdot & \cdot \\ & & & d_{11} & d_{12} & d_{16} & \cdot & \cdot \\ & & & & d_{22} & d_{26} & \cdot & \cdot \\ & & & & & d_{66} & \cdot & \cdot \\ & \text{Symmetric} & & & & & c_{55} & c_{45} \\ & & & & & & & c_{44} \end{bmatrix} \begin{Bmatrix} \varepsilon_s \\ \varepsilon_\theta \\ 2\varepsilon_{s\theta} \\ \kappa_s \\ \kappa_\theta \\ 2\kappa_{s\theta} \\ 2\varepsilon_{s3} \\ 2\varepsilon_{\theta 3} \end{Bmatrix} \qquad \text{(A10)}$$

where c, f and d are shell stiffness coefficients. The nonorthotropic (anisotropic) terms are circled and the dots refer to zero terms.

APPENDIX B - FORMULAS FOR THE ELEMENTAL ARRAYS $[Q]$, $[R]$ AND $\{g_o\}$

The explicit forms of the elemental arrays $[Q]$, $[R]$ and $\{g_o\}$ are given in this appendix. For convenience, each array is partitioned into blocks according to contributions from displacement and contact nodes. The expressions of the typical partitions (or blocks) are given in Table B.1. In Table B.1, N^i and N^j are the shape functions for the Lagrange multipliers and generalized displacements; m is the number of displacement nodes in the element; c is the number of nodal points in contact in the element; and $\Omega^{(e)}$ is the element domain. The range of the indices i and j is from 1 to c, and the range of the index i' is from 1 to m; and $<\tilde{g}>$ is the unit ramp (or singularity) function defined as follows:

$$<\tilde{g}>^n = \begin{cases} \tilde{g}^n & (\tilde{g} > 0) \\ 0 & (\tilde{g} \le 0) \end{cases}$$

where $\tilde{g} = -\bar{g}$ and n=0 or 1.

Table B.1 - Explicit form of the typical partitions of the arrays $[Q]$, $[R]$ and $\{g_o\}$

Array	Number of Partitions or Blocks	Formula for Typical Partition
$[Q]$	$m \times c$	$\int_{\Omega^{(e)}} N^{i'} N^j <\tilde{g}>^o \, d\Omega$
$[R]$	$c \times c$	$-\int_{\Omega^{(e)}} N^i N^j <\tilde{g}>^o \, d\Omega$
$\{g_o\}$	c	$\int_{\Omega^{(e)}} N^i <\tilde{g}_o> d\Omega$

APPENDIX C - TRANSFORMATION OF THE ELEMENTAL ARRAYS FROM SHELL COORDINATES TO GLOBAL CARTESIAN COORDINATES

The transformation of the displacement components from the shell coordinates (s, θ, x_3) to the global Cartesian coordinates (x, y, x) is expressed by the following equation:

$$\{X\}^{(e)} = [T]\,\{\bar{X}\}^{(e)}$$

where $[T]$ is a block-diagonal transformation whose submatrix at each node is given by

$$\underset{(5x5)}{[T]^{(n)}} = \begin{bmatrix} \vec{e}_s & \vec{e}_\theta & \vec{e}_s \times \vec{e}_\theta & \mathbf{0} & \\ & \mathbf{0} & & 1 & \\ & & & & 1 \end{bmatrix}$$

$\vec{e}_s$ and $\vec{e}_\theta$ are the tangential unit vectors in the s- and θ-directions, respectively; $\{X\}^{(e)}$ and $\{\bar{X}\}^{(e)}$ are the generalized displacement vectors in the shell coordinates and the global Cartesian coordinates. Note that the rotation components, ϕ_s and ϕ_θ, are not transformed since the outer surface of the tire was chosen as the reference surface and, therefore, ϕ_s and ϕ_θ do not appear in the contact conditions.

The elemental matrices, $[S]$ and $\left[\frac{\partial M}{\partial X}\right]$, and the external load vector $\{P\}$ are transformed from the shell coordinates to the global Cartesian coordinates as follows:

$$[S] \rightarrow [S][T]$$

$$\left[\frac{\partial M}{\partial X}\right] \rightarrow [T]^t \left[\frac{\partial M}{\partial X}\right][T]$$

$$\{P\} \rightarrow [T]^t\{P\}$$

The nonlinear vectors, $\{G(X)\}$ and $\{M(H,X)\}$, are evaluated with displacement vector $\{X\}$ expressed in terms of $\{\overline{X}\}$ at the end of each iteration cycle.

APPENDIX D - SUMMARY OF THE COMPUTATIONAL PROCEDURE

The computational procedure used in the present study is summarized in this appendix.

PREPROCESSING AND INITIAL CALCULATION PHASES

*Step 1. Model the tire geometry; evaluate stiffness coefficients (19); and generate input data including transformation matrices.

*Step 2. Select estimates for the penalty parameter; and assume the contact status at the selected contact nodes.

*Step 3. Generate linear element arrays.

Solution phase:

*Step 4. Solve inflation pressure case without contact against a pavement using Newton-Raphson iteration scheme.

*Step 5. Generate initial gap between the inflated configuration and the pavement at designated contact nodes.

*Begin displacement incrementation loop.

*Begin Newton-Raphson iteration and contact combined loop.

*Step 6. Generate nonlinear elemental arrays; eliminate the internal forces and the Lagrange multipliers for the elemental equations; and assemble the left- and right-hand side of the equations.

*Step 7. Solve Eqs. (7) for the incremental displacements.

*Step 8. Update the response vector for displacements, internal forces, and the Lagrange multipliers:

$$\{Z^{(r+1)}\} = \{Z^{(r)}\} + \{\Delta Z^{(r)}\}$$

*Step 9. Check the contact status (and modify the contact conditions) at each contact node as needed: If $\overline{g} > 0$, and $\lambda > 0$, then the contact constraint is inactive; if $\overline{g} < 0$, then the constraint is active. If contact status is the same as previously assumed, then continue. Otherwise, add the contact contributions for nodes with active constraint and delete these contributions for nodes with inactive constraint; then return to step 6.

*Step 10. Check convergence of Newton-Raphson iterations; that is:

$$e = \frac{\left[\{\Delta Z\}^t\{\Delta Z\}/\{Z\}^t\{Z\}\right]^{1/2}}{n} < \text{Tolerance}$$

where n is the total number of degrees of freedom in the model; and the tolerance is prescribed. If convergence is achieved, then compute the contact forces at each contact node by

$$P_n^i = \int_{\Omega^{(e)}} N^i\, \overline{\lambda}\, d\Omega$$

and continue. Otherwise, return to step 6.

*Step 11. If the prescribed displacement is greater than the specified maximum displacement, then stop; otherwise, add additional displacement incrementation and return to step 6.

Table 1 - Characteristics of Mixed Finite Element Models Used in the Numerical Studies

Designation	Number of Displacement Nodes	Maximum Number of Lagrange Multiplier Nodes	Number of Parameters Per Internal Force	Number of Quadrature Points*
M9-4	3×3	3×3	2×2	3×3

*All elemental arrays are evaluated using Gauss-Legendre quadrature formulas except for $[Q]$, $[R]$ and $\{g_o\}$ which are evaluated using Newton-Cotes formulas.

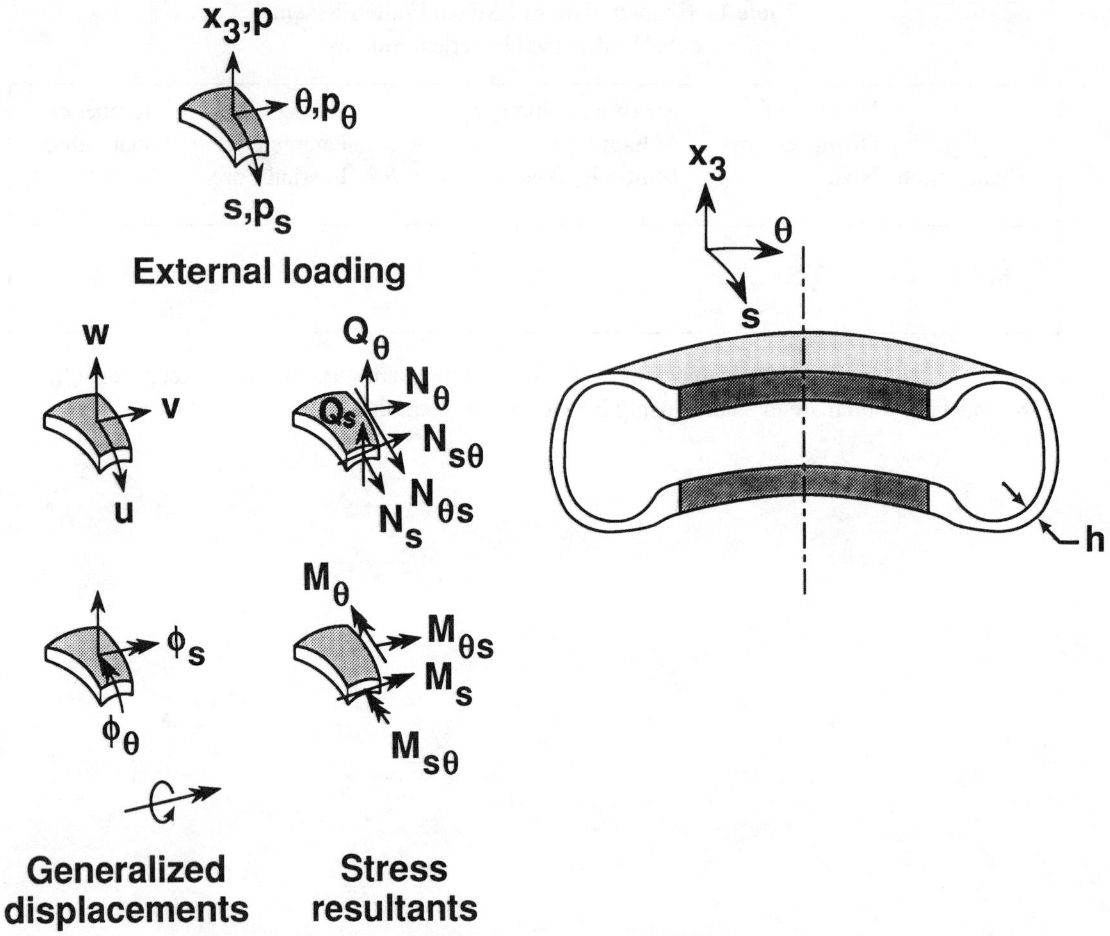

Figure 1 - Two-dimensional shell model of the tire and sign convention for the external loading, generalized displacements and stress resultants.

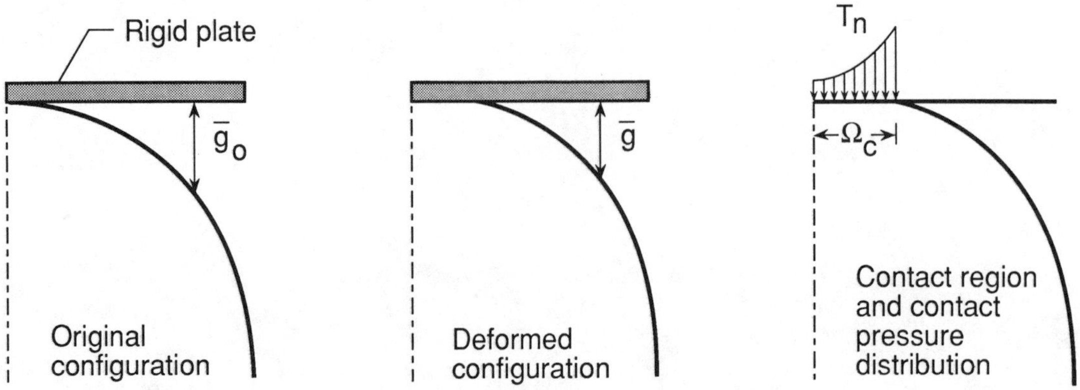

Figure 2 - Characteristics of frictionless contact of a shell pressed against a rigid plate.

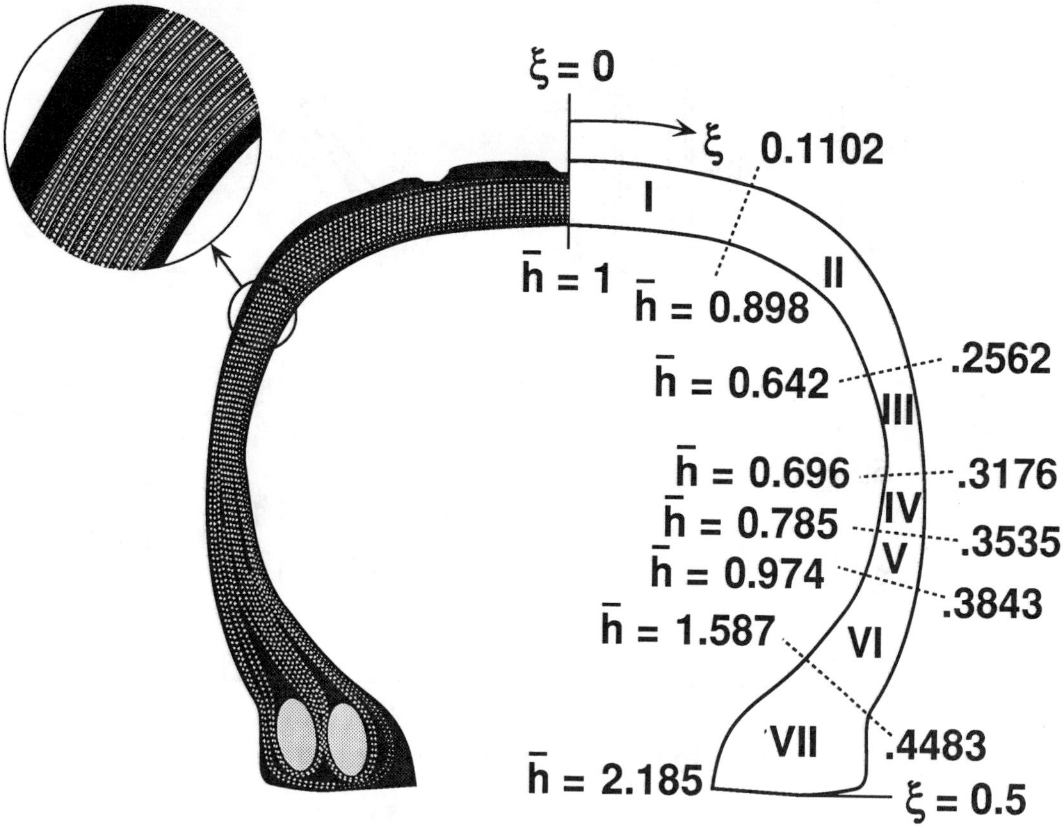

Figure 3 - Cross-section of the space shuttle orbiter nose-gear tire used in the present study.

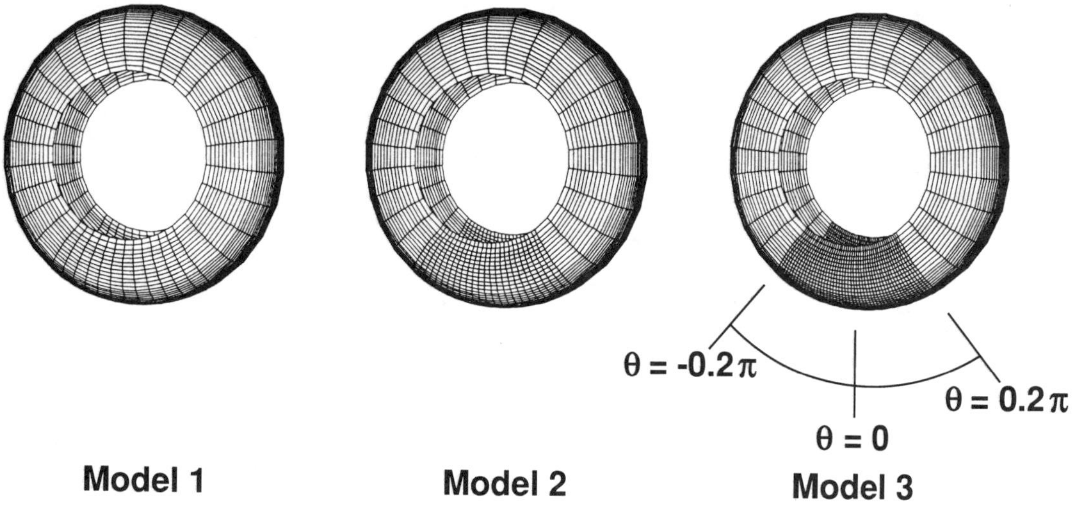

No. of elements (MD9-4)*	Model 1	Model 2	Model 3
$-.2\pi \leq \theta \leq .2\pi$	180 (30 x 6)	360 (30 x 12)	720 (30 x 24)
$\theta < -.2\pi, \theta \geq 2\pi$	360 (30 x 12)	360 (30 x 12)	360 (30 x 12)
Total	540 (30 x 18)	720 (30 x 24)	1080 (30 x 36)

***Meshes are shown in the figure as 4-node elements instead of MD9-4**

Figure 4 - Finite element models used in the present study.

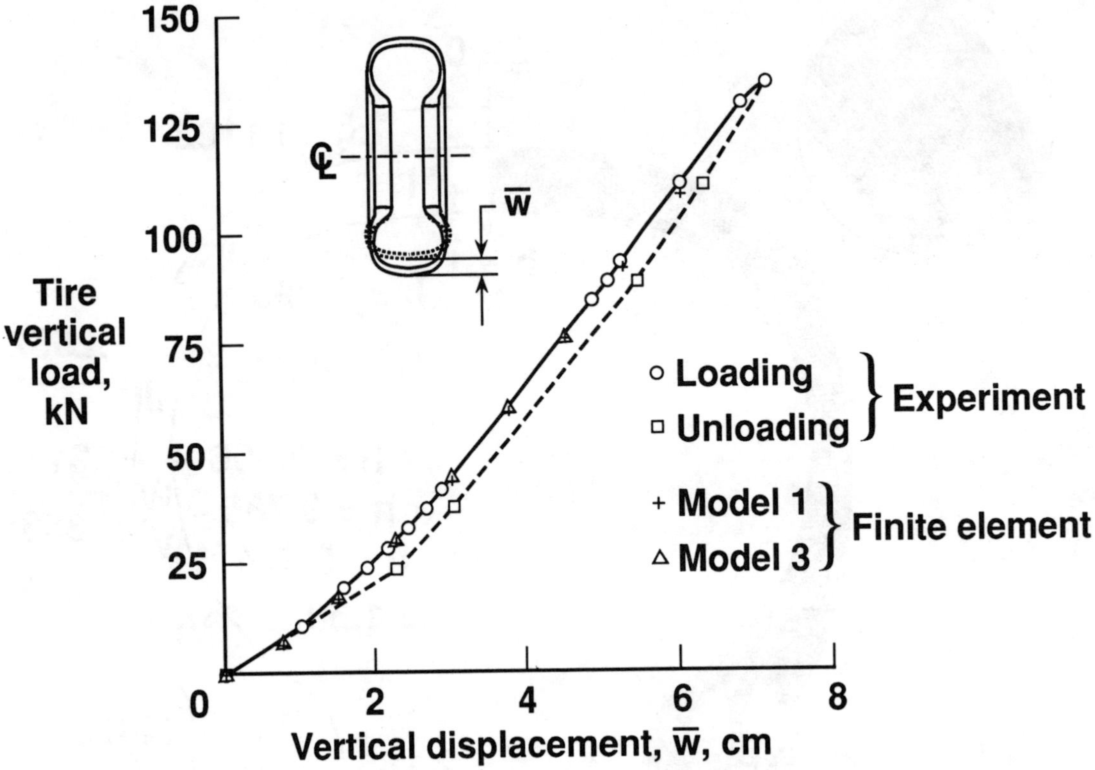

Figure 5 - Load deflection curve for the space shuttle orbiter nose-gear tire shown in figure 3.

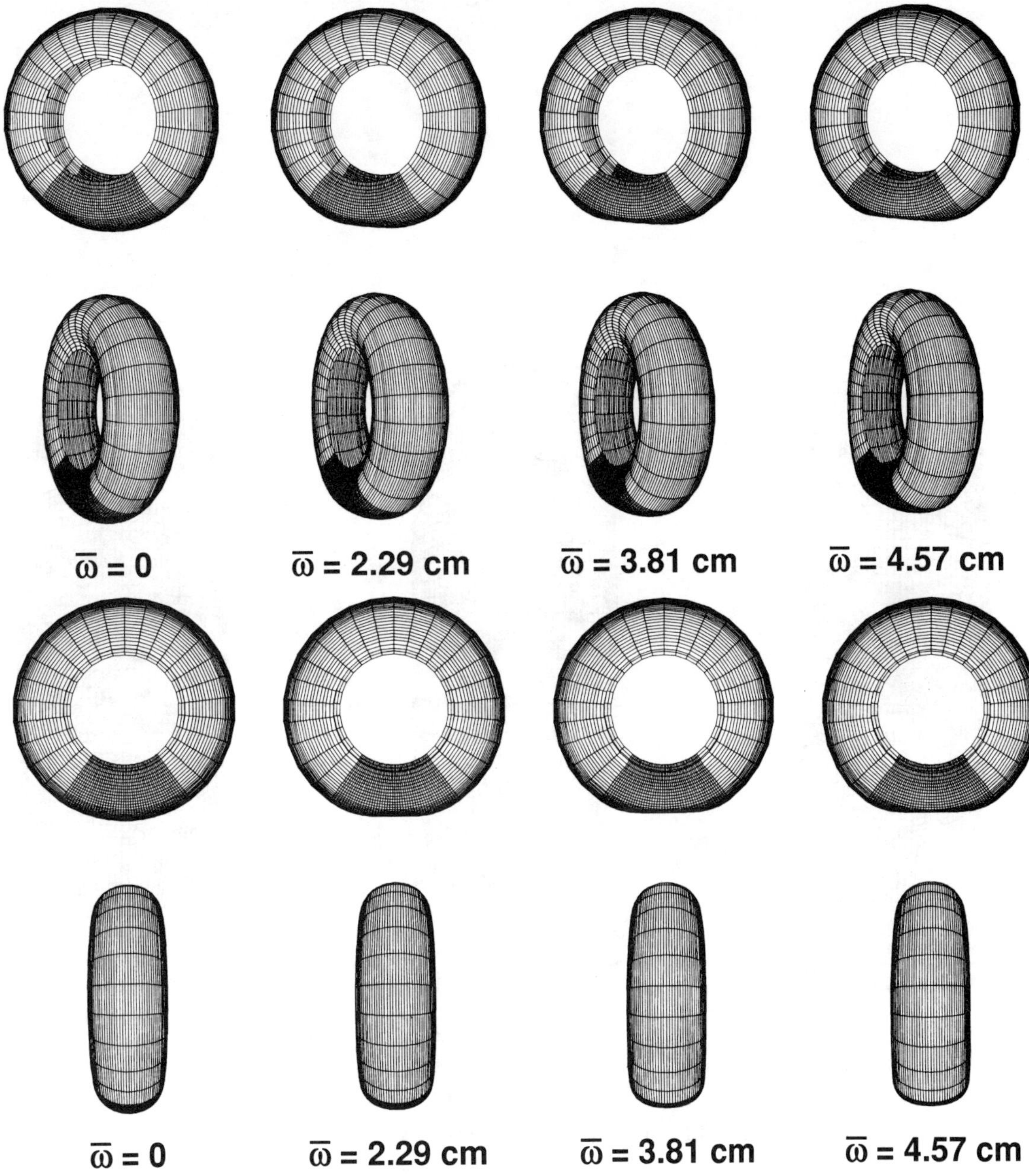

Figure 6 - Deformed configurations of the space shuttle orbiter nose-gear tire shown in figure 3. The tire is subjected to uniform inflation pressure p_o=2.2 M Pa and is pressed against a rigid pavement.

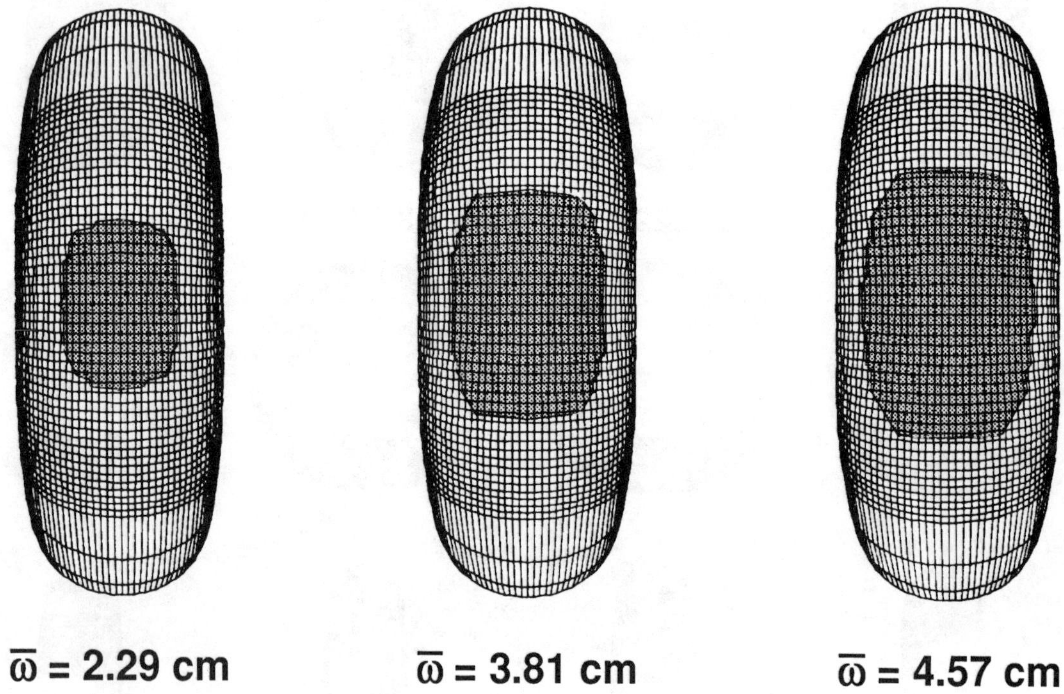

Figure 7 - Variation of footprint areas with applied displacements, space shuttle orbiter nose-gear tire shown in figure 3. The tire is subjected to uniform inflation pressure p_o=2.2 M Pa and is pressed against a rigid pavement.

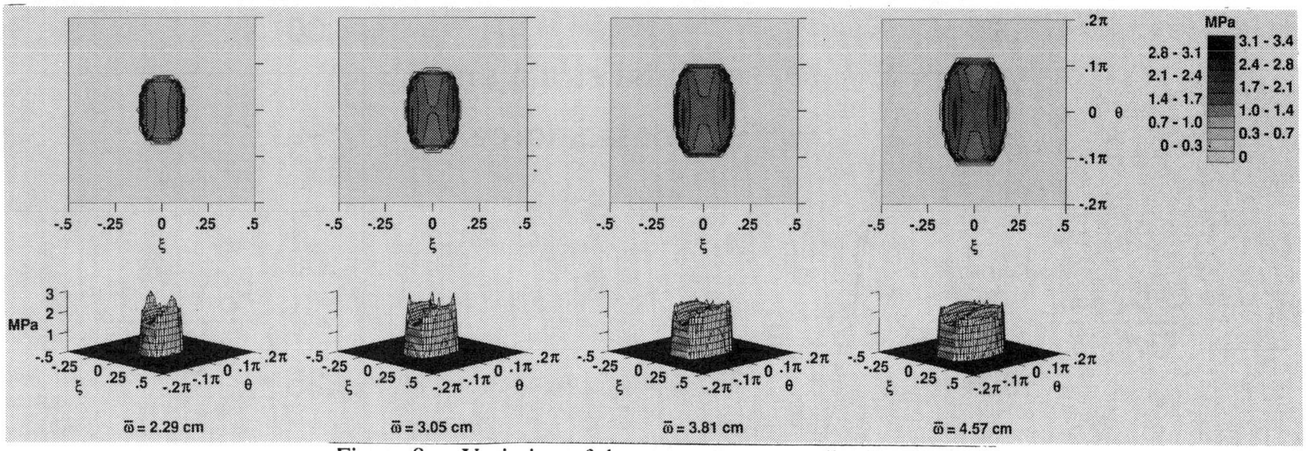

Figure 8 - Variation of the contact pressure distribution with applied displacements, space shuttle orbiter nose-gear tire shown in figure 3. The tire is subjected to uniform inflation pressure p_o=2.2 M Pa and is pressed against a rigid pavement.

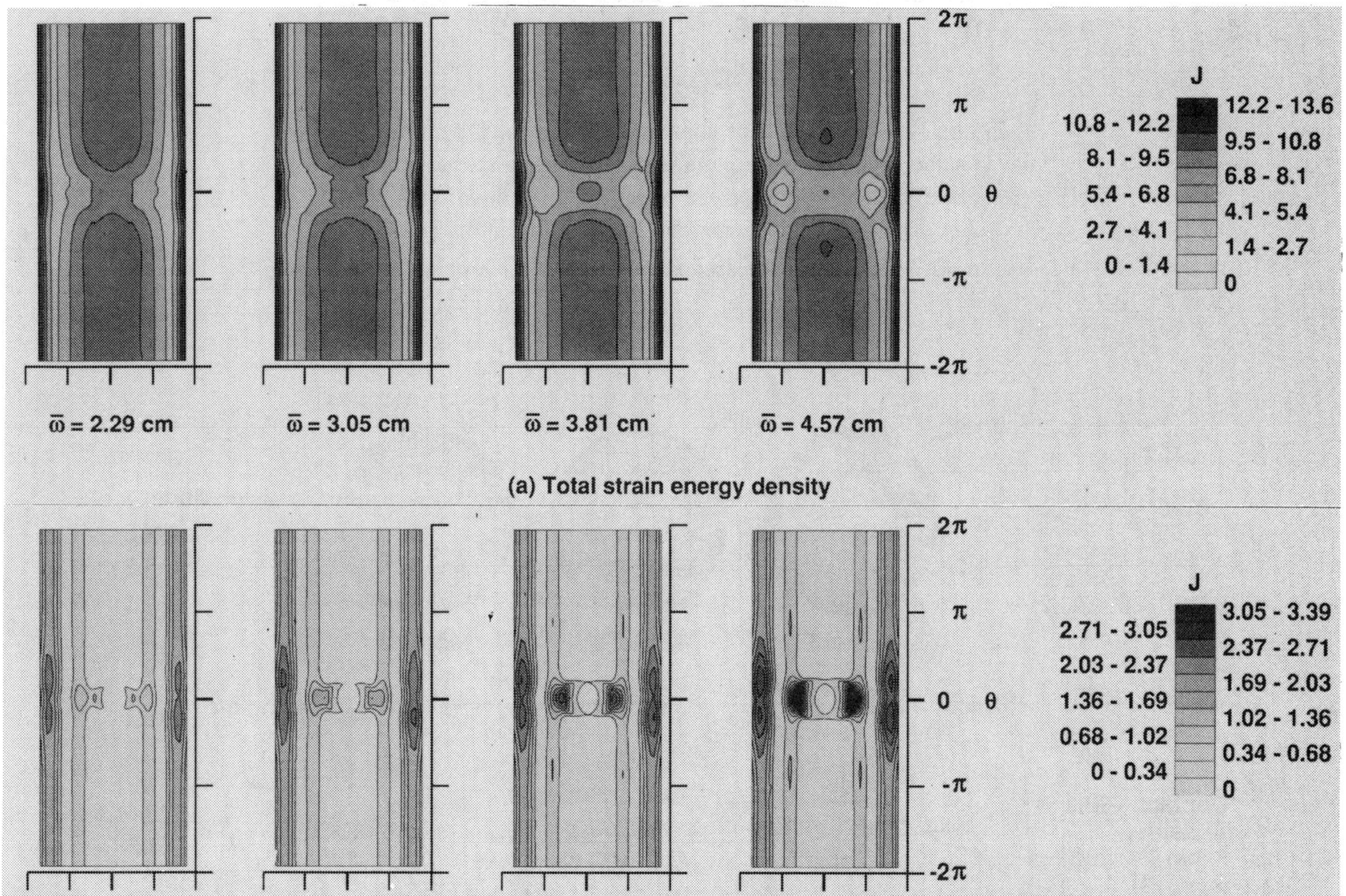

Figure 9 - Variation of the strain energy density distribution with applied displacements, space shuttle orbiter nose-gear tire shown in figure 3. The tire is subjected to uniform inflation pressure p_o=2.2 M Pa and is pressed against a rigid pavement.

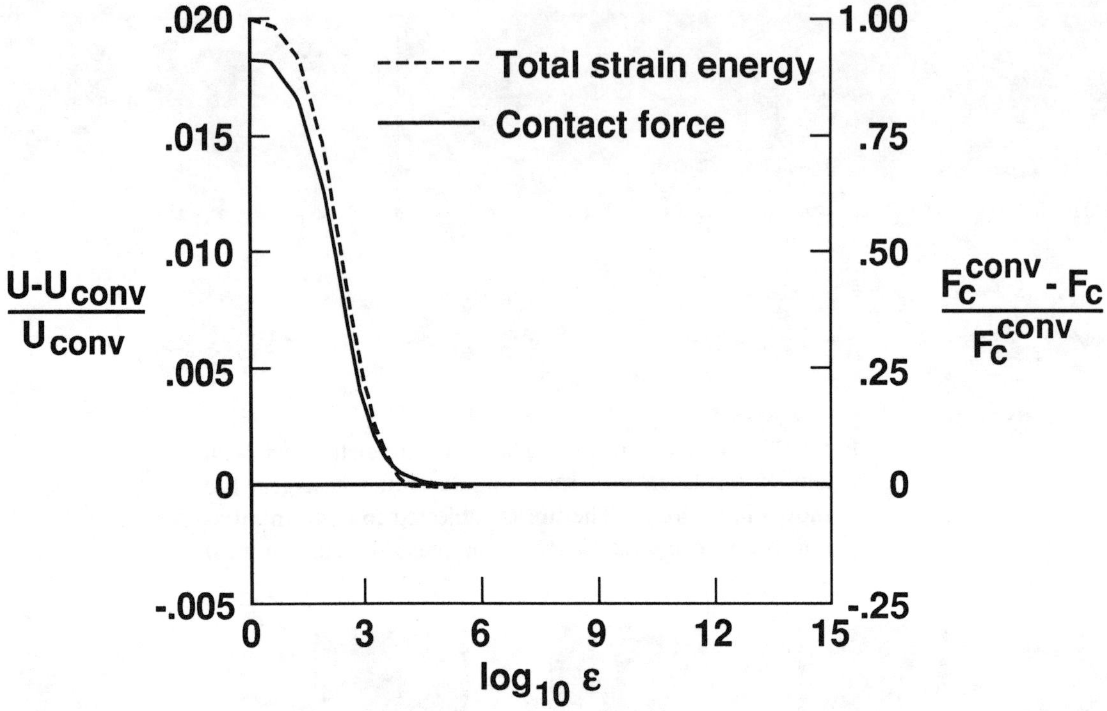

Figure 10 - Effect of the magnitude of penalty parameters on the accuracy of the total strain energy and the contact force. Space shuttle orbiter nose-gear tire shown in figure 3. The tire is subjected to uniform inflation pressure p_o=2.2 M Pa and is pressed against a rigid pavement.

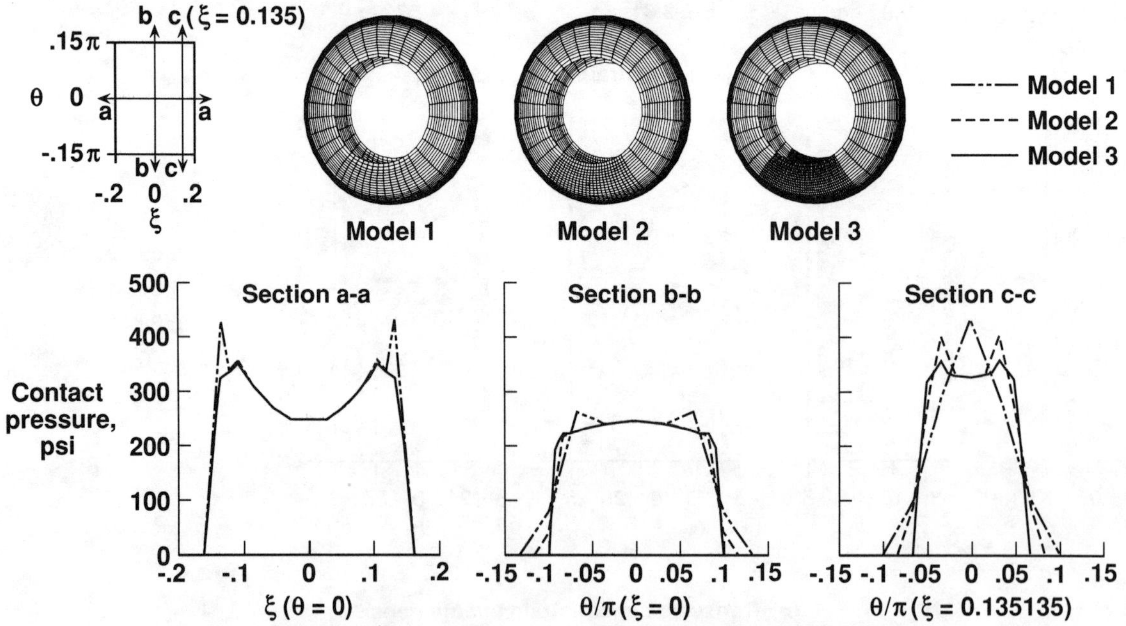

Figure 11 - Convergence of contact pressure distribution with mesh refinements. Space shuttle orbiter nose-gear tire shown in figure 3. The tire is subjected to uniform inflation pressure p_o=2.2 M Pa and is pressed against a rigid pavement.

Aircraft Tire/Pavement Pressure Distributions

John T. Tielking
Texas A&M University

** Paper 892351 presented at the Aerospace Technology Conference and Exposition, Anaheim, California, September, 1989.

ABSTRACT

A finite element tire model has been used to examine pavement pressure distributions developed by statically loaded aircraft tires. This paper briefly describes the tire model's characteristics and presents load versus deflection curves and pavement pressure distributions calculated for three different aircraft tires. The tire model shows that the nonuniformity of the tire/pavement pressure distribution increases when the tire is operated at an inflation pressure or tire load that is above or below the design value for that tire. The data calculated by the tire model are compared with load-deflection curves and contact pressure data measured for the tires modeled.

AIRCRAFT TIRE CONTACT PRESSURES are of interest to tire designers and have recently attracted the interest of airfield pavement engineers. Fighter aircraft takeoff weights have been significantly increased, without increasing the number of tires on their landing gear. The additional load is carried by higher tire inflation pressures, now well over 200 psi (1.4 MPa), with 300 psi (2.1 MPa) tires expected to become common. The extreme stiffness of the relatively thick aircraft tire carcass, at high inflation pressure, is responsible for the considerable nonuniformity found in aircraft tire/pavement pressure distributions. This is of concern in pavement design as it is now known that interior strains in a pavement are sensitive to surface pressure nonuniformities (1)*.

*Numbers in parentheses designate references at end of paper.

Apart from the effect on a pavement, the tire contact pressure distribution has long been recognized as a key indicator of tire performance, with regard to braking/steering and wear.

The purpose of this paper is to present selected results from investigations made with a finite element tire model and compare, whenever possible, the calculated results with measurements made on the tire being modeled.

FINITE ELEMENT TIRE MODEL

The tire model used to calculate the contact pressure distributions shown in this paper was developed earlier for the purpose of investigating tire-pavement interaction during vehicle maneuvering (2). A comprehensive description of tire constructional details is input to the model, thereby permitting study of the influence of factors such as tire materials and size on the pavement contact pressure distribution.

The tire is modeled by an assembly of axisymmetric shell elements positioned along the carcass mid-ply surface. Figure 1 shows an assembly of 15 elements used to model a typical aircraft tire that has a 10 ply carcass. The geometric and material properties of each ply are specified separately in determining the equivalent stiffness of the finite element model. Node 16 (Fig. 1) is a fixed node representing the tire bead. The tire is modeled in the mounted, uninflated configuration.

The tire calculations begin by specifying the inflation pressure. The computer program calculates the inflated shape of the tire and the structural stiffness of the inflated tire. A pavement contact load is then applied by specifying the loaded radius, R_ℓ shown in Figure 2, which is equivalent to specifying tire deflection against a pavement. At present, the tire model is statically deflected

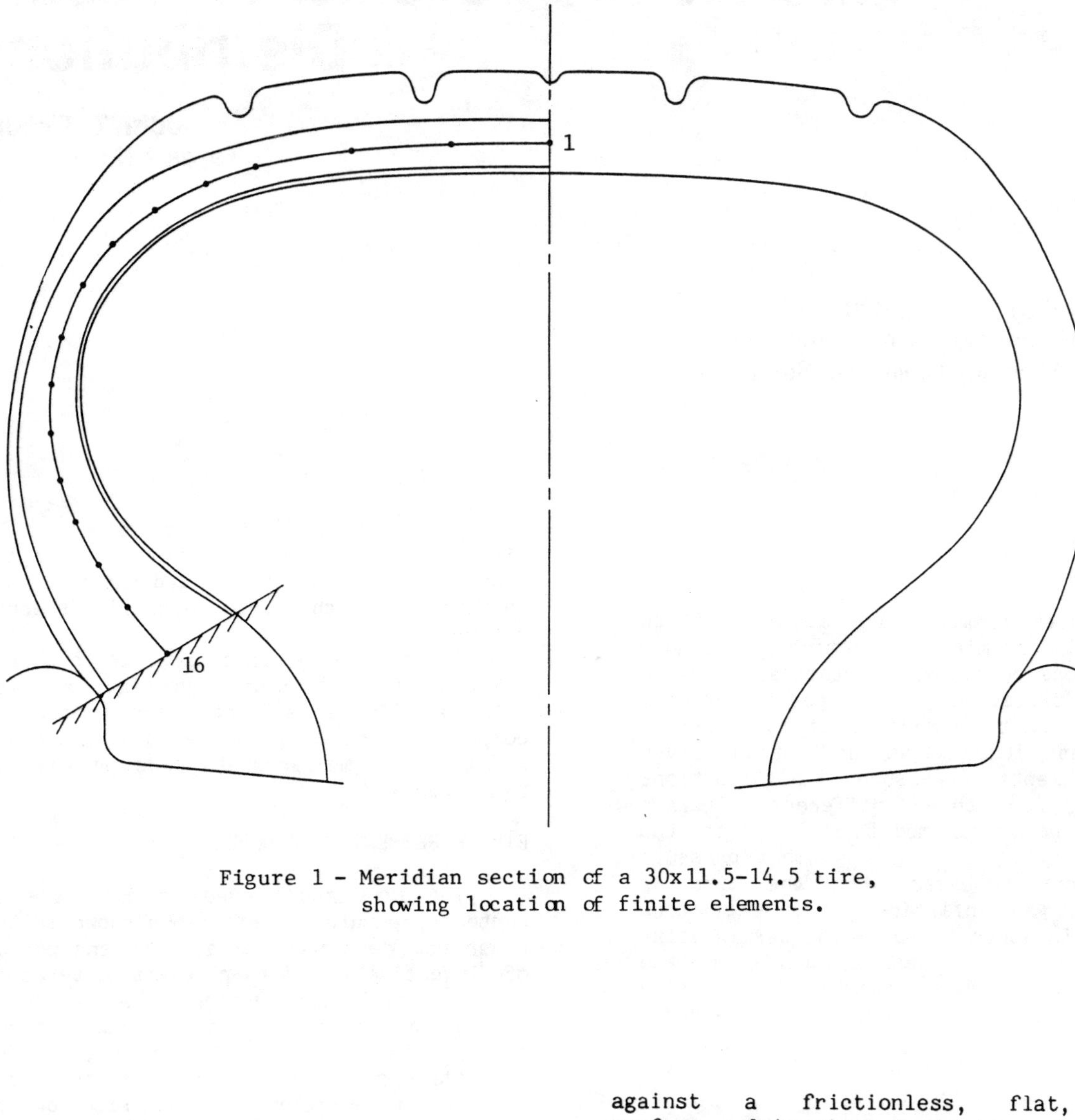

Figure 1 - Meridian section of a 30x11.5-14.5 tire, showing location of finite elements.

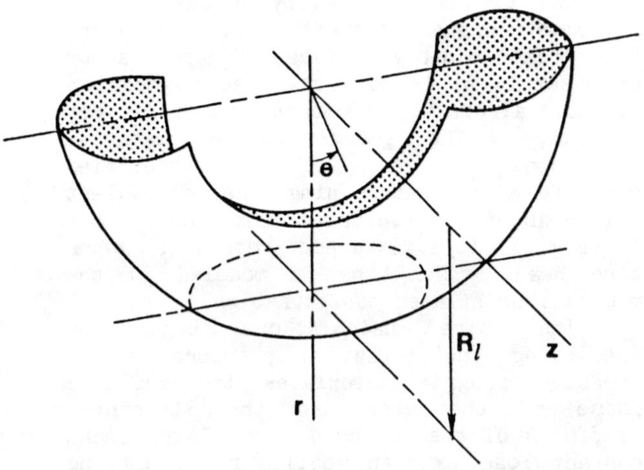

Figure 2 - Loaded radius, R_ℓ, specifying tire deflection.

against a frictionless, flat, rigid surface. Laboratory tests have shown that interfacial friction has very little effect on the normal component of the contact pressure of a standing tire. Also, the contact pressure is only slightly affected by the traveling velocity of a free-rolling tire.

Neither the contact pressure distribution nor the contact boundary are known a priori by the tire model. They are determined by the computer program during the contact solution procedure (3). The calculated pressure distribution is integrated over the contact area to find the tire load that would be necessary to maintain the specified tire deflection. The calculated load-deflection data are compared with measured data as a check on the veracity of the tire model. A good comparison of calculated and measured load-deflection data indicates (but does not guarantee) a realistic calculation of pavement contact pressures.

TIRES STUDIED - Table 1 gives the sizes and design operating data for the three tires in this paper. Tire A is the nose gear tire of the space shuttle, B is the main gear tire of F-4C/G fighter aircraft, and C is representative of those used on the main landing gear of B-737 and DC-9 commercial airplanes. All are bias-ply tires, with nylon cords; cord modulus E_c = 200,000 psi (1.38 GPa) was used for all tires. Cord angles and end counts vary with tire size, and are different in different plies of the same tire; these data were obtained from the tire manufacturers. Ply gages were measured from a tire section. A tire section was also used to locate the mid-ply surface of the mounted, uninflated tire. As contact pressure distributions are very sensitive to tire construction delails, it is essential that tire model data be taken from the tire that is tested.

TIRE A (32X8.8/20)

The space shuttle's tires have been extensively tested at the Landing and Impact Dynamics Branch of NASA-Langley Research Center. The test data have been made available for tire model verification purposes. Figure 3 shows the loading and unloading curves obtained when a vertical load is applied to the space shuttle nose gear tire (Tire A). Load-deflection data points calculated by the tire model are indicated in Fig. 3 by the + symbol. The calculated data are seen to follow the measured data up to a deflection of 1.5 inches (3.8 cm). Beyond this deflection the tire model appears stiffer than the real tire. This behavior is believed due to the use of tension moduli everywhere in the tire model. As tire deflection increases, a larger portion of the tire goes into compression. The considerably lower compression moduli in the tire prevent the structural stiffening that is exhibited by the tire model.

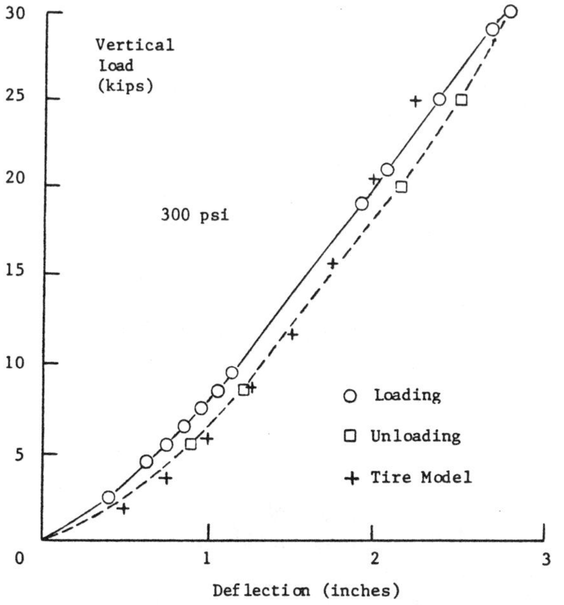

Figure 3 - Measured and calculated load-deflection data for tire A

CONTACT PRESSURES - Tables 2 and 3 give the calculated contact pressures in one-quarter of the footprint of tire A at two different tire loads. Node 1 follows the center line (equator) of the tire, and circumferential position 0.00 is on the meridian passing through the center of the footprint. The pressure values along nodes 2 and 5 are compared in Tables 2 and 3 with measured values given by an array of force transducers (T1 and T2) used in experiments conducted at NASA-Langley. The footprint force transducer array is described in Reference 4. Figure 4 shows the location of the finite element nodes and force transducers on the tread surface of Tire A. The transducer surface is square, with 0.5 inch (1.27 cm) edge length. A pressure value is obtained by dividing the transducer force measurement by the surface area, 0.25 in^2 (1.61 cm^2). In considering the comparison of calculated and measured pressures, it should be remembered that the tread layer is not included in the finite element tire model.

TIRE B (30X11.5-14.5/26)

The results for this tire are taken from a technical report (5) recently published by the Air Force Engineering & Services Center. The tire section is shown in Fig. 1. It is a low profile design, formerly designated as Type VIII. Figure 5 shows the calculated deflection versus load curve and the corresponding curve measured by the tire manufacturer. Here also, the tire model is somewhat stiffer than the actual tire.

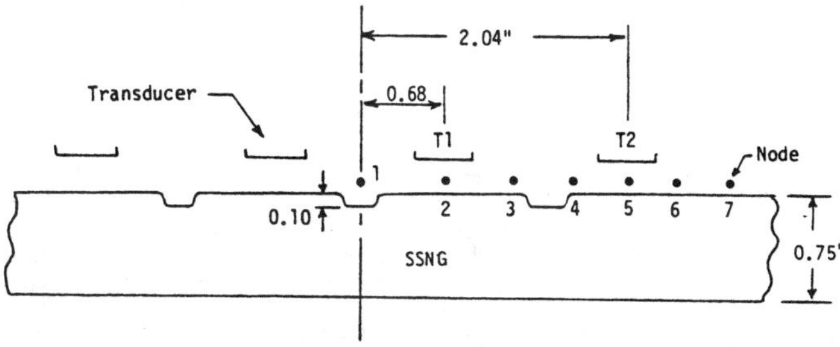

Figure 4 - Location of finite element nodes and force transducers on tire A.

Table 1 - Tires studied, rated inflation pressure and load.

Tire	Size/PR	Pressure (psi)	Load (lb)
A	32x8.8/20	300	23,700
B	30x11.5-14.5/26	265	26,600
C	40x14/28	200	33,100

Table 2 - Contact pressures (psi) at 15,000 lb load.

Node	0.00	1.17	2.33	3.50	4.67	5.84	Circum. Pos. (in)
1	218	210	217	224	220	0	
2	298	298	281	349	183	0	Calculated
T1	498	508	512	520	440	0	Measured
3	369	380	290	555	0		
4	403	452	627	510	0		
5	893	825	392	0	0		Calculated
T2	475	468	432	214	0		Measured
6	0	0	0				

Table 3 - Contact pressures (psi) at 25,000 lb load.

Node	0.00	1.17	2.33	3.50	4.67	5.84	7.00	Circum. Pos. (in.)
1	244	205	255	180	360	0	0	
2	370	274	371	229	505	0	0	Calculated
T1	491	499	500	504	510	504	0	Measured
3	450	369	429	343	476	0		
4	475	435	453	379	775	0		
5	654	648	615	999	0	0		Calculated
T2	464	484	496	500	480	120		Measured
6	610	714	861	106	0			
7	754	614	59	0				
8	0	0	0					

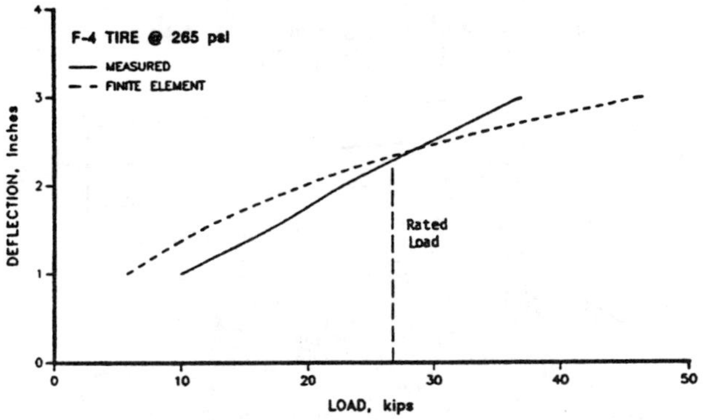

Figure 5 - Comparison of calculated with measured deflection-load data for tire B.

CONTACT PRESSURES - Table 4 gives the location and value of contact pressure calculated in the footprint of tire B at rated inflation pressure and load. The column labeled NCP (Normalized Contact Pressure) gives the pressure values divided by the inflation pressure. The location of each point where a pressure is found is given by Cartesian coordinates, x and y, which originate at the center of the footprint. Figure 6 shows the complete set of points where contact pressures are obtained for tire B. Here, the finite element nodes are the lines parallel to the x-axis, in the force-aft direction of the footprint. The tire model calculates pressures at points 1 through 29 (listed in Table 4) for this tire. The pressures at the unnumbered points in Fig. 6 are identified by reflection across the footprint axes of symmetry (x and y).

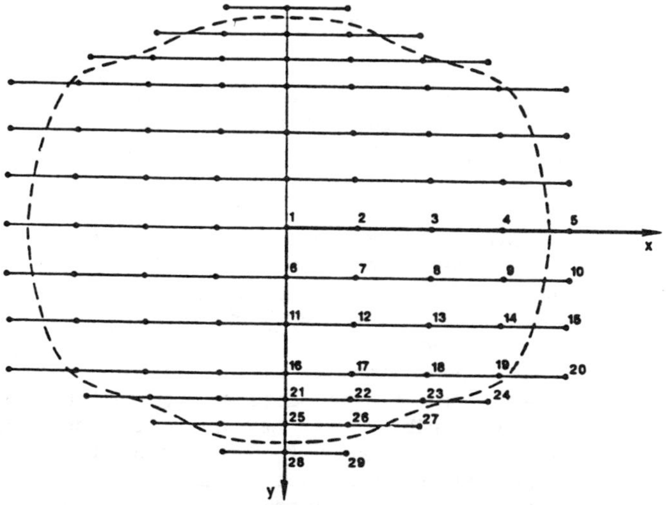

Figure 6 - Calculated contact pressure locations in footprint of tire B.

TABLE 4. CONTACT PRESSURES, p, IN FOOTPRINT OF 30X11.5-14.5 (F-4C/G) TIRE WITH 265 PSI INFLATION PRESSURE AND 26,600 POUND LOAD.

Point	x (in)	y (in)	p (psi)	NCP
1	0.0	0.0	141.9	0.54
2	1.53	0.0	149.1	0.56
3	3.05	0.0	146.9	0.55
4	4.58	0.0	149.9	0.57
5	6.10	0.0	0.0	0.0
6	0.0	1.00	367.2	1.39
7	1.52	1.00	344.1	1.30
8	3.04	1.00	338.8	1.28
9	4.56	1.00	298.8	1.13
10	6.08	1.00	0.0	0.0
11	0.0	2.00	328.6	1.24
12	1.50	2.00	305.0	1.15
13	3.00	2.00	307.1	1.16
14	4.50	2.00	241.4	0.91
15	5.99	2.00	0.0	0.0
16	0.0	3.00	462.2	1.74
17	1.46	3.00	423.9	1.60
18	2.92	3.00	512.8	1.94
19	4.39	3.00	197.7	0.75
20	5.85	3.00	0.0	0.0
21	0.0	3.52	387.5	1.46
22	1.43	3.52	451.2	1.70
23	2.87	3.52	451.3	1.70
24	4.30	3.52	0.0	0.0
25	0.0	4.06	558.0	2.10
26	1.40	4.06	490.7	1.85
27	2.80	4.06	0.0	0.0
28	0.0	4.65	0.0	0.0
29	1.35	4.65	0.0	0.0

CONTACT BOUNDARY - The zero pressure points are included in Table 4 so that the contact boundary can be approximately located. It is shown, for rated inflation pressure and tire load, by the dashed oval curve in Fig. 6. The predicted contact length and width are 10.7 and 8.7 inches, respectively, and agree well with footprint measurements. A 3-D plot of the pavement pressure distribution found at rated inflation pressure and tire load is shown in Fig. 7. This plot was made with the SAS graphics package on a mainframe computer.

EFFECT OF TIRE LOAD - The 3-D pressure plots provide a convenient way to ascertain the effect of tire design and operating variables on the overall pavement pressure distribution. As an example, Figs. 8 and 9 show the distributions found when tire load is increased 25 percent and decreased 25 percent from the rated value. As may be expected a higher tire load is carried by a more nonuniform pressure distributions (cf. Figs. 7 and 8), with shoulder pressure becoming significantly higher. Reducing the tire load reduces the contact area and, comparing Figs. 7 and 9, also appears to produce a slightly more nonuniform pressure distribution than that found for the tire operated at rated load.

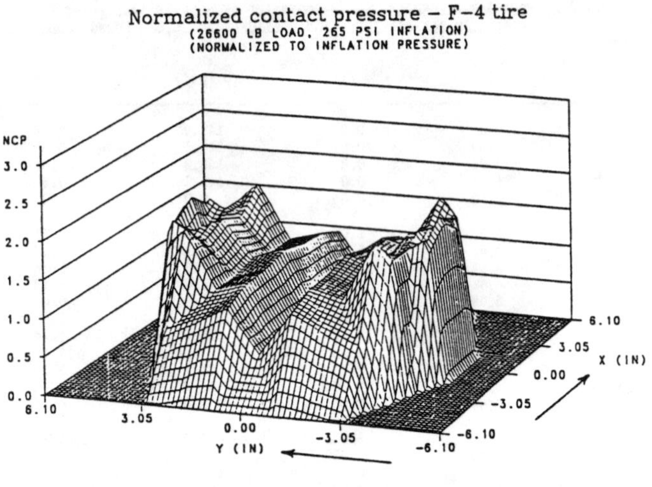

Figure 7 - Pavement pressure distribution for tire B at rated inflation pressure and load.

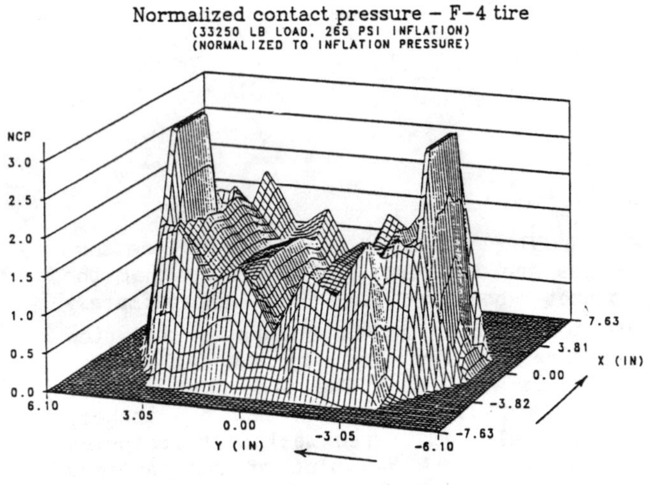

Figure 8 - Pavement pressure distribution for tire B at rated inflation pressure and 25% overload.

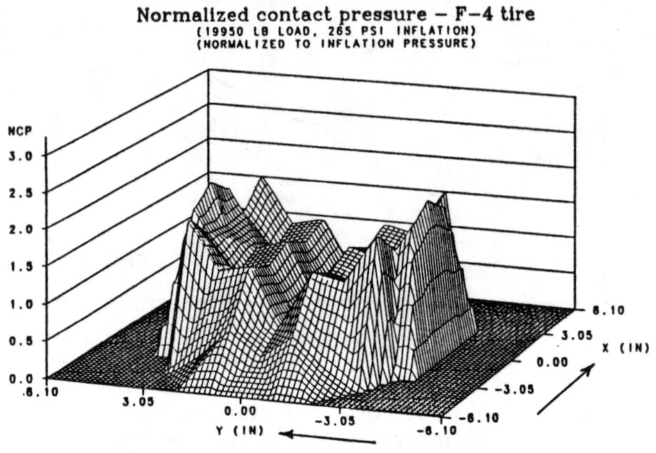

Figure 9 - Pavement pressure distribution for tire B at rated inflation pressure and 25% underload.

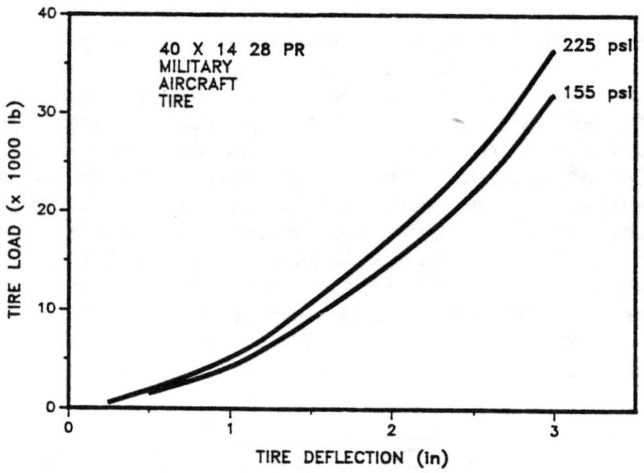

Figure 10 - Load-deflection data calculated for tire C at two inflation pressures.

TIRE C (40X14/28)

The 40X14 is a large tire used on the main gear of commercial and military transport aircraft. Figure 10 shows load versus deflection data calculated for this tire at two inflation pressures.

CONTACT PRESSURES - Figures 11 and 12 show the effect of inflation pressure (155 vs. 225 psi) on contact pressures calculated for tire C at two tire loads (16,000 and 30,000 lbs). For a comparison with measured data, the normalized pressures calculated along the meridian passing through the center of the footprint are shown in Fig. 13(a,b) along with normalized vertical forces measured with the footprint force transducer at NASA-Langley Research Center (4). The location and magnitude of the peak pressures agree fairly well with the footprint force data. The deviation of calculated from measured data at the center of the footprint may be due to the substantial tread layer (~ 0.5 in.) on the test tire, which had been recapped. The NASA data show that maximum force at the center of the footprint is reached at a low tire load, and is maintained throughout the range of loading (4). This finding is confirmed by the tire model contact pressure calculations (cf. Figs. 11 and 12).

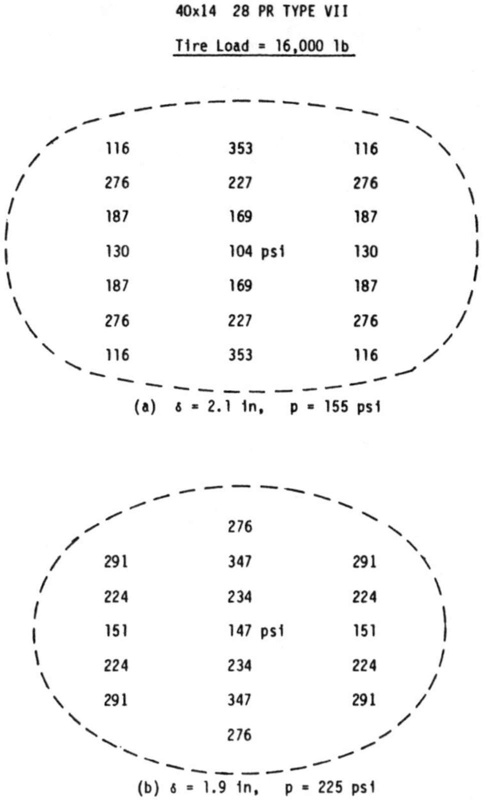

Figure 11 - Calculated pavement pressures for tire C with 16,000 lb load.

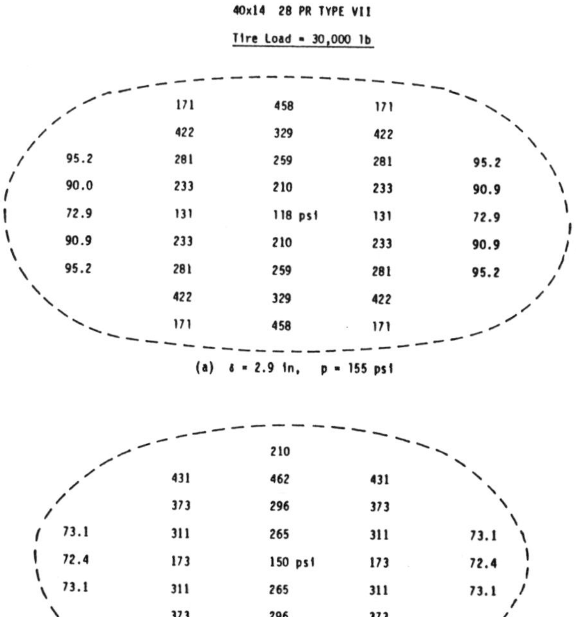

Figure 12 - Calculated pavement pressures for tire C with 30,000 lb load.

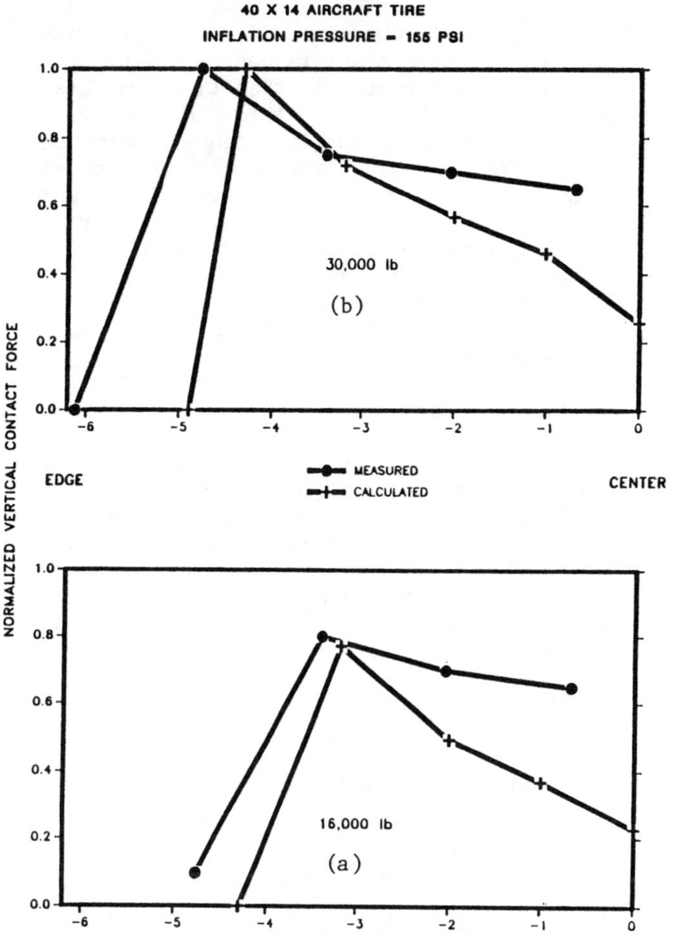

Figure 13 - Comparison of calculated with measured footprint forces at two tire loads.

CONCLUSIONS

Aircraft tire/pavement pressure distributions are characteristically nonuniform. The highest pressures are usually found in the shoulder regions of the footprint. The peak pressures are intensified when the tire load is increased, or the inflation pressure reduced. At rated load and inflation pressure, peak contact pressures seldom exceed twice the inflation pressure. Tires operating at inflation pressures above 300 psi (2.1 MPa) are exceptions. The space shuttle tire (Tire A, Fig. 3) and the main gear tire for the F-16, which operates at 310 psi (5), were found to have peak pavement pressures over three times the value of the inflation pressure. High gradients in the contact pressure distributions are detrimental to pavements (cause rutting and cracking), accelerate tire wear, and may influence other aspects of tire performance.

The nonuniformity in tire/pavement pressure distributions is due to the bending stiffness inherent in the tire carcass, which is relatively thick in aircraft tires. The nonuniformity can, however, be controlled by tire design. A tire model study of the effect of tire load and inflation pressure on an aircraft tire/pavement pressure distribution (5) shows that the most uniform pavement pressure is obtained when the tire is operated at the design values of load and inflation pressure.

ACKNOWLEDGEMENTS

The tire contact pressure calculations presented in this paper were sponsored by NASA-Langley (Tires A&C) and by the Air Force Engineering & Services Center, Tyndall AFB, FL (Tire B). The support and encouragement of John Tanner (NASA) and James Murfee (Air Force) are greatly appreciated.

REFERENCES

1. Tielking, J.T. and Roberts, F.L., "Tire Contact Pressure and Its Effect on Pavement Strain," Journal of Transportation Engineering, 113(1), pp. 56-71, January 1987.
2. Schapery, R.A. and Tielking, J.T., Investigation of Tire-Pavement Interaction During Maneuvering, Vol. 1, Report No. FHWA-RD-78-72, Federal Highway Administration, Washington, DC, June 1977.
3. Tielking, J.T. and Schapery, R.A., "A Method for Shell Contact Analysis," Computer Methods in Appl. Mech. and Engrg., 26(2), pp. 181-195, 1981.
4. Howell, W.E., Perez, S.E., and Vogler, W.A., "Aircraft Tire Footprint Forces," The Tire Pavement Interface, ASTM STP 929, pp. 110-124, 1986.
5. Tielking, J.T., Aircraft Tire/Pavement Pressure Distributions, Report No. ESL-TR-89-01, Air Force Engineering & Services Center, Tyndall AFB, FL, August 1989.

Braking and Steering Systems

Review of NASA Antiskid Braking Research

John A. Tanner
NASA Langley Research Center

** Paper 821393 presented at the
Aerospace Congress and Exposition,
Anaheim, California, October, 1982.

ABSTRACT

NASA antiskid braking system research programs are reviewed. These programs include experimental studies of four antiskid systems on the Langley Landing Loads Track, flight tests with a DC-9 airplane, and computer simulation studies. Results from these research efforts include identification of factors contributing to degraded antiskid performance under adverse weather conditions, tire tread temperature measurements during antiskid braking on dry runway surfaces, and an assessment of the accuracy of various brake pressure-torque computer models. This information should lead to the development of better antiskid systems in the future.

MODERN AIRCRAFT ANTISKID SYSTEMS are sophisticated skid control devices designed to provide maximum braking effort while maintaining full antiskid protection under all weather conditions. The several million landings that are made each year in routine fashion with no serious operational problems demonstrate the effectiveness and dependability of these skid control devices. However, both flight tests and field experience suggest that the performance of these antiskid braking systems is degraded when the runway becomes slippery and this degradation can lead to dangerously long roll-out distances and reduced steering capability during some airplane landing operations (1 to 3)*. Thus, there is a need to study various types of antiskid braking systems to find the reasons for the degraded braking performance that sometimes occurs under adverse weather conditions. A need also exists to update the computer simulations that are used to develop today's advanced skid control devices to insure safer ground handling operations under all weather conditions in the future.

To meet these needs, an experimental research program was undertaken by NASA with support from the Federal Aviation Administration to study the single-wheel behavior of several different airplane antiskid braking systems under the controlled conditions afforded by the Langley Landing Loads Track. The types of skid control devices studied included a velocity-rate-controlled system, a slip-ratio-controlled system with ground speed reference from an unbraked nose wheel, a slip-velocity-controlled system, and a mechanical-hydraulic system. The investigation of all four systems was conducted with a single main wheel, brake, and tire assembly of a McDonnell Douglas DC-9 series 10 airplane (4 to 7). Additional investigations were conducted to study tire tread temperatures during antiskid braking (8) and to study brake and tire dynamics during antiskid braking (9).

The purpose of this paper is to present some of the significant findings from these various research programs. Topics of discussion will include techniques for evaluating braking performance, comparisons between track and flight tests, the reasons for performance degradation under adverse weather conditions, tire tread temperatures during antiskid braking, and current computer simulation studies.

APPARATUS AND TEST PROCEDURE

TEST FACILITY - The Landing Loads Track tests were performed using the test carriage shown in figure 1. Also shown in figure 1 is

---

*Numbers in parentheses designate references at end of paper.

a close-up view of the test wheel and instrumented dynamometer which was used instead of a landing-gear strut to support the DC-9 wheel, tire, and brake assembly because it provided an accurate measurement of the ground reaction forces. The test tire was a 40x14, type VII retreaded tire inflated to 0.98 MPa.

The test runway can also be seen in figure 1. Approximately 244 m of the flat concrete test runway was used to provide braking and cornering data on dry and wet runway surfaces. Test speeds ranged from approximately 40 to 100 knots and tire yaw angles up to 9 degrees were examined; tire vertical loadings ranged from approximately 58 kN to 124 kN.

SKID CONTROL SYSTEMS - Typical brake system hardware used in the track test program is shown in figure 2. The brake system components were a pilot metering valve, brake selector valve, and a hydraulic fuse, all DC-9 aircraft components. The antiskid control valve is peculiar to each antiskid system investigated. The line sizes and lengths were those of the DC-9 but line bends were not simulated.

Four antiskid systems were tested during the track test program. These systems included a velocity-rate-controlled and a slip-velocity-controlled system each with pressure-bias modulation as a key element in their logic circuits, hereafter referred to as systems A and B respectively; a slip-ratio controlled system with input from an unbraked nose wheel to obtain aircraft ground speed, system C; and a hydromechanically controlled system which relied on the relative velocity between the braked wheel and a small flywheel to control the braking effort, system D. Detailed descriptions of these four systems are included in references 4 to 7.

FLIGHT TESTS - A flight test program was conducted using a McDonnell-Douglas DC-9 series 10 airplane, shown in figure 3, equipped with antiskid systems A and B. Tests were conducted on a number of different runways throughout the United States under both dry and wet surface conditions. Time histories of the brake pressure for each of the four main gear brakes were used to measure the braking performance of the two antiskid systems during a number of tests on an ungrooved asphalt runway at Lubbock Regional Airport and on an ungrooved concrete runway at Edwards AFB. A detailed description of the flight test program is included in reference 10.

Figure 1.- Landing Loads Track and test carriage.

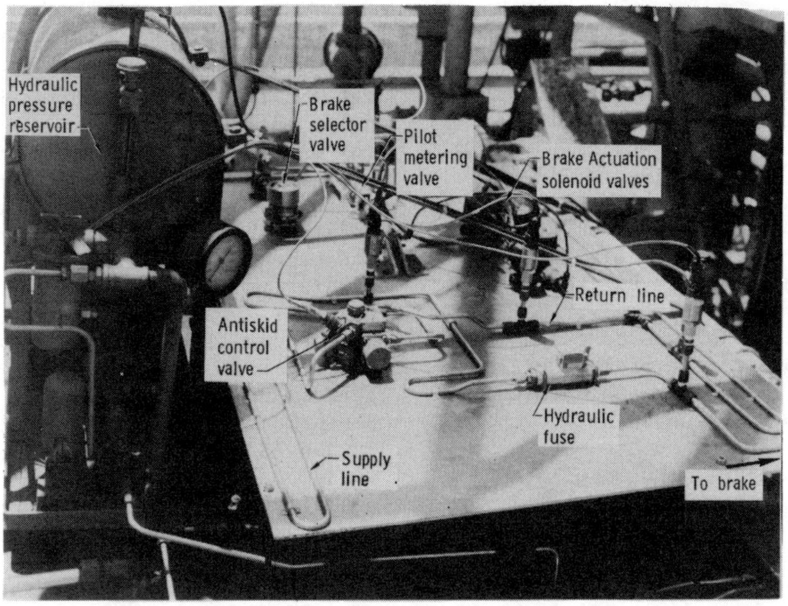

Figure 2.- DC-9 brake system simulation.

Figure 3.- DC-9 test airplane.

COMPUTER SIMULATION STUDIES

Computer simulation studies were conducted to investigate the responses of various brake pressure-torque models and to examine the effects of tire spring and damping characteristics on antiskid braking performance. The brake pressure-torque models investigated include an undamped nonlinear spring, a linear spring with viscous damping, and a variable nonlinear spring with hysteresis memory function. Equations describing these pressure-torque models are included in reference 9. To assess the accuracy of these computer models, the torque output from each model was compared with the actual torque output of the brake for the same pressure input. The difference between the actual torque output of the brake and the model torque response defined an error term which gave a direct indication of model accuracy (9). To examine the effects of tire spring and damping characteristics on the braking performance of antiskid braking systems, the wheel and tire were modeled as two rigid, concentric disks joined by a linear spring and viscous damper. Values for the tire spring rate and damping ratio were computed from free-vibration tests (9). This computer model was used to generate plots of the friction coefficient as a function of the wheel slip ratio during simulated cyclic braking operations. The characteristics of these computer generated computer plots were compared with the actual friction slip curves developed during antiskid braking tests on the Landing Loads Track.

RESULTS AND DISCUSSION

The paragraphs that follow will discuss the use of pressure, torque, and friction ratios to estimate braking performance; the comparison between track tests and flight tests of two different antiskid braking systems, the reasons for degraded antiskid performance under adverse weather conditions, tread temperature measurements during antiskid braking and their relationship with rubber deposits on the runway, and recent results from computer simulation studies.

ESTIMATION OF BRAKING PERFORMANCE - References 11, 12, and 13 discuss several different sources from which antiskid-system performance can be computed. Ideally antiskid performance should be based upon the friction developed between the tire and the runway surface. However, friction measurements are not always available in practice and other measurements such as brake torque or brake pressure must be substituted. Figure 4 illustrates the first step of the two-step procedure used to estimate the braking performance of antiskid systems. Shown in the figure are time histories of the brake pressure, brake torque, and friction coefficient from a typical antiskid braking test on the Langley Landing Loads Track. These data are from a test with antiskid system B on a dry runway. Brakes were applied about two seconds into the run as indicated by the rapid buildup of pressure, torque, and friction. During the braking portion of the run, the antiskid system modulated the braking effort to prevent

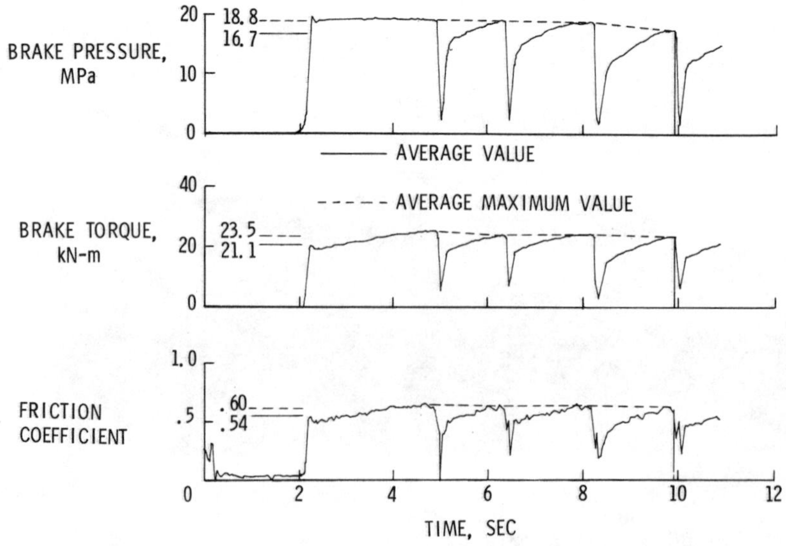

Figure 4.- First step for estimating antiskid braking performance. Antiskid system B on a dry runway.

excessive wheel skidding which resulted in a rapid decay in the three time histories. When the antiskid system sensed that the danger of excessive tire skidding had passed, the brakes were reapplied and the cycle was repeated four times in figure 4. Numerical integration techniques were used to compute both the average values of pressure, torque, and friction during antiskid braking and the average maximum values of these parameters as defined by the dashed curves connecting the local maximum values reached during the brake cycles. The average values computed from the data in the figure are listed on each time history and illustrated graphically by the solid and dashed line segments located near the brake application point.

The second step in the braking performance estimation procedure is illustrated in figure 5. The average values of brake pressure, torque, and friction coefficient obtained from track tests of antiskid system A are plotted as a function of their respective average maximum values. In each case the dry data (fig. 5(a)) and the wet data (fig. 5(b)) are plotted separately. The solid line in each plot represents the line of ideal performance and has a unit slope. The dashed line in each plot is the linear least-squares fit passing through the plot origin. The slope of each dashed line β represents the average braking-performance index or performance estimate for each data set. Values of β obtained from the track tests of all four antiskid systems are summarized in table I.

On the dry runway surface, the value of β determined from pressure, torque, and friction values for antiskid system A varied between 0.85 and 0.81, a difference of approximately 5 percent. On the wet runway surfaces, the variation in β was between 0.69 and 0.66 a difference of about 4 percent. Hence the estimates of braking performance from the three data sources are nearly identical. A similar

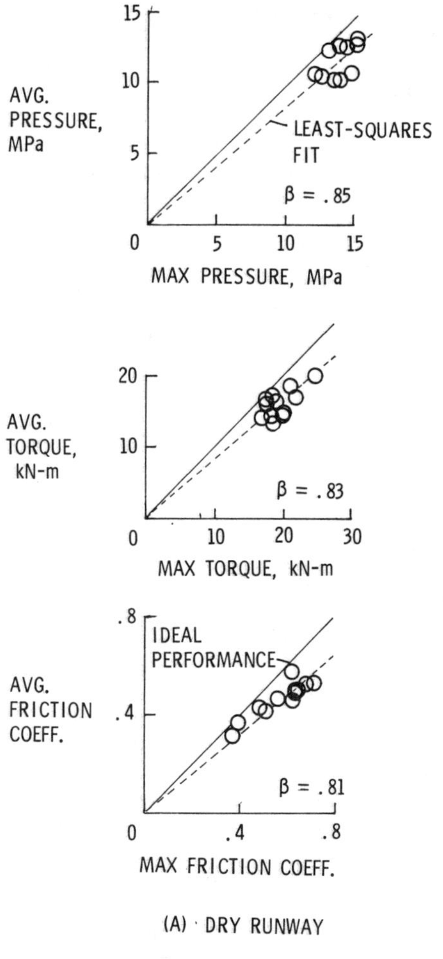

Figure 5.- Second step for estimating antiskid braking performance. Antiskid system A.

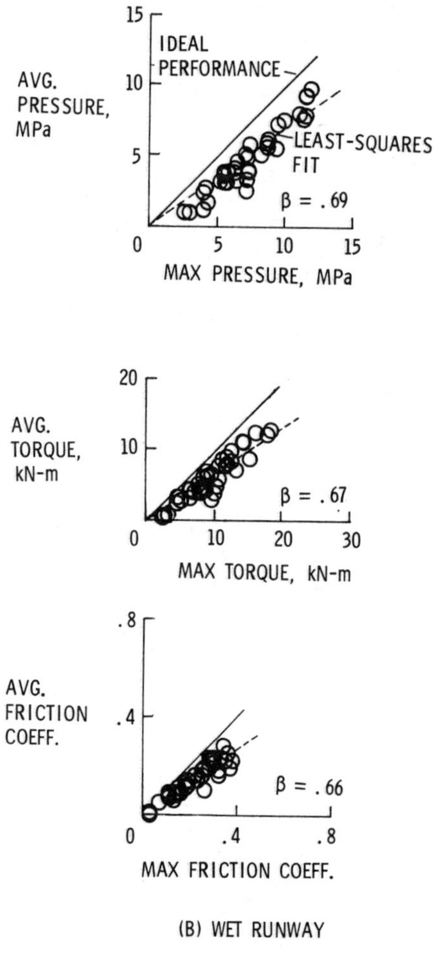

Figure 5.- Concluded.

Table I.- Summary of antiskid performance estimates from track tests.

System	Runway Condition	Performance estimates		
		Pressure data	Torque data	Friction data
A	Dry	0.85	0.83	0.81
	Wet	.69	.67	.66
B	Dry	.91	.93	.91
	Wet	.70	.71	.68
C	Dry	.89	.93	.94
	Wet	.70	.75	.81
D	Dry	.81	.76	.74
	Wet	.71	.78	.69

trend was noted for antiskid system B and estimates of its braking performance varied by 2 percent and 4 percent on the dry and wet runway surfaces, respectively. However, the estimates of braking performance for systems C and D showed considerably larger variations, especially on the wet runway surfaces. These results indicate that estimates of braking performance based upon the three data sources can be used interchangeably for antiskid systems A and B but not for systems C and D. For antiskid systems C and D the estimates of braking performance based upon friction measurements should be considered more accurate than the estimates based upon pressure or torque measurements. Finally, a comparison of the brake performance estimates on the dry runway surfaces with those on the wet runway surfaces indicates that all four antiskid systems suffer performance degradation on the wet surfaces in addition to the obvious reduction in available friction coefficient.

COMPARISONS WITH FLIGHT TESTS - The results of the comparison between the braking performance estimates from the flight tests with the estimates from the track tests are presented in table II. The comparison is limited to antiskid systems A and B, the only systems installed on the test airplane and the wet data from the track tests included only the damp runway wetness condition which closely approximated the wet runway condition during the flight test program. The estimates listed in the table were based upon the brake pressure measurements since the DC-9 airplane was not instrumented to measure brake torque or friction coefficient. It is worth repeating, however, that results from the track tests of antiskid systems A and B revealed that braking performance estimates from pressure, torque, and friction measurements can be used interchangeably for the two systems. The data in the table indicate generally good agreement between the flight tests and the Landing Loads Track tests although track test results tend to be slightly lower than those from the flight tests. These small differences were attributed, in part, to the relatively short distance of the test track. The test track was about 244 m long whereas the airplane stopping distances during the flight test program ranged up to 1890 m. The flight test results also indicate that antiskid systems A and B suffer degraded braking performance when the runway surface becomes wet.

PERFORMANCE DEGRADATION UNDER ADVERSE WEATHER CONDITIONS - Both track and flight tests have indicated that antiskid braking systems are subject to degraded performance under adverse weather conditions and this degradation is attributed to several factors. Most aircraft brakes are designed to provide sufficient torque to produce acceptable airplane deceleration during a rejected takeoff. Under dry runway conditions this maximum brake torque is nearly balanced by the ground reaction torque and the antiskid system can modulate the braking effort at pressures which are near the maximum operating pressure of the brake. Under wet runway conditions, however, the maximum available brake torque greatly exceeds the available ground reaction torque and the antiskid system is forced to modulate the braking effort at greatly reduced pressure levels. Unfortunately the response characteristics of most airplane braking systems are diminished at these lower pressure levels thus leading to degraded braking performance. Furthermore, the reduced friction coefficients on wet runways cause tire spin-up accelerations to be lower than those

Table II.- Comparison between flight tests and track tests.

System	Runway Condition	Pressure performance estimates	
		Flight tests	Track tests
A	Dry	0.91	0.85
	Wet	.66	.65
B	Dry	.92	.91
	Wet	.78	.72

observed on dry runways. These reduced spin-up accelerations increase the time required for the tire to recover from a deep skid and leads to further degradation in braking performance. These torque and spin-up acceleration factors affected the wet runway braking performance of all four antiskid systems tested.

Figure 6 illustrates a case when pressure-bias modulation can also contribute to degraded braking performance under adverse weather conditions. Pressure-bias modulation is an antiskid logic design feature used in systems A and B to enhance performance by increasing the operating time on the front side of the friction coefficient - slip ratio curve. This design feature may be satisfactory under constant friction conditions, but, as shown in figure 6, can be less satisfactory when the friction conditions change rapidly. Time histories of wheel speed, skid signal, brake pressure, and friction coefficient are presented in the figure for a test of system B on a dry runway that has one damp spot about 0.6 m long approximately midway down the test section. The figure shows that the wheel speed is cycling as designed on the dry surface such that wheel spin-down as at A causes a skid signal buildup B which closes the antiskid control valve, thereby reducing the brake pressure C. When the wheel recovers from the skid D, the skid signal reduces and the brake pressure is reapplied. At approximately 6 seconds into the test, the wheel encounters the damp spot on the runway and immediately goes into a deep skid causing a saturated skid signal E and a corresponding reduction in brake pressure. When the wheel recovers from this deep skid, the skid signal is only slightly reduced F because the pressure-bias-modulation system

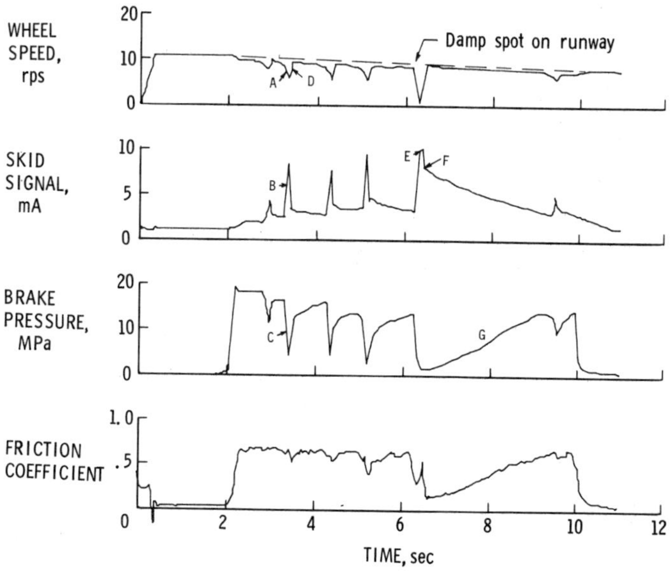

Figure 6.- Antiskid response with pressure-bias modulation. Antiskid system B.

causes a slow reduction in skid signal and, consequently a slow reapplication of brake pressure G. The resulting friction coefficient trace indicates that a friction level of about 0.6 is maintained on the dry runway, but the friction level drops abruptly upon reaching the damp spot on the runway and remains below 0.6 for some time because of the slow rate of brake pressure reapplication following the deep skid. Hence the slow rate of pressure reapplication contributes to degraded system performance.

Figure 7 shows the same type of test conducted with system C which does not employ pressure-bias modulation. When the wheel reaches the damp spot on the runway, the wheel speed once again drops suddenly, the skid signal reaches its maximum value, and there is a corresponding drop in brake pressure. When the wheel recovers from the skid, however, the skid signal rapidly drops to zero and brake pressure is quickly reapplied. The friction coefficient indicates good antiskid action since the friction level drops only momentarily when the damp spot is encountered.

TREAD TEMPERATURES DURING ANTISKID BRAKING - During the course of the Landing Loads Track tests a recording optical infrared thermometer was used to measure tire tread temperatures during antiskid braking on a dry runway. The infrared thermometer was focused on a center rib of the tire in a region about 1/6 revolution aft of the tire-pavement contact area. Time histories of wheel speed, brake pressure, brake torque and tread temperature from a track test of system D is presented in figure 8. The data indicate that brakes were applied approximately 2 seconds into the test and there was no modulation of the braking effort for about 9 seconds. The tire entered into a locked-wheel skid about 11 seconds into the test, the antiskid system quickly released the brake pressure to reduce the brake torque which allowed the wheel to recover from the skid and then modulated the reapplication of the brake pressure, and the brakes were released about 13 seconds into the test. Figure 9 is a photograph of the runway from the point of the locked-wheel skid to the end of the test section. The skid pictured in the figure was 3.7 m long and the tread temperature was 200°C (see fig. 8). Following recovery from the skid, the hot portion of the tread that was in contact with the pavement during the skid continued to deposit rubber on the runway on each wheel revolution thereafter until the final release of brake pressure. These rubber deposits, called tire stamping, correspond to the cyclic peaks observed in the tread temperature time history (fig. 8) which ranged from 184°C to 114°C. Following the test, the tire skid patch (fully cooled) was found to be tacky; thus, indicating that the rubber in this portion of the tread had undergone a change in chemical composition.

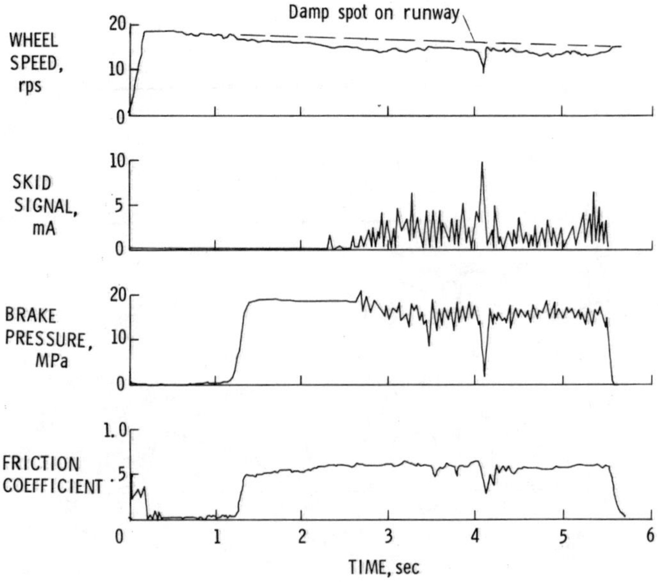

Figure 7.- Antiskid response without pressure-bias modulation. Antiskid system C.

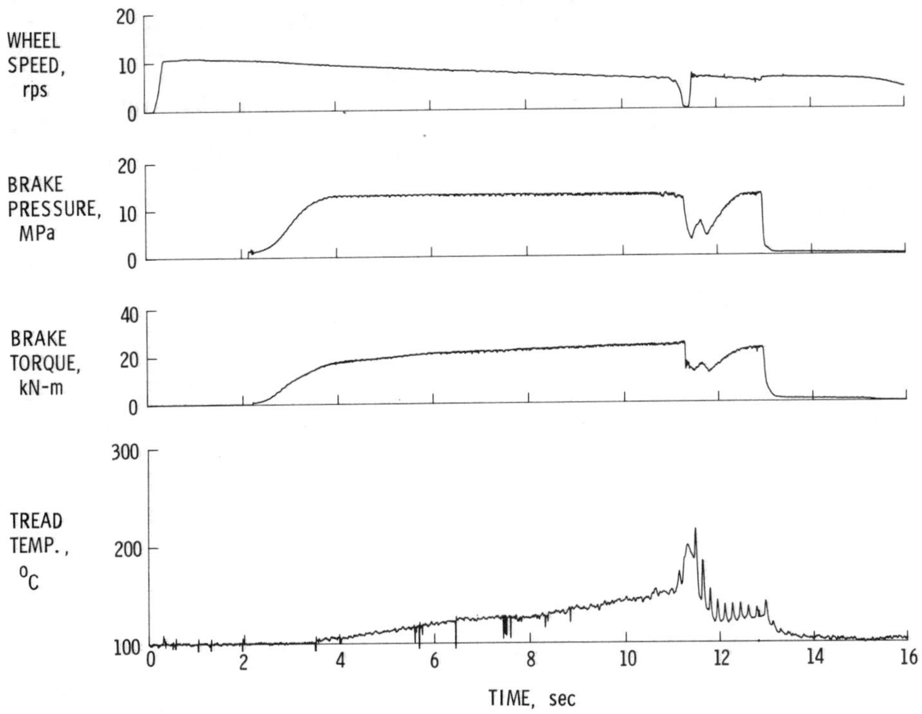

Figure 8.- Time histories from antiskid braking test on a dry runway. Antiskid system D.

Analysis of tire tread temperature data from several tests revealed that tread temperatures never exceeded 142°C when the braking effort was too weak to allow the tire to reach incipient skid conditions. When braking forced the tire beyond the incipient skid point, the peak tread temperatures ranged from 208°C to 265°C. During unyawed runs, these peak tread temperatures were highly transitory and were usually diminished within the time span of 1 wheel revolution, whereas, during runs at high yaw angles the tread temperatures stayed elevated for longer periods of time; thus, the average tread temperatures were increased. Finally, tire tread temperatures above 200°C usually resulted in rubber being deposited on the runway surface.

COMPUTER SIMULATION STUDIES - Accurate models of brake and tire dynamic behavior are necessary to enable computer simulations to tune existing antiskid systems to optimize their braking performance for specific aircraft applications and to aid in the design of future systems. Many response characteristics of the brake can be duplicated by including the braking hardware in a "breadboard" simulation, but the pressure-torque response of the brake and the spring and damping characteristics of the tire must be modeled mathematically.

The relationship between brake pressure and torque defines the brake behavior during antiskid operation and plays a critical role in establishing the braking performance of antiskid systems. The torque from a typical

Figure 9.- Runway rubber deposits following antiskid braking test.

antiskid braking test on the Landing Loads Track is plotted as a function of brake pressure in figure 10. The relationship depicted in the figure is characterized by a number of hysteresis loops, and these loops are developed in a counterclockwise sense during the course of the test. Hence the torque developed as a result of a given pressure input can vary over a large band and is dependent upon previous braking history. The outer boundary of the hysteresis loops shown in the figure generally encompasses the measured pressure-torque responses of the DC-9 brake used in the track tests. Instantaneous pressure-torque responses of the brake within this envelope are a function of runway friction level and the response characteristics of the individual antiskid systems. The extent of the hysteresis envelope is a function of brake temperature, fade, and stiffness characteristics; back pressure in the hydraulic lines; low-speed torque peaking; brake-lining friction characteristics; and, to a lesser extent, brake wear.

Most current computer simulations model the brake pressure-torque response as an undamped nonlinear spring (14) or as a linear spring with viscous damping coupled with a nonlinear static torque gain (15). During the course of the NASA simulation studies, a third brake pressure-torque model was developed which can be characterized as a variable nonlinear spring with hysteresis memory function. Figure 11 presents the pressure-torque responses of these models to the measured brake pressure input from the experimental data shown in figure 10. If the models

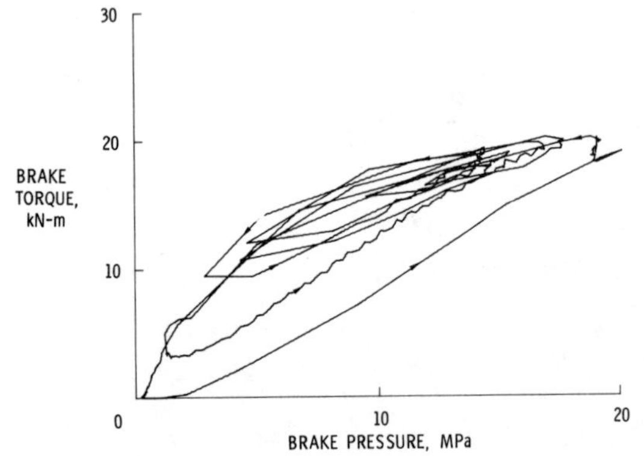

Figure 10.- Typical brake pressure-torque response during antiskid braking.

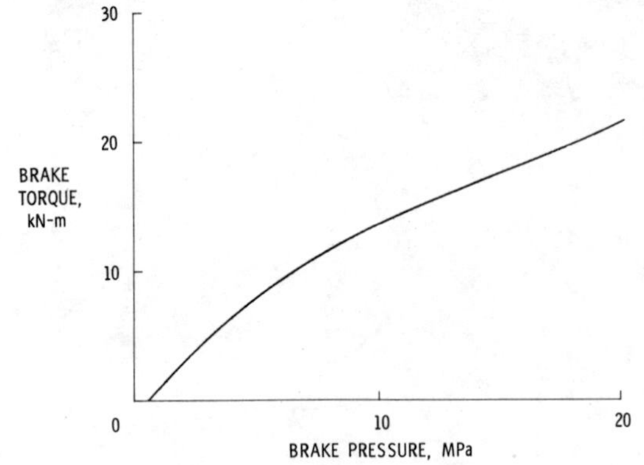

(A) UNDAMPED, NONLINEAR SPRING

Figure 11.- Computer model pressure-torque response.

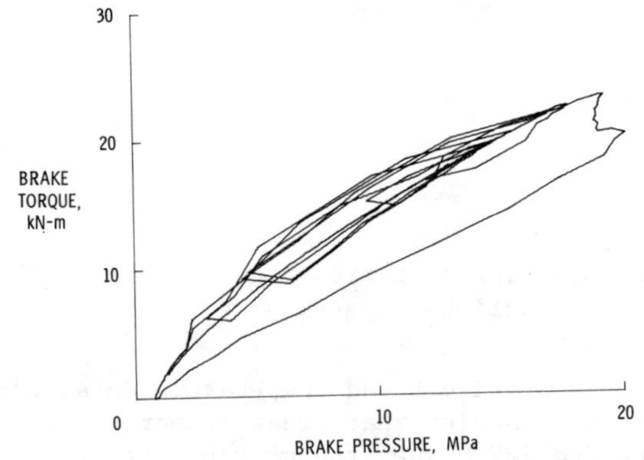

(B) LINEAR SPRING WITH VISCOUS DAMPING

Figure 11.- Continued.

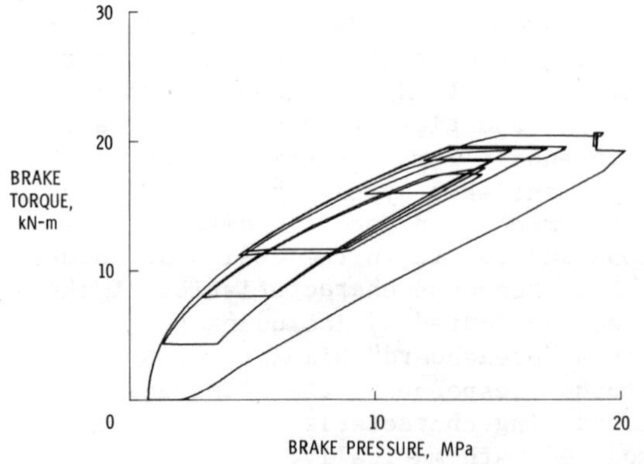

(C) VARIABLE NONLINEAR SPRING WITH HYSTERESIS MEMORY FUNCTION

Figure 11.- Concluded.

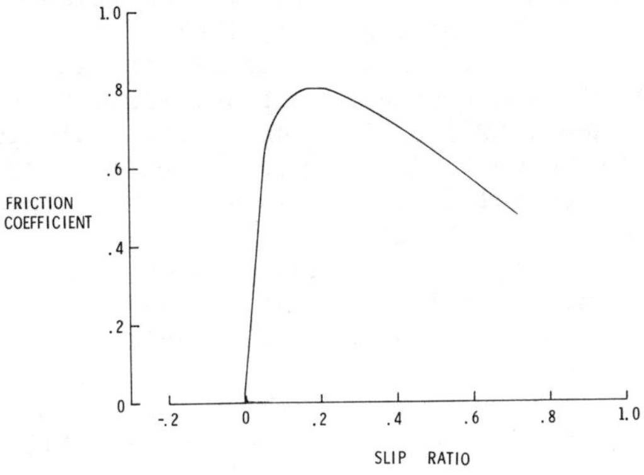

Figure 12.- Computer simulation of friction coefficient - slip ratio relationship. Tire spring and damping characteristics not included.

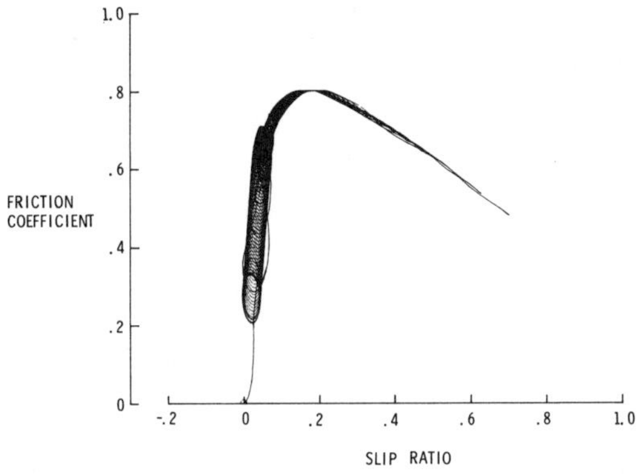

Figure 13.- Computer simulation of friction coefficient - slip ratio relationship. Tire spring and damping characteristics included.

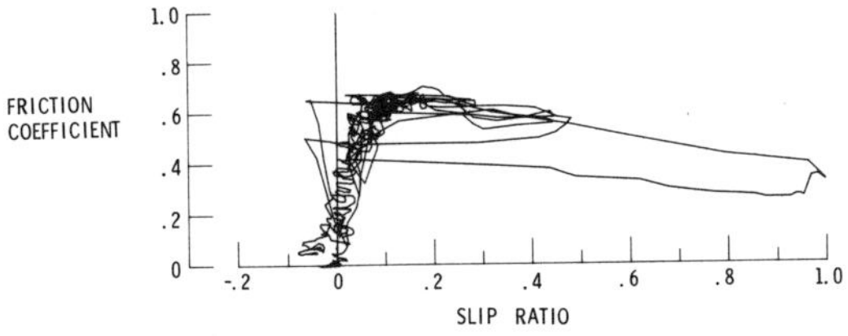

Figure 14.- Measured friction coefficient - slip ratio relationship from typical antiskid braking test.

performed perfectly, then the model responses in figure 11 would duplicate exactly the measured pressure-torque response shown in figure 10. The response of the undamped nonlinear spring is shown in figure 11(a) and can be characterized as a nonlinear, single-valued curve. This curve provides an accurate fairing of the actual torque response of the brake but does not exhibit the pronounced hysteresis loops observed in the measured response. The responses of both the linear spring with viscous damping (fig. 11(b)) and the variable nonlinear spring with hysteresis memory function (fig. 11(c)) exhibit hysteresis characteristics that are similar to the measured brake response. An assessment of the accuracy of the various pressure-torque models based upon the torque error criterion indicates that the variable nonlinear spring with hysteresis memory function is about 40 percent more accurate than the undamped nonlinear spring and the linear spring with viscous damping. Antiskid computer simulations which fail to model accurately the hysteresis characteristics of the brake pressure-torque response generally underestimate (1) the severity of tire skidding which occurs during antiskid braking and (2) the time required for skid recovery. Such simulations produce unconservative estimates of antiskid braking performance.

Computer simulations of antiskid braking systems can also be made more accurate by including the effects of tire friction, spring, damping, and inertia characteristics. Most computer simulations have the ability of modeling friction and inertia characteristics, but many simulations do not model the spring and damping characteristics. Figure 12 shows a simulation of friction coefficient variation with wheel slip ratio that does not include the spring coupling between the wheel and the tire-pavement contact area. Figure 13 shows the same simulation, but with tire spring and damping characteristics included and figure 14 shows a measured friction coefficient variation with wheel slip ratio during a typical antiskid braking test. Without these additional refinements, the tire dynamic characteristics are not adequately represented in the computer simulation, and the antiskid model could produce misleading results.

CONCLUDING REMARKS

Recent NASA antiskid research including Landing Loads Track tests, flight tests, and computer simulation studies have been reviewed, and the techniques for estimating antiskid braking performance based upon pressure, torque, and friction data have been outlined.

The results of the track tests of four different antiskid systems indicate that each system is subject to degraded braking performance when the runway becomes wet. These track results were corroborated by flight tests of two of the antiskid systems.

Measurements of the tire tread temperature during antiskid braking on dry runway surfaces indicate that the peak temperatures observed are a function of the tire skidding allowed by the antiskid system.

The computer simulation studies demonstrated the need for accurate modeling of the brake pressure-torque response and the dynamic characteristics of the tire to insure that antiskid computer simulations provide the correct information for designing and tuning antiskid braking systems.

REFERENCES

1. Tracy, William V., Jr.: Wet Runway Aircraft Control Project (F-4 Rain Tire Project). ASP-TR-74-37, U.S. Air Force, Oct. 1974. (Available from DTIC as AD A004 768).

2. Danhof, Richard H.; and Gentry, Jerauld R.: RF-4C Wet Runway Performance Evaluation. FTC-TR-66-6, U.S. Air Force, May 1966. (Available from DTIC as AD 486 049).

3. Horne, Walter B.; McCarty, John L.; and Tanner, John A.: Some Effects of Adverse Weather Conditions on Performance of Airplane Antiskid Braking Systems. NASA TN D-8202, 1976.

4. Stubbs, Sandy M.; and Tanner, John A.: Behavior of Aircraft Antiskid Braking Systems on Dry and Wet Runway Surfaces - A Velocity-Rate-Controlled, Pressure-Bias-Modulated System. NASA TN D-8332, 1976.

5. Tanner, John A.; and Stubbs, Sandy M.: Behavior of Aircraft Antiskid Braking Systems on Dry and Wet Runway Surfaces - A Slip-Ratio-Controlled System With Speed Reference From Unbraked Nose Wheel. NASA TN D-8455, 1977.

6. Stubbs, Sandy M.; Tanner, John A.; and Smith, Eunice G.: Behavior of Aircraft Antiskid Braking Systems on Dry and Wet Runway Surfaces - A Slip-Velocity-Controlled, Pressure-Bias-Modulated System. NASA TP-1051, 1979.

7. Tanner, John A.; Stubbs, Sandy M.; and Smith, Eunice G.: Behavior of Aircraft Antiskid Braking Systems on Dry and Wet Runway Surfaces - Hydromechanically Controlled System. NASA TP-1877, 1981.

8. Tanner, John A.; Dreher, Robert C.; Stubbs, Sandy M.; and Smith, Eunice G.: Tire Tread Temperatures During Antiskid Braking and Cornering on a Dry Runway. NASA TP-2009, 1982.

9. Tanner, John A.; Stubbs, Sandy M.; Dreher, Robert C.; and Smith, Eunice G.: Dynamics of Aircraft Antiskid Braking Systems. NASA TP-1959, 1982.

10. Horne, Walter B.; Yager, Thomas J.; Sleeper, Robert K.; Smith, Eunice G.; and Merritt, Leslie R.: Preliminary Test Results of Joint FAA-USAF-NASA Runway Research Program, Part II - Traction Measurements of Several Runways Under Wet, Snow Covered, and Dry Conditions With a Douglas DC-9, a Diagonal-Braked Vehicle, and a Mu-Meter. NASA TM X-73910.

11. Skid Control Performance Evaluation. ARP862, Soc. Automot. Eng., Mar. 1, 1968.

12. Lester, W. G. S.: Some Factors Influencing the Performance of Aircraft Anti-Skid Systems. Tech. Memo. EP 550, British R.A.E., July 1973.

13. Brake Control Systems, Antiskid, Aircraft Wheels General Specifications for Mil. Specif. MIL-B-8075D, Feb. 24, 1971.

14. McGowan, J. A.: Expansion of Flight Simulator Capability for Study and Solution of Aircraft Directional Control Problems on Runways - Appendixes. NASA CR-145281, 1978.

15. Straub, H. H.; Yurczyk, R. F.; and Attri, N. S.: Development of a Pneumatic-Fluidic Antiskid System. AFFDL-TR-74-117, U.S. Air Force, Oct. 1974. (Available from DTIS as AD A009 170).

European Aircraft Steering Systems

Donald W.S. Young
Dowty Rotol Ltd. UK
Burkhard Ohly
Messerschmitt-Bolkow-Blohm GmBh
West Germany

* Paper 851940 presented at the Aerospace Technology Conference and Exposition, Long Beach, California, October, 1985.

ABSTRACT

The paper discusses system and hardware design characteristics of the steering for several current aircraft and aircraft under development, covering both mechanically and electronically controlled systems. Possible future developments for steering systems are discussed briefly.

STEERING SYSTEM DEVELOPMENT AND DESIGN has evolved from relatively simple systems which cater only for taxi operation to those which accommodate steering over the whole range of ground speeds and the transition to flight.

The purpose of this paper is to review European designs in this field to date and to comment on aims for future development. The subject is not covered exhaustively, but picks out salient developments using several examples, concentrating on the mechanisms and systems directly associated with the nose gear rather than the total system.

BASIC REQUIREMENTS

All steering systems must meet certain design requirements:-

- Sufficient torque to steer under all normal ground conditions.
- Controlled rate of steering with a maximum rate adequate for all defined ground manoeuvres and forward speeds.
- Shimmy stability under all modes of operation.
- Input/output hysterisis adequate for good control.
- Provide centring for retraction or is compatible with mechanical centring provision.
- Does not inhibit rudder control motion.
- Castors freely when the steering is not activated.

Steering torque is not defined by standards in Europe as in the US. There is no requirement to move the tyres with the aircraft at static load with μ of .8 as defined in MIL-S-8812. Steering torques with much lower friction co-efficients have been used and historically μ values of as low as .375 have proved to be adequate where space or other contraints have prevented a larger steering actuator being fitted.

In addition to the above requirements some systems need to be protected against damage resulting from external inputs such as one tyre bursting, hitting obstructions, or rolling over the runway edge, any of which may castor the nose gear at a high velocity and generate high pressures in the actuator and associated valves.

DESIGNS 1950 - 1960

Steering systems and their associated components have developed much from those fitted to the early post-war aircraft as a result of the need for improvements in performance, increased reliability and the need to castor freely with the minimum of ground crew attention during towing.

Several basic steering actuation mechanisms were fitted during this era including multiplying linkages, direct operation of a crank with an oscillating actuator, moving body actuator and twin actuator designs, as shown diagrammatically in Figure 1. Each type has different torque to angular movement relationships, as shown in Figure 2, which compares some typical characteristics.

It is apparent that mechanisms of these types must be designed to provide the torque needed at extreme steering angles, yet for strength considerations must accommodate the maximum forces due to 'oversized' actuators at the central position, with a consequent adverse effect on mechanism weight.

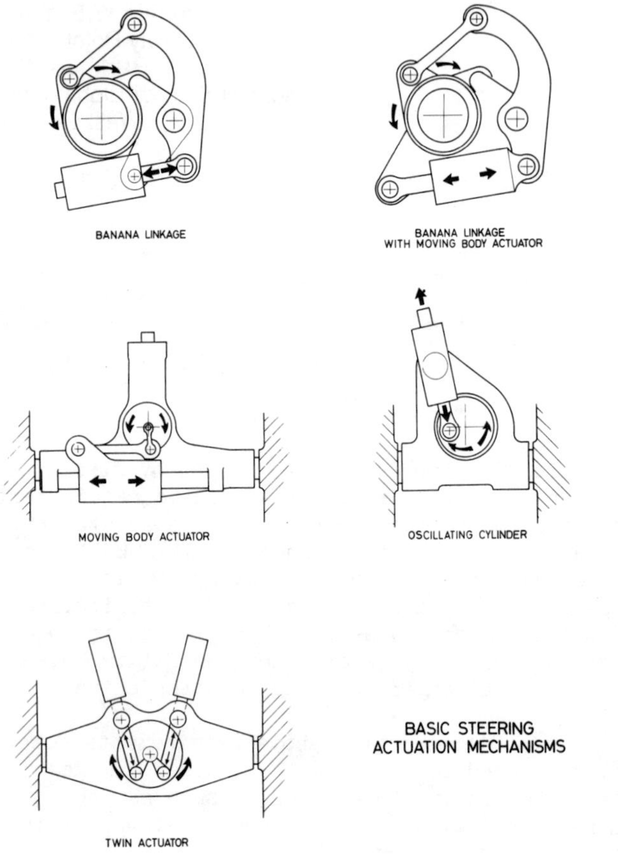

FIGURE 1

COMPARISON OF STEERING ANGLE CAPABILITY

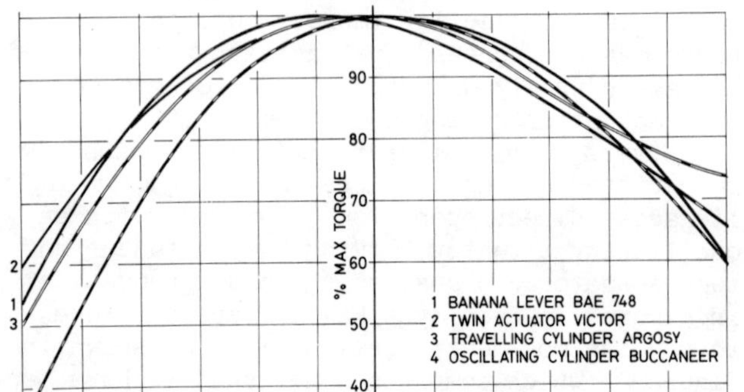

FIG.2

BAe 748 - During the 1950's aircraft frequently used a 'Banana' multiplying linkage to drive a cuff fitted around the main cylinder in order to achieve steering angles of up to $\pm 50^{o}$ with a significant fall off of output torque at extreme angles. Typical of this arrangement is that of the BAe 748, illustrated in Figure 3, which has the pilot's input fed through a torque tube on the nose gear with the complete hydraulic control system mounted close to the actuator.

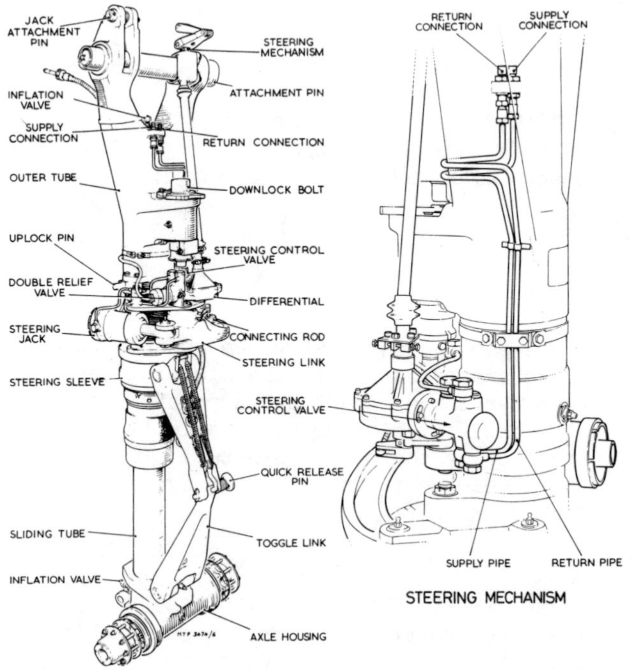

NOSE LANDING GEAR
BAE 748

FIG.3

On aircraft fitted with simple steering systems similar to that of the HS 748, shown diagrammatically in Figure 4, castoring when the steering is inactive is by fluid transfer between actuator volumes through a by-pass valve which closes when system pressure is present. Shimmy suppression on twin wheeled aircraft of this era was often by means of coupled wheels using a live axle with only incidental coulomb friction and inherent hydraulic damping assisting.

An alternative 'Banana' linkage sometimes used - typically on the Trident and HS 125 aircraft - has a moving body actuator, which eliminates the need for a differential linkage to provide mechanical error signals to the valve. Any power mechanism with a number of joints and a large number of load cycles, such as the multiplying linkage, is susceptible to fatigue problems. Efforts should always be made to reduce the effect of fatigue by simplifying the mechanism to reduce the number of joints necessary.

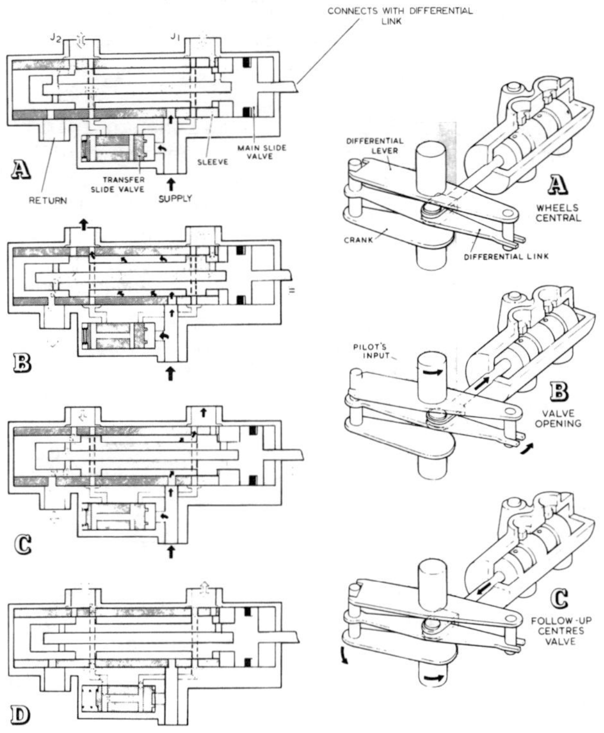

STEERING CONTROL SELECTOR DIFFERENTIAL ASSEMBLY

STEERING COMPONENT OPERATION
BAE 748 FIG 4

BAe BUCCANEER - Lever suspension nose landing gears, because of their short length and the internal turning member, which is often but not invariably fitted, lend themselves to the location of the steering mechanism above the leg. The oscillating cylinder mechanism provides a limited effective steering range, similar to that of the 'Banana' linkage, it has relatively few joints and also allows the leg to castor through $\pm$ 360 ^{o} without ground crew intervention. For deck landing aircraft with lever suspension nose gears this is an important feature. Typical of this type is the Buccaneer nose gear with a system fitted in a very constrained space as shown in Figure 5.

For this aircraft the simple electro-hydraulic system shown in Figure 6, was chosen to control nose wheel steering angle as a function of rudder bar displacement. The velocity of nose wheel rotation is not proportional to error signal since an electro-selector is used to control fluid flow to the steering actuator. This simple concept is not so crude as it first seems since a spool valve in a rudder controlled system operates at full flow for much of the time during coarse movements and low flows are only achieved with small error signals, which are only present on initial selection and when the nose wheel approaches the commanded steering angle. On the Buccaneer aircraft the lack of space means that the steering actuator is cantilever mounted from the main fitting and the crank pin is also cantilever mounted on the turning member, an arrangement that requires careful design to minimise deflections and achieve an adequate fatigue life. The steering control switch is mounted on the airframe above the nose gear and the input signal. A mechanical input signal rotates a contacting brush to energize the appropriate sector for selector operation. Feedback from the nose gear turning tube rotates the sectors to cancel the error signal.

Centring for retraction can be a problem with lever suspension nose gears, but the Buccaneer is provided with steer-to-centre by means of centralizing sectors inside the control switch. These are energized on 'gear retract' selection.

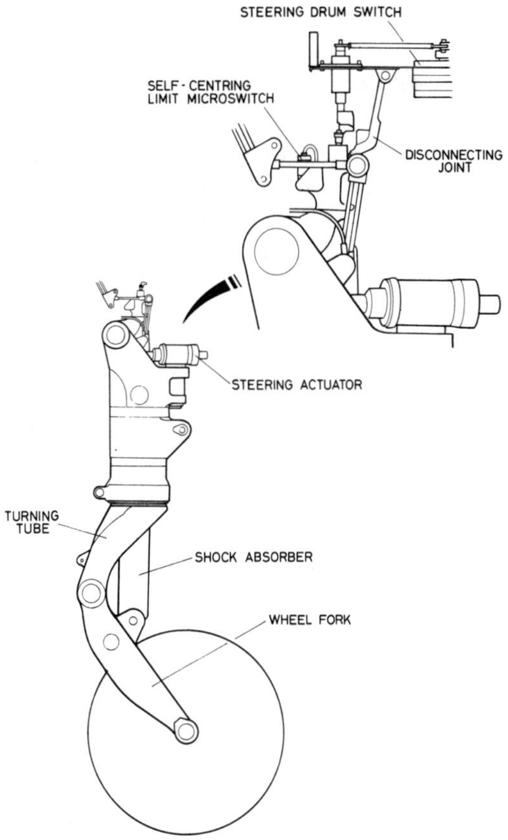

BUCCANEER NOSE GEAR

FIG.5

The control switch is disconnected from the landing gear on retraction and is maintained centralized by a simple spring-loaded lock lever.

Shimmy suppression for this landing gear is by coulomb friction at the castor axis bearings assisted by a friction clutch energized by the vertical wheel load mounted adjacent to the upper castor axis bearings. Over-pressure protection is given by cross line relief valves fitted to the nose gear adjacent to the actuator. An electrically closed by-pass valve permits castoring when the steering is inactive. The Vulcan aircraft has a similar steering actuator arrangement but in this case centring is effected by a separate actuator which is powered from the gear retract line.

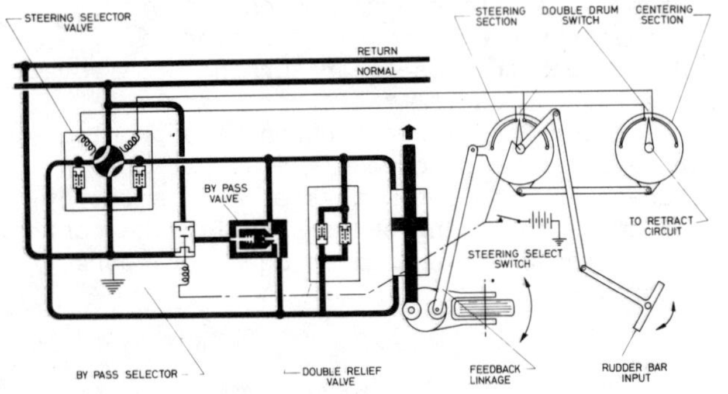

BUCCANEER NOSEGEAR STEERING SYSTEM

FIG.6

BAe ARGOSY - The travelling actuator design, fitted to the BAe Argosy aircraft, permits a simple telescopic nose gear design with the minimum number of large parts. The connecting link design extends the effective steering range and can simplify installation as flexible fluid supply pipes are not needed, since fluid supply can be through the fixed piston rod. The landing gear pivot arms can be used as convenient actuator attachments, although this may have an adverse effect on fatigue strength of the arms.

Figure 7 illustrates this solution which includes a serrated torque shaft which rotates the sliding member directly, so eliminating a second rotating tube and torque arms. This design is satisfactory

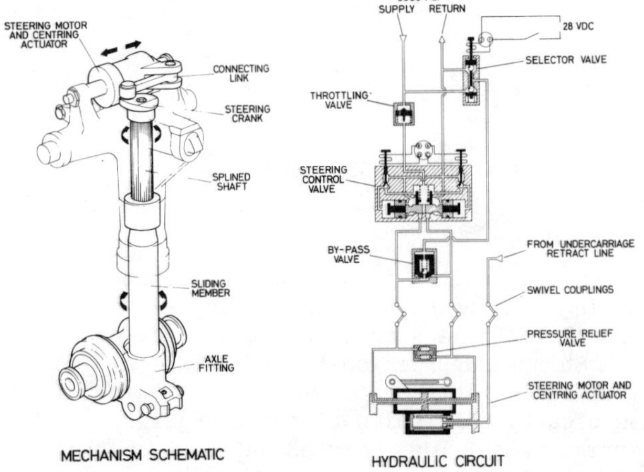

ARGOSY NOSEGEAR STEERING

FIG.7

where wheels are coupled to provide shimmy suppression, but care must be taken to provide good torsional stiffness, particularly if tyres of low stiffness and high enertia are fitted.

TWIN ACTUATOR DESIGNS - Although twin actuator designs which power a steering cuff low on the nose gear, as fitted to a number of US aircraft, have not been used extensively in Europe some nose gears have incorporated twin actuators fitted at the top of the leg, acting directly on a turning tube with limited steering angles have been produced. Unlike some US designs with twin actuators, trunnion sequenced changeover valves to enhance the steering power curve have not been fitted.

ACTUATION SYSTEM MECHANISM DESIGNS evolved up to 1960 continue in use to this day and some designs for landing gears utilize them when positive advantages with respect to weight or installation can be demonstrated. For example, a variant of the Argosy arrangement is used on the De Haviland Canada Dash 8 nose gear and a 'Banana' linkage is fitted to the nose gear for the GP-180 aircraft.

DESIGNS 1960 - 1970

This period brought the rack and pinion actuator into general usage with the advantages of lower hydraulic volume requirements, constant steering torque and the possibility of effective steering at greater steering angles.

JAGUAR - The Anglo-French Jaguar aircraft's lever suspension nose gear is fitted with a hydro-mechanical steering control system, as shown in Figure 8, which has several points of interest.

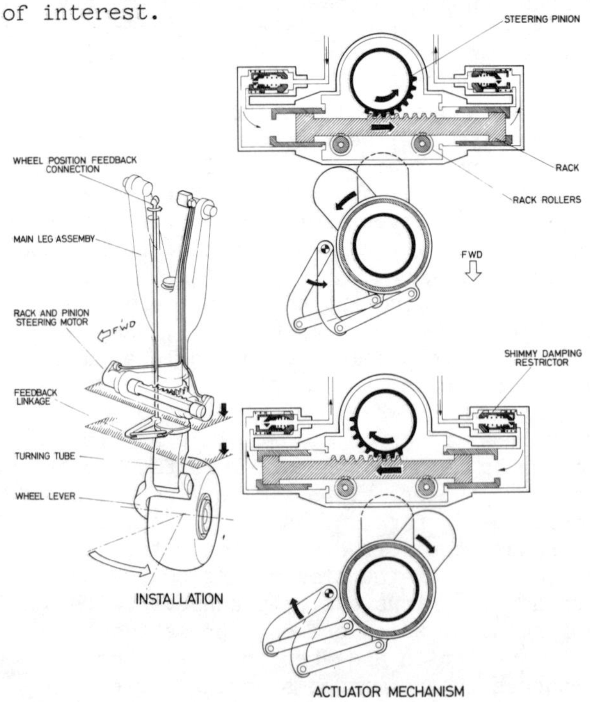

JAGUAR STEERING

FIG.8

The rack and pinion steering motor is an integral part of the main fitting structure. The rack is powered by hydraulic cylinders, each of which contain a 'normal' steering piston, within a co-axial centring piston. In normal operation, this assembly moves together when steering towards the centre position; when steering out from the centre position, effort is from the inner piston only. When retraction is selected the steering cylinders are pressurized equally at a reduced pressure, the outer centring pistons bottom on stops to provide a positive central position of the wheel for retraction. The feedback linkage geometry is arranged to raise slightly during nose gear retraction and engage a dog clutch which prevents rotational movement of the linkage and retains the wheel centralized mechanically during flight. A retraction inhibiting switch prevents retraction if the wheel is not centred.

The steering valve on the Jaguar is located on the airframe and therefore the anti-shimmy restrictor valves are included in the actuator cylinder assemblies to minimize the oil volume under compression.

The steering system schematic of Figure 9 shows the pressure reducing valve with its associated accumulator which supplies centring pressure to the steering actuator through the steering valve. An hydraulically engaged clutch unit is interposed between the steering valve and the pilot's rudder so that the steering valve is only connected when the steering is switched on.

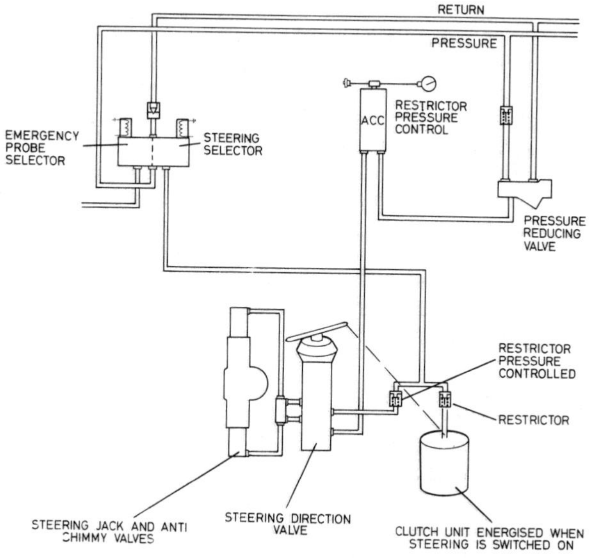

JAGUAR STEERING SYSTEM SCHEMATIC

FIG. 9

BAe HARRIER - The Harrier (AV-8B) aircraft is now as well known in the US as Europe and while many changes have been made to enhance the original aircraft, the steering concept has not been changed significantly from the design which evolved in the latter '60's'. It has an unusual quadricycle landing gear arrangement where the nose gear carries approximately 40% of the aircraft weight. Since the steering is effectively like that of a bicycle and there is no possibility of steering by differential braking, integrity and accuracy are both vital aspects. The steering of this compact nose gear shown in Figure 10, also has to accommodate a shortening mechanism, which passes through the centre of the steering pinion.

For integrity, the system fails 'on' in the event of electrical system failure and emergency accumulators are provided to allow steering control during landing in the event of hydraulic failure. To minimise loss of fluid from the accumulators, the steering valve has overlapped lands. Careful attention to detail ensures that the accuracy of resolution compares well with a 'modern' electrically controlled system.

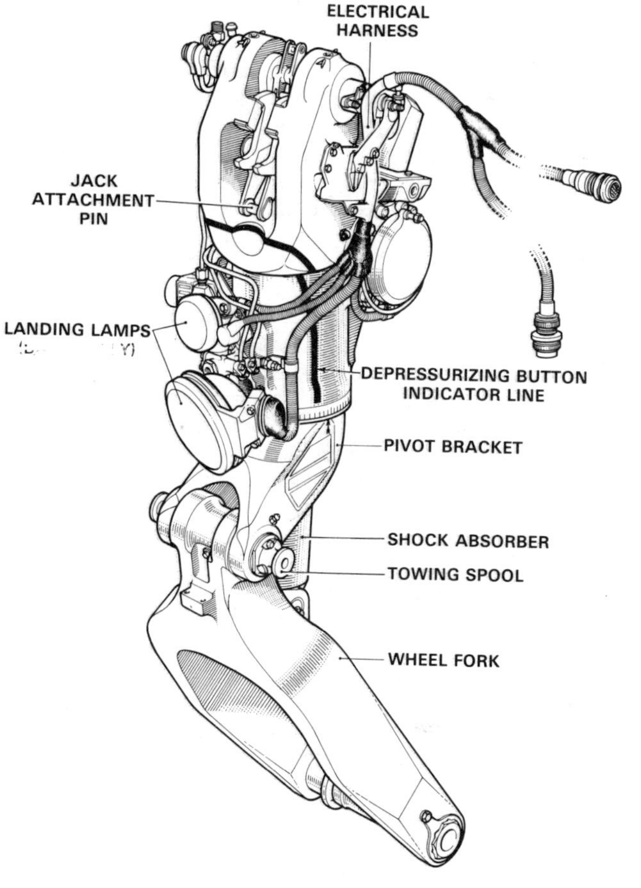

HARRIER NOSE LANDING GEAR

FIG 10

The steering system components directly associated with the nose gear are shown in Figure 11 and includes a by-pass valve to permit free castoring for towing and a relief valve to prevent system overloads. Before towing the aircraft at nose gear angles outside the normal steering range, the system must be de-pressurized and a manually operated valve is fitted for this purpose.

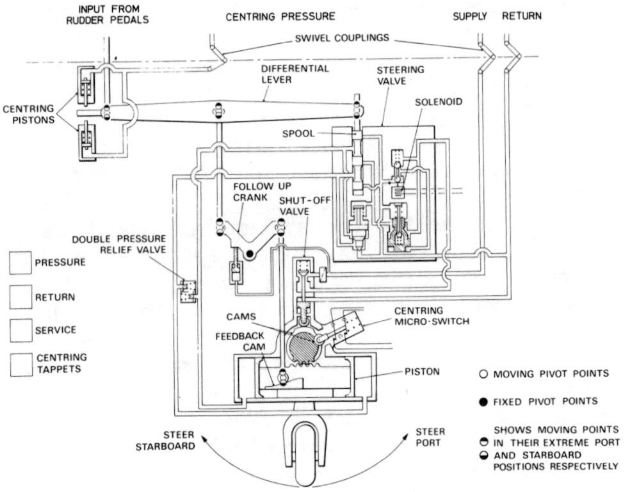

HARRIER STEERING SCHEMATIC

FIG 11

A 'Vernier' device incorporated in the input control linkage provides two gain modes at any angle of steer. The 'fine' control mode provides low gain with limited angular motion for accurate steering. The second 'coarse' control mode is achieved if the demand is greater than the 'fine' control limit of authority by overriding a friction clutch in the variable gain mechanism.

The rack and pinion steering motor has power steering to $\pm 45^{\circ}$ and features an automatically disengaging and re-engaging mechanism which permits free castor to $\pm 179^{\circ}$ which is accomplished by the pinion pushing the rack out of engagement, as shown in Figure 22. The rack is held out of engagement by means of a baulking ring which permits re-engagement under the influence of return pressure in the steer actuator only when the gear teeth are in correct alignment. This design permits a short rack assembly which fits into the very narrow landing gear bay. The nose gear centres prior to retraction from any angle within the steering range by means of centralizing the input lever.

The gain of the system also varies with the steered position with a 3:1 gain difference between the central position and the extreme steering angles, as shown in Figure 12. This gain characteristic is provided by a profiled feedback cam fitted to the steering rack.

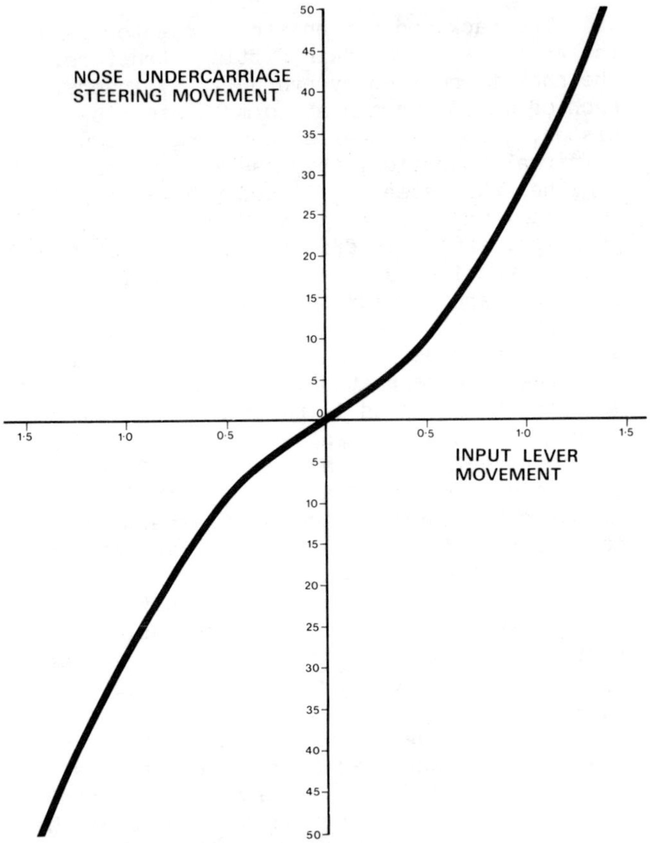

HARRIER STEERING GAIN CHARACTERISTICS

FIG 12

SAAB VIGGEN - The SAAB Viggen steering is unusual in that limited authority force feedback is employed. Originally the steering of the Viggen's cantilever type nose gear was of a conventional type, of good response and accuracy, with the moving body steering actuator rotating the nose wheel through a multiplying 'Banana' linkage. The aircraft has a large delta wing and it was found that the use of reverse thrust on landing destabilized the aircraft due to wing lift, which reduces the load on the main gears, and therefore their ability to provide stabilizing drag forces, compounded by aerodynamic asymmetry due to thrust reverser aircraft attitude and configuration effects which tend to turn the aircraft.

After careful consideration of the steering characteristics under all conditions, it was decided to incorporate a limited authority force feedback element in the system so that over a controlled angle about any steer position the pilot is able to feel the forces generated at the same time that positive feedback rotates the nose gear so that it tends to follow the ground force vector and minimises any de-stabilizing side force at the ground. Figures 13 and 14 respectively illustrate the mechanism layout in principle and the system schematic.

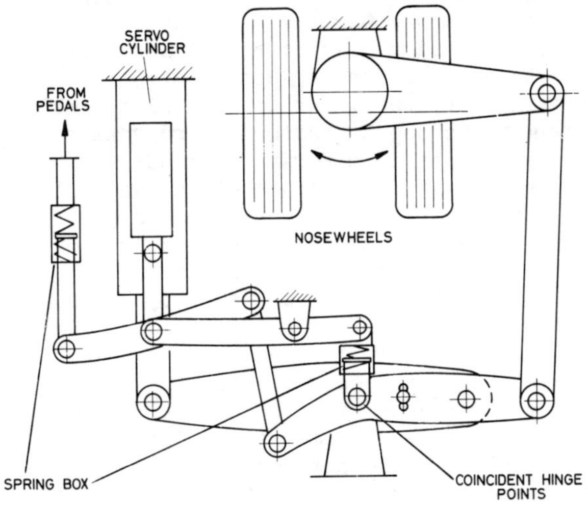

SAAB VIGGEN STEERING MECHANISM

FIG.13

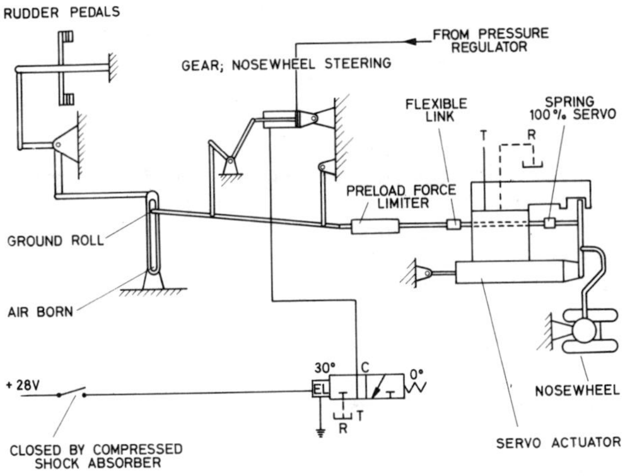

SAAB VIGGEN STEERING SCHEMATIC

FIG 14

DESIGNS 1970 - 1980

This period saw the introduction in Europe of steering systems for military aircraft which used steer-by-wire techniques. These, for military aircraft at least, have now virtually replaced mechanically controlled steering systems. The advantages of electrically controlled steering are mainly with respect to space and weight savings. Also of considerable significance, particularly for small aircraft, is the elimination of mechanical backlash and the ease with which electrical wiring can be routed by comparison with cable runs or push/pull rods. The ability to change characteristics by changing the control module circuitry can be advantageous for experimental purposes. Because of the large number of components involved, reliability has not improved over hydro-mechanical systems and it has been necessary to incorporate additional circuits to improve safety, which further degrades the reliability of the steering system.

ALPHA JET - This aircraft's nose gear is a lever type with a single wheel as shown in Figure 15. The steering control system has a single mode which is engaged either by nose gear weight on wheels or pilot selection. This mode allows steering to $\pm 50^{\circ}$ at rates of up to 15° per second with a dual slope steering positional response which permits sensitive steering up to approximately 7° for compatibility with landing and high speed taxi operations. Figure 15 also shows these characteristics.

ALPHA JET STEERING SYSTEM
PEDAL STROKE V.S. STEERING ANGLE

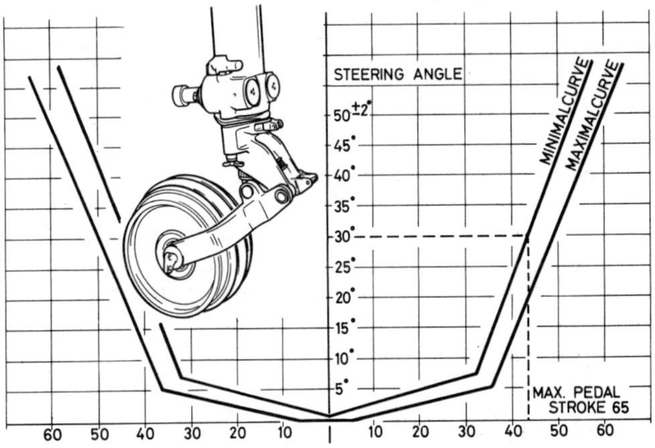

FIG. 15

Centring for retraction is by a mechanical cam and the steering may be disconnected by manual withdrawal of a connecting pin to permit free castoring through 360° ground handling.

The dual rack and pinion steering actuator was probably selected because of space constraints, but although undoubtedly heavy, has a minor advantage as gear clearances are not important for shimmy suppression since the active gear tooth profiles are always in engagement due to pressure in the steering cylinders.

The control system illustrated in Figure 16 is simple in concept, with separate selectors in series controlling pressure to the steering selector. One selector accepts nose gear switch inputs and the other accepts pilot control units.

The steering control valve is mounted on the airframe and shimmy damping is by one-way restrictors within the steering actuators.

ALPHA JET STEERING SCHEMATIC

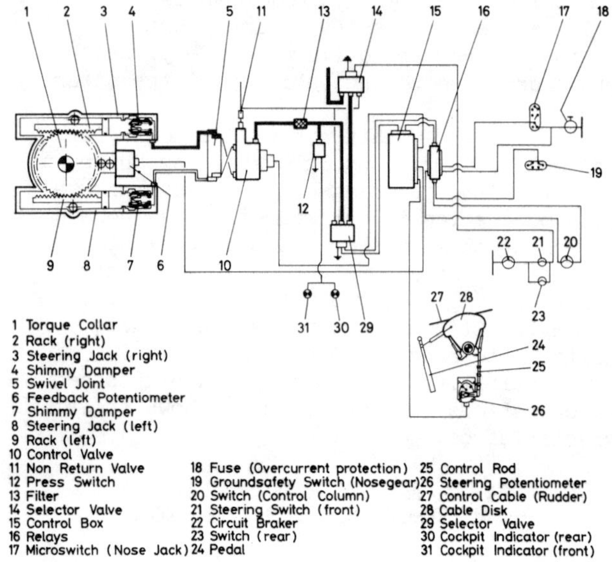

FIG. 16

TORNADO - The nose gear of this aircraft is a cantilever type with twin uncoupled nose wheels mounting 18 x 5.5 tyres. The original concept for the steering system was driven by the fact that F4 Phantom aircraft were in service with the UK and Germany, two of the partners in this trinational programme.

It was wished to use the maximum number of 'off the shelf' components and the F4 steering system was recognized as a source that offered the potential of a steer-by-wire system in quantity production with appreciable service experience, low development costs and the possibility of taking advantage of the 'lessons learned' information which was available.

To cater for emergency high speed landings with the Tornado's variable geometry wing fully swept back, a two mode system was made a requirement to cover high (steering range $\pm 30^\circ$) and low speed (steering range $\pm 60^\circ$) conditions.

As the F4 hydraulic system pressure is 3000 psi and the Tornado's is 4000 psi, equipment requalification was necessary. Changes were also made to improve sealing against moisture ingress on electrical components and dual track command and feedback potentiometers with duplicated control lines were introduced to provide failure protection to avoid the possibility of hardover failures.

Initial trials indicated that the steering system had a peaceful nature, but it was soon found that unexpected yaw moments, which were difficult to compensate by steering inputs, occurred during thrust reverser application at high rolling speeds. As noted earlier, SAAB had experienced similar problems with their Viggen and advantage was taken of their experience to design a solution. A mechanical solution to overcome the effect on stability of low main gear loads, due to wing lift combined with yaw motions due to taileron aerodynamic loads, was not possible for the Tornado.

No effective rudder control is available when reverse thrust is active and the steering system is therefore the only active source of directional control available, although MBB concluded that pilot reaction was too slow to form part of a control loop which could counteract the aggressive yaw movements experienced. It was decided to connect the steering system with the yaw gyro function of the CSAS (CSAS = Command Stability and Augmentation System) lateral computer to detect deviations from zero acceleration and compute an automatic steering correction to compensate for, and damp, the aircraft's yaw motion. This change was 'breadboarded', tried successfully on a pre-series aircraft, and incorporated in a redesigned electronic control box.

At the same time automatic system test facilities were incorporated to indicate proper functioning to the pilot or to indicate and revert the system to free castor in the event of a system malfunction. In this event the pilot must not use the thrust reverser when landing.

The automatic in-flight test steers the nose gear $\pm 5^\circ$ for error detection. Figure 17 illustrates the Tornado nose gear configuration. The general mechanical and hydraulic layout of the system is shown in Figure 18 and its positional gain characteristics are shown in Figure 19.

TORNADO STEERING SYSTEM

- o Steer-by-wire system — Electrical Control / Hydraulic Powered
- o System pressure = 4000 psi
- o Electrical supply = 28V/DC
- o Two steering modes (Non linear)
  Low $\pm 30^\circ$ for TO
  High $\pm 60^\circ$ for taxi and speed operations
- o Steering speed 20° per second under static load.
- o Control box connected to the lateral FCS computer in order to automatically control the aircraft on the ground if excessive yaw rates are experienced.
- o In-flight automatic $\pm 5^\circ$ steering test after nose gear extension.
- o Failure detection system with automatic reversion to free castor. No single failure to cause hardover steering.
- o BITE facility for ground checks.
- o Position monitoring to prevent retraction with non-centred nose gear.
- o Duplexed electrical control circuit.
- o Nose wheel centring after TO if nose landing

gear and one main landing gear 'weight off' indicated and uplock engaged.

- Automatic engagement system on NW touch-down with 1,5 second fade in (low gear).
- 360° free castoring without disconnection when steering is de-energized.
- Shimmy damping when steering is de-energized
- Automatic low mode selection after touchdown with automatic yaw control.
- Alternative manual mode selection by the pilot.

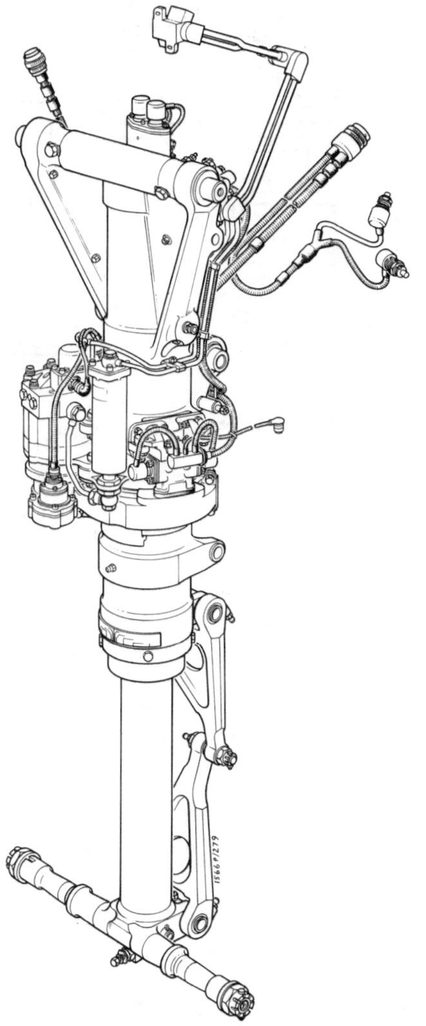

TORNADO NOSE GEAR

FIG 17

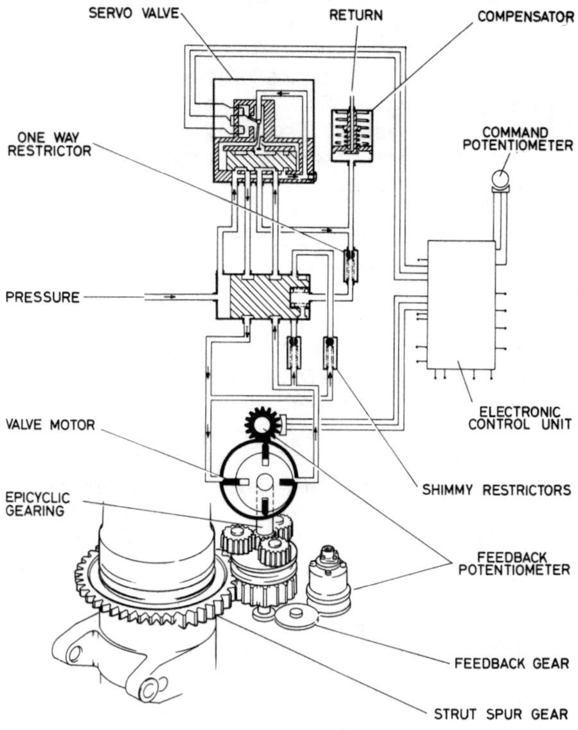

TORNADO STEERING SCHEMATIC

FIG.18

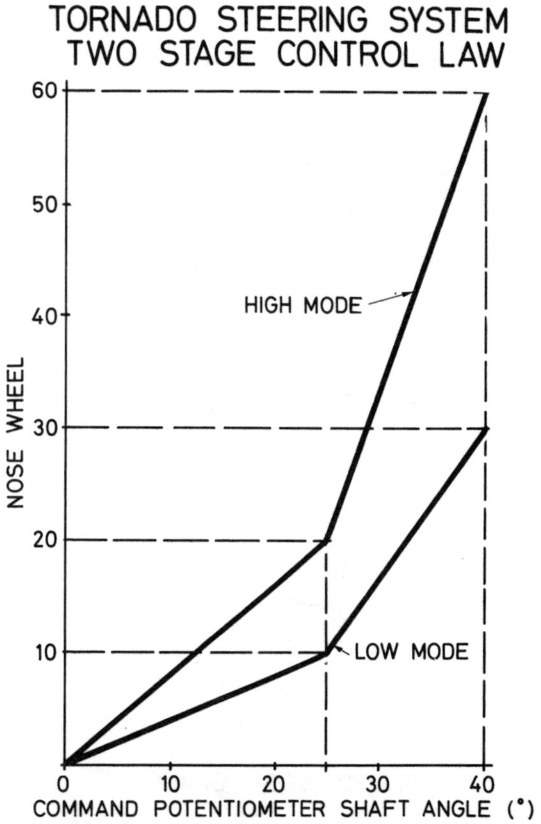

FIG 19

A300 AND A310 - The steering for these aircraft is virtually identical and is of a dual rack and pinion type with a steering range of $\pm 65^{\circ}$. Control is hydro-mechanical with either full range from the pilot's hand-wheel or $\pm 6^{\circ}$ by use of the rudder pedals which are disconnected from the steering input mechanism automatically when the nose gear is retracted.

The cantilever nose gear has a dual rack and pinion steering mechanism with four cylinders as shown in Figure 20. Two cylinders are supplied from the green system and provide the normal means of steering. In the event of a system problem, a completely separate control valve is supplied from the yellow system as shown in Figure 21. This valve supplies a completely separate pair of steering cylinders to actuate the rack and pinion. The steering valve incorporates shimmy damping restrictors and a compensating accumulator provides a return line back pressure to prevent hydraulic cavitation in the sytem.

Both steering valves are activated by a rocker beam which is controlled by differential tension in the steering control cables, which also form the feedback as they connect with a pulley on the nose gear.

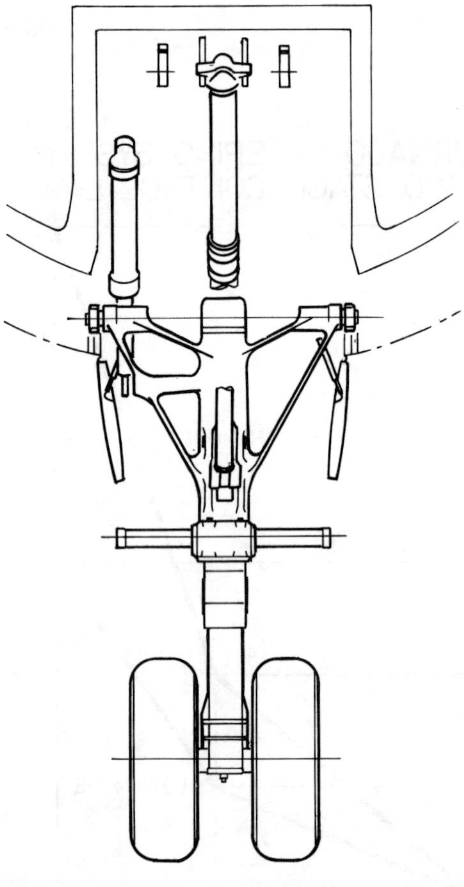

A300/310 NOSE GEAR

FIG 20

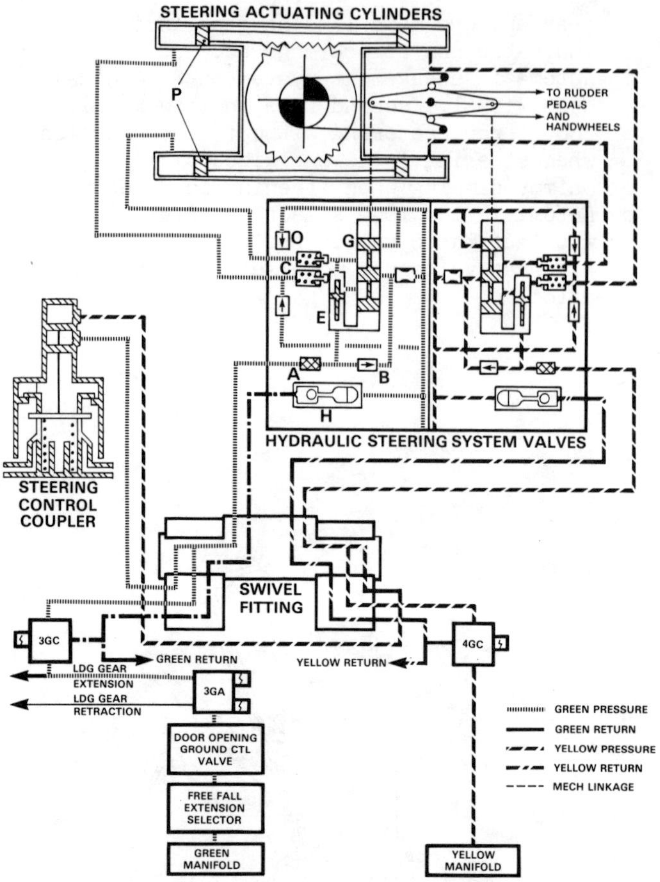

A300/310 STEERING SCHEMATIC

FIG 21

BAe 146 - This aircraft's cantilever nose gear has a rack and pinion hydro-mechanical system with cable input,chosen for overall simplicity and reliability. The cable is routed through the landing gear retraction trunnions and connects directly to an input quadrant on the steering valve differential mechanism. Steering by hand-wheel control is $\pm 70^{\circ}$ with $\pm 180^{\circ}$ available without manual disconnection for ground manoeuvring by differential braking or towing. A rudder bar controlled 'fine' mode of $\pm 7^{\circ}$ is available as a customer option.

The rack and pinion steering motor, shown in principle in Figure 22, disengages automatically past the $\pm 70^{\circ}$ normal steering angle to permit free castor. Re-engagement is only possible in the correct position as a baulk ring prevents inward motion of the rack under the influence of return line pressure until the appropriate gear teeth are in line.

The leg mounted steering system components are shown in Figure 23, no additional hydraulic units are needed.

To prevent application of steering pressure when the steering is being disconnected or is within the free castor range,a valve operated by a cam concentric with the steering collar shuts off the system pressure to the steering valve when the normal steering angle is exceeded.

Feedback to the steering valve is provided by an eccentric circular cam on the steering collar operating a spring loaded cam roller. The cam eccentricity gives a gain characteristic variable with the steering angle as shown in Figure 24.

Coulomb damping for shimmy suppression is provided by a disc spring loaded friction pack mounted below the steering motor which engages with the steering housing and the steering collar.

Similar steering systems are fitted to the Fokker F28, BAe ATP and SD3-60 aircraft nose gears. These are designed to suit individual installation and system requirements with variations with respect to the need for coulomb friction and design of feedback mechanisms. Some incorporate electrical switches to shut off hydraulic supply at the extremes of steering angle. On the BAe 146 hydraulic pressure is available for steering at all times.

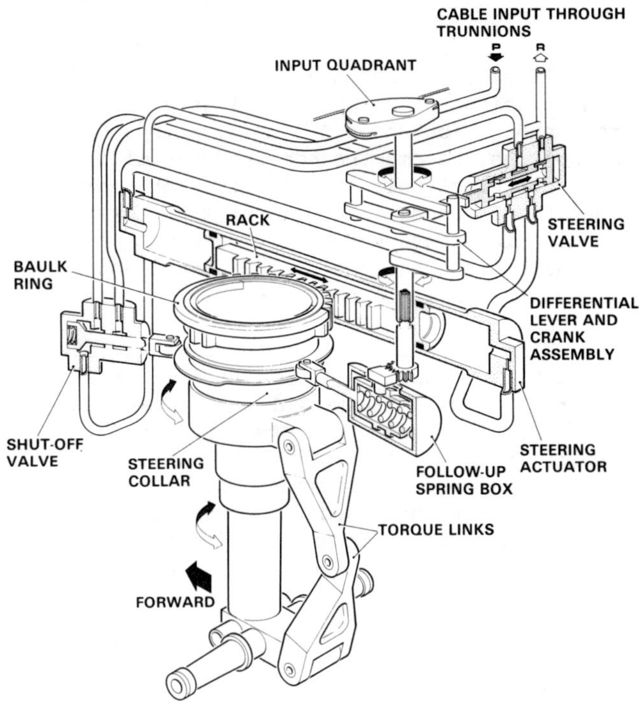

BAe 146 STEERING SYSTEM

FIG 23

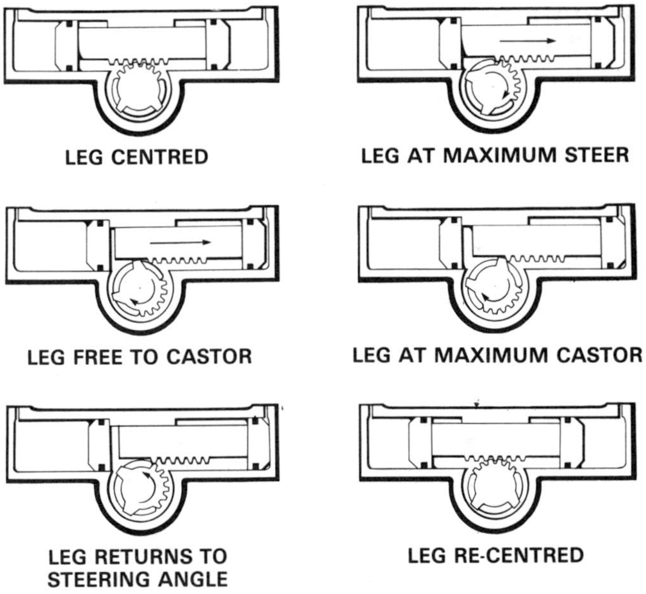

AUTO-DISCONNECTING RACK AND PINION

FIG 22

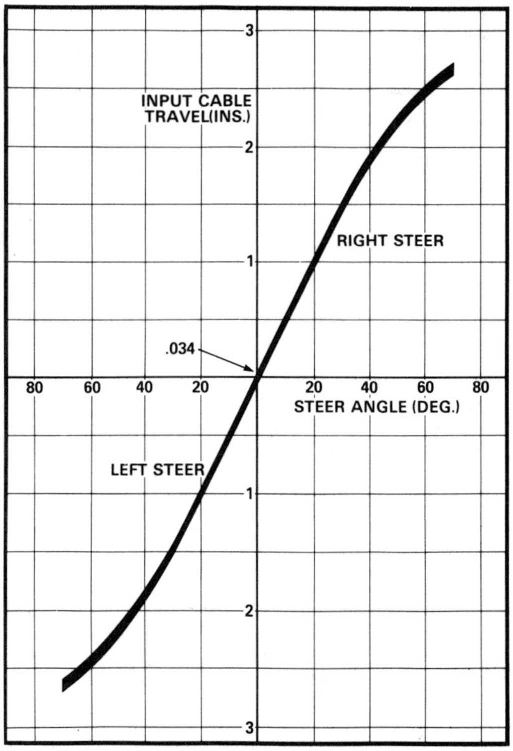

BAe 146 STEERING GAIN CHARACTERISTICS

FIG 24

DESIGNS 1980 -

This decade has seen consolidation of the techniques developed previously. In Europe rack and pinion actuators have, with a few exceptions, become the norm and for military aircraft electronic steering is nearly always specified. There is a general move towards electronic steering for civil aircraft and, to quote extremes, the GP-180 Avanti and the Airbus A320 will both have electronic steering control.

Figure 25 illustrates the simple system which will be used on the 'Avanti' aircraft. In order to minimise weight and volume and to eliminate some connectors, the control amplifier is combined with the command potentiometer. Dual command and feedback potentiometers are used to provide a non-linear characteristic and provide runaway failure protection. Each channel drives one half of the servo valve and the centre tap voltage deviation from zero is monitored to shut off the fluid to the servo valve and revert to the castor mode should an excessive error occur.

A 'Banana' linkage actuation system has been designed to suit this particular installation and the linear feedback potentiometer is located within the steering actuator.

This type of control system will see increasing use on relatively small aircraft due to the weight, space and cost reductions achieved when considering the total steering system. The compatibility with rudder pedal control, because no additional pedal loads are involved and the absence of direct mechanical connection to the steering valve, are additional favourable factors.

Despite these advantages, many aircraft intended for use from remote airfields with minimal servicing facilities will still require mechanically commanded steering systems.

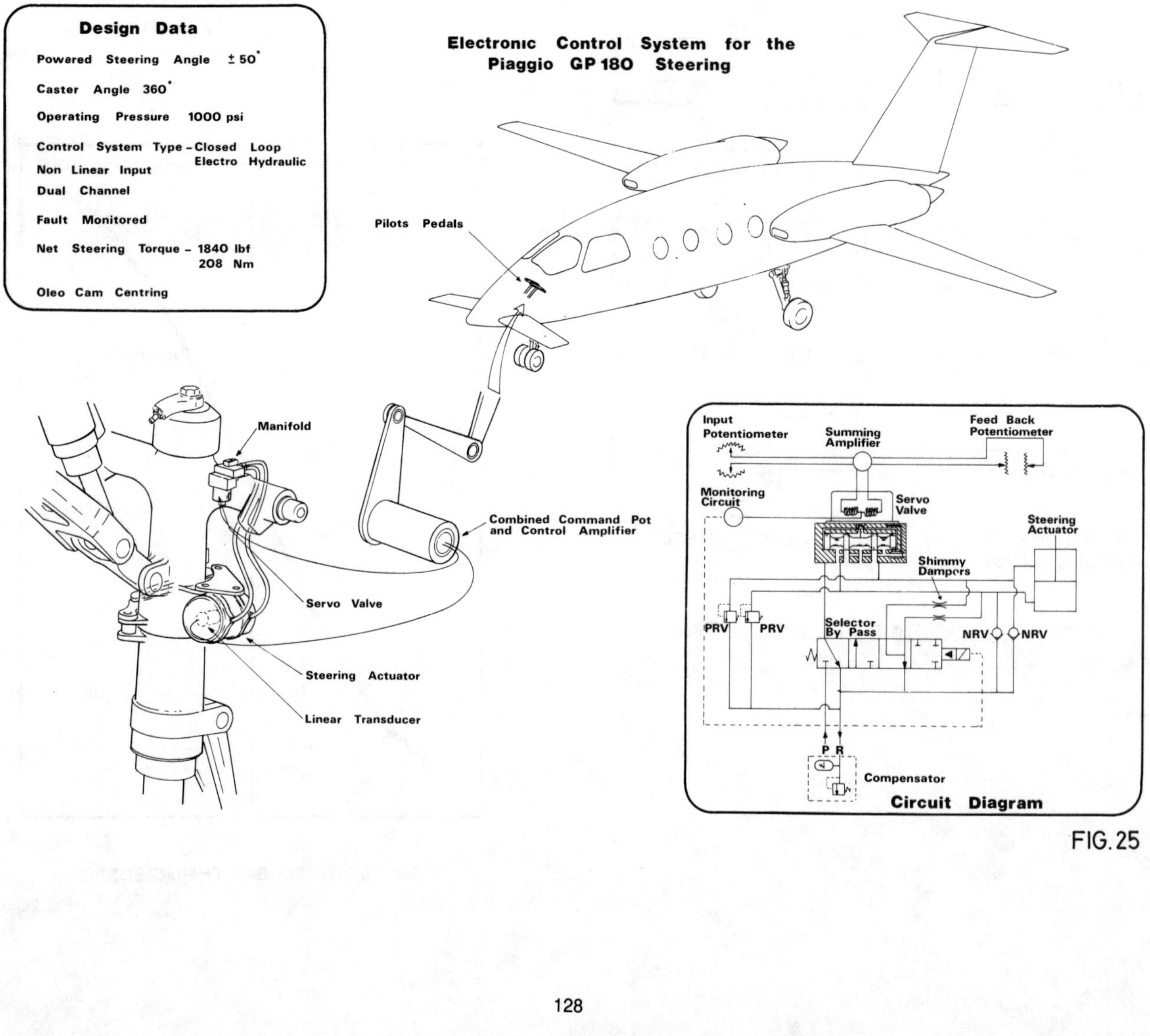

FIG.25

There is little doubt that, properly designed, this type of control provides maximum possible reliability with ease of trouble shooting and rectification by ground engineers, since seals can be changed and mechanical problems can be dealt with without, generally speaking, needing replacement units. The repair of electronic control modules needs to be carried out by specialists using facilities which will not be available with the smaller operators.

For military aircraft, where there is a tendency to centralize flight control and service management computing functions, comprehensive control of the steering system by the utility systems management computer, and the flight control management system computer will, in the future, be realized economically. This will allow inputs from various sources to be evaluated to determine if the pilot's command input requires modifying in order to achieve control harmonization for optimum stability under all aircraft control and aerodynamic configuration conditions.

Inputs from the flight control gyros and the antiskid system may be utilized, the antiskid input may provide ground speed measurements which can be used to modify steering gain progressively to suit landings, taxi and manoeuvring speeds.

The antiskid system may also provide an indication of the amount of differential braking used and it would be possible to use this information to control the steering to provide neutral steering characteristics.

The block diagram in Figure 26 indicates the degree of integration with the central computers which will be achieved on the UK EAP aircraft. On this aircraft, apart from command and input potentiometers, the only other units involved which are dedicated solely to the steering function are the steering valve module and the steering actuator.

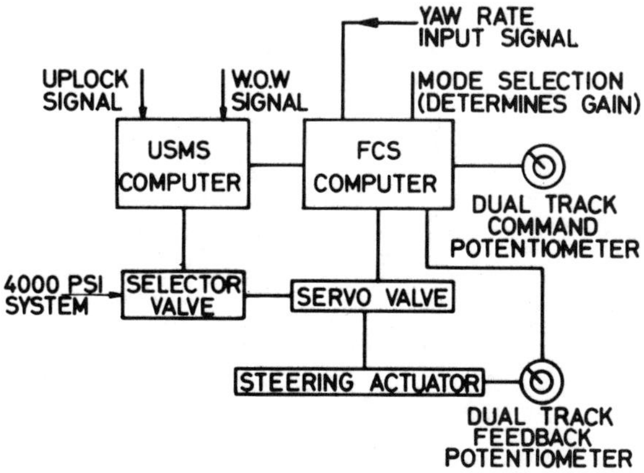

E.A.P. STEERING SYSTEM BLOCK DIAGRAM

FIG. 26

The use of potentiometers on this system complicates matters since analogue to digital convertors are needed for compatibility with the computer. For the future, it is envisaged that absolute value digital transducers will be used for command and feedback functions, which will also allow fibre-optic signal transmission rather than hardwiring to improve electro-magnetic compatibility of the steering system.

ACKNOWLEDGEMENTS

The authors wish to thank their colleagues at Dowty Rotol Ltd, Gloucester, England and MBB, Munich for the help given in preparing the paper, and their respective Directors for permission to publish and present it.

Thanks are also given to Messier Hispano Bugatti of Paris for providing information for inclusion in this paper.

APPENDIX 1

STEERING SYSTEM DATA

Aircraft	Max TO WT - lbs	Max Steer Angle °	Castor Angle	Gross Steering Torque lbs ins	Centring	Actuator Type	Input Type	Means of Shimmy Suppression	No of Tires
Alpha Jet	15,430	± 50 °	360 ° (1)	12,880	Mech Cams	Rack & Pinion	Elct	Hydraulic	1
Argosy	90,000	± 55 °	360 ° (2)	54,300	Actuator	Moving Cylinder	Elct/ Mech	Coupled Wheels	2
ATR 42	34,725	± 60 °	± 95 °		Oleo Cams	Rack & Pinion	Mech	Hydraulic	2
Avanti (GP 180)	10,200	± 50 °	360 ° (1)	2,000	Oleo Cams	'Banana' Linkage	Elct	Hydraulic	2
BAe 146-200	88,250	± 70 ° or ± 7 °	± 180 ° (3)	38,170	Oleo Cams	Rack & Pinion	Mech	Coulomb Friction & Hydraulic	2
BAe 748	44,495	± 45 °	360 ° (1)	28,400	Oleo Cams	'Banana' Linkage	Mech	Coupled Wheels	2
BAe ATP	49,500	± 47 °	± 180 ° (3)	26,150	Oleo Cams	Rack & Pinion	Mech	Hydraulic	2
Buccaneer	59,350	± 45 °	360 ° (2)	14,300	Steer to Centre	Oscillating Cylinder	Elct/ Mech	Coulomb Friction	1
EAP	44,000	± 45 °	± 120 ° (3)	12,300	Oleo Cams	Rack & Pinion	Elct	Hydraulic	1
F50	45,900	± 73 °	± 138 ° (3)	21,680	Steer to Centre	Rack & Pinion	Mech	Coulomb Friction	2
F100	95,000	± 76 °	± 135 ° (3)	37,770	Steer to Centre	Rack & Pinion	Mech	Coulomb Friction	2
Harrier	21,810	± 45 °	± 179 ° (3)	66,000	Steer to Centre	Rack & Pinion	Mech	Coulomb Friction	1
Jaguar	33,070	± 55 °	360 ° (1)	22,850	Actuator	Rack & Pinion	Mech	Hydraulic	1
SD 3-60	26,450	± 55 °	± 180 ° (3)	26,850	Oleo Cams	Rack & Pinion	Mech	Hydraulic	1
Tornado	58,546	± 60 ° or ± 30 °	360 °	16,200	Steer to Centre	Rotary Vane & Epicycle Gearing	Elct	Hydraulic	2
Vulcan	210,000	± 47 °	± 47 °		Actuator	Oscillating Cylinder	Mech	Coulomb Friction & Hydraulic	2

(1) = With Manual Disconnection

(2) = Without Disconnection

(3) = Automatic Disconnection

KEY: Mech = Mechanical
Elct = Electronic
Elct/Mech = Electro-mechanical

Integrated Braking and Ground Directional Control for Tactical Aircraft

Kevin L. Smith and Calvin L. Dyer
Air Force Wright Aeronautical Laboratories
Steven M. Warren
Boeing Military Airplane Co.

* Paper 851941 presented at the Aerospace Technology Conference and Exposition, Long Beach, California, October, 1985.

ABSTRACT

An autonomous ground handling system is being developed that integrates and automatically controls selected aircraft functions, such as the rudder, brakes, and nose wheel steering, to prevent Air Force aircraft from veering off runways during adverse weather operations. The system uses only aircraft-generated signals and pilot inputs to control the lateral forces on the aircraft and keep it on the runway centerline. Piloted simulations of an F-4 equipped with the system indicate that it can extend the crosswind landing capability of the aircraft into a region designated "unsafe" by the flight manual.

THE AIR FORCE is currently conducting research concerning the problem of stopping and maintaining directional control of fighter aircraft during high crosswind, low runway friction conditions. This research was initiated as a result of numerous runway departure accidents directly attributable to adverse weather, and has led to the development of an integrated system which automatically controls the brakes, rudder, and nose gear steering during takeoff and landing. To date, all of the work has been performed by the Boeing Military Airplane Company under three contracts to the Air Force Wright Aeronautical Laboratories of the Aeronautical Systems Division. The first contract was a study effort to determine the feasibility of an Advanced Brake Control System (ABCS) designed to prevent runway departure accidents. During this program, several control concepts were defined, developed, and briefly evaluated. These control concepts were later refined and optimized for the F-4 aircraft during the second contract. Piloted simulations flown in the Air Force's Large Amplitude Multimode Aerospace Research Simulator (LAMARS) established the adverse weather control limits of an unassisted pilot and provided a baseline for comparing the adverse weather capability of an ABCS controlled aircraft. Under the third contract, a digital controller will be designed, fabricated, and simulator tested. Integrating the controller with current aircraft subsystems will also be investigated.

This paper will focus on the refinement, optimization, and simulation of the Advanced Brake Control System. The development method and tools will be presented along with the results from the piloted simulations.

THE PROBLEM

The problem a pilot faces when landing during adverse weather (defined here as the combination of high crosswind and wet or icy runways) is decelerating the aircraft while preventing it from skidding off the side of the runway. The deceleration problem has been solved by the development of modern aircraft brake systems with autobrake control that automatically apply the brakes after touchdown and antiskid control constantly adjusts the brake pressure to maximize the braking force and prevent excessive wheel lock-up. The task of directional control, however, has generally been left up to the pilot.

Maintaining proper directional control requires that the pilot recognize any deviation and promptly take corrective action. While most directional control problems occur during landing, a refused takeoff (RTO) can be just as difficult because of the high aircraft weight and speed associated with takeoff conditions. Directional control during a normal takeoff is somewhat easier due to the availability of engine thrust to counteract the crosswind. Due to the high speed usually involved in a landing or RTO situation, a slight delay in recognizing or correcting the problem can put the pilot in an unrecoverable situation. Even if he recognizes a problem in time, the pilot's corrective action can often make the situation worse.

High speed, however, is not the only time that ground control problems can occur. In fact, problems can occur at practically any speed and at any time during the rollout. During the touchdown phase, alignment of the aircraft with the runway, from a crabbed or wing-low attitude, can introduce both yaw and lateral motions which can contribute to the control problem. As the ground speed decreases, rudder authority also decreases and control must be transitioned to the nose gear steering (unless the steering is a full-time system). When and how to make this transition can create confusion and result in improper control inputs. As the speed decreases further, the aerodynamic sideslip angle increases and the effect of the crosswind becomes greater. Coupled with a lack of rudder control and very low runway friction, this can also cause directional control problems even at relatively low speeds.

The current technique for reducing the potential occurrence of these problems is to restrict operations during adverse weather. During wartime, such restrictions could limit or delay the number of sorties that could be launched. Additionally, wartime sortie generation requires that bomb damaged runways be repaired as rapidly as possible. To reduce runway downtime, not all of the craters will be repaired. Instead, a minimum operating strip (MOS), possibly as small as 50 feet (15.2 meters) wide by 5000 feet (1524 meters) long, will be laid out on the damaged runway so that only a minimum number of craters will have to be repaired. Requiring aircraft to operate from such small strips can aggravate directional control in several ways. First of all, with only 50 feet of runway width, the pilot's margin for error will be greatly reduced. For example, the distance between the main landing gears on the F-4 is approximately 18 feet (5.5 meters). On a 50 foot wide runway, this leaves the pilot with a useable portion of only 32 feet (9.8 meters). Secondly, the MOS may not be aligned with the prevailing winds, requiring the pilot to operate in higher than normal crosswinds. Finally, the MOS may contain bomb-damage repair mats and other roughness which the pilot may not be accustomed to operating on.

To address these problems, the Air Force began studying the feasibility of an integrated braking and ground directional control system that would assist the pilot in successfully launching or recovering his aircraft during adverse weather.

EARLY INTEGRATED CONTROL CONCEPTS

During the Advanced Brake Control System feasibility study, two different integrated control concepts were proposed for solving the runway departure problem - an automatic system and a semiautomatic system. Both systems control the lateral aerodynamic and ground forces acting on the aircraft by controlling the aircraft's yaw angle. The ground track of the aircraft is then determined from motion sensors and compared with the desired ground track to generate a course error signal. Commands are then sent to the rudder, brakes, and nose gear steering to drive the course error to zero. The difference between the automatic and semiautomatic concepts is the way the desired ground track is obtained.

With the semiautomatic system, the desired direction of travel is continuously supplied by the pilot with the rudder pedals. At landing impact, the weight-on-wheels switch (squat switch) signals the ABCS to disconnect the rudder pedals from their normal function. The pilot then indicates his desired direction of travel by using the rudder pedals to aim at a point some distance down the runway (Fig. 1).

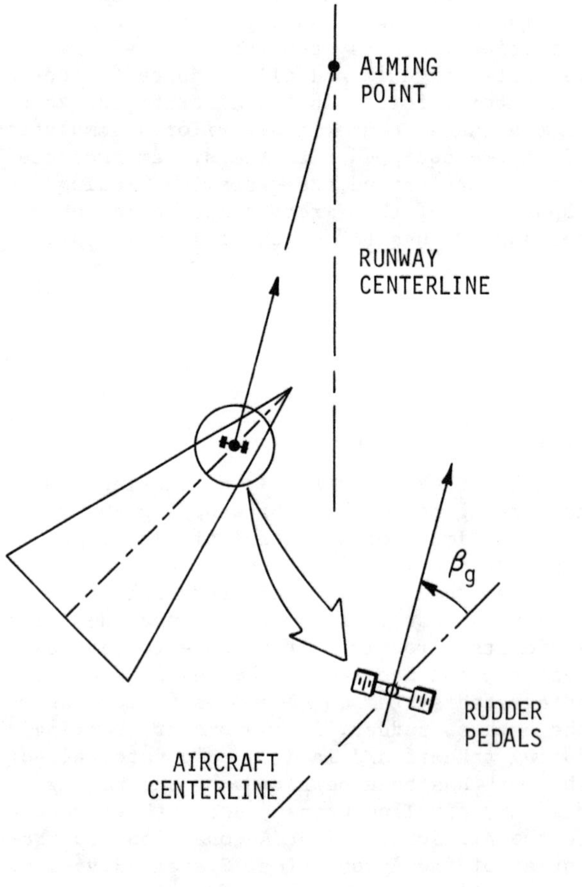

Fig. 1 - Semiautomatic ABCS

The control system uses this signal as input to generate rudder, steering, and braking commands to direct the aircraft to the aiming point. In addition to landing, this system could also be used for taxi or takeoff by simply indicating the desired direction of travel with the rudder pedals.

The fully automatic ABCS is essentially a "hands off" system. During final approach, the pilot initializes the ABCS by inputting the desired runway heading and anticipated friction condition through a cockpit panel (Fig. 2).

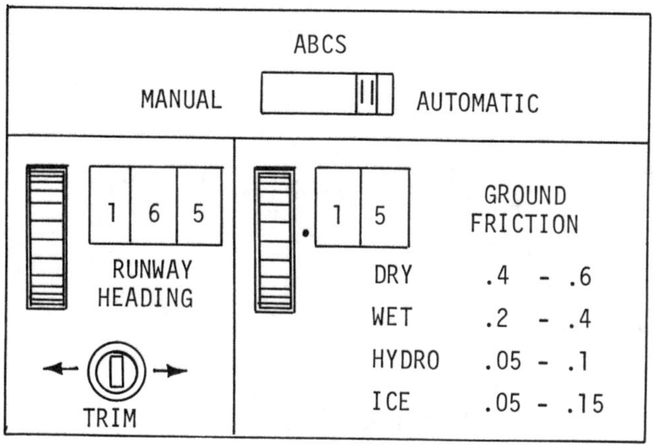

Fig. 2 - Automatic ABCS Cockpit Panel

His task is then to land the aircraft as close to the runway centerline as possible. An ABCS runway centerline, which passes through the touchdown point and is parallel to the pilot's input runway heading, provides the desired ground track. At landing impact, the squat switch signal engages the ABCS and the aircraft is then under automatic control throughout the landing rollout. Deviation of the aircraft from the desired ground track is then determined from the Inertial Navigation System (INS). Again, rudder, steering, and braking commands are generated to control the available lateral forces and keep the aircraft on the centerline. Should the pilot land off-center, a lateral trim function allows him to adjust the desired ground track and direct the aircraft back to the runway centerline. When the pilot is ready to turn off the runway, he disengages the ABCS and resumes manual control of the aircraft.

An analysis of the equations of motion showed that for both concepts, all of the required signals and control variables could be obtained from onboard aircraft instruments such as the INS, yaw rate gyro, and accelerometers. This was very desirable from an operational standpoint in that neither system would have to rely on signals from a beacon or microwave landing system which may not be available during wartime. This ability to use only aircraft-generated signals became a design requirement for all later ABCS work.

Once obtained, the necessary signals and control variables are processed by the three controller components shown in Figure 3.

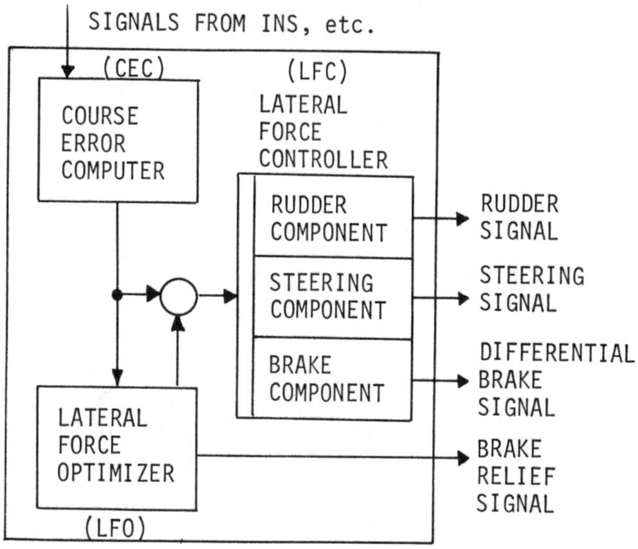

Fig. 3 - ABCS Components

The Course Error Computer (CEC) calculates a course error angle and lateral deviation using the aircraft signals previously mentioned. This tracking error information is then used by the three subcomponents of the Lateral Force Controller (LFC) to generate corrective rudder, steering, and differential braking commands. The Lateral Force Optimizer (LFO), considered an optional component, maximizes the available lateral forces during severe control conditions.

To evaluate the effectiveness of the controller, the F-4 was chosen as the study aircraft and a real-time, unpiloted, triple hybrid simulation was developed at Boeing's Hybrid Brake Control Laboratory. The simulation consisted of analog and digital computers as well as a complete mockup of the F-4 brake hydraulic system. Because of the highly nonlinear characteristics of both the hydraulic system and the antiskid control box, using actual aircraft hardware assured increased accuracy in predicting the aircraft's braking performance. The mathematical models incorporated into the simulation included the following: six degree-of-freedom rigid body dynamics; longitudinal and lateral aerodynamics; landing gear shock strut response; engine, rudder, and steering system response; brake torque; wheel dynamics; longitudinal, lateral, and normal tire/ground force prediction; and the ABCS control laws. Figure 4 shows the major components of the simulation. Because no pilot inputs were available, only the fully automatic ABCS was simulated. Normal pilot actions such as retarding the throttles and

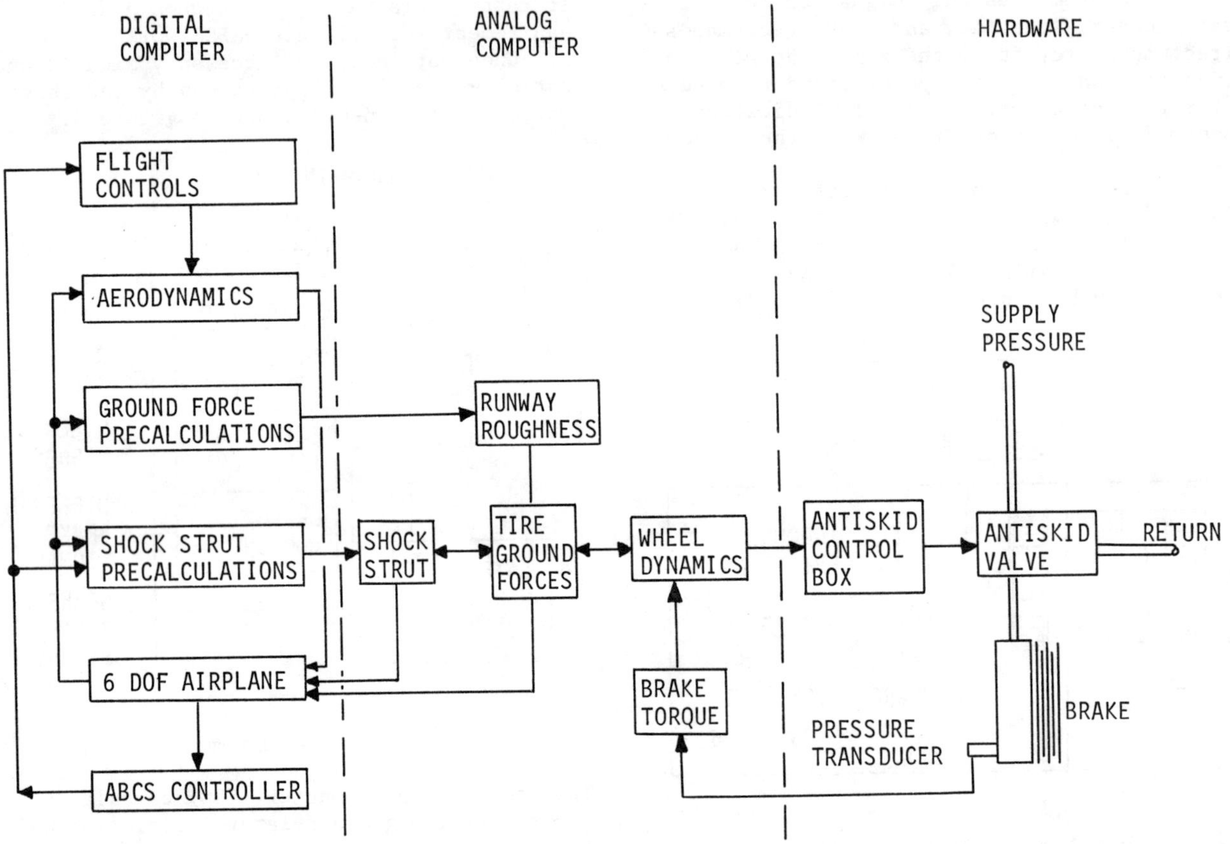

Fig. 4 - HYBCOL F-4 Simulation Schematic

applying the brakes were programmed as simple time dependant functions consistent with normal operating procedures. Also, incorporating the Lateral Force Optimizer routines and differential braking would have required extensive antiskid control box modification and retuning. Because of the limited scope of the initial feasibility study, these two components were not simulated.

The results of the initial simulations were very encouraging. With only a course error computer and a lateral force controller, the ABCS was able to safely control an F-4 under weather conditions considered to be "unsafe" by the flight manual. The limits in capability were determined, not by the ABCS, but by the antiskid system being used in the simulation (due to availability, a Mark II system was used rather than the higher capability Mark III system). At low aircraft speeds, generally below 40 feet per second (12.2 meters per second), the main gear wheels would lock up under low friction conditions and loss of directional control would occur. Overall, however, the simulation demonstrated that an Advanced Brake Control System was feasible and that such a system could potentially extend the F-4's adverse weather landing capability.

SYSTEM REFINEMENT AND OPTIMIZATION

Recognizing the limitations of the results obtained in the feasibility study, the Air Force initiated the Advanced Brake Control System Development Program. The objectives of this effort were to refine the ABCS control concepts and optimize them for the F-4, incorporate the Lateral Force Optimizer and differential braking concepts into the hybrid simulation and evaluate their effectiveness, and conduct piloted adverse weather simulations of an F-4 equipped with the "best" ABCS configuration.

Refinement and optimization of the control concepts were performed with the hybrid F-4 simulation developed during the feasibility study. First, however, several modifications were made to the simulation to improve its accuracy and expand its capability. The aerodynamics model was updated to more accurately predict the effects of ground proximity, large aerodynamic sideslip angles, large rudder deflection, and positive (trailing edge down) stabilator deflections. Rudder and steering system response data were obtained and included in the rudder and steering models. To permit simulated operation over rough runway surfaces, a terrain model was added (multiple bomb-damage repair mats and long wavelength bumps and dips) and the landing gear model was changed to include the effects of unsprung mass response and tire deflection. A simplified drag chute model was added to determine the effect of chute deployment on ABCS control. In addition to landing capability, the simulation was upgraded to include takeoffs and RTOs. The necessary pilot actions during an RTO were programmed as a function of aircraft velocity and time to represent recommended flight manual procedures. Finally, a new brake hydraulic system mockup was constructed (Fig. 5) incorporating a Hydro-Aire Mark III antiskid control box.

Fig. 5 - F-4 Brake Mockup

The first major task during the ABCS refinement effort was to refine the original ABCS concepts with the new Mark III antiskid control system included. This task consisted of retuning the existing components and adding several new components to improve the overall directional control. A wash-in filter (function EASY-IN) was added to gradually ramp on the steering and rudder commands after touchdown. This allowed the aircraft to decrab and stabilize prior to ABCS control, and made the transition from pilot to automatic control much smoother. Initial simulations also indicated that misalignment of the nose gear tire with the ground track at nose gear touchdown and a nose gear steering rate limit of 12 degrees/second were causing degraded directional control. The steering rate limit was gradually increased and a minimum rate of 40 degrees/second was found to provide adequate control. This increased rate was used for all subsequent ABCS simulations.

Retuning of the controller components was done at both high and low speeds because of the large variation in rudder and steering effectiveness over the speed range from touchdown to taxi. The rudder is most effective at high speed (touchdown to approximately 70 knots (130 kph)) and it was tuned, with the nose gear commanded to the castor angle, by varying the system gains until the best performance was obtained. High speed tuning of the steering system was accomplished with the rudder system active because of the steering system's inability to adequately control the aircraft at high speed with no help from the rudder. At low speed, however, rudder effectiveness decreases and directional control is provided by the nose gear steering. Therefore, low speed tuning of the steering was performed with the rudder controller inactive. A recheck of the high speed rudder and steering control following low speed tuning revealed that, while the addition of steering improved high speed control, the combination of rudder and steering produced a two Hertz oscillation in the aircraft's yaw rate. Since yaw rate was being used as one of the primary control variables, a filter was added to the yaw rate signal which eliminated the oscillation.

To allow for off-center landings, a lateral trim function was developed and tested. When the ABCS is activated at touchdown, it assumes that the pilot has landed the aircraft on the runway centerline. It uses this actual lateral touchdown point and the desired runway heading to define a reference runway centerline. Any lateral deviation of the aircraft from this reference centerline causes a control command to be generated so that the deviation will be reduced. Should the pilot land 10 feet (3.1 meters) to the right of the actual runway centerline, the ABCS would command the aircraft along a path that was parallel to, but 10 feet to the right of, the actual runway centerline. The lateral trim function corrects this situation by redefining the reference centerline based on input from the pilot. The aircraft trim switch, a self-centering thumb switch located on the control stick, was chosen for this purpose rather than adding a panel-mounted switch as shown in Figure 2. Each time the switch is moved right or left, the reference runway centerline is shifted 5 feet (1.5 meters) to the right or left. With the lateral trim function, if the pilot lands 10 feet to the right of the centerline, he can press the trim switch twice to the left and the ABCS reference centerline will be shifted to correspond to the actual centerline. In simulations, the aircraft responded to the lateral trim commands in approximately 15 seconds.

The second major refinement task was developing and integrating differential braking into the ABCS and evaluating its effectiveness. An early design requirement imposed on the ABCS was that only the pilot should be able to apply brake pressure. Any braking commands generated by the ABCS for directional control purposes should only be able to reduce brake pressure. For this reason, ABCS-commanded differential braking would have to be achieved by relieving the pilot-commanded brake pressure on one side of the aircraft and leaving the other side unaffected. The F-4 brake system, however, contains a single antiskid valve that relieves brake pressure on both sides of the aircraft upon receiving a signal from the antiskid control box. To incorporate ABCS differential braking, a second antiskid valve was added to the mockup along with an additional electronic circuit. This allowed the ABCS to provide individual commands to each antiskid valve which would, in turn, generate individual right and left brake pressures. A servo-controlled pressure equalization line was also added between the left and right brake metering valves so that, with the ABCS engaged, the pilot would not be able to generate differential brake pressures.

A series of landing simulations established that differential braking for directional control would only be required during extremely severe weather conditions (i.e. runway friction coefficients of 0.05 and crosswinds between 60 and 80 feet/second (18 and 24 meters/second)). However, two problems became immediately apparent. The differential braking control logic developed during the feasibility study commanded continuous or alternating right and left brake pressure dumps. This constant loss of braking greatly increased stopping distance. Secondly, the conditions for which differential braking was required were all low friction. Under these conditions, the brake pressure required to lock up the wheels was approximately 300 pounds/square inch (2068 kPa). Hysteresis within the two-stage electro-hydraulic antiskid servo valve, designed for operation at 3000 psi (20.7 MPa), made modulating the brake pressure between 300 psi and the retractor spring pressure of 150-200 psi(1034-1379 kPa) virtually impossible. The only reasonable solution would have been to use differential braking in a full on/full off mode. Several other differential brake control logic concepts were investigated but none significantly improved directional control without degrading stopping performance.

The final major task in the ABCS refinement effort was the investigation of several Lateral Force Optimizer concepts and modification of the Mark III antiskid control box to accept the LFO commands. The function of the LFO is to modify the basic control commands so as to maximize the landing gear and rudder forces available for directional control. Since the basic ABCS could control the F-4 in most weather conditions, the objective of the LFO development effort was to improve the marginal directional control encountered during severe conditions (icy runways with high, gusty crosswinds).

Three types of LFO concepts were investigated during this effort. The first of these, Brake Relief, was based on the behavior of the braking system and the tire forces.

Both lateral and longitudinal tire forces are functions of tire slip which is the relative ratio between the actual angular wheel speed (ω) and the synchronous (free rolling) wheel speed (ω_0) as shown in Equation 1.

$$1.\quad \text{TIRE SLIP (S)} = \frac{\omega_0 - \omega}{\omega_0}$$

A free rolling, unbraked tire yields a slip value of zero ($\omega = \omega_0$) while a locked up, skidding tire ($\omega = 0$) yields a slip value of one. The variation of lateral (side) and longitudinal (braking) tire force as a function of tire slip is shown in Figure 6.

Modern antiskid systems attempt to maintain the maximum braking force by keeping the tire slip at or near the optimum value. In a desperate ground control situation, however, maintaining lateral control of the aircraft is much more important than achieving maximum deceleration. The problem is that with the antiskid operating at the optimum slip value, very little lateral force is available for directional control. To improve this situation, three Brake Relief configurations were investigated that reduce the braking force in favor of increased lateral force. Gear Force Brake Relief would monitor the side load on the main landing gear tires. During maximum braking, the peak available side load on the tire is relatively low and would be used as a threshold. When this threshold is approached, signaling a need for more side force capability, the braking would be reduced so that the side force could increase. When considering the mechanization of this concept, however, it became immediately obvious that accurately monitoring the tire side load would be virtually impossible under operational conditions. Therefore, the Gear Force Brake Relief concept was dropped from further consideration.

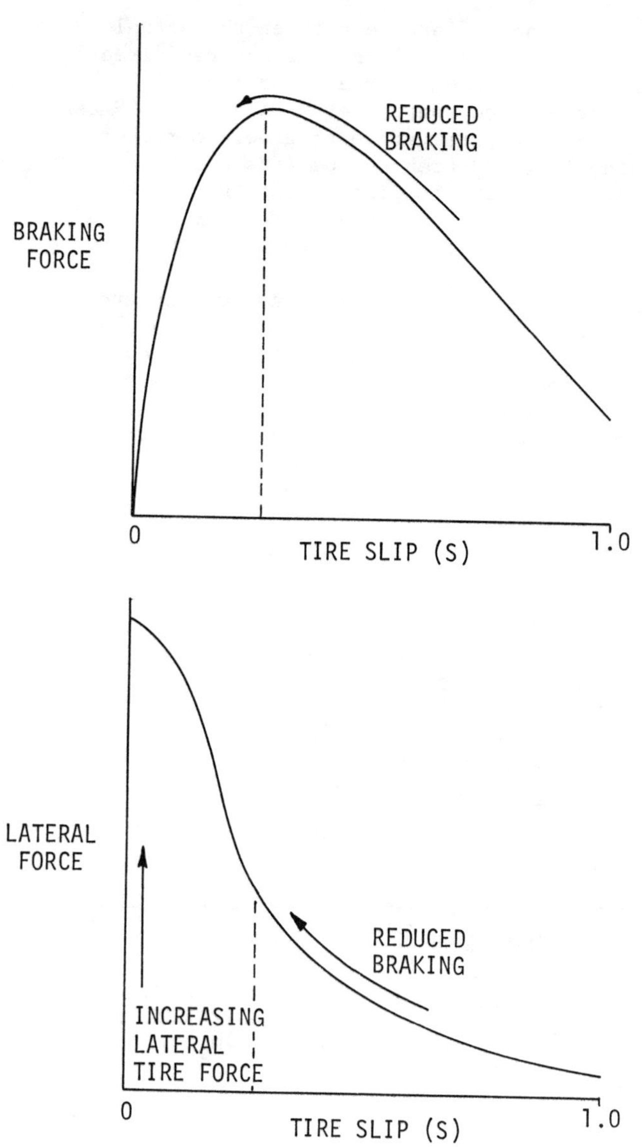

Fig. 6 - Tire Forces vs. Slip

In addition to tire slip, the lateral tire force is also a function of the slip angle (Figs. 7, 8). For a constant braking effort (constant tire slip) the lateral force increases with slip angle to its maximum value at some optimum slip angle. Any further increase in slip angle causes a reduction in the lateral force. A second brake relief concept, Slip Angle Brake Relief, would relieve braking whenever a certain slip angle threshold was exceeded. Simulation of this concept with the slip angle threshold set to 15 and 20 degrees showed that the yaw angle and lateral deviation were reduced slightly, but as a result of the reduced braking, the stopping performance was degraded significantly.

The final brake relief concept investigated, Velocity Brake Relief, was designed to

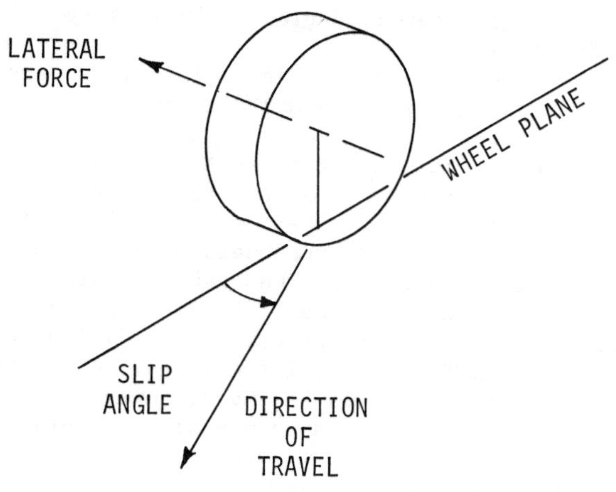

Fig. 7 - Tire Slip Angle

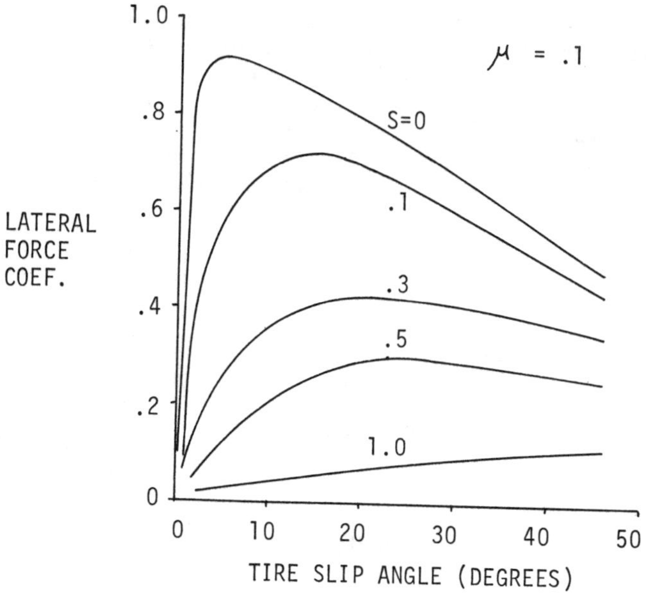

Fig. 8 - Lateral Force vs. Slip Angle

relieve brake pressure whenever the aircraft was accelerating away from the runway centerline. A lateral displacement threshold of ± 10 feet was used and the simulation results indicated that the concept could slightly reduce the yaw angle and lateral deviation during gusty winds and step friction conditions (wet/dry patches). During steady wind and friction conditions, however, the concept allowed the aircraft to oscillate back and forth across the threshold. As with the other brake relief concepts, Velocity Brake Relief significantly degraded the braking performance.

Before the brake relief simulations were actually conducted, the Mark III antiskid control box had to be modified to accept the brake relief command from the ABCS. The antiskid controller circuitry was studied and three modification concepts designed. The best of these was chosen and breadboard hardware was constructed and tested. The results of the tests indicated that the Mark III antiskid control box could be successfully modified to accept an ABCS brake relief command. When the brake relief logic was included, however, the F-4's stopping performance was significantly degraded. This was not a fault of the logic, but, as previously stated, was due to the inability to proportionally control the brake pressure on low friction surfaces.

The second type of LFO investigated was a Slip Angle Limiter. This concept was designed to optimize or limit the course error angle so that the peak lateral force available for directional control is not exceeded. Figure 9 shows an aircraft aligned with the runway but moving away from the centerline with a total velocity V_T at some ground slip angle β_g. However, the desired direction of travel (back toward the centerline), is shown as the desired velocity V_T' at the desired ground slip angle β_g'. The presence of a crosswind component V_w creates an aerodynamic sideslip angle β between the relative aerodynamic velocity V_a and the aircraft's longitudinal

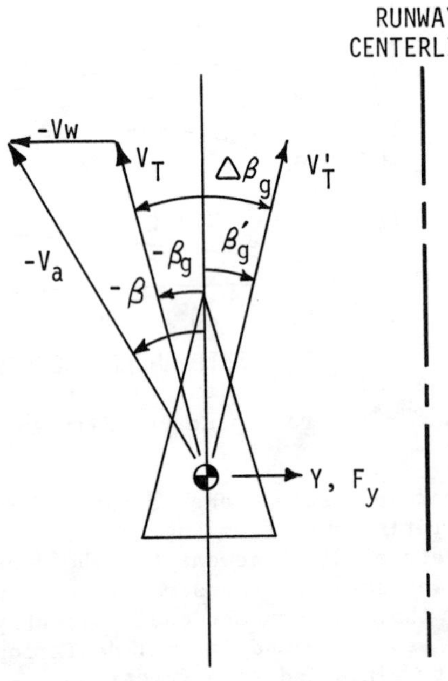

Fig. 9 - Angle Definition

axis. The difference between the actual direction of travel and the desired direction of travel is the course error angle $\Delta\beta_g$. The Lateral Force Controller within the ABCS uses the course error angle to generate rudder, steering, and braking commands which result in the aircraft traveling at some ground slip angle β_g. If the course error angle is large, the LFC would command an equally large ground slip angle to reduce the course error as rapidly as possible. In doing so, however, the ground slip angle β_g for peak available lateral force (Fig. 10) may be exceeded. The function of the Slip Angle Limiter is to determine when that point has been reached and modify the course error angle signal to prevent a further increase in β_g.

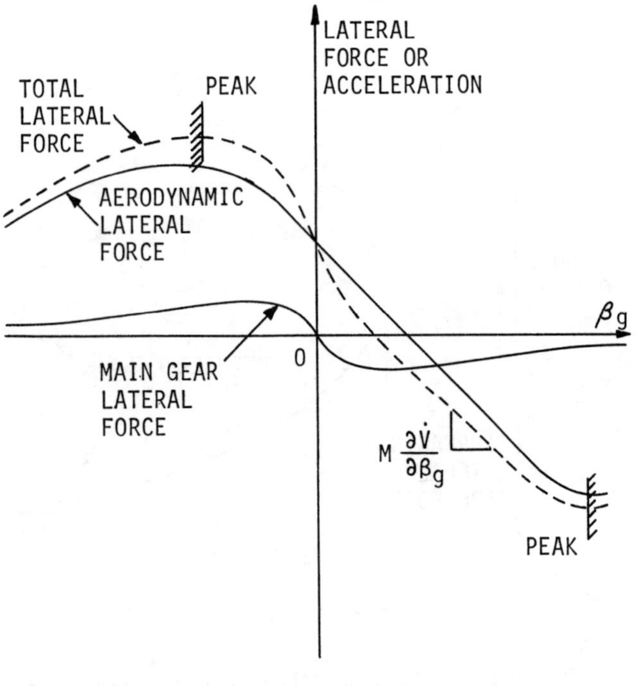

Fig. 10 - Slip Angle Limiter

To implement this concept, the total lateral force, which is a combination of the aerodynamic and tire forces, was replaced (Eq. 2) with the lateral acceleration ($\dot{V}$) which was available from the INS.

$$2. \quad \dot{V} = \frac{\Sigma F_y}{M}$$

This signal could then be differentiated with respect to the ground slip angle to locate the peak. When the concept was installed in the ABCS, however, it was discovered that extensive signal conditioning was required to attenuate the high frequency noise generated by differentiating V. The filters required to attenuate this noise reduced the Slip Angle Limiter signal to a level which made this LFO concept ineffective.

The final LFO concept investigated was a Steering Angle Limiter. The ABCS steering controller generates two steering angles which are summed to yield a single steering command. The first angle, the castor angle, commands the nose gear to steer into the skid which reduces the nose wheel slip angle to zero. This also reduces the lateral nose wheel force to zero. When lateral nose gear forces are needed for directional control, a second steering angle is calculated, summed with the first angle, and a nonzero steering slip angle results. The maximum lateral force that can be generated by the nose gear is a function of the steering slip angle and the peak runway friction coefficient as shown in Figure 11. The function of the Steering Angle Limiter was to maximize the nose gear lateral force by limiting the steering slip angle to the values along the Maximum Force curve. To simplify this non-linear curve for the simulations, a linear limit function was used which calculates the optimum steering angle as a function of peak runway friction coefficient.

The results from simulations including this concept showed that the Steering Angle Limiter was active for a brief time immediately after touchdown for severe weather conditions.

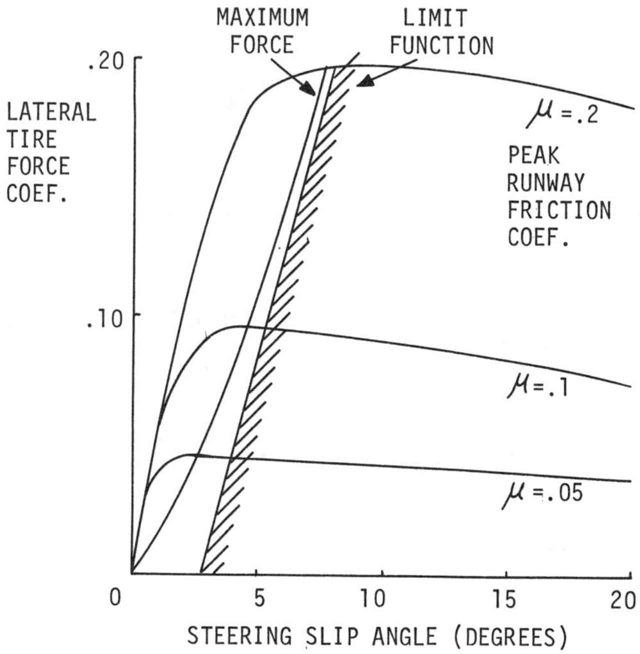

Fig. 11 - Steering Slip Angle Limiter

As previously mentioned, however, the nose gear steering is not very effective at the high speeds immediately following touchdown. Even though the limiter was active, the reduced steering effectiveness resulted in no overall improvement in directional control, so the concept was dropped. Deleting this LFO concept also eliminated the need for the runway friction switch on the ABCS cockpit panel.

BEST F-4 ABCS

To determine which ABCS components would yield the "best" system, each component's effect on the F-4's overall directional control and stopping capability were evaluated. To aid in this evaluation, a set of performance requirements were developed based on the objectives of the program. These requirements were:

1. The ABCS shall provide safe aircraft operation (takeoff and landing) for the following conditions:

a. Constant runway surface friction down to 0.05

b. Hydroplaning conditions,

c. Step surface friction (icy/wet/dry patches),

d. Constant crosswind up to 35 knots (65 kph), and

e. Gusty crosswind up to 35 knots;

2. The ABCS shall allow no more than ± 15 feet (4.6 meters) of lateral deviation and ± 40 degrees of ground slip (βg);

3. Surface roughness shall not negate normal ABCS performance;

4. The ABCS shall not significantly degrade aircraft stopping performance;

5. The ABCS shall control the aircraft in a stable and well behaved manner; and

6. The ABCS shall not adversely affect pilot workload.

With these requirements in mind, the best ABCS configuration for the F-4 contained the following components and features:

1. Initialization

2. Engage ABCS signal

3. Individual rudder and steering Course Error Computers

4. Lateral trim

5. Yaw rate filter

6. Individual rudder and steering Lateral Force Controllers

7. EASY-IN function

A schematic of these components is shown in Figure 12.

PILOTED SIMULATION

With the ABCS refined and optimized for the F-4, the next step in the ABCS program was to conduct piloted simulations to evaluate the improvement in directional control provided by the system under adverse weather conditions. This involved developing an F-4 ground handling simulation and conducting piloted takeoffs, landings, and refused takeoffs under a wide range of simulated adverse weather conditions with and without the ABCS.

Based on an industry-wide survey of aircraft simulators conducted during the feasibility study, the Air Force's LAMARS facility was chosen for conducting the ABCS piloted simulations. Several requirements which led to the choice of this simulator were: digital/analog computer capability, F-4 simulation experience, wide angle field-of-view visual capability, brake system hardware interface capability, and a fighter cockpit. Analog computer capability was required for real-time simulation of the high frequency response of the brake system and landing gear (over runway roughness). Wide angle field-of-view was required because of the large yaw angles that would be encountered during the adverse weather simulations. Brake system hardware interface capability would allow the F-4 hydraulic brake system mockup to be included for accurate modelling of the brake system response. Finally, a fighter cockpit would add to the realism and increase pilot acceptance of the simulation.

The LAMARS simulator (Fig. 13) is a five degree-of-freedom motion system carrying a single-place cockpit and ten-foot radius, spherical projection screen on the end of a 30 foot beam. The visual display used for the

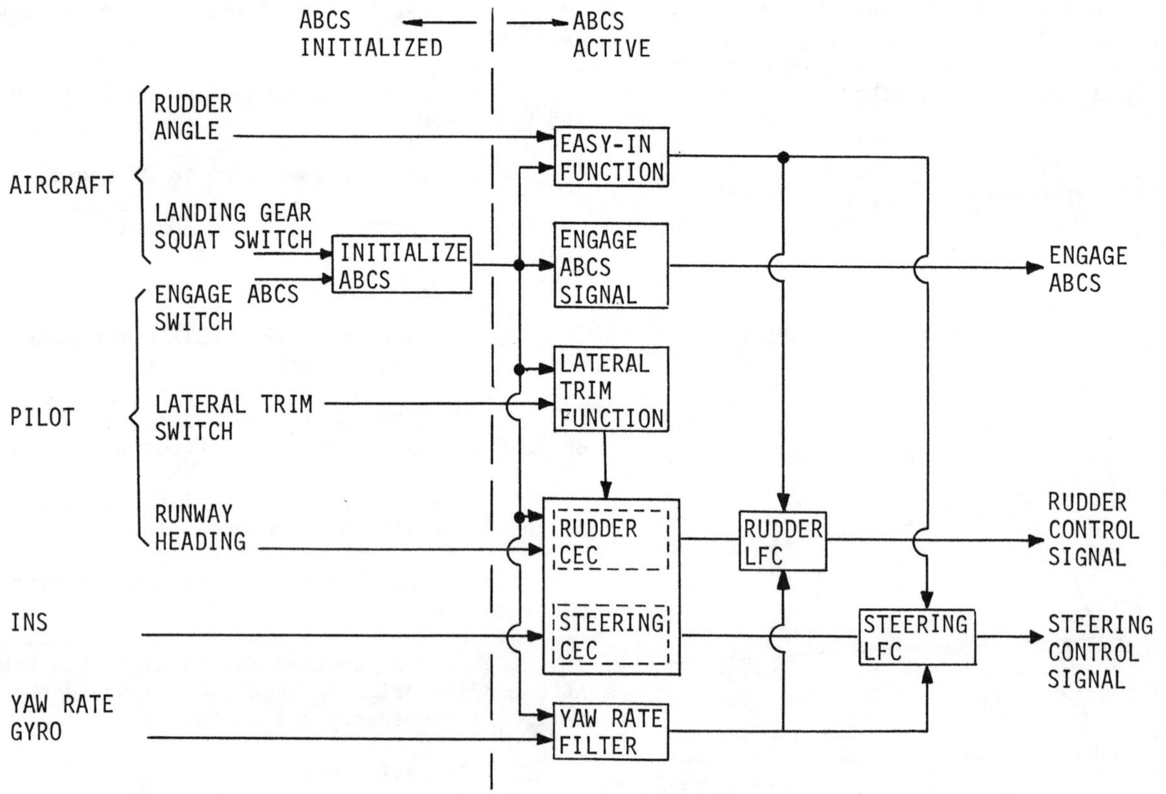

Fig. 12 - Best ABCS for the F-4

Fig. 13 - LAMARS

Fig. 14 - 1500:1 Terrain Board

ABCS simulations consisted of a projected image from a 1500:1 scale terrain board (Fig. 14). The generic fighter cockpit was equipped with a center stick and programmable feel system, and a simulated head-up display for providing the pilot with ground speed and sink rate information. To simulate both standard runway and MOS operations, two separate runways were required. The primary runway on the 1500:1 terrain board was used for standard runway simulations. As no 50 by 5000 foot runway was available on the terrain board, an MOS was simulated by changing the terrain board scale factor to 750:1 and using a 100 by 10000 foot taxiway. To lessen the impact of the scale change on the pilot, the MOS was approached from the opposite end of the terrain board.

Development of the F-4 simulation was a joint effort. The Air Force developed the basic F-4 simulation which included the rigid body equations, aerodynamics, flight controls, and cockpit configuration, and was responsible for operating the simulator during the testing. Boeing developed the landing gear, ground force, and ABCS models, and provided the brake system mockup and interface hardware. A schematic of the LAMARS F-4 simulation is shown in Figure 15.

The objectives of the testing were to establish the adverse weather limits of a pilot controlled F-4 and determine if these limits

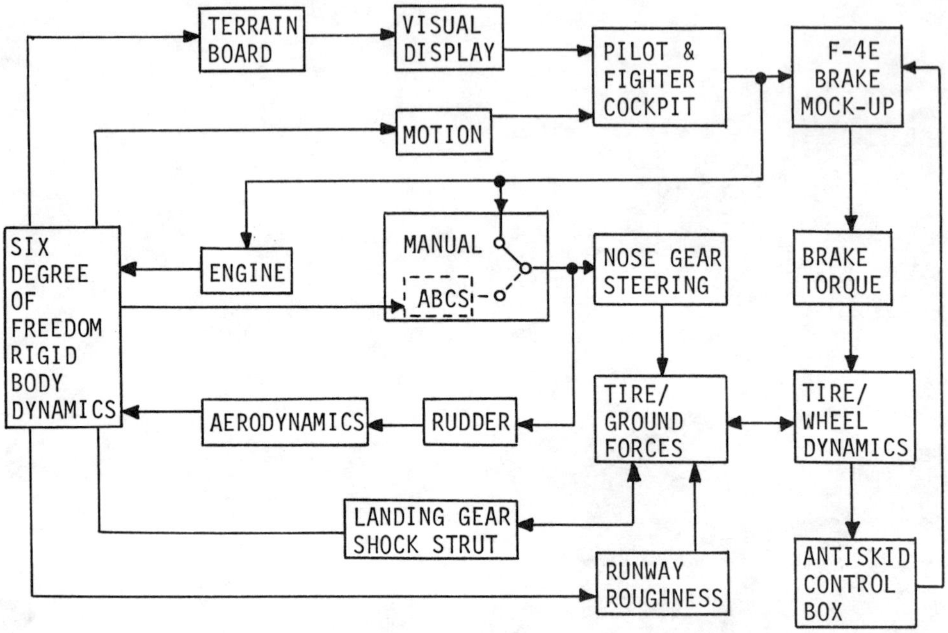

Fig. 15 - LAMARS F-4 Simulation Schematic

could be safely extended with the Advanced Brake Control System. To meet these objectives, the testing was divided into three phases: baseline, tuning, and ABCS. During Phase I, simulations were conducted to establish a baseline of the pilot's ability to control an F-4 under various weather and runway conditions. In Phase II, a smaller number of runs were made to ensure that the ABCS components were properly tuned. Finally, during Phase III, the adverse weather limits of an ABCS controlled F-4 were determined.

Three engineering test pilots, each with at least 500 hours in the F-4, were supplied by Boeing for the testing. Engineering test pilots were chosen because of their experience with simulator testing, their ability to adapt to differences between simulators and actual aircraft, and their understanding of aircraft control and dynamics. Numerous other Air Force and civilian pilots flew the F-4 ABCS simulation and provided valuable comments on both the simulation and the ABCS. However, the actual testing was done by the Boeing pilots only.

An extensive test matrix was developed that resulted in more than 1100 runs being conducted. During the baseline and ABCS phases, each pilot performed tests in the following five categories:

1. Landing Rollout, Standard Field, No Drag Chute;

2. Landing Rollout, Standard Field, Drag Chute Conditional;

3. Landing Rollout, Temporary Field, Drag Chute Conditional;

4. Takeoff, Temporary Field; and

5. Refused Takeoff, Standard Field, Drag Chute Conditional.

Within each category, steady crosswinds were varied from 0 to 80 feet/second (0 to 24.4 meters/second) and constant friction coefficient conditions were varied from 0.5 to 0.05. Gusty crosswinds from 40 to 60 feet/second (12.2 to 18.3 meters/second), icy patches (step friction change from 0.4 to 0.075), and hydroplaning (linearly varying friction from 0.4 to 0.05) were also simulated. Sinusoidal surface roughness (bumps and dips) was simulated in part of the Category 1 runs and bomb-damage repair roughness was included in Categories 3 and 4. The pilot's ability to control the F-4 under each of the above conditions was determined during the baseline tests. These conditions were then repeated with the ABCS controlling the aircraft. Only the results from Category 1 are presented here.

The landing simulations were begun with the aircraft trimmed at an altitude of approximately 390 feet (118.9 meters) at a distance of approximately 6400 feet (1950.7 meters) from the intended touchdown point. For the crosswind conditions, the aircraft was trimmed by yawing it into the wind. Crosswind gusts were introduced only after the aircraft was on the ground. The simulations were ended when the aircraft had decelerated to a speed of 30 knots (55.6 kph).

A great deal of time history and post-simulation data was obtained during the testing. However, of particular importance were the maximum peak-to-peak lateral deviation (SWING), the maximum yaw angle (PSIMAX), and the stopping distance. Comparison of these values for the pilot controlled F-4 versus the ABCS controlled F-4 gave an excellent indication of the ABCS's control capability.

Figure 16 is a comparison of the SWING (maximum peak-to-peak lateral excursion) versus crosswind velocity for a dry (mu = .5) runway.

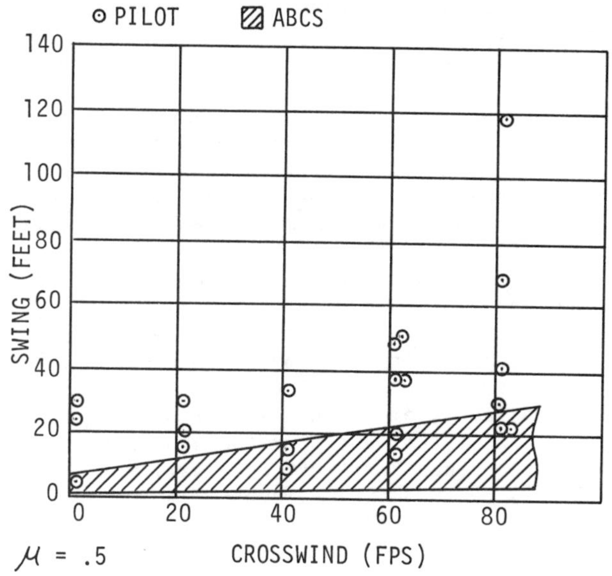

Fig. 16 - SWING vs. Crosswind

The data points are the maximum values obtained from individual simulation runs during the baseline (pilot controlled) tests. The values are widely scattered and increase to a lateral excursion of over 100 feet (30.5 meters) at a crosswind of 80 feet/second. The shaded area shows the maximum SWING obtained with an ABCS controlled F-4. When compared to the pilot's ability, the ABCS allowed a maximum SWING of only 30 feet (9.1 meters) with a crosswind of 80 feet/second. Figures 17 and 18 show similar comparisons for wet (mu = .3) and icy (mu = .15) runway conditions, respectively. Again, the shaded areas of ABCS performance show significant improvement in lateral control over an unassisted pilot.

The maximum yaw angle (PSIMAX) encountered during the landing simulations is shown in Figure 19 for an icy (mu = .15) runway. Again the data points represent the pilot's ability while the shaded area is the performance of the

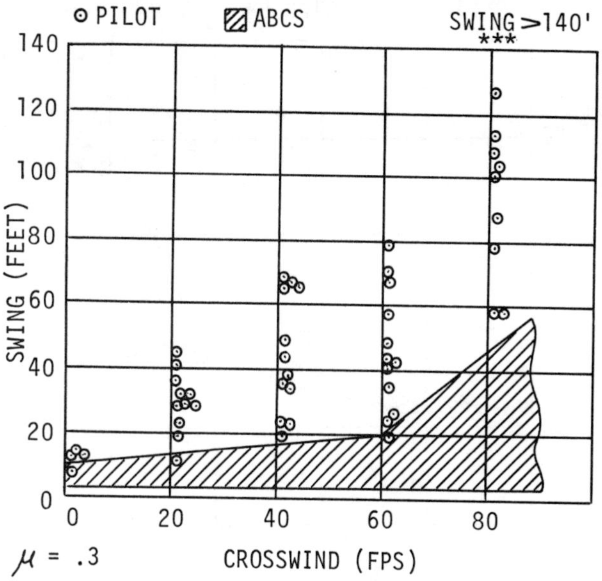

Fig. 17 - SWING vs. Crosswind

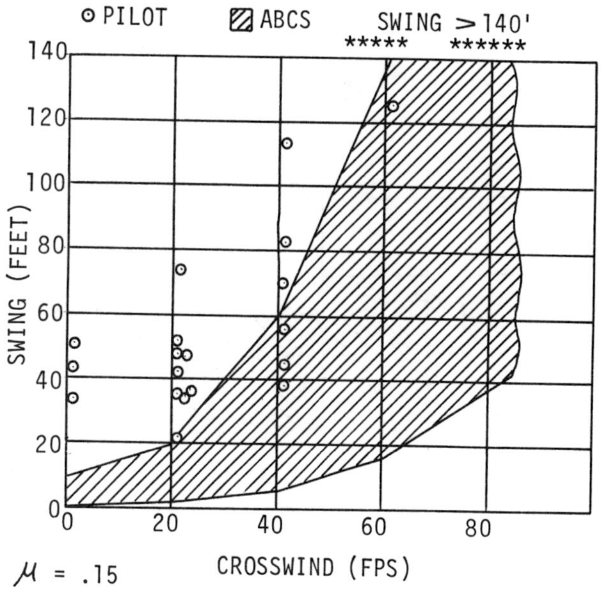

Fig. 18 - SWING vs. Crosswind

ABCS. As can be seen, the ABCS consistently controls the maximum yaw angle to within a relatively narrow margin. This demonstrated that the system was, in fact, well behaved.

Finally, the stopping distance versus crosswind for an icy (mu = .15) runway is shown in Figure 20. This data indicates that, since the performance for the piloted and ABCS controlled F-4 are nearly identical, the ABCS has very little effect on stopping performance.

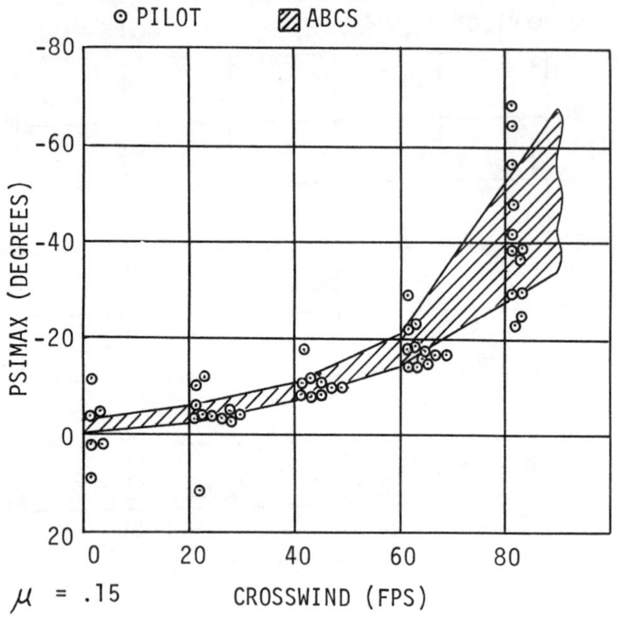

Fig. 19 - PSIMAX vs. Crosswind

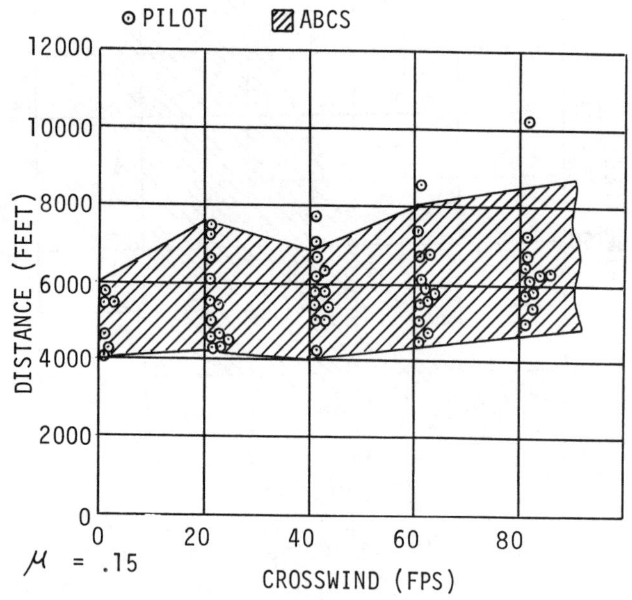

Fig. 20 - Stopping Distance vs. Crosswind

SIMULATION CONCLUSIONS

After reviewing all of the simulation data, several conclusions were drawn concerning the ABCS. The most important of these were:

1. The ABCS provides stable, well behaved aircraft control during landing rollout, takeoff roll, and refused takeoff operations;

2. The ABCS assisted F-4 performance is consistent and repeatable;

3. Stopping performance is not affected by the ABCS; and

4. The ABCS improves directional control for all crosswind and runway friction conditions.

CURRENT WORK

To answer several important questions generated by the Advanced Brake Control System Development Program, and to further develop the ABCS, the Air Force initiated the Integrated Aircraft Brake Control System (IABCS) program. The objectives of this effort, which is also being conducted by Boeing, are to design, fabricate, and laboratory demonstrate a reprogrammable, digital controller that can be used for ground directional control on several aircraft. Designing such a system will require a detailed investigation of how to interface the controller with current aircraft subsystems (rudder, steering, brakes, etc.) which can range in complexity from mechanical to electromechanical to electrical.

Several other important issues will also be addressed by this program. During the previous LAMARS F-4 simulations, the aircraft velocity and position information used by the ABCS to calculate control commands was obtained, not from an actual Inertial Navigation System, but from the simulator computers. Such "perfect" data will not be available on an actual aircraft. A specific objective of the IABCS program will be to determine the effect that INS error has on the controller's performance. Real aircraft also operate with systems that are subject to failure. Detecting the failures, determining the criticality of a particular failure, and deciding how to handle the failure will also receive heavy emphasis during this effort. This work is scheduled for completion in December, 1986.

Performance Testing on an Electrically Actuated Aircraft Braking System

Douglas D. Moseley
Loral Aircraft Braking Systems
Thomas J. Carter
Air Force Flight Dynamics Laboratory

* Paper 881399 presented at the Aerospace Technology Conference and Exposition, Anaheim, California, October, 1988.

ABSTRACT

The concept of utilizing an electrically actuated aircraft braking system could result in greater fire safety, the elimination of centralized hydraulics, and compatibility with an all-electric aircraft. Using the Air Force A-10 as a test bed, the first fully functional electric brake was laboratory tested, qualified, and installed on an aircraft for testing. On-aircraft testing was curtailed due to a dynamic instability between the brake and landing gear. An extensive laboratory dynamometer test program was substituted. The prototype electric brake demonstrated performance nearly equivalent to the production hydraulic brake with a potential for more accurate torque control.

BACKGROUND

For many years, people have been encouraged to "live better electrically." Someday, airplanes may be taking that advice too. "Fly-by-wire" or electronic controls have already taken over in controlling flight surfaces, engines, and other systems in some aircraft. The concept of an "All Electric Airplane" in which all hydraulic systems are replaced with electric components could result in weight savings and a more maintainable, efficient, and available aircraft; however, the successful development of this concept relies on the ability to produce subsystems capable of performing specific tasks using electric power.

Electrically actuated subsystems could also provide greater safety for the aircraft. During the years 1970-1975, the U.S. Air Force attributed a $20 million annual loss to hydraulic fluid fires, of which a substantial portion resulted from fluid leaks in the area of hot brakes. Elimination of hydraulics in the braking system would greatly enhance overall fire safety - it also poses great challenges to the brake system designer.

The subject of this paper, electrically actuated aircraft brakes, is a developmental subsystem to be matured before the All-Electric Airplane can become a reality. Loral Aircraft Braking Systems (formerly Goodyear Aerospace) began looking at the concept in 1979 and by late 1982, had built and tested a pre-prototype electric brake system (EBS) based on their A-10 aircraft brake. The A-10 was chosen as a test bed to evaluate the design concepts because of its availability for a U.S. Air Force sponsored on-aircraft test, its large power reserve, and Loral's familiarity with the existing hydraulic brake system.

OBJECTIVE

The purpose of this paper is to provide a brief review of the proof-of-concept testing for the first electromechanical braking system for a high performance aircraft. The design concept, laboratory testing, and on-aircraft demonstration will be covered in this paper.

DESIGN CONSIDERATIONS

The A-10 aircraft, weighing approximately 40,000 lbs, requires its braking system to absorb kinetic energy of about 7 million foot-pounds per brake during a normal stop and up to 21 million foot-pounds per brake for a Refused-Take-Off (RTO) stop. The existing hydraulic brake system consists of two multiple-disk brakes, one on each main gear wheel. Each brake assembly (See Figure 1) consists of four rotating disks (rotors) and three stationary disks (stators) which are clamped between the pressure and back plates by seven hydraulic pistons carried in the brake housing. The rotors are keyed to and rotate with the wheel; the stators are keyed to the axle. Clamping force is controlled by metering the high pressure hydraulic fluid supply via variations in the brake pedal pressure (See Figure 2). In addition, skid protection is provided by a conventional aircraft antiskid valve installed upstream of the brakes.

The electric brake assembly (See Figure 3) is identical to the standard hydraulic brake with the exception that the hydraulic housing subassembly is replaced by an electromechanical actuator subassembly. The electromechanical actuator consists of nine direct current motors driving a ring gear which in turn drives a pressure ram that exerts a clamping force on the brake disks in a fashion analogous to the pistons in the hydraulic system. The transformation, within the actuator, of the rotational motion of the samarium cobalt motors to the axial compressive motion of the ram is accomplished via a large diameter ball screw mechanism. The ring gear is under positive, geared control at all times, thus to release the brakes, the motors must be reversed.

Commands are sent to the motors by a central controller (Refer to Figure 2). The controller requires three inputs: pedal signal, antiskid signal (EBS retains the standard A-10 antiskid control unit), and brake torque feedback. The controller balances the pedal signal input and antiskid signal input with a torque feedback signal to determine power level and motor direction.

QUALIFICATION TESTING

Prior to installation and testing on an actual aircraft, all brake systems must be subjected to exhaustive dynamometer qualification testing to ensure integrity and performance for aircraft usage. Since the EBS on-aircraft demonstration was a non-flying test (ground taxi at various speeds), approval could be obtained to qualify the brake to Safety of Flight requirements, a less extensive qualification program. The qualification program consisted of 27 normal energy stops, 3 overload energy stops (225% of normal), a structural torque test, a landing sequence test, and an endurance test. In addition, to qualify the electronic controller, antiskid and degraded system stops were made. The average deceleration requirement for the normal and overload stop conditions is 10 feet per second^2. All stops were conducted on inertia-type landing dynamometers as shown in Figure 4. The brake, wheel and tire are mounted on the test axle and installed in the dynamometer carriage as shown. Weight disks are added to the large flywheel to achieve the desired A-10 kinetic energy requirements when rotated at the specified landing speed.

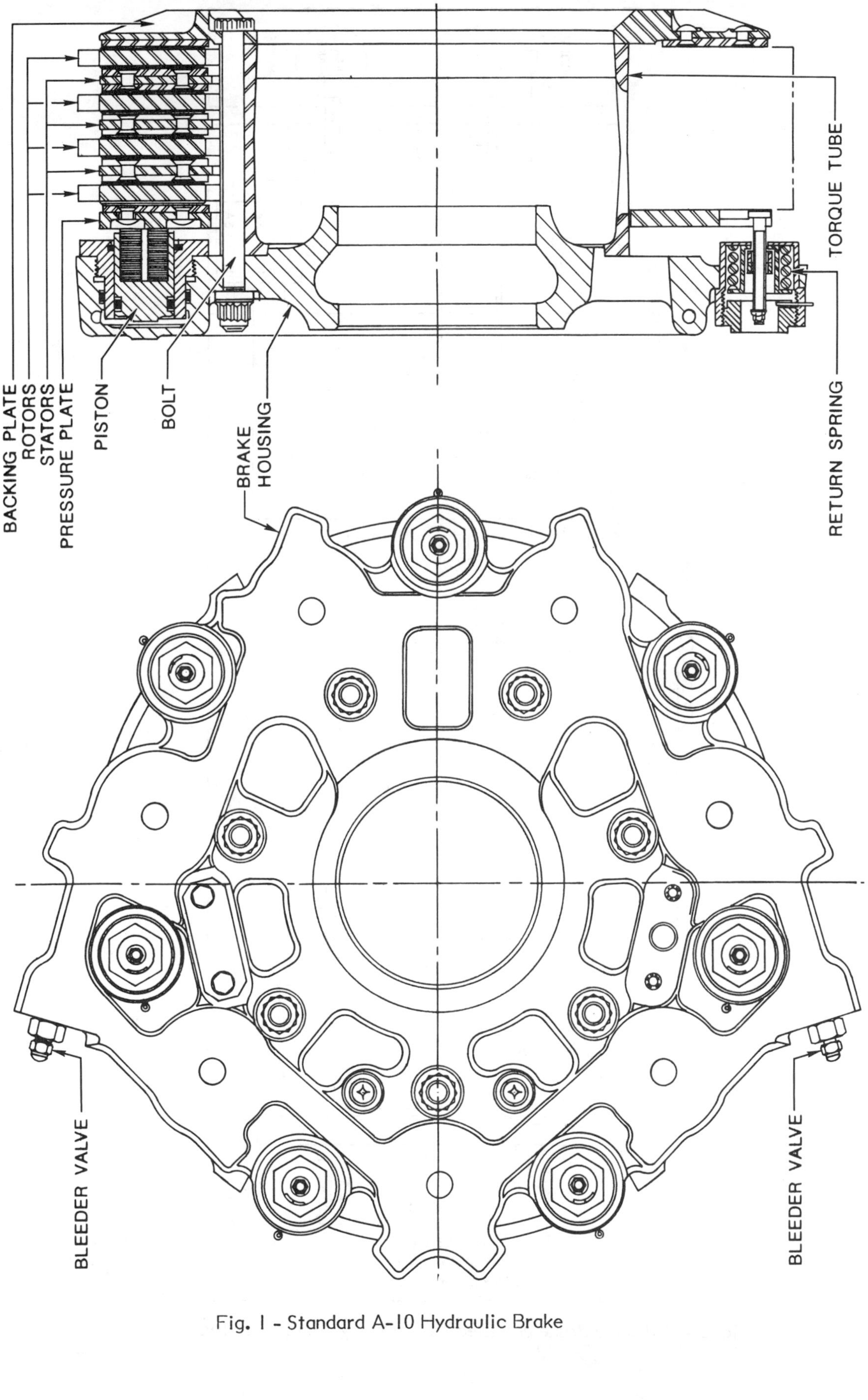

Fig. 1 - Standard A-10 Hydraulic Brake

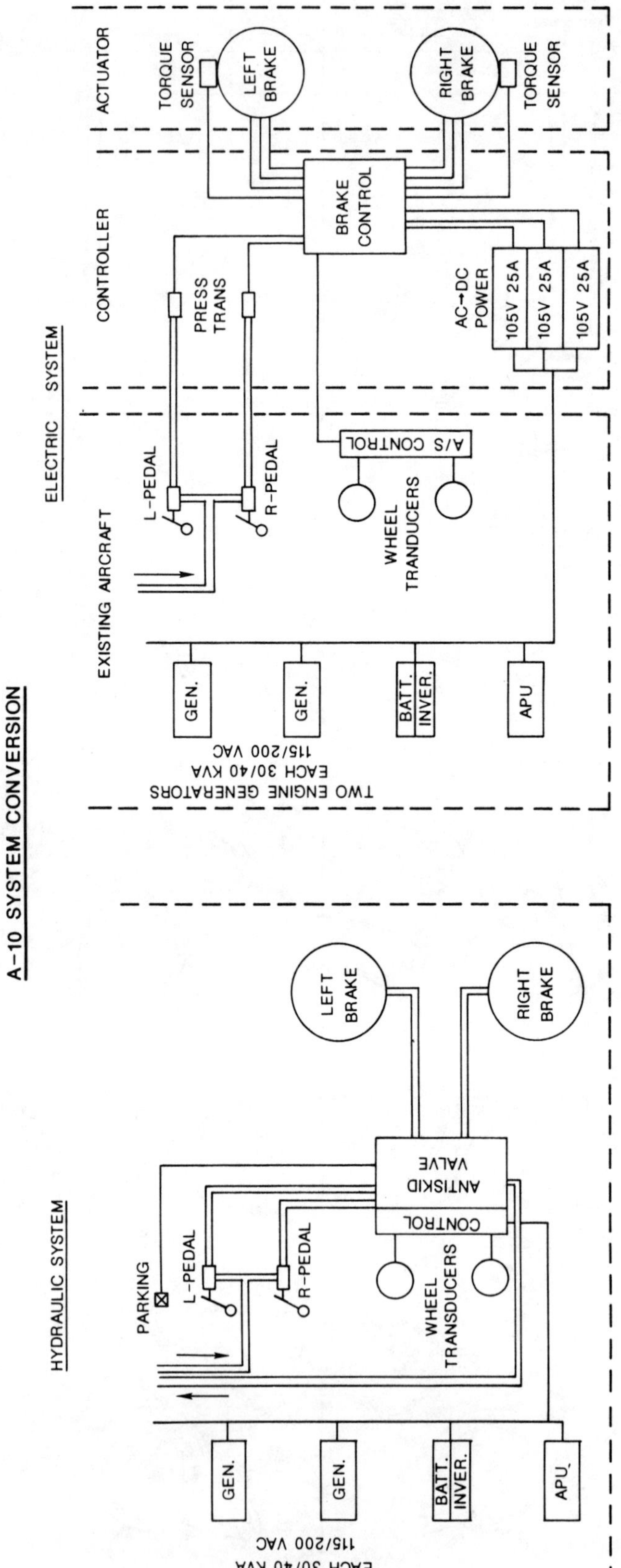

Fig. 2 - Hydraulic Brake and Electric Brake System Schematics

Fig. 3 - Electric Brake Actuator

Fig. 4 - Electric Brake Dynamometer Test Set-Up

At the proper speed, the flywheel is allowed to "free wheel" and the wheel and brake unit is then "landed" against it and loaded to simulate the equivalent portion of the aircraft weight (normally 21,000 lbs). The aircraft brake is applied and stops the flywheel, thus absorbing the same energy as it would in actual practice. Antiskid stops were accomplished by reducing the wheel load, thereby lowering the tire drag force the flywheel surface will support without skidding a tire. In effect, lowering the wheel load simulates lowering the tire to runway coefficient which induces wheel skids.

QUALIFICATION TEST RESULTS

The normal stop deceleration averaged 10.38 ft/sec^2 which exceeded the requirement. The overload deceleration averaged 9.19 ft/sec^2 which was within the acceptance range. Time versus brake torque and motor voltage curves are shown in Figures 5 and 6 for a normal and overload stop respectively. For these stops, a laboratory power supply was manually controlled to achieve the target decelerations. Later stops, including antiskid stops, employed the EBS controller to automatically adjust the motor current level. Note that the motor voltage was applied only as necessary to reach or maintain the desired brake torque level. Once the desired torque level is achieved, the voltage is released until more torque is desired. In both of the stops shown, the brake torque level exceeded deceleration requirements. No excessive temperatures were recorded in any component during the qualification testing. For the antiskid stops, brake application speed was varied between 50 miles per hour (mph) and 117 mph (the normal energy brake application speed). Wheel load was varied between about 3,000 lbs and 7,000 lbs. All stops were successfully completed and demonstrated skid release at simulated runway coefficients as low as 0.16.

The landing sequence as performed on the dynamometer included two taxi stops simulating taxiing to the active runway. A 15-minute wait period was then included to simulate a delay for runway clearance. A normal energy stop was then performed followed by two taxi stops simulating the return to the hangar. The goal was to achieve a 10 ft/sec^2 deceleration for the normal stop, however, a slightly lower deceleration of 8.93 ft/sec^2 was achieved. A slight increase in power would have raised the deceleration to the desired level. The EBS passed both the structural torque and endurance tests.

ON-AIRCRAFT DEMONSTRATION

The on-aircraft test schedule was developed to evaluate EBS operations over an envelope of conditions. These included brake application speeds between 40 and 120 knots (kts), aircraft weights between 32,000 and 40,000 lbs, wet and dry runways, and light to full brake pedal force. Brake stops were scheduled in a build-up manner starting with low speed, aircraft weight, and braking force and proceeding gradually to higher levels. Instrumentation was installed for wheel speed, brake pedal, brake torque, antiskid signal, power supply status, aircraft acceleration and velocity, and brake component temperatures.

Installation of the EBS involved removal of the main landing gear wheel/tire and the hydraulic brake assembly. The hydraulic brake housing was replaced by the electromechanical actuator and the brake and wheel/tire reinstalled on the landing gear. Shown in Figures 7 and 8 are the hydraulic and EBS components as installed. To provide an electrical pedal signal, a closed manifold containing pressure transducers is installed in place of hydraulic antiskid valve. Hydraulic lines downstream of the valve are capped. The EBS controller, emergency controller, and three power supplies (necessary for conversion of aircraft alternating current to direct current for the brake motors) are housed in a large centerline pod secured to the aircraft by means of a modified bomb rack.

The first test point was an operational check conducted at a brake application speed of 40 kts. Light pedal pressure was used. The pilot noted a shake in the aircraft during braking. Examination of the strip chart data showed indications of vibration in the brake torque and wheel speed traces. The antiskid voltage was elevated above zero indicating antiskid action. This was unexpected as the aircraft deceleration and tire drag forces were well below the skid threshold.

A second run was conducted at 40 kts with medium brake pedal pressure. Upon brake application, the pilot noted a stronger aircraft vibration. Analysis of the strip chart data showed oscillations in the brake torque, wheel speed, and antiskid traces at frequencies between 11 and 17 hz. Observers reported a sustained fore-aft motion of the aircraft landing gear struts. This motion is known as "gear walk" and is caused by the brake torque frequency nearly matching the gear natural frequency of 12 to 16 hz. Sustained gear walk can induce high structural loads into the landing gear.

Figure 9 shows the data traces for the first two seconds of the stop and reveals the initiation of the gear walk. Brakes are applied at zero time as shown on the bottom trace which is brake motor current. At approximately 0.25 seconds, the left and right brake torque traces raise above zero indicating brake actuation. The torque traces are instrumented from the torque feedback sensors in the brake actuators. This sensor is unique to the EBS and provides closed loop brake control to achieve a torque feedback signal equivalent to the brake pedal signal. The torque traces remain approximately level to 0.65 seconds and then show sharp increases. Note that both the left and right wheel speed traces show some oscillations between 0.3 and 0.6 seconds but then

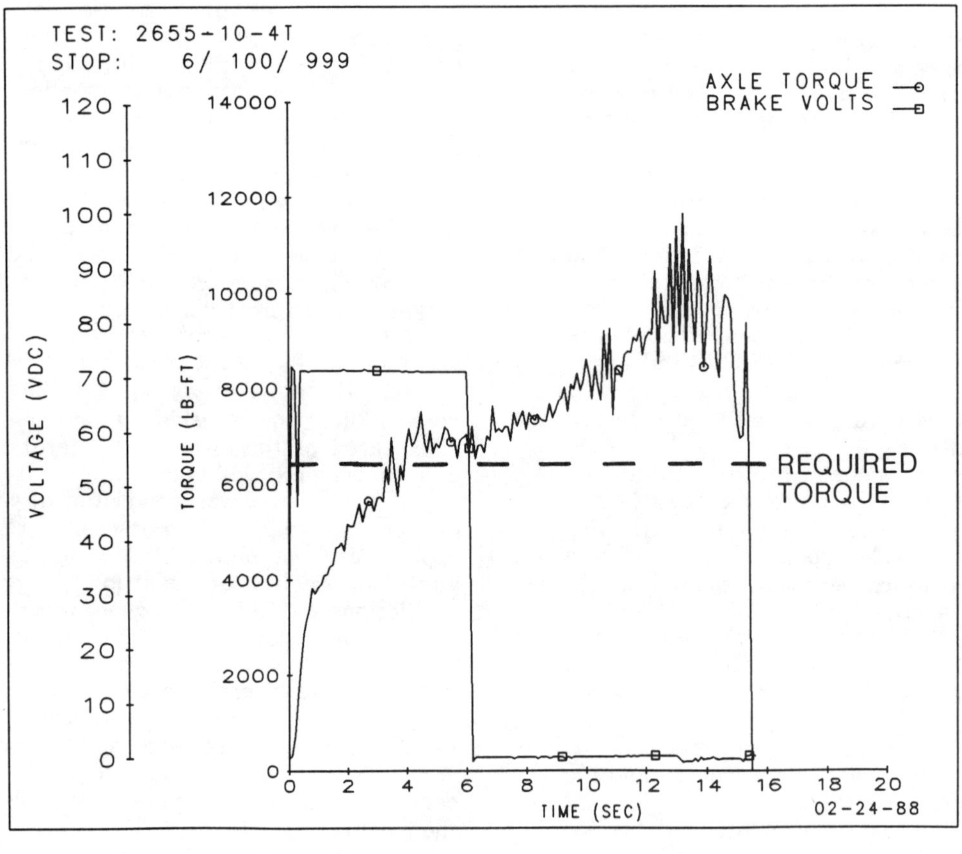

Fig. 5 - Normal Energy Stop Torque, Voltage versus Time

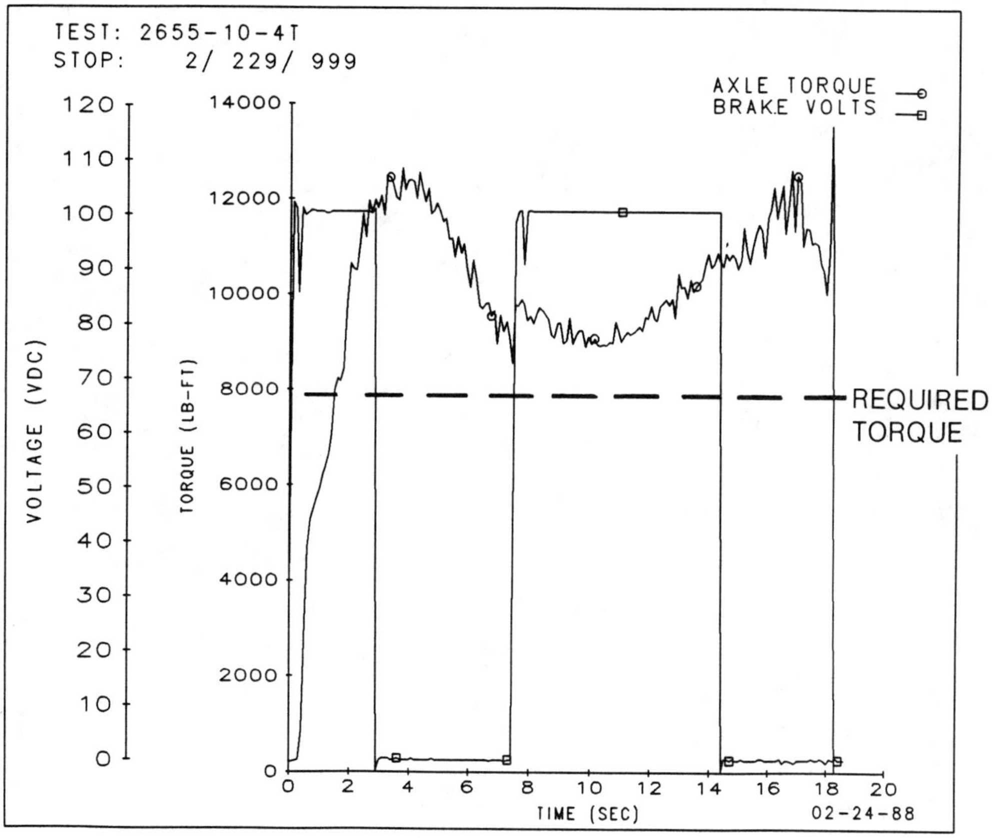

Fig. 6 - Overload Energy Stop Torque, Voltage versus Time

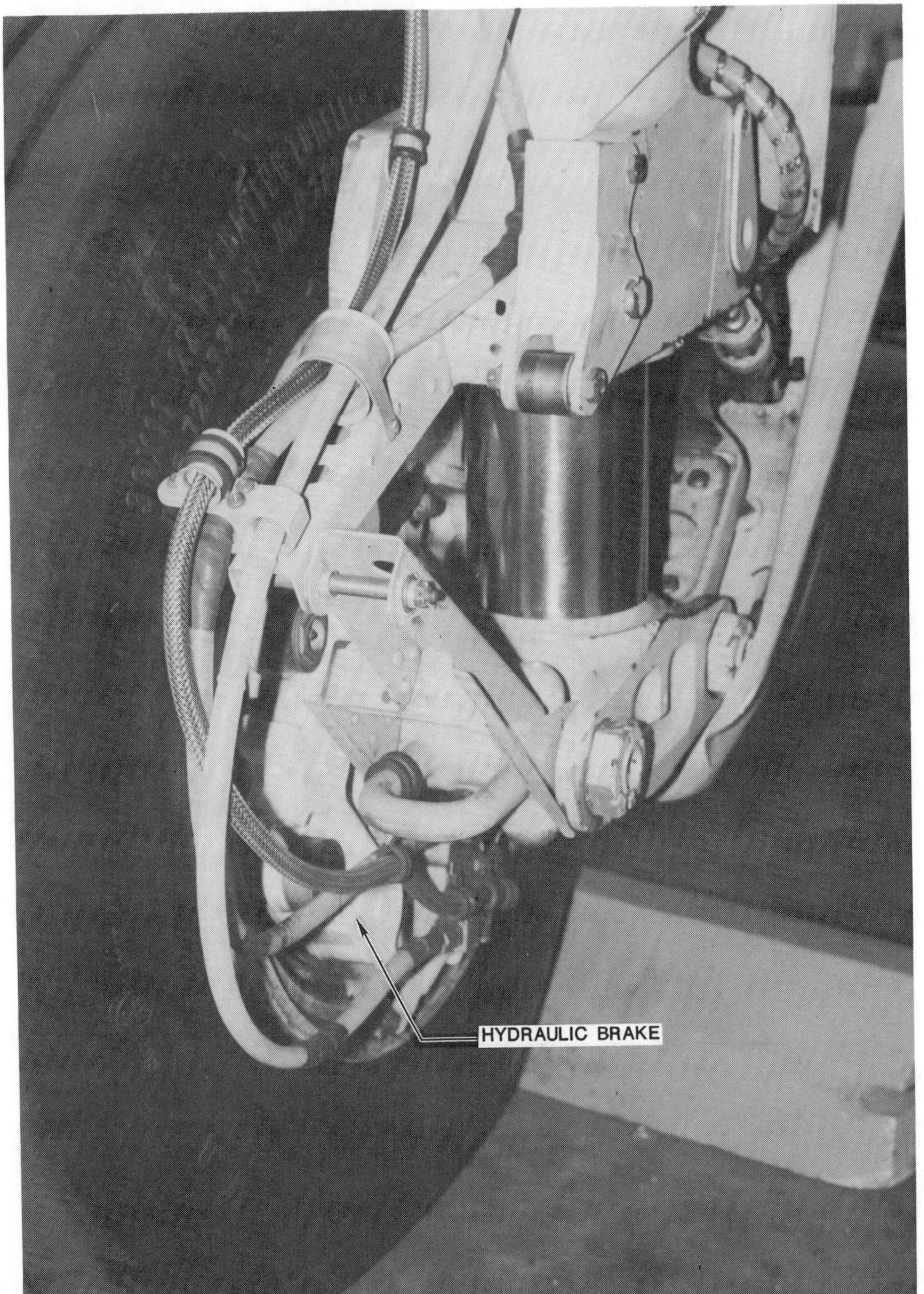

Fig. 7 - A-10 Hydraulic Brake Installation

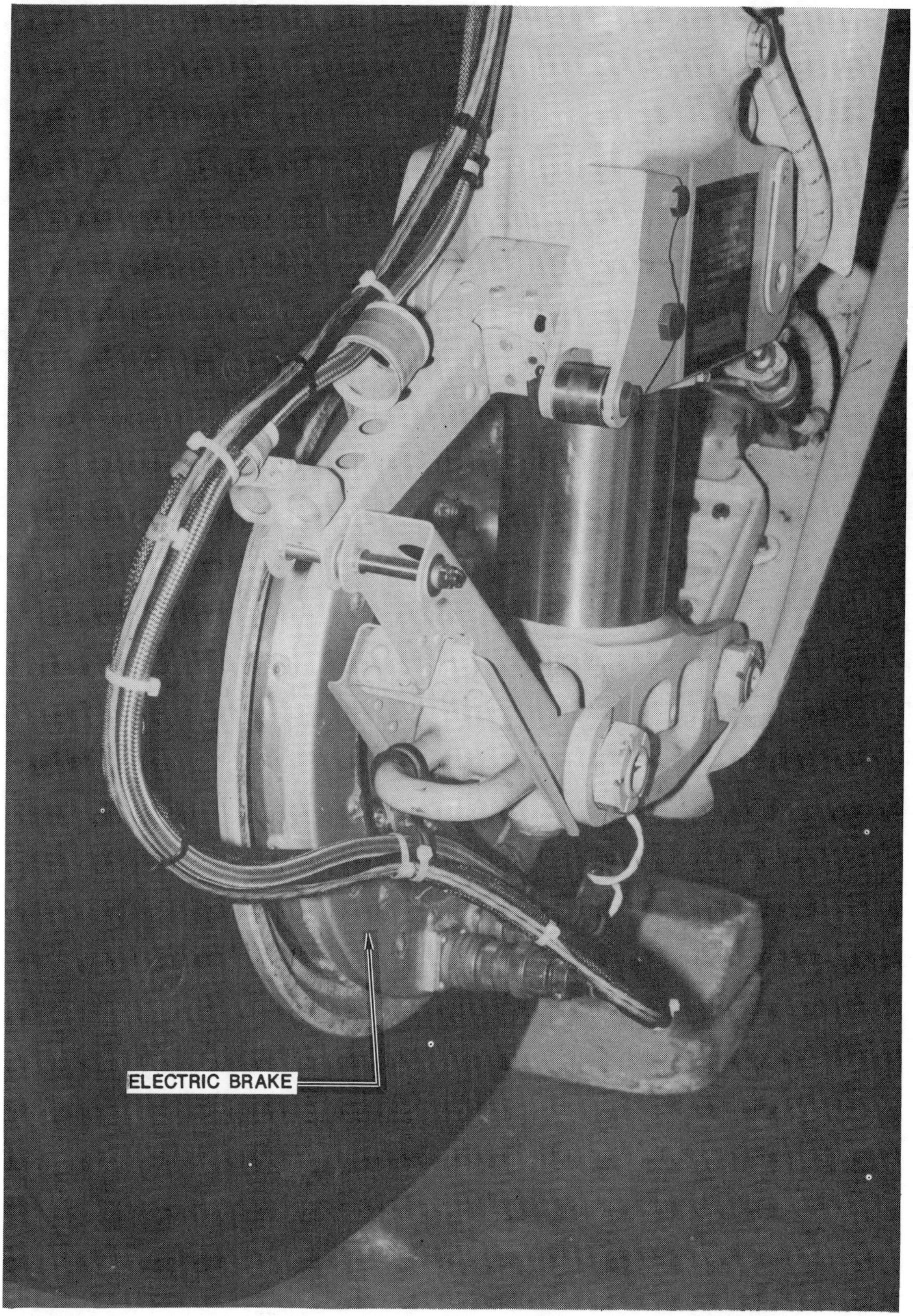

Fig. 8 - A-10 Electric Brake Installation

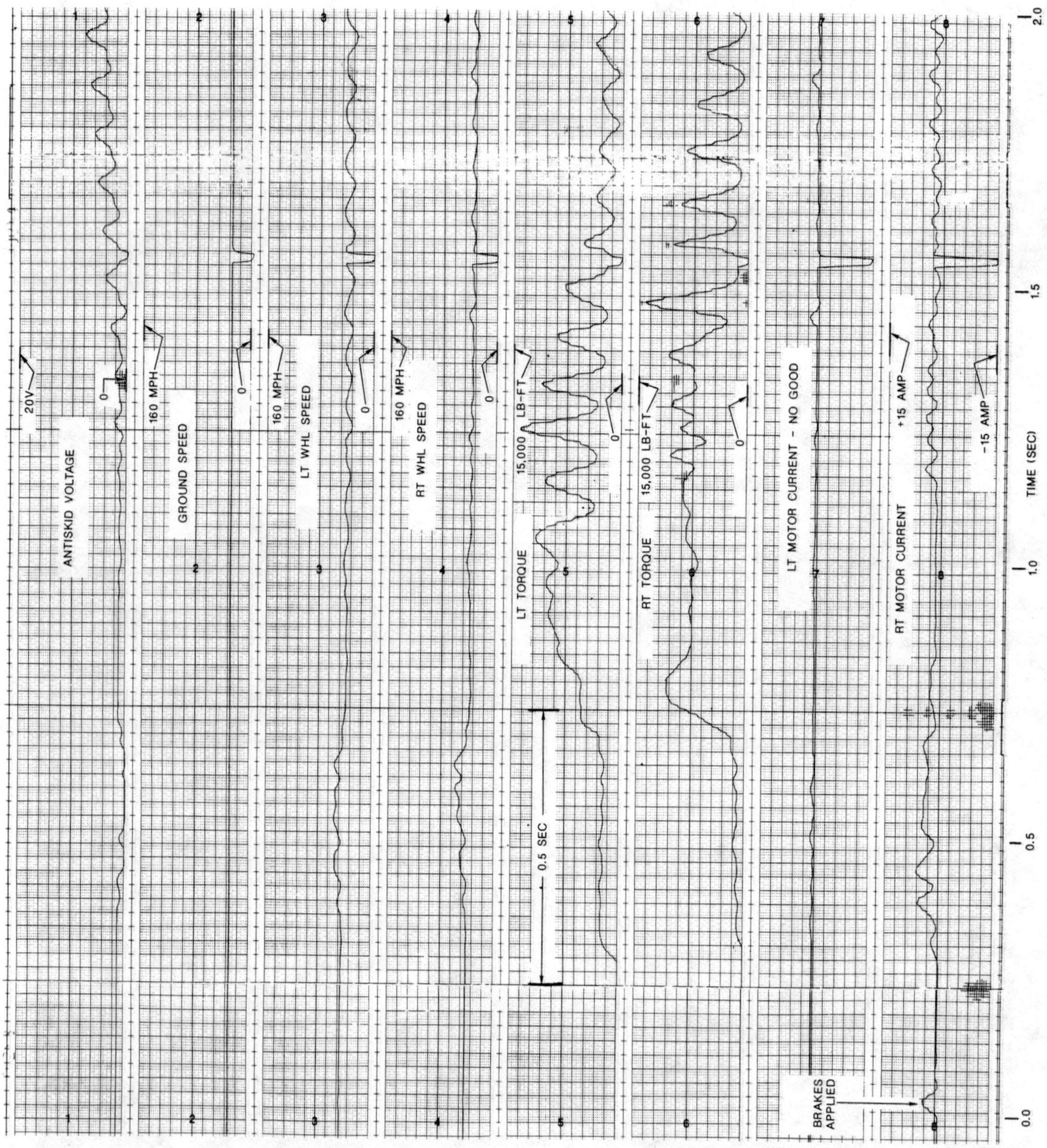

Fig. 9 - Aircraft Taxi Test Data Run 2

settle out as the brake torque increases. At about 1.0 seconds, the left brake torque begins to show an oscillatory shape. The torque feedback signal is overshooting and undershooting the desired value relative to the pedal deflection. At the same time, both wheel speed traces begin to show an oscillatory shape indicating fore-aft gear motion. The gear aft motion creates high wheel speed deceleration that the antiskid system interprets as an impending wheel skid. Beginning at about 1.1 seconds, elevated antiskid voltage is shown with peaks occurring in response to the highest wheel speed decelerations. From about 1.3 seconds on, the brake torque traces reduce in average value in response to the elevating antiskid voltage.

The oscillations in the torque feedback signals were analyzed and thought to be caused by too much clearance between the sensor and the torque take-out lug on the landing gear. As a first attempt to eliminate the gear walk, the torque sensors were shimmed to achieve a tight sliding fit on the torque lug. The third run was then conducted at 40 kts with light to medium brake pedal pressure. The gear walk phenomena was again exhibited but the amplitude of fore-aft deflection was noticeably reduced. The wheel speed traces showed the fore-aft gear motion, but at lower levels with the frequency increased to 15 to 18 hz. Figure 10 shows the data traces for a portion of this stop. Analysis of the data shows that the antiskid influence had been greatly reduced, but the brake torque oscillations remained.

The next adjustment made in the interest of brake torque stability was widening the controller error bandwidth (allowable difference between torque feedback and pedal signals). A fourth run was scheduled for 40 kts brake application speed. The run was aborted, however, when gear walk was displayed during the pretest taxi to the active runway. The source of the gear walk problem was now considered to be a basic control instability or a dynamic incompatibility with the A-10 landing gear. The solution would require alteration of the EBS control circuitry which could not be efficiently accomplished on-site. At this point, further testing was postponed to consider control circuit refinements to eliminate the gear excitation tendency.

ANALYSIS OF PROBLEM

The EBS taxi test showed brake to landing gear instability in the form of brake torque cycling. Brake cycling, which normally occurs only during antiskid stops, is an indicator of the response speed of the brake actuator. Cursory examination of the data concluded that the electric brake cycles much faster than the A-10 hydraulic brake. In order to quantify the response difference, the EBS cycling rate was compared with hydraulic brake antiskid data. The frequency for one torque release and reapply cycle was calculated for both brakes at several amplitudes. The hydraulic brake has a maximum cycling rate of about 5 hz at low torque amplitudes and a minimum rate of about 1 hz at high amplitudes. The electric brake by comparison has a relatively constant torque cycling rate of about 10 to 14 hz.

Closer examination of the brake cycling function shows that both brakes release torque at about the same rate, but the electric brake reapplies torque several times faster than the hydraulic brake. Figure 11 compares the brake torque reapplication for both brakes at a net torque increase of about 4,000 lb-ft. The much slower torque application function for the hydraulic brake is due to the combination of fluid flow to fill the piston cavities and orifices installed in the fill lines to purposely slow the brake fill rate. The orifices were incorporated to enhance the hydraulic brake stability on the A-10 landing gear.

Therefore, to improve dynamic stability on the A-10 landing gear, the electric brake application rate should be slowed similarly to the orificed hydraulic brake. The brake release rate, however, must be maintained as fast as possible to prevent wheel skids.

TORQUE APPLICATION CONTROL CIRCUIT

One of the control signals generated by the electric brake controller is directly related to the amount of current sent to the actuator motors. The magnitude of the motor current is directly related to the brake clamping force which in turn determines the brake torque. Therefore by controlling this signal in the presence of the increasing brake torque, the torque application rate can be limited. By eliminating control in the brake release mode, the brake release speed will be maintained to prevent wheel skids. A circuit of this nature was designed and incorporated into the EBS controller. Bench testing demonstrated that it effectively limited the brake torque application rate to about 17,000 lb-ft/sec similar to the orificed hydraulic brake.

TEST RESULTS

The torque control feature was dynamometer tested in February, 1987. Figure 12 shows the pressure and torque curves for one stop. The slope or rise rate of the curves were compared to evaluate the effect of the torque control circuit. This procedure was repeated with all available data and the resulting pressure rate versus torque rate points are plotted in Figure 13. Note that all of the test points except one fall within the assumed torque control limit of 17,000 lb-ft/sec,

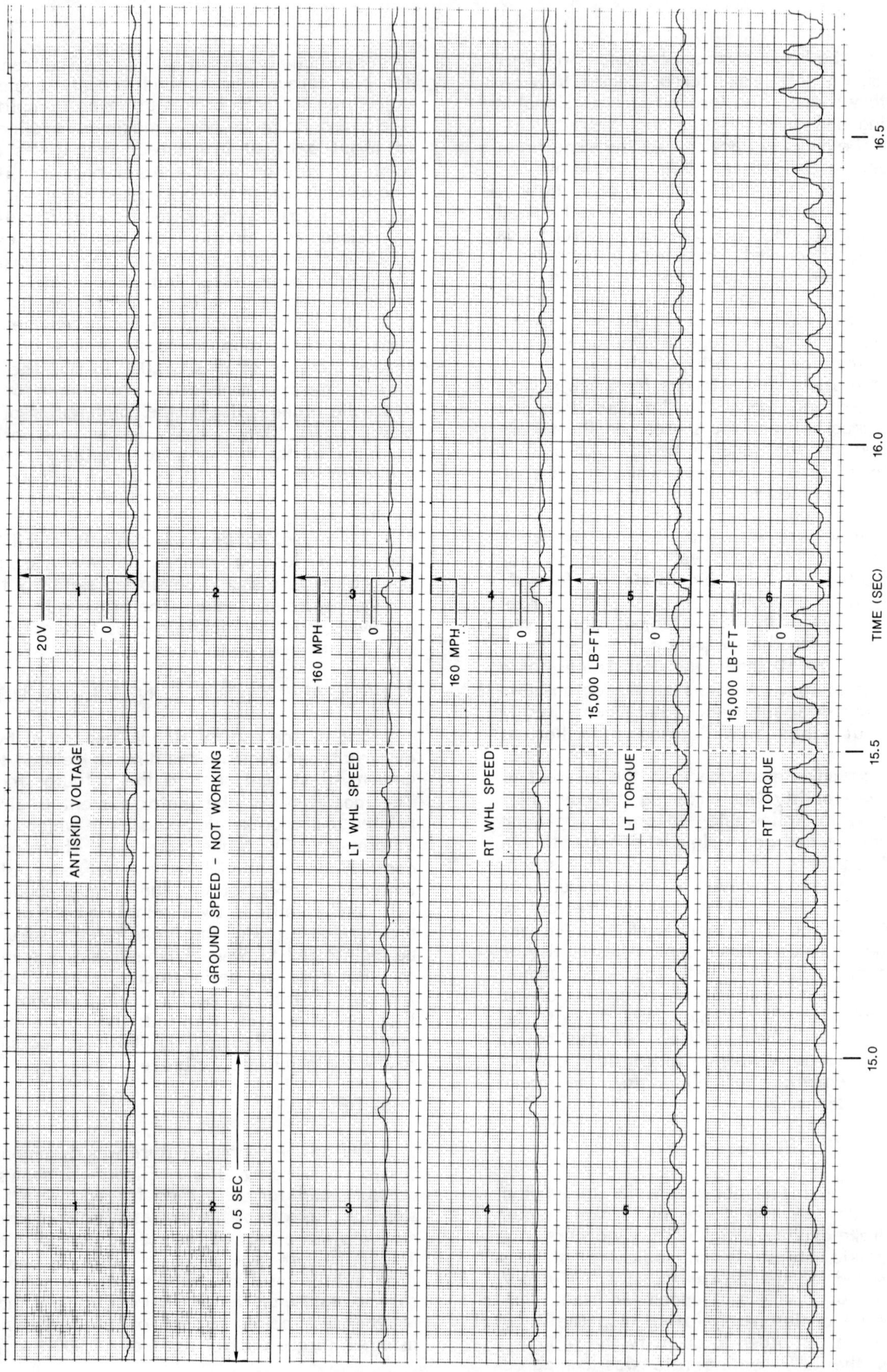

Fig. 10 - Aircraft Taxi Test Data Run 3

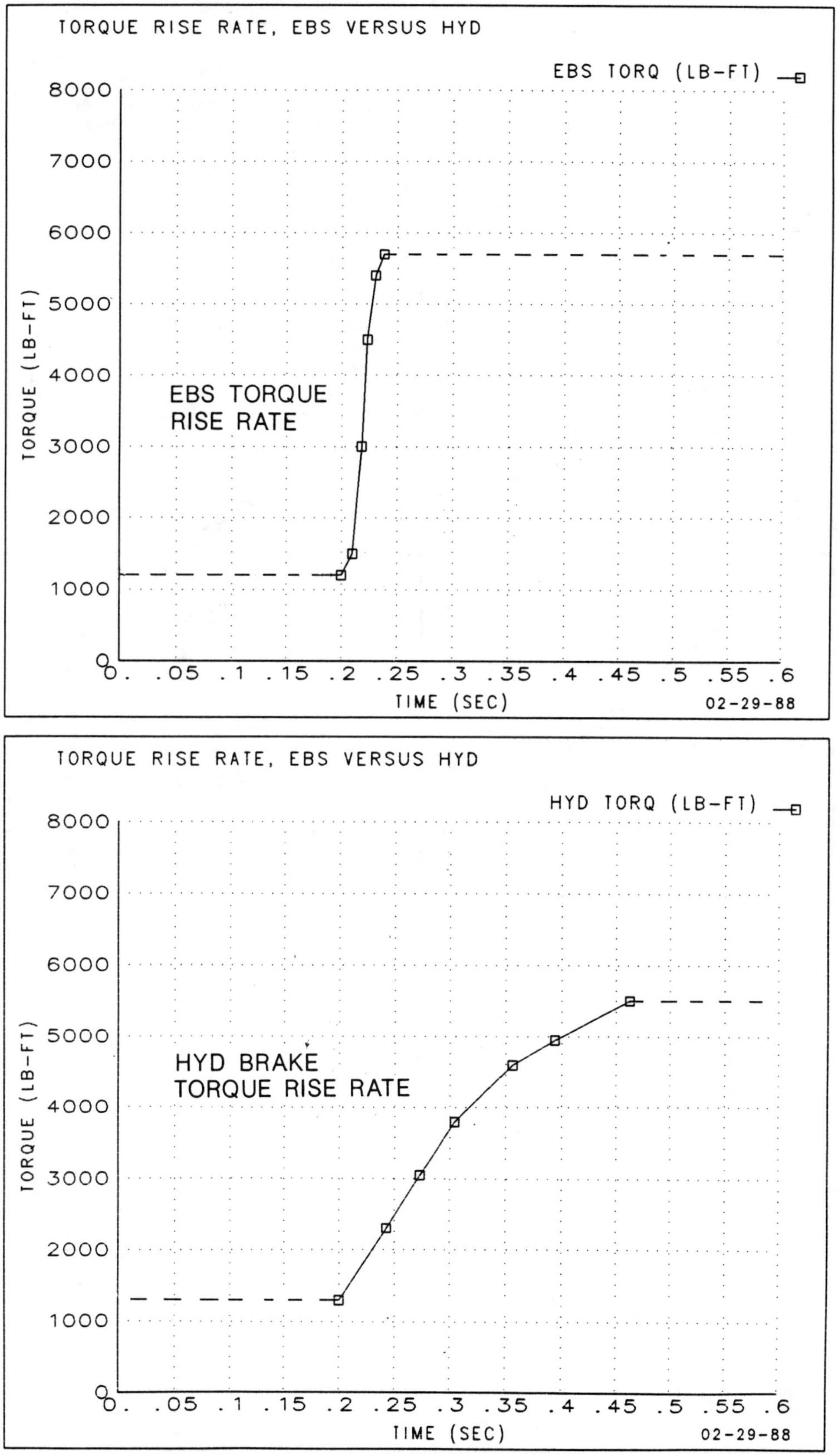

Fig. 11 - Electric versus Hydraulic Brake Torque Application Rate

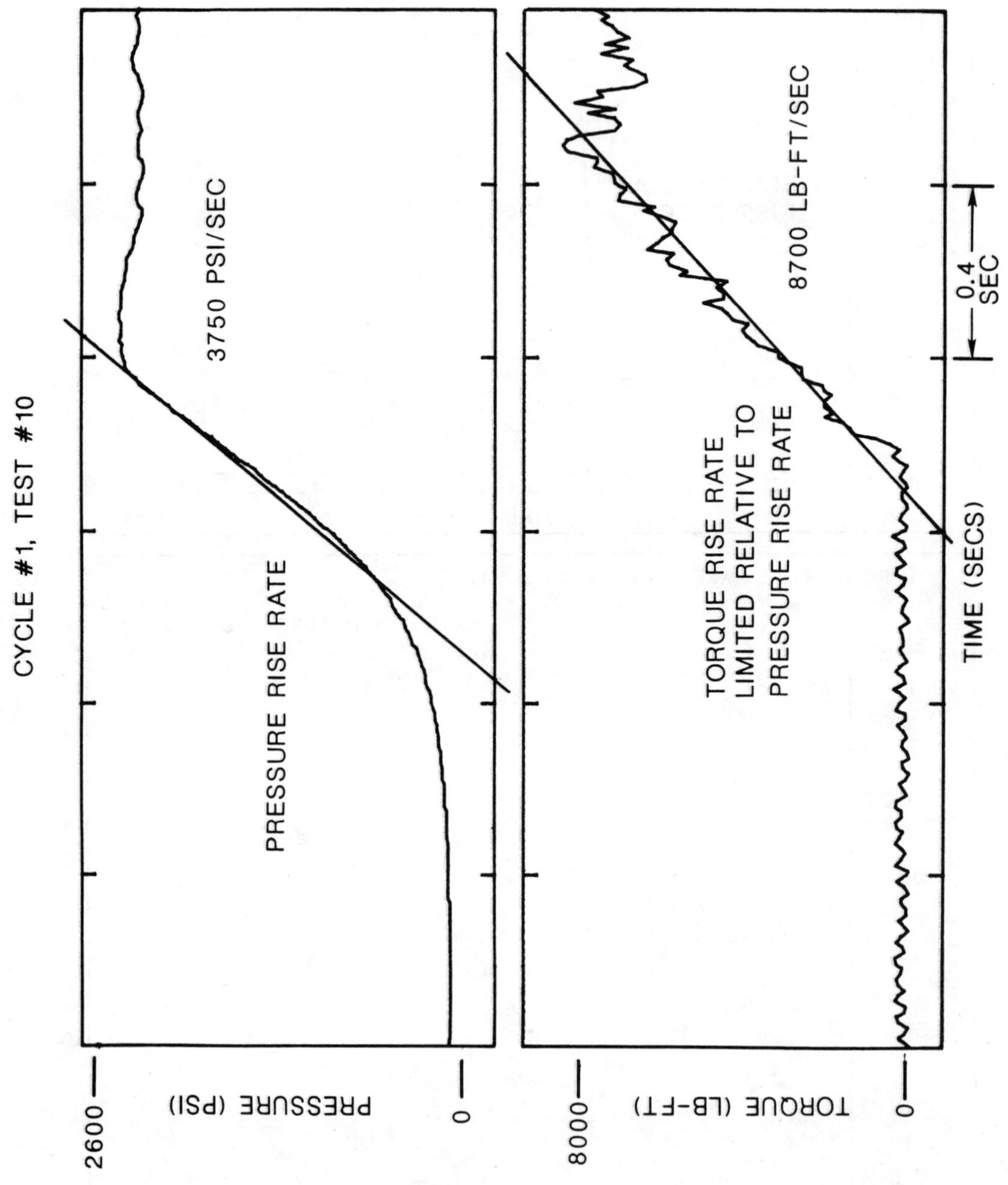

Fig. 12 - EBS Torque Rise Rate Limited by Torque Rate Control Circuit

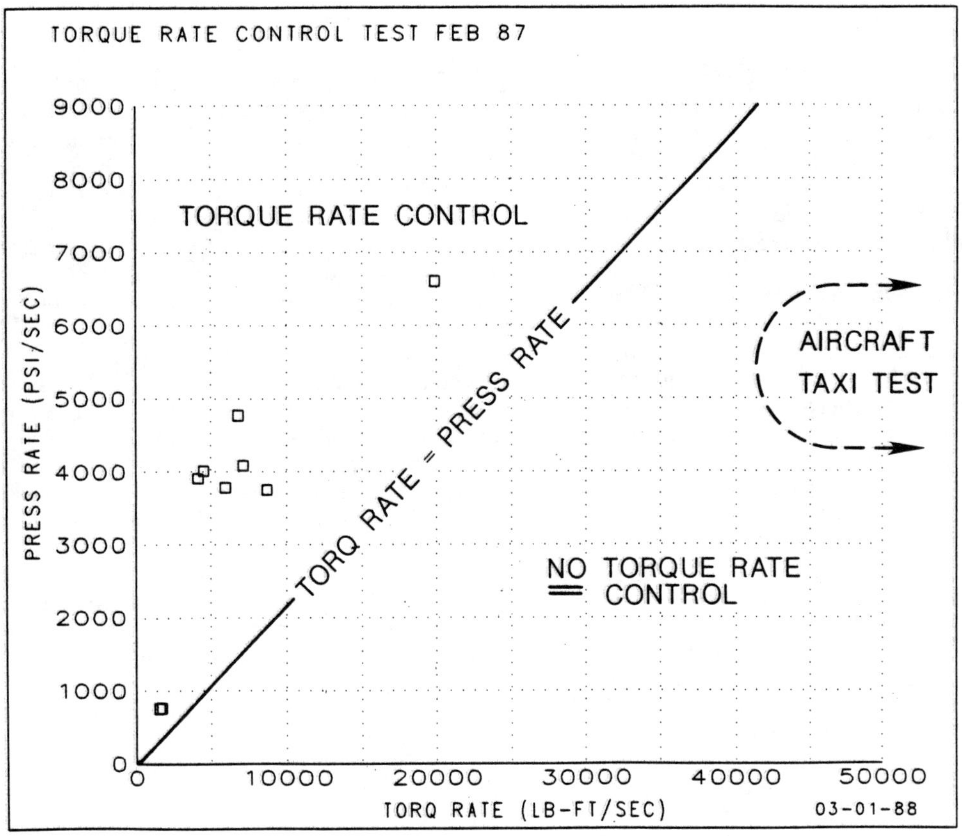

Fig. 13 - Torque Rate Control Test Results

with most points clustered between 4,000 and 9,000 lb-ft/sec. The torque control feature reduces the motor current when the torque is rapidly increasing. Once the torque rate is reduced, the motor current and the torque are allowed to increase again and the cycle may be repeated. Because of this action, the actual torque curve may then resemble a pseudo-stair step function, and the average torque increase rate may be less than the assumed limit of about 17,000 lb-ft/sec. The highest torque rise rate point was about 20,000 lb-ft/sec which still shows torque limiting. Therefore, the torque control feature functioned as designed.

DYNAMOMETER EVALUATION

Aircraft taxi testing with the Torque Rate Control installed was curtailed when inspections revealed the accumulation of testing had worn the electric brake mechanical components to the point that further aircraft testing was not recommended. An extensive dynamometer comparison between the A-10 electric and hydraulic brake systems was conducted instead.

The test objective was to thoroughly assess the performance of the electric brake and compare it to the standard A-10 hydraulic brake over identical operating conditions. Testing evaluated brake torque, deceleration, energy capability, and antiskid response. The effectiveness of the torque control circuit to reduce the tendency for landing gear excitation was examined further.

NORMAL DECELERATION STOPS

Normal deceleration stops were completed at a brake energy of 6.9×10^6 ft-lbs with a brake application speed of 119 mph. A constant brake pressure of 2,300 psi was established prior to the stop to achieve approximately 10 ft/sec^2 deceleration. Recal! that for the electric brake, pressure is converted to a proportional electrical signal by pressure transducers installed in a closed manifold.

Figure 14 shows the brake pressure and torque curves for a normal stop. The actual brake pressure average for this stop was 2,227 psi. The brake torque average was about 5,700 lb-ft which achieved a deceleration of 9.81 ft/sec^2. The deceleration was below 10 ft/sec^2 because the brake pressure was slightly below the target of 2,300 psi.

TORQUE SURVEY STOPS

Torque survey stops were conducted to establish the relation between applied brake pressure and torque for the electric brake. Since the brake pressure is converted to a proportioned electrical signal for transmission to the actuator motors, a relationship can be drawn between electric brake torque output and the brake pressure applied.

The torque feedback sensor output is nonlinear with respect to the physical brake torque. Therefore, high brake pressure is required, compared to the hydraulic brake, to achieve moderate torque levels. Full pressure will, however, generate the full electric brake torque.

The data is plotted in Figure 15. The solid line shows a maximum brake torque of about 10,500 lb-ft. This is an acceptable maximum torque, but it shows evidence of ball screw efficiency losses. Dynamometer testing earlier in the electric program had shown maximum brake torque exceeding 12,000 lb-ft.

ANTISKID STOPS

Five antiskid stops were conducted at wheel loads ranging from 10,000 lbs to 3,000 lbs.

The stop conducted at 6,000 lbs wheel load (simulated coef. = 0.33) produced clear antiskid action throughout the stop. The stop time was 28.40 seconds and deceleration was 6.15 ft/sec^2. The stop time and deceleration are indicative of the antiskid action to reduce brake torque to prevent wheel skidding. The average brake torque was about 4,250 lb-ft.

Figure 16 shows the stop data. The brake pressure is constant at about 2,300 psi as in all antiskid stops. The wheel speed curve now shows deep dips which show rapid deceleration and are the onset of wheel skids. The antiskid nominal voltage is elevated above zero which reduces the average brake torque. High spikes in the antiskid voltage command brake release at the points where the wheel deceleration is the greatest. Corresponding large reductions in brake torque can be seen as the brake releases in response to the spikes in the antiskid voltage.

The antiskid action shown in the stop indicates the ability of the electric brakes to prevent wheel skids. At several points, however, the wheel speed nearly decelerates to zero before the brake is released. The indicates that the ball screw action is sluggish as the brake release is too slow to prevent the wheel speed from decaying to near zero. An alternate ball screw was installed into the actuator for the 5,000 lbs wheel load antiskid stop.

The stop conducted at 5,000 lbs wheel load (simulated coef. = 0.28) produced improved antiskid action compared to the previous stop. The stop time was 28.65 seconds and the deceleration was 6.09 ft/sec^2. The average brake torque was reduced to about 3,650 lb-ft.

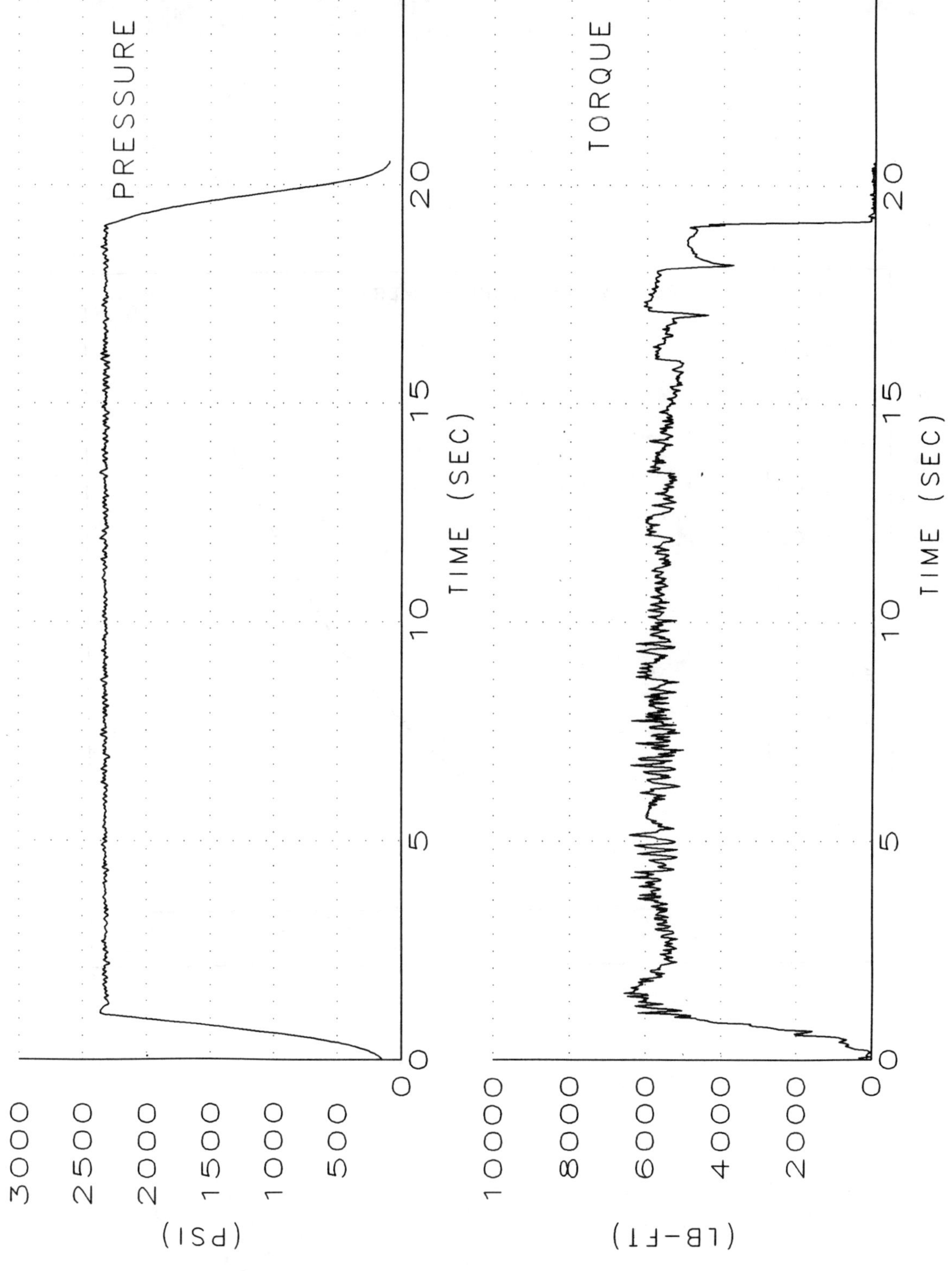

Fig. 14 - EBS Normal Deceleration Stop

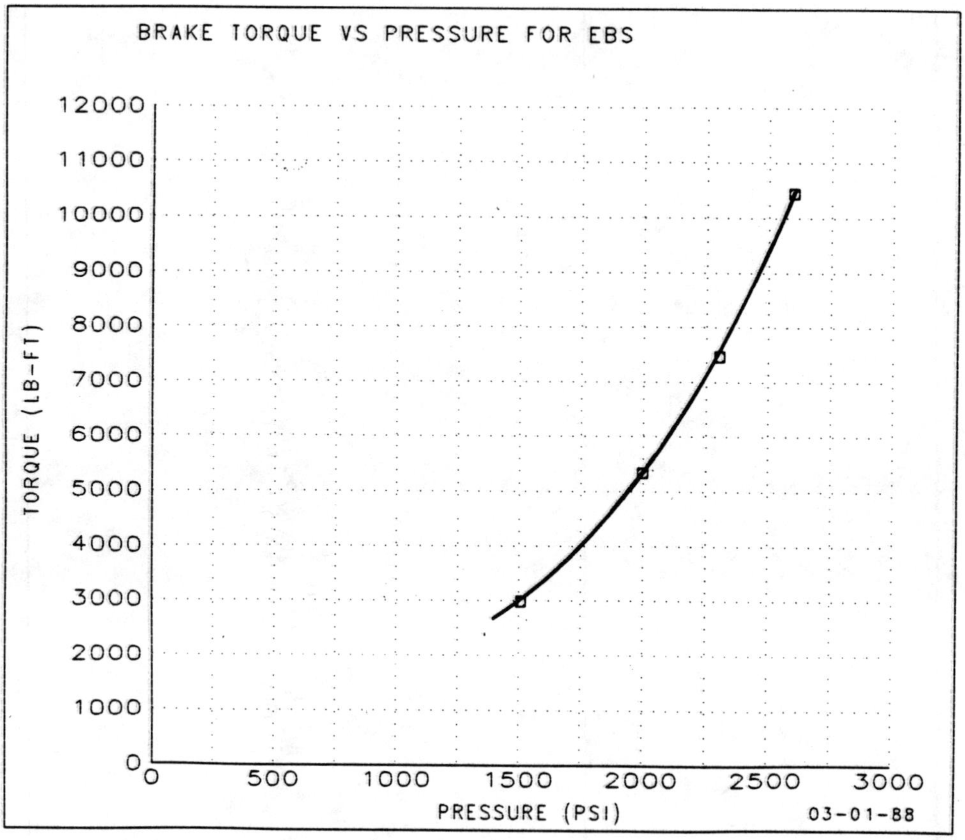

Fig. 15 - Brake Torque versus Pedal Pressure for EBS

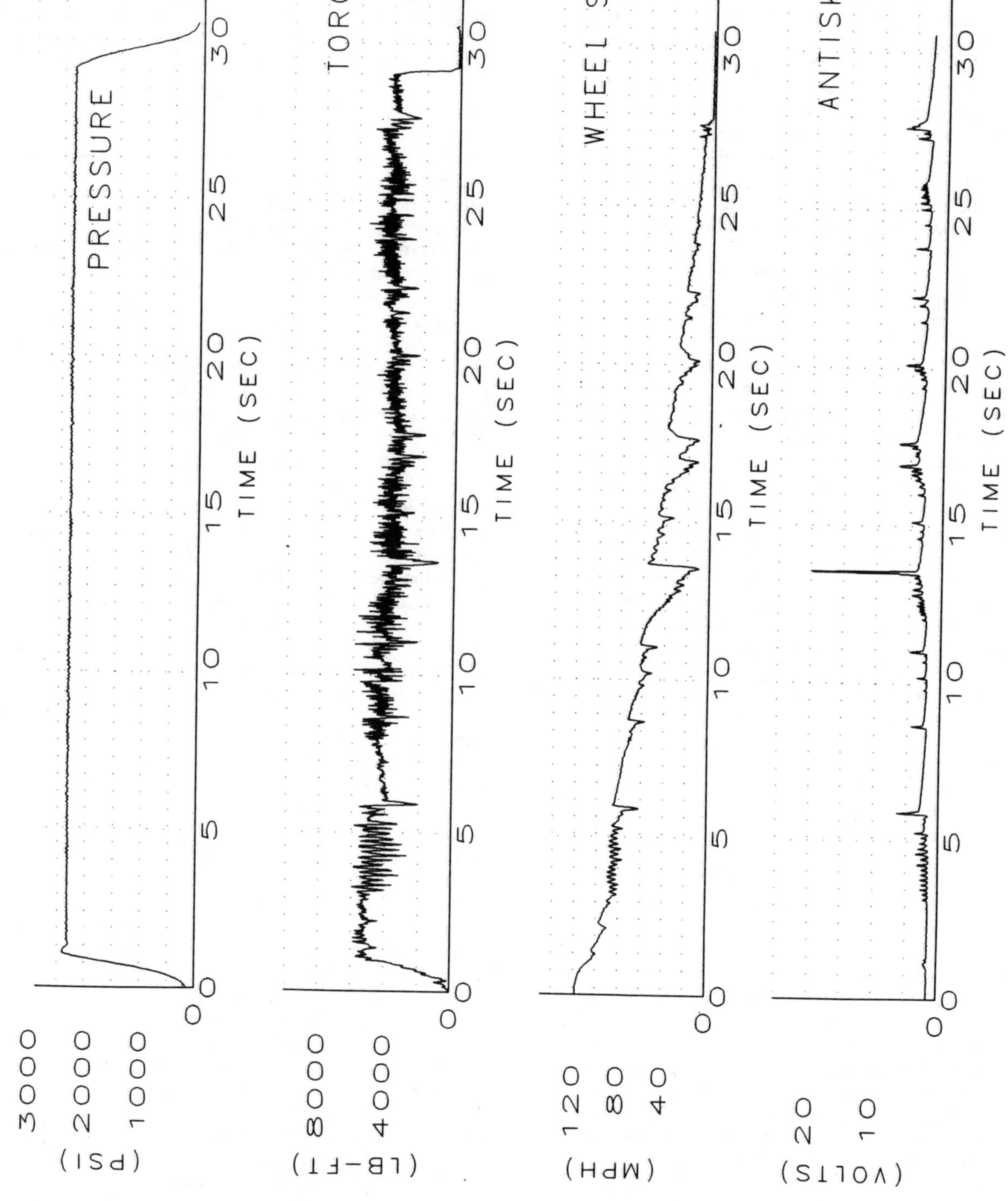

Fig. 16 - EBS Antiskid Stop, 6,000 Pounds Wheel Load

Figure 17 shows the stop data. Compared to 6,000 lbs wheel load stop, major differences are shown in wheel speed and antiskid voltage traces. The wheel speed trace shows numerous shallow dips in wheel speed. This indicates that brake release is much faster with the alternate ball screw and skids are prevented without a significant decay in wheel speed. The antiskid voltage shows a nominal elevation above zero and lower peaks that command brake release. This is also evidence of a lower friction ball screw as lower antiskid voltage peaks commanded effective brake torque release.

This stop clearly illustrates the significance of ball screw response in antiskid stops. A lower friction screw greatly improves the control of wheel skid by providing faster brake release.

Antiskid stops with wheel loads greater than 7,000 lbs showed no appreciable skid indications and the attempt at 3,000 lbs wheel load (simulated coef. = 0.17) resulted in unrecovered skids.

When comparing electric and hydraulic brake antiskid response, the best comparison is made for the 5,000 lbs wheel load case. Figure 18 shows an overlay of the wheel speed traces for the electric and hydraulic brakes for the first five seconds of the stops. A general similarity is shown as both traces show oscillations in the wheel speed beginning almost immediately.

Closer examination of the data reveals that the electric brake wheel speed trace shows more frequent and smaller peak to peak excursions in the wheel speed. This indicates that the brake torque, which drives the wheel speed changes, is being more accurately controlled by the electric brake. More accurate torque control is probably due to the torque feedback sensor which provides closed loop torque control not available with the hydraulic brake. More accurate torque control leads to higher deceleration and shorter stop distance during antiskid operation. The electric brake antiskid capability demonstrated at this wheel load is an improvement over the hydraulic brake.

Further evidence of the more accurate torque control of the electric brake is shown in Figure 19. The brake torque traces are shown for both brakes for the first five seconds of the stop. The electric brake torque excursions are about half as large as those for the hydraulic brake.

Although the above two figures are plotted for only the first five seconds of the stops, these trends continued throughout the stop and were even more pronounced at lower speed. Figure 20 shows both electric and hydraulic torque traces from brake application speed down to 20 mph. The higher frequency lower amplitude torque oscillations for the electric brake are clearly maintained. By virtue of a higher average torque, the electric brake decelerated the flywheel in about 25 seconds compared to about 30 seconds for the hydraulic brake. This yields a deceleration of 5.8 ft/sec^2 for the electric brake compared to 4.8 ft/sec^2 for the hydraulic brake.

ON-OFF RESPONSE STOPS

On-off response stops were conducted to evaluate the electric brake reaction to abrupt changes in brake pressure. The torque control feature, which limits the torque increase rate, was explicitly tested here. In the presence of a pseudo-step increase in pressure, the corresponding torque increase should be slowed similarly to the orificed hydraulic brake. During brake release modes, the electric brake actuator speed should be at a maximum to provide antiskid protection. Previous dynamometer testing indicated that the torque control feature was functioning as designed. For this test, it was further examined by instrumenting the voltage signal that limits the motor current.

The on-off response stop was conducted at the normal energy condition. The brake pressure was manually cycled on and off during the stop. For each cycle, the brake pedal pressure is increased at maximum rate to 2,300 psi (recall a proportional pedal signal activates the EBS), held steady for one to two seconds, and then released to back pressure. The brake torque rises in response to the pressure command and returns to zero with pressure release.

Figure 21 shows the brake pressure, torque, and motor current control voltage from a representative brake pressure cycle. The plot of the current control signal clearly illustrates the effect of the torque control feature. Brake application occurs at 8.3 seconds as shown by the step rise in the signal to 100 percent motor current. No current limiting occurs initially as the ball screw is translating through the brake clearance and is not yet applying clamping force to the brake stack. As soon as the torque begins to increase, motor current is regulated between about 30 percent and 90 percent to control the torque rise rate. When the torque is low, the current is reduced to about 30 percent of full value. As the torque level raises, the current is limited much less as low current would no longer support the desired torque increase. At 9.1 seconds, the motor current is essentially unlimited as the brake torque has reached its desired value. The controller interprets a steady state condition (torque magnitude equal to pedal) and allows the current to fold back to a low level to simply maintain this brake torque. The motor current pulses shown afterwards are corrections to achieve a relatively constant brake torque. At about 10.3 seconds, the brake pressure is released and a constant reverse motor current of 85 percent is shown. The controller has internally switched the motor voltage polarity (not shown) for brake release to running clearance. Note that there is no current control during brake release. This verifies the proper operation of the torque control feature as brake release must be unlimited to maintain antiskid performance.

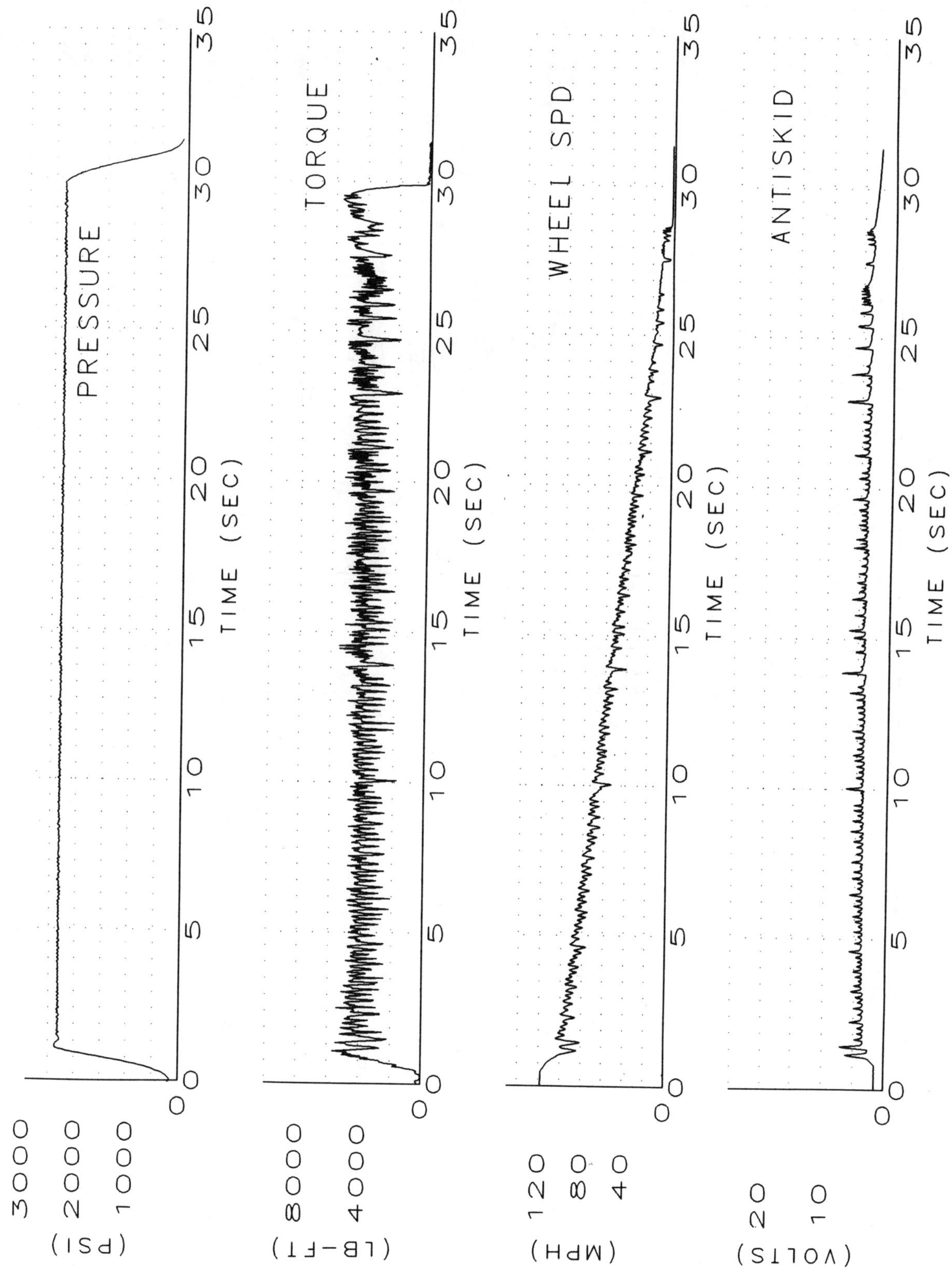

Fig. 17 - EBS Antiskid Stop, 5,000 Pound Wheel Load

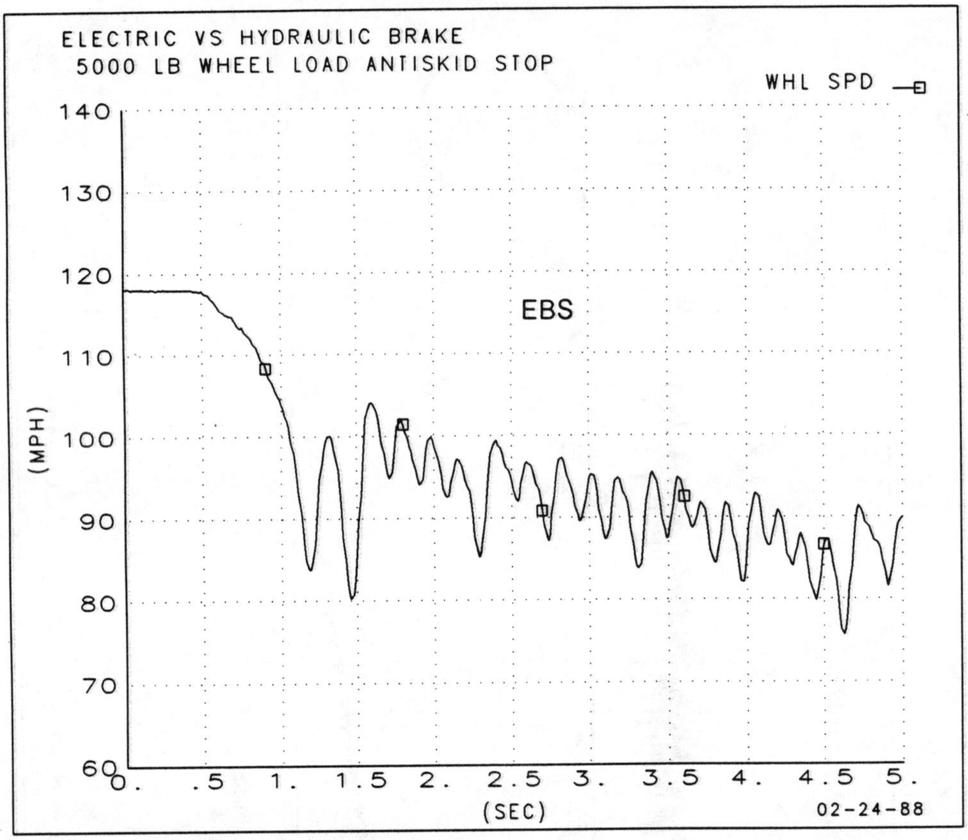

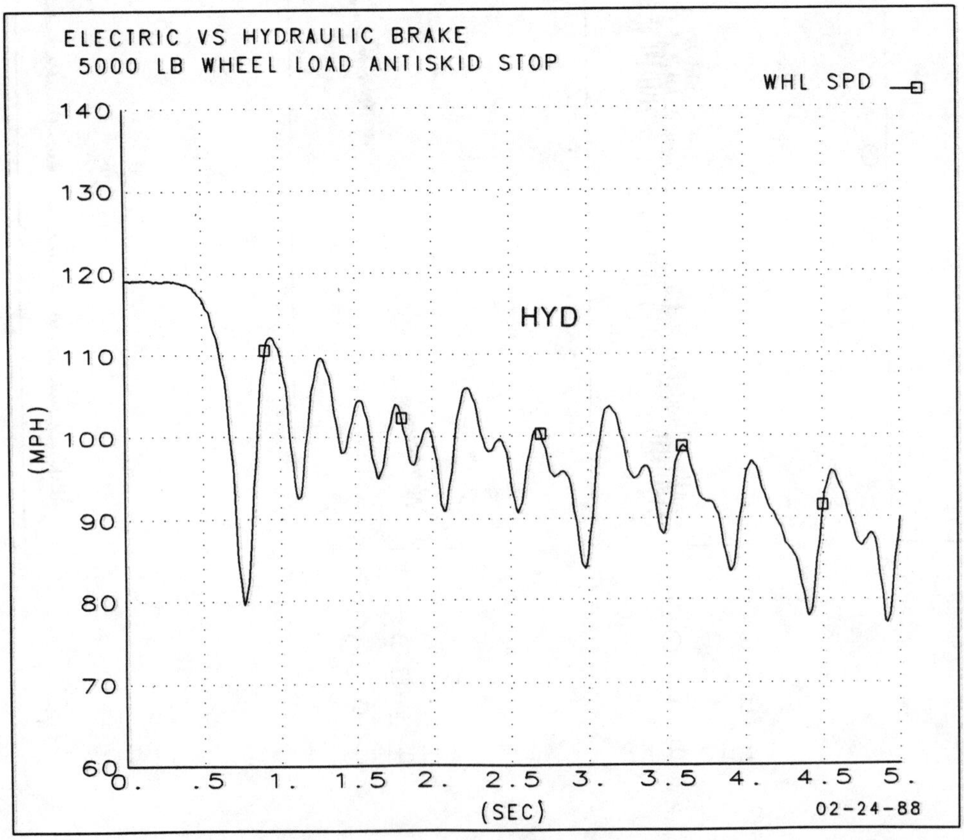

Fig. 18 - Antiskid Wheel Speed Control, EBS versus HYD

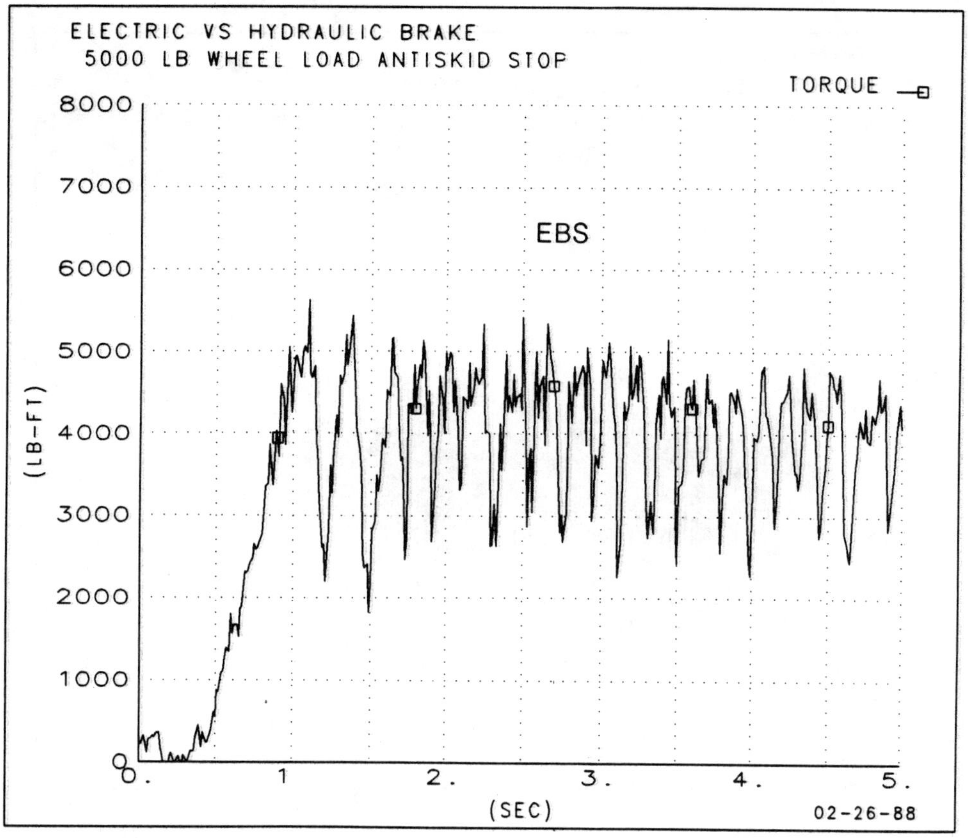

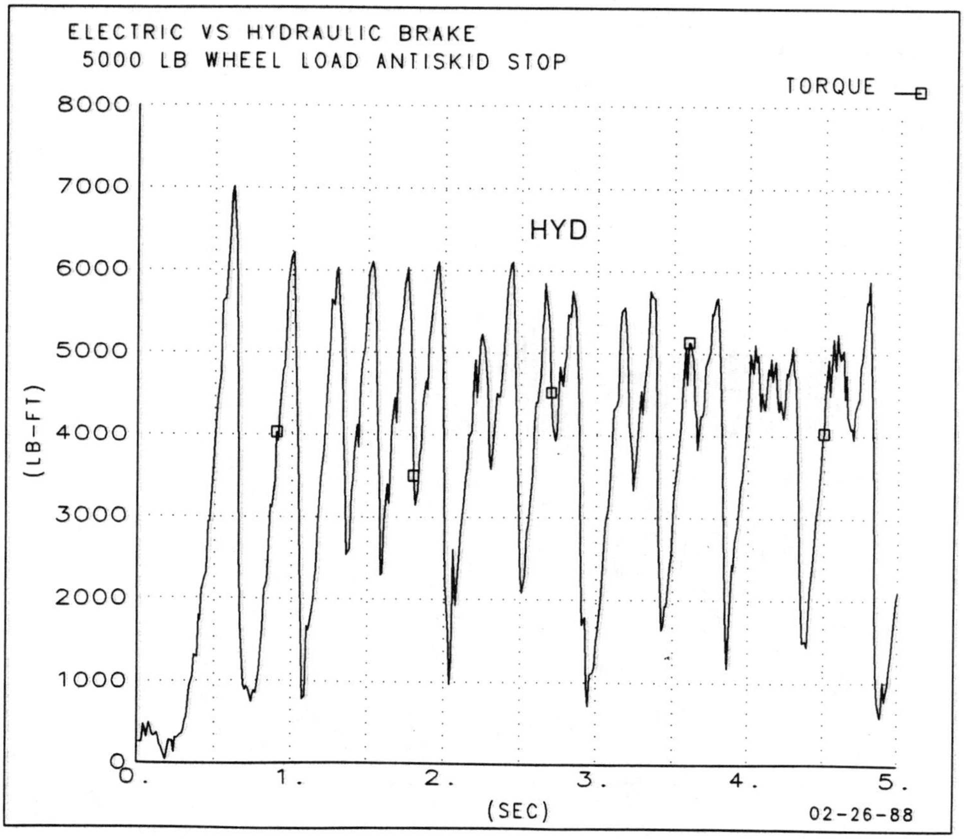

Fig. 19 - Antiskid Torque Control, EBS versus HYD

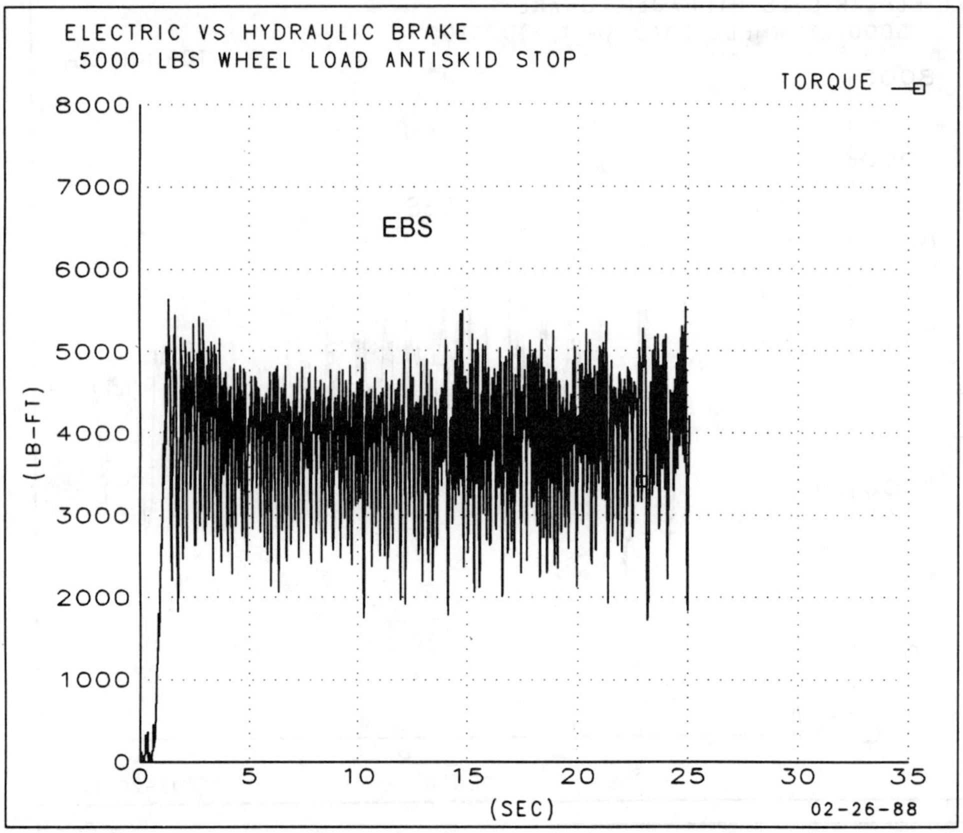

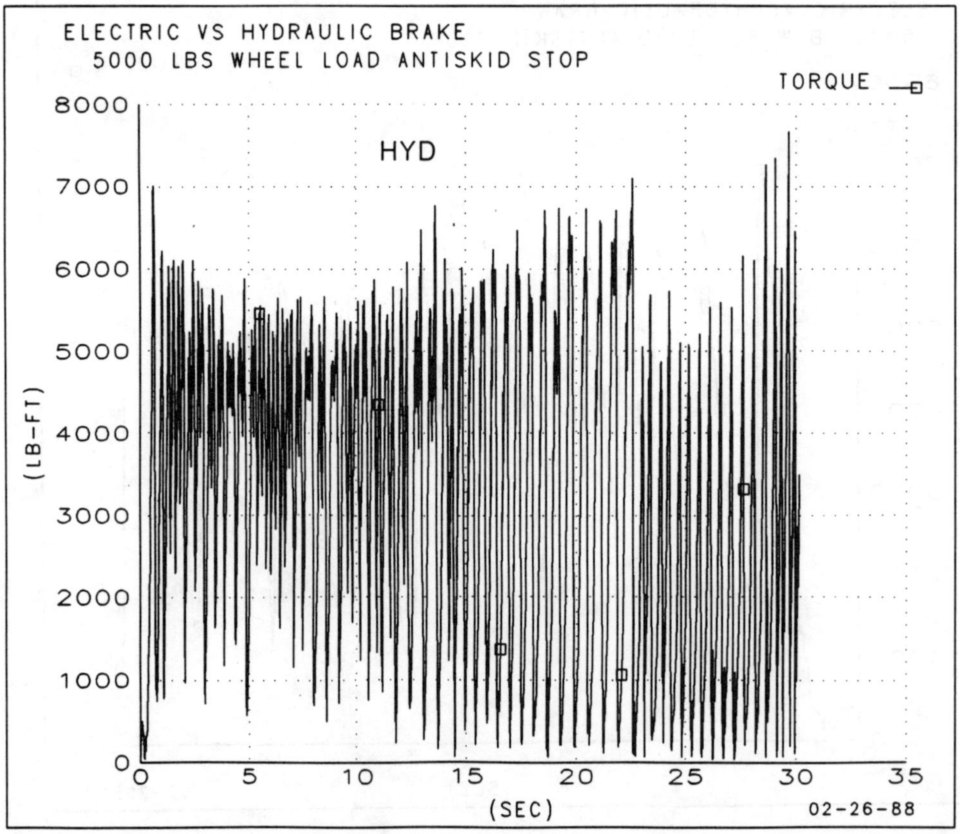

Fig. 20 - Whole Stop Torque Control, EBS versus HYD

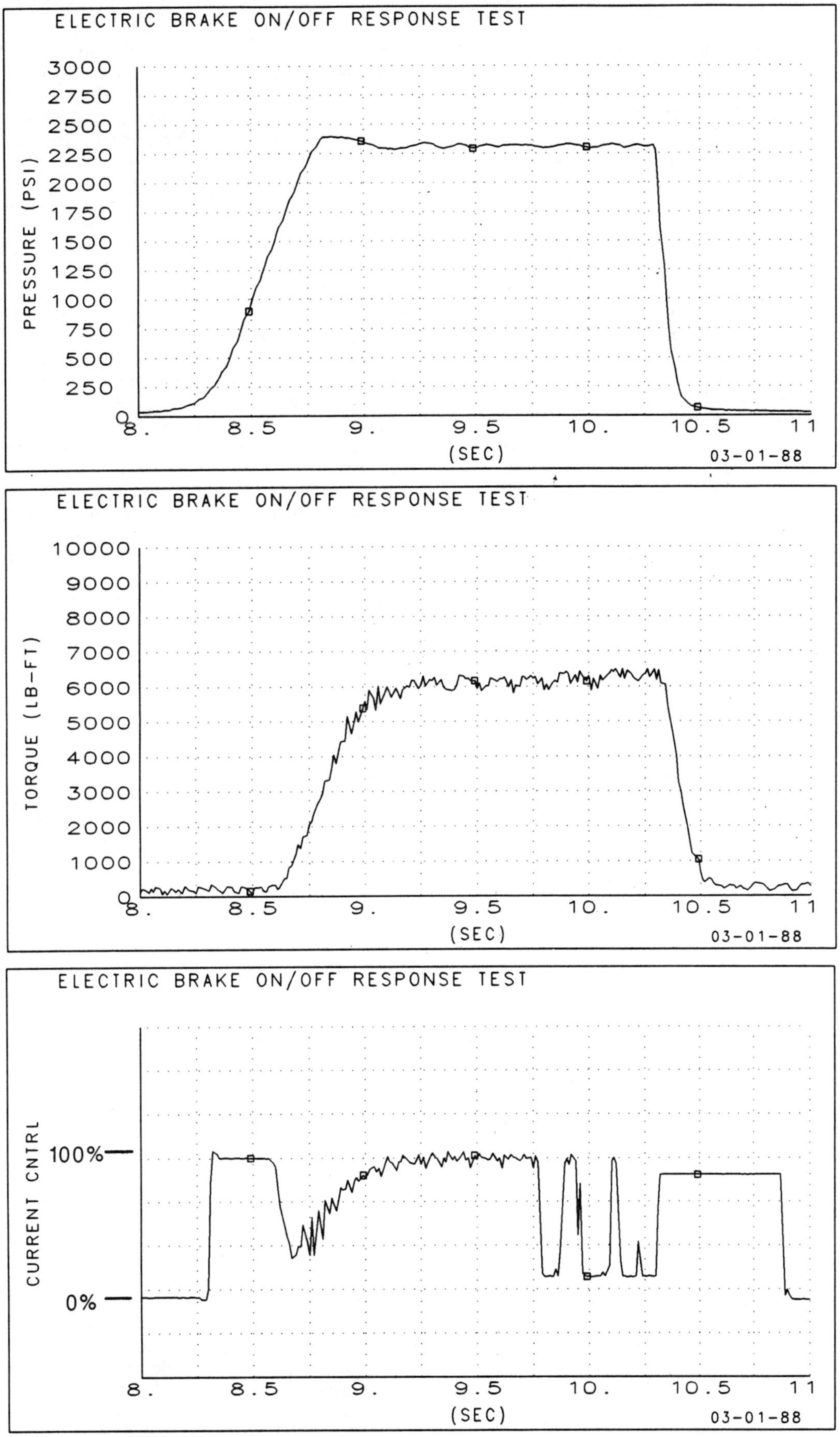

Fig. 21 - Electric Brake, On/Off Response Illustrating Motor Current Control

The on/off response test was also conducted with the hydraulic brake. Shown in Figure 22 is the electric brake pressure curve superimposed on the hydraulic brake curve. There are no brake piston cavities to fill with the electric brake and as a result, the pressure reaches its target maximum about twice as fast. The pressure decay rate is also faster, especially below 500 psi. Because these pressure curves are so different, a precise comparison between the hydraulic and electric brake actuator speed cannot be made. One effect will be maintained with electric brakes, the brake release can be faster than hydraulics at the low pressure end. Hydraulics will tend to release based on the pressure difference between the brake cavity and the return. At low cavity pressure, the release is much slower than at high cavity pressure. Electric brakes, however, can release at the maximum rate irrespective of the "pressure head." This could lead to improved antiskid response at very low runway coefficients where high speed torque release is required.

OVERLOAD ENERGY DECELERATION

An overload energy stop was completed with the EBS at a brake energy of 15.8×10^6 ft-lbs. The brake application speed was 164 mph and a constant brake pressure of 2,600 psi, which is the full system pressure, was used for this stop. The deceleration goal was the A-10 specification deceleration of 10.0 ft/sec^2. Relative to the normal energy stop, the overload condition is a much more severe test of the brake actuator as the overload stop energy is about 2.3 times that of the normal stop.

Figure 23 shows the brake pressure and torque curves for this stop. The brake pressure average was about 2,600 psi. The brake torque peaked at about 10,240 lb-ft early in the stop, fell off to about 6,830 lb-ft about 15 seconds into the stop, then recovered to about 9,600 lb-ft at the end of the stop. This was in proportion to the brake disk coefficient which degraded during the middle of the stop due to the high energy input rate. The brake torque average for the entire stop was about 8,100 lb-ft and the stop time was 21.95 seconds. The stop deceleration was 10.96 ft/sec^2 which surpasses the 10.0 ft/sec^2 goal.

REFUSED TAKEOFF

The maximum energy RTO stop was completed at a brake energy of 20.6×10^6 ft-lbs, some three times the energy of a normal stop. The brake application speed was 175 mph and a constant brake pressure of 2,600 psi was used. The deceleration goal was again 10.0 ft/sec^2. A new brake stack was installed prior to this test.

Figure 24 shows the brake pressure and torque curves for this stop. The brake torque peaked at about 10,440 lb-ft about 2 seconds into the stop. The torque began to decay rapidly at that point and eventually fell off to a minimum of about 3,000 lb-ft about 33 seconds into the stop, then recovered to a value of about 5,400 lb-ft at the very end of the stop. The average torque for the entire stop was about 4,950 lb-ft. The deceleration was low at 6.24 ft/sec^2.

Two reasons combined to cause the brake torque and deceleration to fall short. The biggest factor was low ball screw efficiency. The ball screw design is experimental and did not meet expectations for efficiency and life. The accumulated wear on the screws reduced their efficiency which directly reduces the brake clamping force and torque capability. A deficiency in maximum clamping force will show most clearly in the RTO condition. The brake energy is so high that localized melting off the rubbed surfaces occurs. To maintain torque under this condition requires a significant increase in clamping force, over and above that required for normal operation, which exceeded the capability of the electric brake actuator.

One other factor may have impacted the brake performance. The initial brake torque of 10,440 lb-ft exceeds the required average torque of 8,300 lb-ft to achieve a 10 ft/sec^2 deceleration. This high torque in combination with the high speed at the beginning of the stop yields an exceptionally high brake energy rate. The energy rate is a function of the brake torque multiplied by the instantaneous velocity and other relatively constant terms. For the electric brake, the energy rate reached a maximum of 2.05×10^6 ft-lbs/sec at the maximum torque. By comparison, the hydraulic brake qualification test run in 1977 has a maximum energy rate of 1.48×10^6 ft-lbs/sec. Therefore, the electric brake energy rate was almost 40 percent higher than that of the hydraulic brake qualification test. The extremely high energy rate causes rubbed surface melting early in the stop which in turn compounds the need for high clamping force to maintain brake torque. For the electric brake then, a lower initial torque may have achieved higher deceleration by delaying the onset of torque decay.

SUMMARY AND CONCLUSIONS

Dynamometer testing has demonstrated that the electric brake stopping power is equivalent to the hydraulic brake for normal and overload energy stops. The hydraulic brake has superior clamping force and brake torque output, and this enables it to exceed electric brake performance for the RTO condition. The electric brake clamping force deficiency is directly related to the low efficiency of the experimental ball screw design. A more efficient actuator configuration, already under study, would yield clamping force levels equivalent to hydraulic brakes.

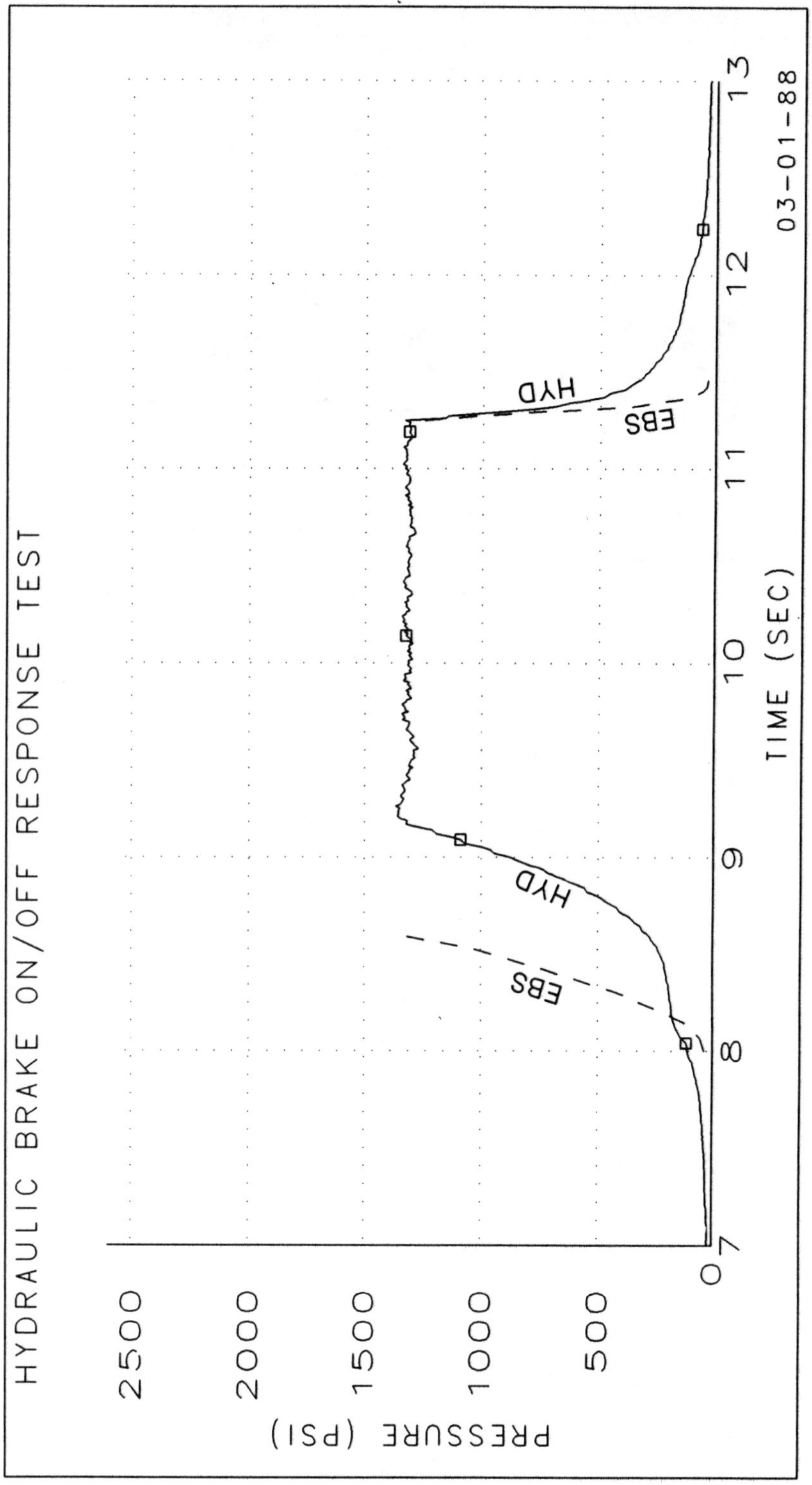

Fig. 22 - Pedal Pressure Response Comparison, EBS versus HYD

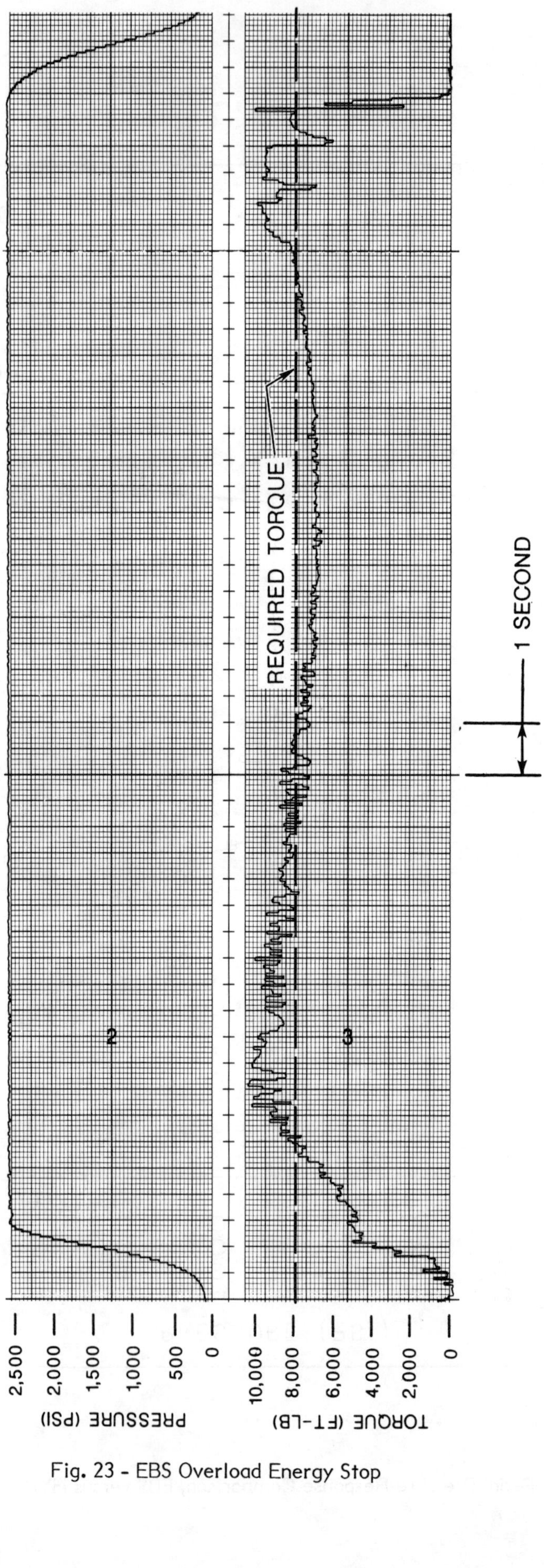

Fig. 23 - EBS Overload Energy Stop

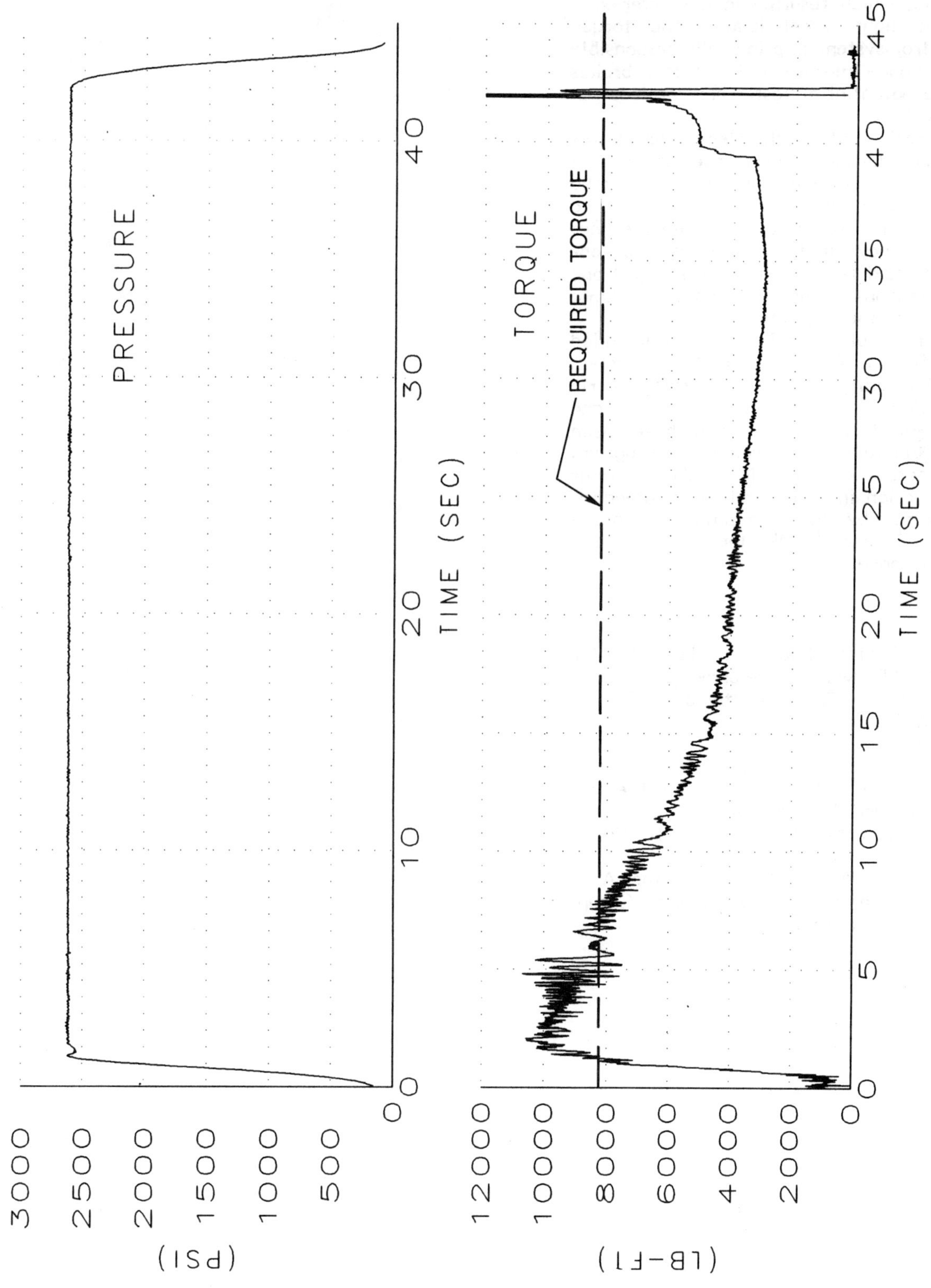

Fig. 24 - EBS Refused Take-Off Stop

The electric brake demonstrated more accurate torque control than the hydraulic brake during antiskid testing which resulted in a shorter stop distance at a light wheel load. The torque feedback control system is principally responsible for the performance improvement. Electric brakes also offer the potential of faster brake release at low torque levels.

The application rate of the electric brake was modified by the addition of the torque rate control circuitry. It effectively limited the brake application to emulate the orificed hydraulic brake. This control would have reduced the gear excitation tendancy that terminated the on-aircraft ground taxi test. This represents just one approach to response tuning to obtain dynamic compatibility with the airframe.

With the first fully functional system installed on an aircraft, the A-10 electric brake system represents a feasibility study into aircraft electric brake actuation. Both strong points and shortcomings of this initial design have been identified in the four-year development program. Potential benefits of electric brake systems include a reduction in brake system maintenance and fire hazards by the elimination of brake hydraulics, and improved brake torque control and antiskid performance.

REFERENCES

1. Capt. Tom Carter, "Laboratory Test Plan for Electric Brake Hardware Development." Flight Dynamics Laboratory, Wright-Patterson AFB, OH, September 16, 1985.

2. D.D. Moseley and R. Morris, "Qualification Test Report for the PD 2655-10 A-10A Electric Brake Assembly and Electric Brake Controller." Loral Systems Group, Akron, OH, August 18, 1986.

3. D.D. Moseley, "Final Report for the A-10 Electric Brake Program." Loral Systems Group, Akron, OH, December, 1987.

Struts, Couplings and Actuators

Aircraft Landing Gears - The Past, Present and Future

D.W. Young
Dowty Rotol Ltd.
Cheltenham, Gloucestershire

* Paper 864752 published by I.Mech.E
(U.K.), 1986.

This paper discusses landing gear basic requirements with a review of historic and current equipment designs, and concludes by pointing to possible future developments to improve functional efficiency.

1 INTRODUCTION

Aircraft take-off and landings are now virtually hazardless due to the improvement of the aerodynamic art, the control exercised by the air traffic authorities and the airworthiness requirements which have evolved. The landing gear has also progressed much since the early days when human legs, skids or spoked wheels were the alternatives, but because of its weight and volume, it has an adverse effect on the aircraft's economy and performance.

Attempts to dispense with the conventional wheeled type of landing gear to improve efficiency have not been successful, except for specialized types, and thus this paper will be confined to the nosegear tricycle wheeled variety with major emphasis given to the structure and suspension design.

2 DESIGN REQUIREMENTS FOR LANDING GEARS

2.1 Basic requirements

While to the fare-paying passengers the landing gear is no more than a set of wheels, it is a sophisticated device with several, sometimes conflicting, requirements.

The landing gear must have an adequate base and track, positioned relative to the aircraft's centre of gravity so that ground manoeuvring characteristics are not limited by the aircraft being easily overturned sideways or by tipping back onto its tail. The geometry must also provide clearance of the aircraft with the ground at all operational conditions with nosegear static ground loads that allow steering and maingear loads high enough for effective braking.

The need for an aircraft to operate from a given strength of runway or unprepared ground dictates the size and number of tyres and their pressures. Many different tyre arrangements are in use with up to eight tyres per landing gear used to provide a footprint which will not damage the ground surface.

There are many methods for assessing the effect of landing gears on rigid runways and semi-prepared surfaces, but these are far too detailed to discuss here. The current system for rigid runways is the ACN (aircraft classification number) and PCN (pavement classification number) system introduced by the ICAO (International Civil Aviation Organization).

This paper was presented at a Joint Meeting of the Institution of Mechanical Engineers and the Royal Aeronautical Society held in Bristol on 17 April 1985. The MS was received on 4 June 1985 and was accepted for publication on 17 July 1985.

The importance of tyre and undercarriage characteristics for satisfactory rough and soft ground compatibility has long been recognized, but weight and installation constraints sometimes dictate landing gear designs with significant limitations when operating from rough surfaces.

The main task of the landing gear is to absorb horizontal and vertical energy. The former is catered for by the brakes and the latter by the shock absorber. Ideally, the shock absorber system should cater for all conditions, but the requirements for good ground ride are neither compatible with the need to absorb landing energy, nor the need to provide space for wheel brakes to absorb the horizontal energy.

Evolution of solutions which minimize the volume of the stowed landing gear and provide a low weight, is a prime task of the landing gear designer. The frontal area necessary for the landing gear, its volume and installed weight all affect the performance of the aircraft adversely, so realistic and thorough evaluation of landing gear solutions is essential at an early stage. Often a good solution for the landing gear can penalize the airframe and the correct solution for the aircraft is not, necessarily, the one which provides the lightest landing gear structural weight.

Retraction of the landing gear before the aircraft achieves critical.speed after take-off is vital, and design of the retraction mechanism and its associated hydraulics system to keep hydraulic demands low is important for minimum system weight. Emergency lowering by some means is always a requirement.

Landing gears have additional tasks placed upon them to ease maintenance and operation. Attachment points for towing, jacking and debogging are provided, and picketing points are sometimes specified on naval aircraft. In addition, limit switches to sequence or inhibit the action of various functions of the landing gear or other aircraft systems are necessary on most aircraft.

2.2 Energy requirements

The landing energy requirement results from the aircraft's landing mass and its vertical velocity. A simple theoretical diagram of the suspension system is shown in Fig. 1.

The reaction factor is the ratio of landing gear reaction to its associated mass and varies from less than unity to about 4, determined by the type of aircraft. The value specified results from consideration of the 'hardness' of the structure and passenger comfort.

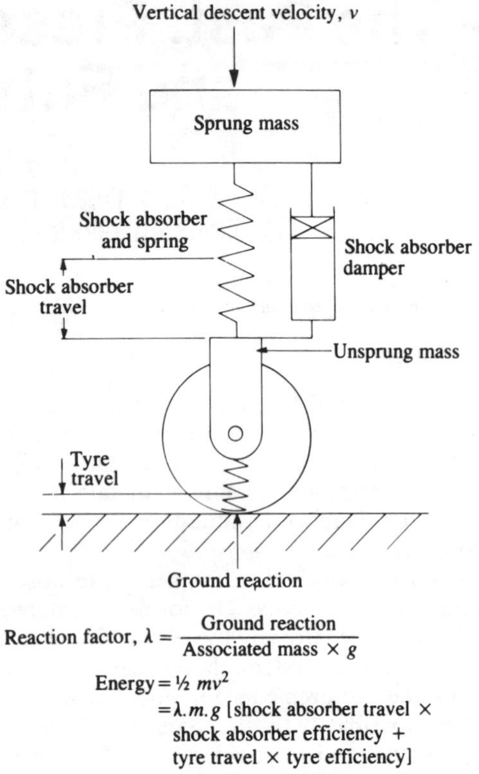

Fig. 1 Shock absorber schematic

The current requirements for vertical descent velocity range from 3.05 m/s for civil aircraft, through 3.66 m/s for fighters and 4.00 m/s for trainers, to above 6.00 m/s for deck landing aircraft.

The combination of reaction factor, vertical descent velocity and shock absorber efficiency determines the vertical axle travel necessary to absorb the landing energy. The tail-up and tail-down attitudes have an effect on the vertical axle travel required, as has the selection of suspension, since lever suspension geometries usually require a greater travel due to changes of mechanical advantage over their travel due to motion of the axle in an arc.

An idealized diagram of shock absorber reaction plotted against travel is shown in Fig. 2. The reaction at any points during the shock absorber travel is the sum of the gas spring force and the oil damping force (which is usually a function of the square of shock absorber velocity). The area under the curve represents the energy capacity of the shock absorber.

On touchdown, the wheel spins up to ground speed under the influence of the ground reaction and tyre/ground friction. The resulting drag force at the axle stores energy in the structure. When the tyre velocity matches that of the ground, the stored energy is released and the structure vibrates half a cycle to produce reverse loads at the axle. This is known as springback.

The shock absorber continues to close until all vertical energy has been absorbed, and the maximum stroke is defined by the polytropic spring curve. It can be seen that the energy is absorbed in a relatively short time. The landing then enters its run-out phase with oscillation about the static position which will be dependent on the degree of airborneness at any point in time.

2.3 Strength–load cases

Strength–load cases include combinations of vertical drag and side loads generated by a number of different landing and ground conditions, which depend on the particular certification requirements for the aircraft.

These cases include wheel spin-up, springback and lateral drift during landing with variations of aircraft attitude in pitch and roll, taxi and ground manoeuvring cases, and towing operation. Burst and unequally inflated tyres are also taken into account.

In general, stresses produced by any load case must

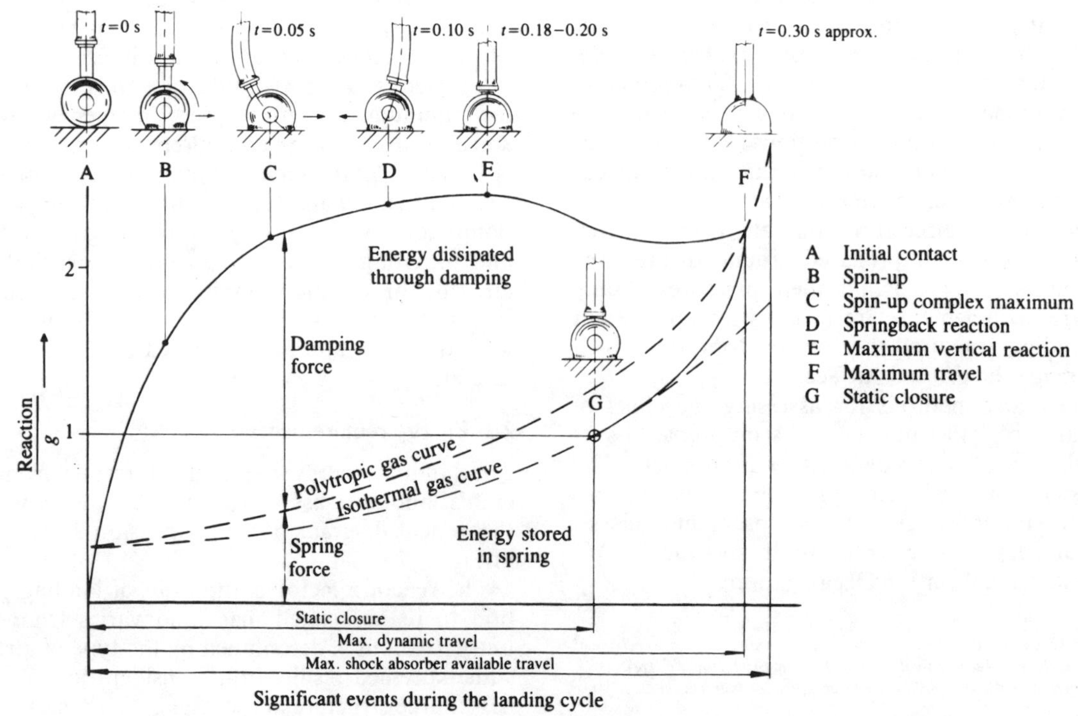

Fig. 2 Significant events during the landing cycle

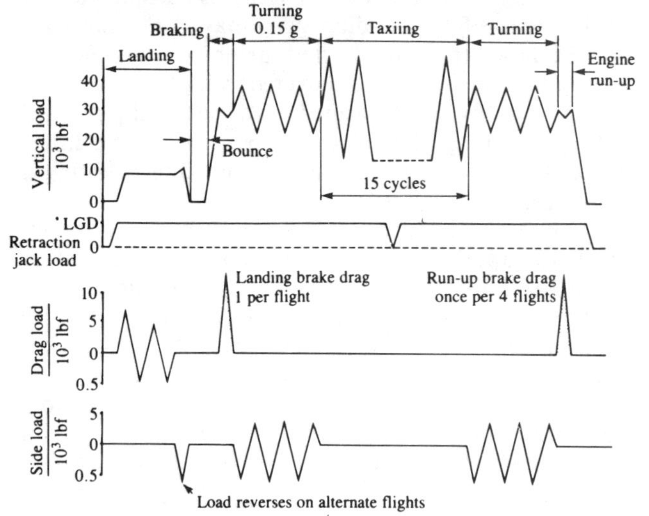

Fig. 3 Typical fatigue test spectrum

be less than the elastic limit of the materials used and no more than two-thirds of their ultimate strength.

2.4 Fatigue

With the ever-increasing length of aircraft operating life, design for good fatigue resistance assumes greater prominence, particularly for civil aircraft with low reaction factors where landing loads assume little significance.

Figure 3 is a typical fatigue test spectrum for an aircraft with a reaction factor of about 1.4. This spectrum is resolved from a much more complex set of conditions to provide equivalent damage for a practical test spectrum. The diagram shows the vertical, drag and side loads generated by various operations. Fatigue case landing loads are much less than those during taxi and the number of cycles is very much less, since an aircraft can taxi 500 000 km or more in its working life, but landings usually occur only once per flight.

Fig. 4 Drop tests of AV-8B nose landing gear

2.5 Performance and strength testing

For certification, it is essential to demonstrate that the equipment meets the design requirements.

Drop testing demonstrates the ability of the landing gear to meet the energy requirements within its available travel, and is performed on a rig which can apply the appropriate energy with the correct vertical velocity at touchdown. Drag forces are simulated by spinning the wheels in the reverse direction to normal motion. Figure 4 shows the AV-8B nose landing gear being prepared for drop testing on a swinging arm rig.

Fig. 5 Strength test BAe 146 main landing gear

Fig. 6 Fatigue test A310 main landing gear

Strength tests are performed using a strong baseplate to apply loads for a number of cases before applying the case which is critical for ultimate strength. Both loads and deflections are recorded. Where it is considered that certain areas of the landing gear may prove strength critical, these are often tested prior to manufacture using aluminium models covered with photo-elastic material. Analysis is performed using reflected polarized light by counting the tints and passage of the interference fringes. Figure 5 shows the BAe 146 main landing gear strength test.

Fatigue tests are carried out using special test frames, with loads applied to servo-valve controlled actuators and load cell feedback to the controlling microprocessor. The processors are programmed for flight by flight application of the test spectra and testing proceeds continuously, stopped only by automatic safeguards and the need for maintenance and inspection. The installation of the Airbus Industry A310 main landing gear in its fatigue strength frame is shown in Fig. 6.

3 HISTORY

The design of landing gears has followed the development of aircraft and their operational requirements. Design development is a continuing process resulting in as great a number of solutions as there are different types of aircraft. Weights of aircraft have increased from 225 kg at the turn of the century to 340 000 kg now.

Table 1 identifies significant landing gear developments against time. For the structure, the main events are the development of various types of shock absorbers, retractable landing gears, the introduction of the multi-wheeled bogie landing gear, and the use of ultra-high tensile strength steels. For rotating

Table 1 Significant landing gear developments

Year	Structure	Wheel equipment
1900		
	Wright Brothers' flight	
	Oil damper shock absorbers	
1910		'Special' wheels and tyres
	Oleo shock absorbers	'Wired on' tyres
		First wheel brakes
	Retractable landing gear	
1920		
		Wheel brakes in general use
	Internally sprung wheel	
1930		Differential brakes demonstrated
	Retractable landing gears for production aircraft, portal frame maingears, liquid spring development	Hydraulic braking systems
1940	Bogie landing gears	First twin-wheeled main and nose-gear aircraft (B29)
	Pressure welding of steel parts	
	Adoption of nosegear layout	
	High strength aluminium alloys	Disc brakes
1950		Anti-skid systems (mechanical Maxaret)
	Ultra-high tensile steels used for landing gears	High-pressure tyres
1960		Tubeless tyres
	Stress corrosion resistant high strength aluminium alloys	Electronic anti-skid systems
1970	Electronic steering control	
	Nitrogen supplemented liquid spring	Carbon disc brakes
1980	Passive shock absorber devices for ground ride	
	Adaptive shock absorber devices for ground ride	
1990		

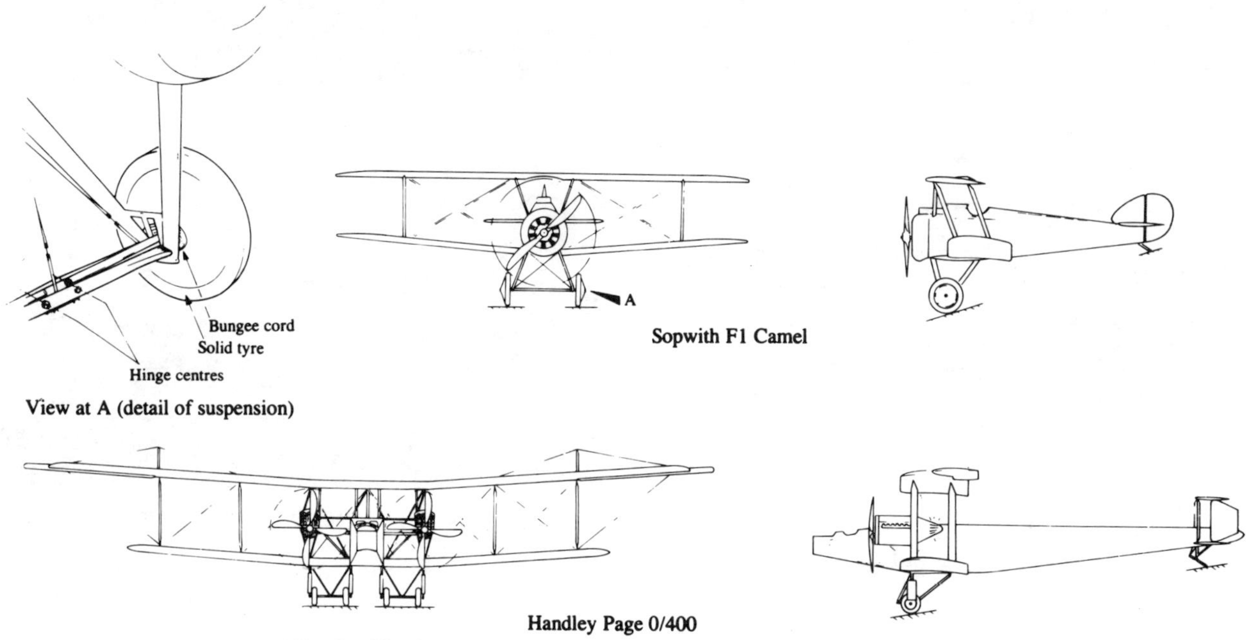

Fig. 7 Single and twin-engined aircraft landing gear, World War I

equipment, the introduction of wheel brakes, disc brakes, high-pressure tubeless tyres, anti-skid development, carbon brakes, and the development of the radial tyre are the most significant events.

3.1 The 'experimental' era

For early aircraft, as shown in Fig. 7, the fragility of wing structure meant that, for single-engined aircraft, the landing gear was attached directly to the fuselage in the region of the engine mounts. For multi-engined aircraft, direct connection to the main masses was also necessary. This figure shows a sideways lever arrangement which is connected to the fuselage and the engine mounting structure.

The 'vee' type landing gear, the usual solution for World War I fighters, often had a one-piece axle sprung by rubber cord wound around the axle and apex of the 'vee', although some aircraft used steel springs with or without damping. Brakes were not fitted and the track was very narrow, with poor landing and take-off stability. This disadvantage was outweighed by its relatively low drag. The split axle, introduced by Sopwith, helped to improve stability by providing positive location, and this design continued into the 1930s. The poor lateral strength of the 'vee' type undercarriage tended to be considered as a crashworthy feature which protected the airframe.

Vertical descent velocities of up to 4.5 m/s for aircraft of this era were not unusual. Reaction factors were similar to the flight factors for airframe design, and varied from 4 to 8 depending upon aircraft type.

During this period, it was appreciated that low shock absorber efficiency due to the absence of damping caused high airframe reactions and undesirable recoil characteristics. Several inventors proposed the use of oleo-pneumatic damping to help overcome this problem, notably Duncan of Vickers, J. D. North of Boulton Paul and, in France, Breguet, who should be credited with the first application of the oleo-pneumatic shock absorber to aircraft. In the UK, oleos were fitted to BE2, F.E2b, Siskin, Avro 504 and Bristol Braemar undercarriages. Some aviators, however, complained that oleos were not so good for 'getting off' because they did not 'bump' at all and full flying speed was needed before one could lift from the ground.

3.2 The 'between wars' period

Increasing cruising speeds made landing gear drag of greater importance. Efforts were made to reduce drag by using streamlined struts with fairings to cover the gap between tyre and wheel hub. Cleaner aerodynamic design and increased landing speeds meant that, by the end of the 1920s, wheel brakes were a positive requirement, and it was found that the conventional 'vee' type undercarriages could not react brake torques, and that redesign with additional members to react the torque imposed significant weight penalties. These factors led to the introduction of the internally sprung wheel (Fig. 8) which allowed a 'rigid' structure of low weight which could react the landing loads and brake drag readily. This development, resulting in an order from Kawasaki of Japan to George Dowty, led to the founding of the Dowty Group. The main disadvantage of this arrangement was the need for large-diameter, low-pressure tyres to permit adequate travel, but significant reductions of drag and overall weight were achieved. The installation on the Gloster Gladiator (Fig. 9) is a typical application which illustrates well the aerodynamic cleanliness of the concept.

Retractable undercarriages were considered from the early days of aviation. Probably the first application was that of the Martin Company in the USA, which proposed a design during World War I which claimed to reduce the aircraft's drag by 10 per cent and increase its speed by 6 per cent. Credit should be given to the Bristol Company who produced one of the first retractable undercarriages for the Type 72 Racer of 1922/23. This particular concept, where the landing gear follows the fuselage profile when retracted, can be seen on the present-day RFB Fan Trainer.

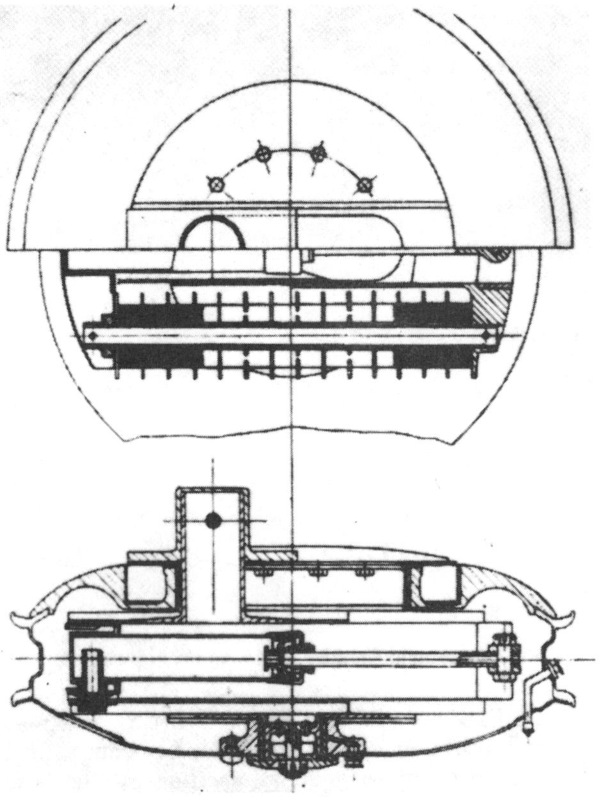

Fig. 8 Internally sprung wheel

The move towards retractable landing gear was of considerable importance, but biplanes did not lend themselves to their general introduction, and many designers considered that speeds would not increase enough to make the weight penalties worthwhile.

Indeed, it was not until 1936 that the Avro Anson monoplane was the first aircraft with retractable landing gears to be put into service with the RAF, although notable efforts had been made by Messier and Lucien in France with the Messier 'Laboratoire' aircraft (Fig. 10). The resemblance to the Harrier landing gear layout is of interest.

During the mid-1930s, retractable telescopic units were designed to fit into the nacelles of bomber aircraft and the wings of fighters. The twin shock absorber portal frame undercarriage became the standard on the heavy and medium bombers as exemplified by Fig. 11, which shows a Lancaster installation. For single-

Fig. 9 Gloster Gladiator

Fig. 10 Messier 'Laboratoire' aircraft

engined fighters, the single telescopic oleo leg was almost universally adopted, exemplified by the Spitfire installation in a relatively thin wing, which had the demerit of a narrow track.

This era also saw the introduction of the first true nosewheel type aircraft. This arrangement protected propellers and enhanced stability during take-off, on landing and when braking.

Steering control was, in general, by differential engine and brake control or limited rudder pedal steering.

3.3 1940 to the present

The advent of the jet engine posed new problems for both aircraft and landing gear designers. The reduction of aircraft height from the ground, resulting from the absence of propellers, led to a new set of landing gear solutions to overcome the new problems.

The mechanically shortening lever main landing gear suspension for the Meteor aircraft (Fig. 12) was designed to stow itself in the wing between engine and fuselage. This arrangement was a forerunner of the geometries used on the later Gnat, Buccaneer and Harrier aircraft.

The Meteor nosegear, although it did not shorten, used a tension shock absorber to provide a short single-

Fig. 11 Lancaster landing gear

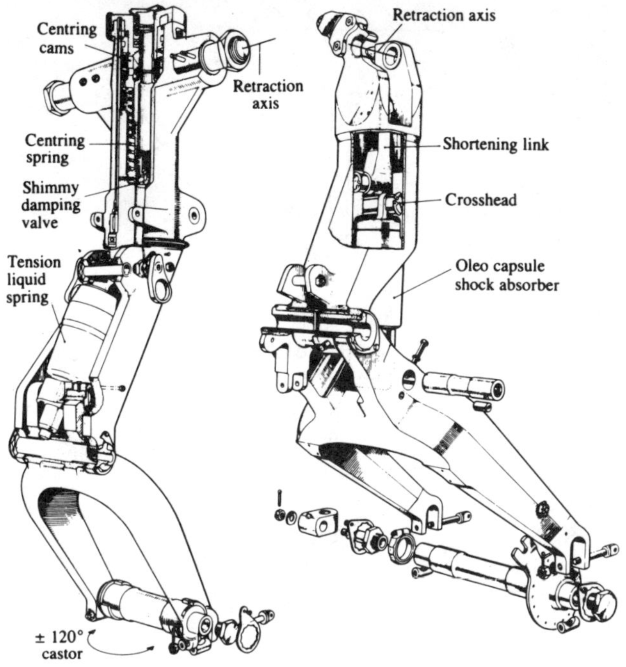

Meteor castoring nose landing gear **Meteor shortening main landing gear**

Fig. 12 Meteor landing gears

wheel lever suspension undercarriage with a relatively small retracted volume, due to the small mechanical trail of the wheel.

Prior to World War II, tyre pressures were restricted to about 0.24 MPa with wheel loads, in general, no more than 7000 kg due to the need to operate from grass runways.

The rapid increase in aircraft weight during World War II caused single-wheel loads in excess of 14 000 kg and made tyre diameters of up to 1600 mm necessary in order to keep tyre pressures below 0.31 MPa. This increase led to the use of concrete or asphalt runways.

At the same time, plans were laid for larger aircraft; both the Bristol Brabazon in the UK and the B36 in the USA had all-up weights of about 140 000 kg, and it became obvious that even twin-wheel undercarriages were not feasible due to pavement strength limitations and the impracticable size of the wheels needed for leg loads of 70 000 kg. These aircraft were the first larger size aircraft with multi-wheel bogie main landing gears which allowed tyre diameters to be reduced to below 1170 mm with maximum single-wheel loads of 22 500 kg, which also helped to extend tyre life and reduced the weight of the rotating equipment considerably.

Following World War II, there was an upsurge of development of different types of aircraft with a wide range of landing gear requirements and geometries. Many of these aircraft were not put into service. From the experience gained in designing, developing and manufacturing landing gears for these aircraft together with the feedback from those in service have evolved the practices now used for landing gear design and manufacture.

The installation of main landing gears into thin wings meant that a number of fighter aircraft, such as the Lightning and F4, had very high tyre pressures—up to 2.75 MPa—which was feasible from good runways and aircraft carriers. These same space constraints also meant that innovative, and often difficult geometries had to be evolved.

High wing propeller driven aircraft required installation of the main landing gear in the engine nacelle with a long leg length, and many ingenious geometries were developed to stow twin-wheel units with minimum impact on the nacelle volume.

The stowage of bogie maingears within nacelles or deep wings, such as the Vulcan and Victor aircraft, similarly produced complex solutions where leg shortening and rotation of the bogie for retraction was necessary.

More recently, the high wing fighter aircraft configuration, sometimes with variable geometry, together with the need for unencumbered under-wing and fuselage surfaces has led to the more general adoption of fuselage-mounted main landing gears for this category of aircraft.

There are almost as many unique landing gear solutions as there are aircraft. Each type presents its own set of geometric and operational conditions, so that the landing gear has to be 'custom built' in order to provide an optimum solution.

4 DESIGN ASPECTS

The economic facts of life have become plain to the aviation business over the last twenty years, and the large investment needed for new aircraft programmes with the emphasis on reducing the cost to the fare-paying passenger, combine to affect the landing gear designer in much the same way as the aircraft manufacturer.

The drive for minimum cost and weight with maximum reliability and maintainability has resulted in design simplification with the use of proved rather than innovative solutions. Advances have been made in detail design and analytical areas to improve functioning and maintainability and improve fatigue resistance to meet the ever-extending aircraft life requirements.

4.1 Suspension geometry

There are two main types of landing gear suspension: telescopic (otherwise known as cantilever) and lever (also known as trailing link). In addition, semi-articulated types are a compromise between the two and have advantages when space is limited. A selection of basic geometries is given in Fig. 13 which can apply to single- or twin-coaxial wheel arrangements.

The telescopic unit generally provides the cost and weight effective solution where it can be employed, but this is not possible if the stowage is restricted in length, or the attachment point positions give large rake angles for a telescopic unit, which results in unacceptably high friction. Lever suspension solutions permit easy shortening for stowage, and their low friction and sensitivity to drag forces show considerable advantages for operation from rough ground or uneven runways.

On multi-wheel bogie landing gears it is essential to prevent overloading of the front wheels due to brake drag and minimize pitch oscillation of the bogie due to the relatively undamped tyre spring. It is also necessary to control bogie motion during landing to prevent high front wheel contact velocities.

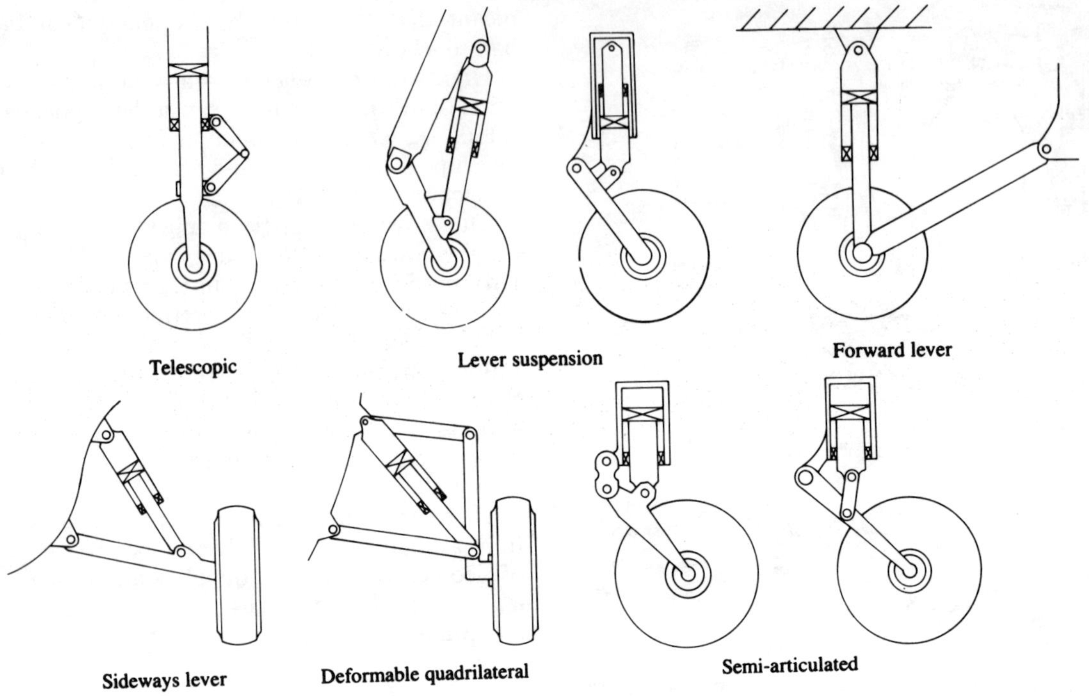

Fig. 13 Landing gear suspensions

It is usual to incorporate reaction rods, which connect the brake to the leg member and provide a moment about the bogie pivot which cancels that due to brake drag (Fig. 14). One or more dampers are fitted which control pitch oscillation during taxi, prevent second contacting wheel overloads on landing, and control the bogie position for retraction. On the innovative Victor main landing gear, the damper also acts as a secondary shock absorber with the rear tyres positioned to contact first and thus act as a lever suspension on initial contact.

Other bogie layouts are sometimes used where the rear wheels contact first; these types have controlled articulation and use a telescopic tie rod attaching near the front wheels to prevent front wheel slam. Again the bogie acts as a lever suspension during the first stage of landing. These types may rotate the bogie for retraction by shortening the tie rod or, in the case of the Vulcan, by pulling up the sliding member. In the latter case, the friction due to the inclined leg meant that a pitch damper was not needed.

Where height is particularly limited, as on the Comet and Brabazon, a lever type solution where the bogie is hinged at its centre point has to be used. On the Comet, a link and beam connects the forward lever to the upper shock absorber attachment so that front and rear wheel

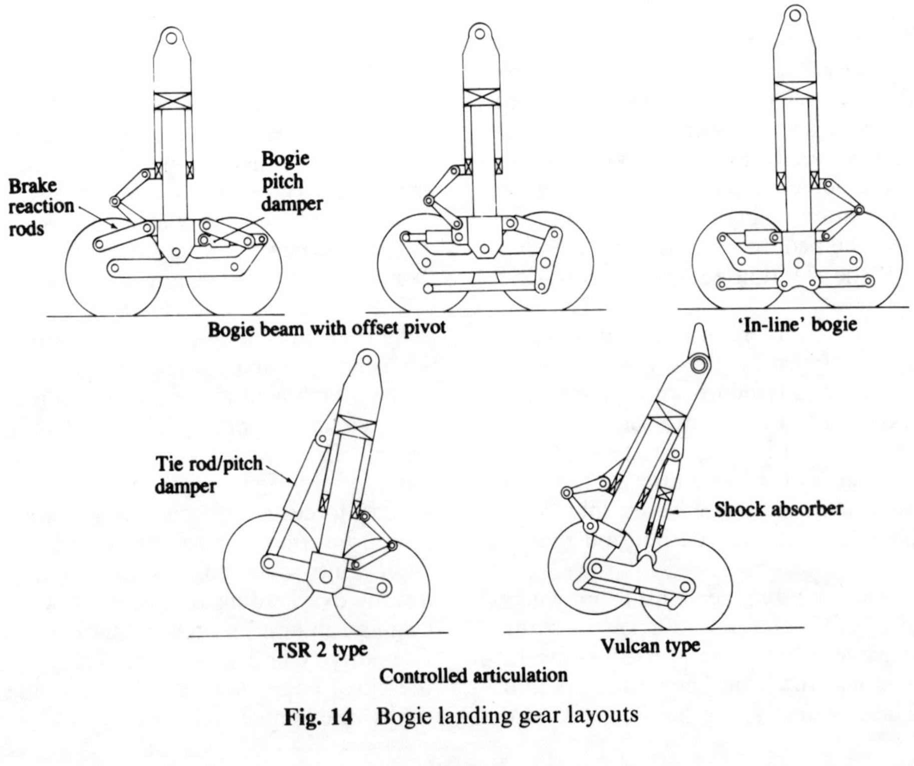

Fig. 14 Bogie landing gear layouts

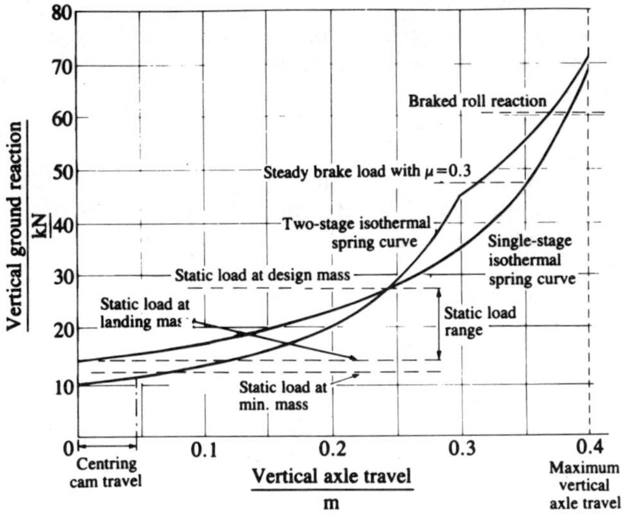

Fig. 15 Nose landing gear spring characteristics

loads are balanced. On the Brabazon, separate shock absorbers were used for forward and aft wheel levers.

4.2 Shock absorbers

The shock absorber is the heart of the landing gear system and its spring characteristics should be designed to prevent full extension or closure under any operational load conditions.

Figure 15 illustrates two types of spring curve for a nosegear. The nosegear has a wider range of conditions to cater for than the maingear, because of the effects of centre of gravity movement and braking reactions. Single-stage gas springs are not always adequate to cover the full range, although they are the normal first choice of the designer. The diagram shows, for this particular example, the need for a two-stage spring to prevent full leg extension under minimum static loads.

There are many types of shock absorber construction, a few of which are shown in Fig. 16:

(a) is a simple unseparated oleo where oil and gas can mix;
(b) shows an unseparated oleo with position dependent damping which is used to improve diagram efficiency;
(c) shows a simple separated oleo which can be mounted in any attitude and has no fluid aeration problems, but which has disadvantages because of its high polytropic gas index when reaction factors are low; and
(d) shows a typical two-stage shock absorber with a 'separated' second stage gas spring. The second stage is inflated to a pressure which corresponds to the spring changeover load. When this load is exceeded the two springs operate in series.

Where high damping at low velocities is necessary, so as to control aircraft pitching during taxiing, a two-level damping valve is often fitted to the nosegear. This opens under the influence of damping pressure in a bi-stable manner to provide landing mode damping when a predetermined closure velocity is sensed.

There are, however, a number of instances where liquid springs or nitrogen supplemented liquid springs provide the most efficient installation because of space or spring curve characteristic reasons.

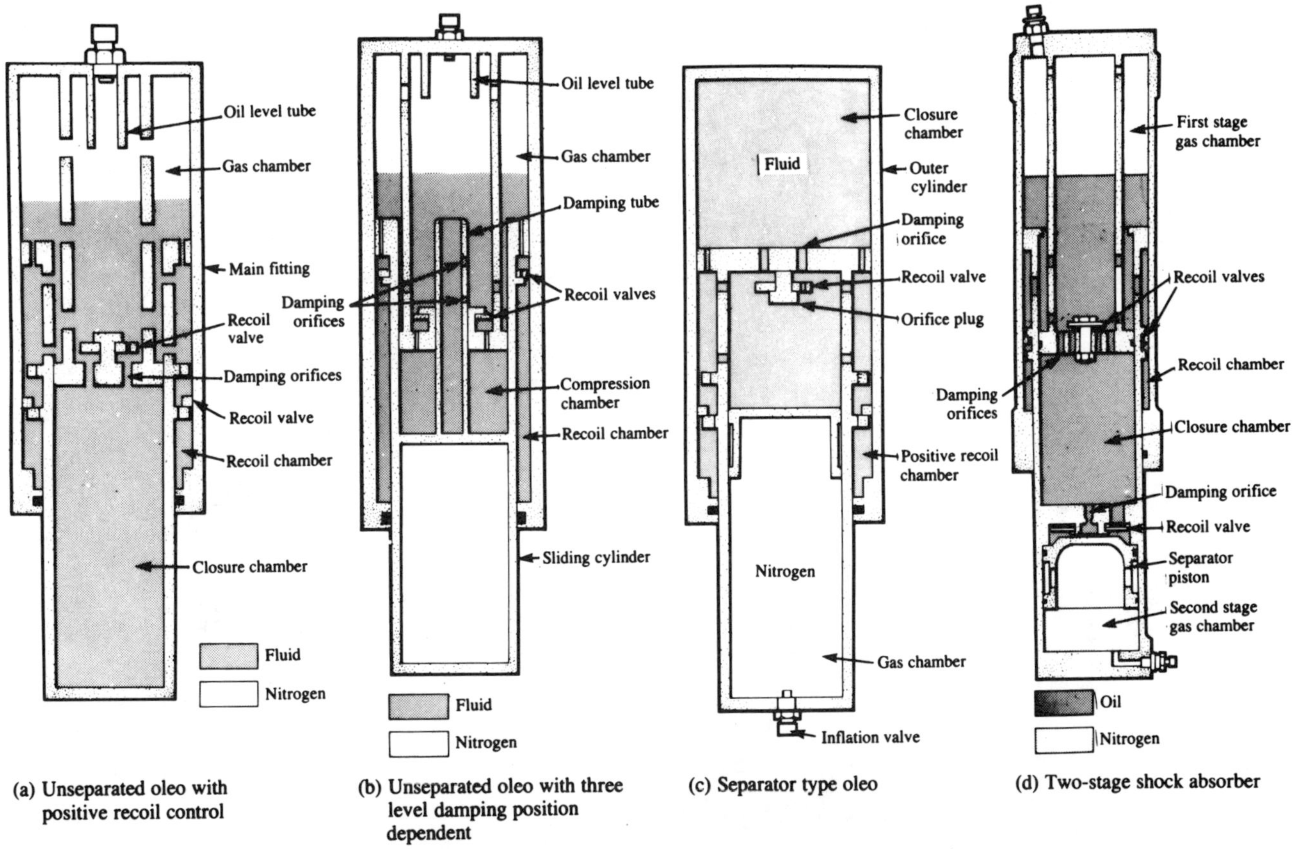

(a) Unseparated oleo with positive recoil control

(b) Unseparated oleo with three level damping position dependent

(c) Separator type oleo

(d) Two-stage shock absorber

Fig. 16 Several types of shock absorber

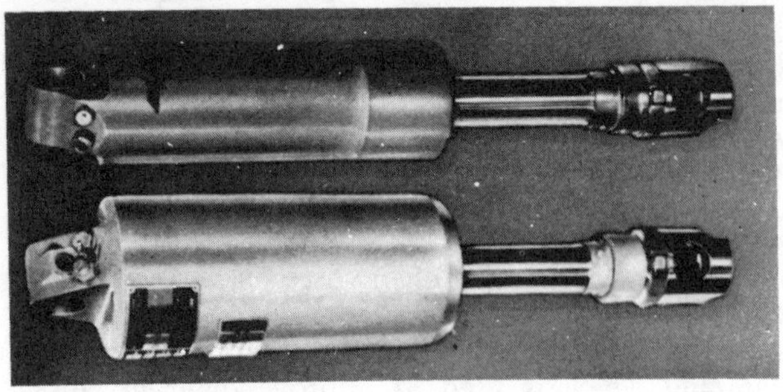

Fig. 17 Liquid nitrogen supplemented springs: *upper*, nitrogen filled; *lower*, standard

Figure 17 shows an example of the high-pressure liquid spring shock absorber which was developed for landing gears by the late Sir George Dowty in the early 1940s. Liquid springs are still in use, but their weight and space effectiveness have been eroded by the much higher oleo shock absorber pressures now used. A significant development of the liquid spring is modification of the fluid bulk modulus by absorbed nitrogen to reduce both the total volume of the spring and its static pressure. This enhances gland life, reduces the volume of the unit, and ensures extension at low temperatures. The liquid spring is most suited to lever suspension units. The example shown in Fig. 17 compares a nitrogen supplemented liquid spring with a 'conventional' unit of the same energy absorption.

4.3 Retraction

The retraction system is one area which has exercised the ingenuity of designers over the years. In order to retract into the smallest space possible, complex motions are often necessary, sometimes including skew rotation axes and two axes of rotation in order to achieve a wheel locus which puts the landing gear into its assigned space and misses the landing gear doors and external stores.

Most retraction mechanisms are powered by a hydraulic actuator which acts about the pivot axis to raise the landing gear against weight and aerodynamic loads.

To illustrate typical retraction systems, a few present-day landing gears are given as examples:

The BAe 146 is a jet-engined, high wing aircraft with a take-off weight of 40 700 kg. For the fuselage mounted maingear, a lever suspension allows shortening through a direction bar mechanism by raising the shock absorber by means of a crank and direction bar. The landing gear is locked down by a folding side-stay which is stabilized by a lock linkage at the centre joint. Emergency lowering is assisted by an actuator acting on the side-stay powered from an emergency hydraulic power source. Figure 18 shows the small space the landing gear retracts into and only about 6 per cent additional fuselage area is needed to accommodate this installation.

The F50 turbo-prop is a high wing aircraft with a take-off weight of 20 860 kg and is a development of the successful Fokker Friendship, which includes the introduction of a telescopic nosegear and the replacement of pneumatic with hydraulic operation of the landing gear. The landing gear is mounted on the forward pressure bulkhead and is locked down by a folding drag-stay with a hydraulically released lock linkage at its centre joint (Fig. 19). Steering is by a rack and pinion mechanism, rotating a turning tube connected to the axle by means of torque arms.

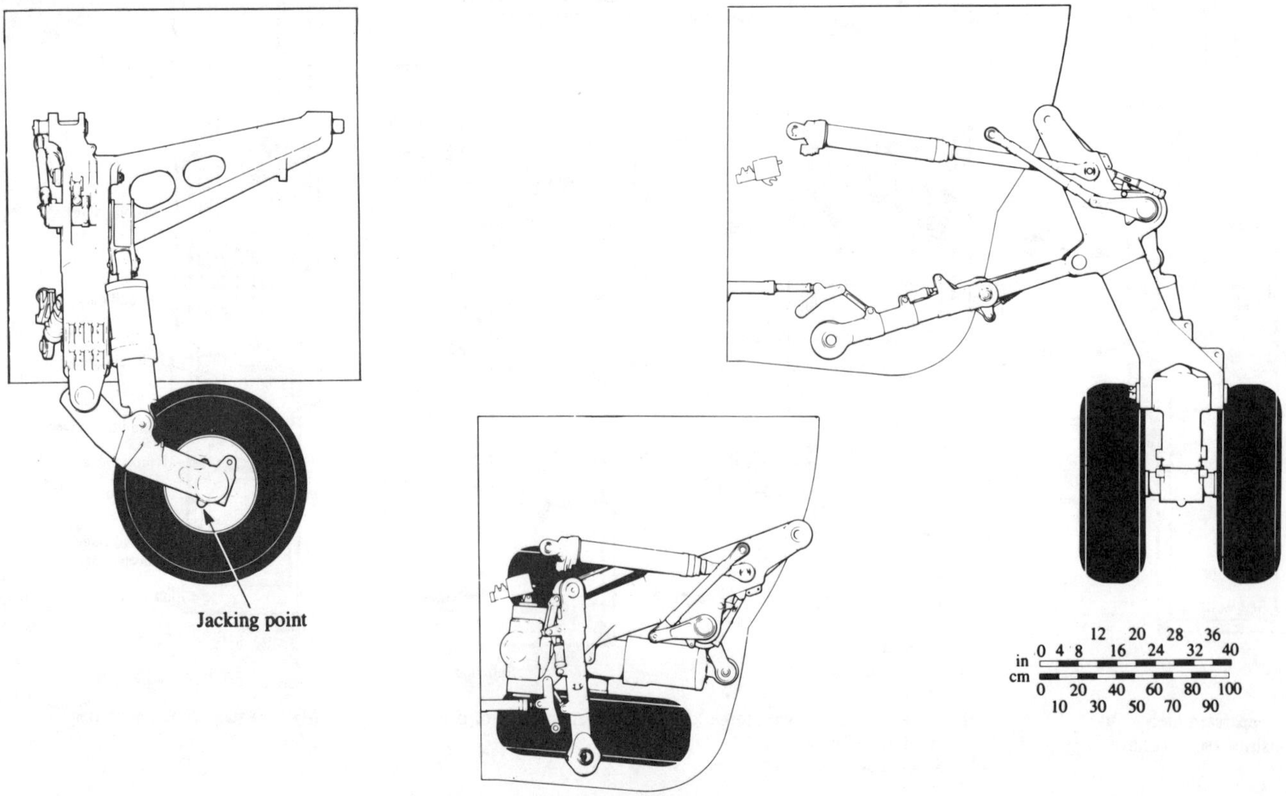

Fig. 18 BAe 146–200 main landing gear

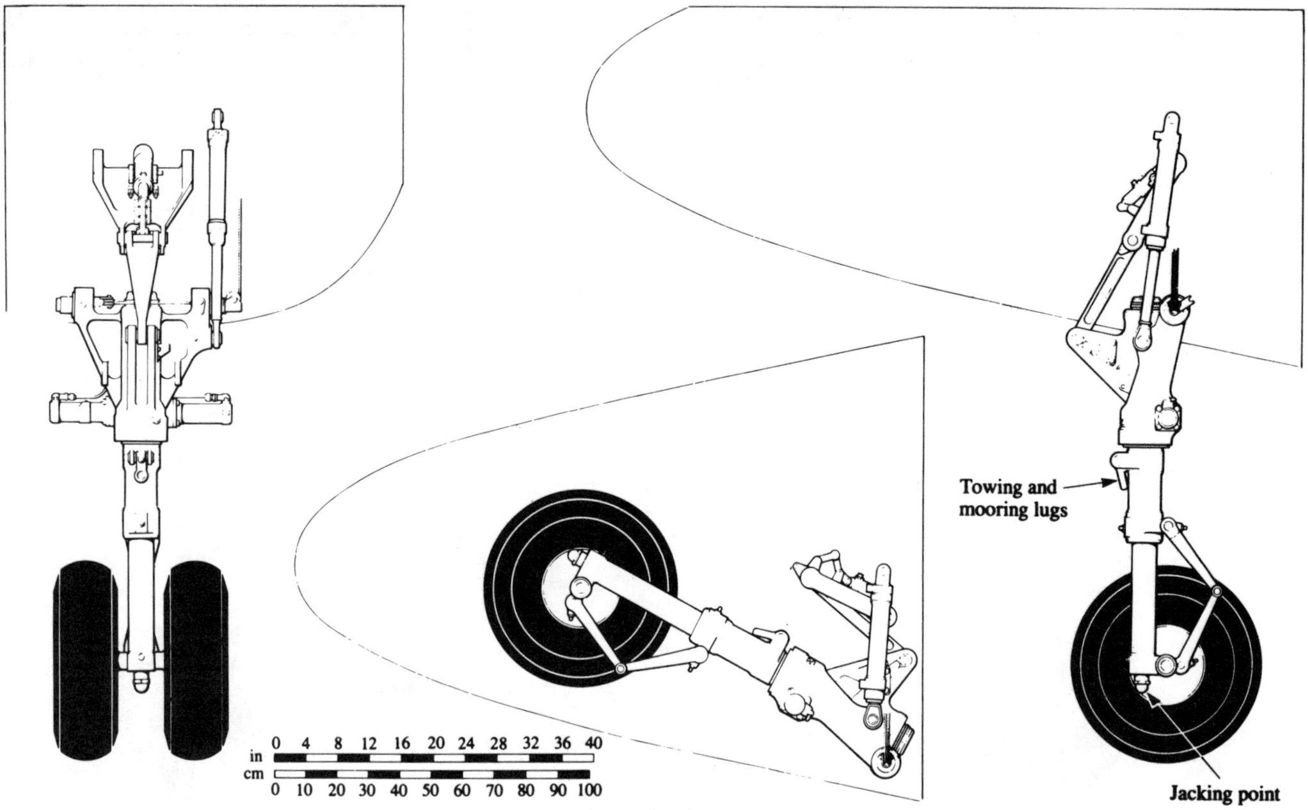

Fig. 19 F50 nose landing gear

The F50 maingear structure is almost identical to that of the Friendship, but has had minor changes to accommodate hydraulic operation with the addition of a springbox to ensure locking after freefall emergency lowering. This is a good example of a leg which folds at a hinge to retract aft into the engine nacelle. In this case, the retraction actuator motion also releases the geometric lock of the downlock links for retraction (Fig. 20).

The Fokker 100 is a low wing aircraft of 41 820 kg take-off weight, with its twin jet engines mounted at the rear of the fuselage, and is a development of the Fokker 28 Friendship. The telescopic nosegear (Fig. 21) resembles that of the Fokker F50 in concept, with an integral steering system which provides a large steering angle with automatic disconnection for ground handling beyond the steering range. The steering valve is mounted on the airframe and is controlled by a mechanical connection to a handwheel with positional follow-up from the nosegear by means of cables.

In this case the undercarriage is locked down by a plunger which fits into a location in the airframe. The plunger is withdrawn by initial travel of the retraction actuator.

The BAe ATP is a re-engined development of the BAe 748. The larger propellers meant that more ground clearance was necessary, and therefore longer landing gears, but no more stowage space for the retracted landing gears was made available.

The solution found for the nosegear is shown in Fig. 22, and attaches to the existing pivot axis. By the addition of a linkage, it is possible to pull the shock absorber capsule up during retraction to stow in the same landing gear bay. Downlock is effected by a plunger lock fitted to the rear of the leg. The plunger is released by a hydraulic actuator.

Similar problems were encountered for the maingear but, in this case, it is possible to use the same pivot point without shortening the leg by adjusting its geometry. Downlock is by means of a tumbler latch fitted to the airframe engaging a pin on the end of a hinged strut. A guide link positions the strut to engage the downlock when lowered and locate it within the nacelle profile when retracted (Fig. 23).

The Tornado nosegear (Fig. 24) is a telescopic design with steering effected by a vane motor geared down by an epicyclic track to drive a cuff which rotates the axle via the torque arms. The motor is servo-valve controlled by an electronic control amplifier with two gain levels—one for landing and the other for taxi operation. Input and feedback for the amplifier are by potentiometers fitted to the rudder pedal assembly and steering motor respectively.

An interesting point about the retraction mechanism is that the drag-stay and its lock linkage re-erects in the retracted position to form the uplock. Lock release for down- and uplock is by initial motion of the retraction actuator, which is a considerable design simplification with positive advantages for both weight and reliability.

The Tornado maingear is also an interesting unit is that it employs two axes of rotation to achieve retraction into the bay. Figure 25 shows three views of this gear in the down position. The primary rotation is about a skew pivot axis on the leg member which retracts the leg into the bay. Simultaneously, a direction bar mechanism rotates the shock absorber assembly to

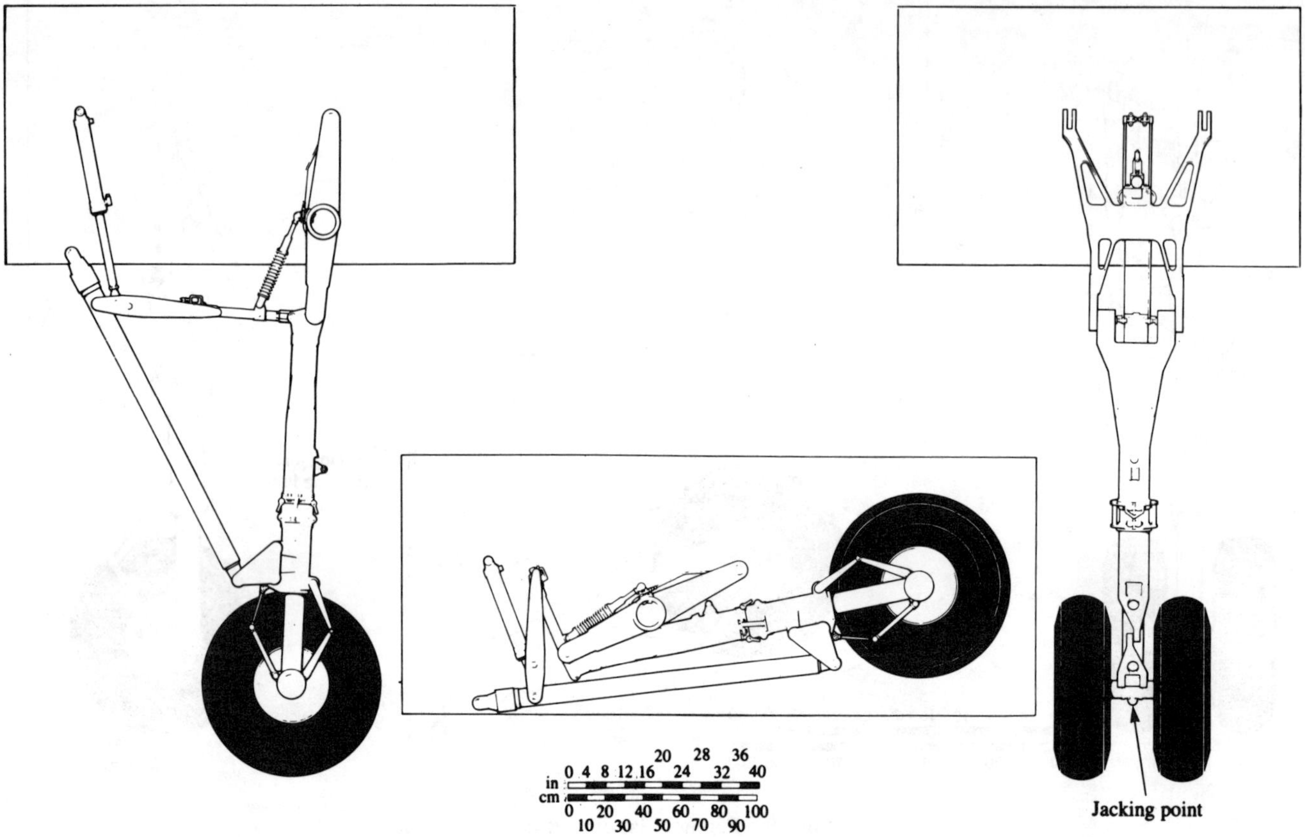

Fig. 20 F50 main landing gear

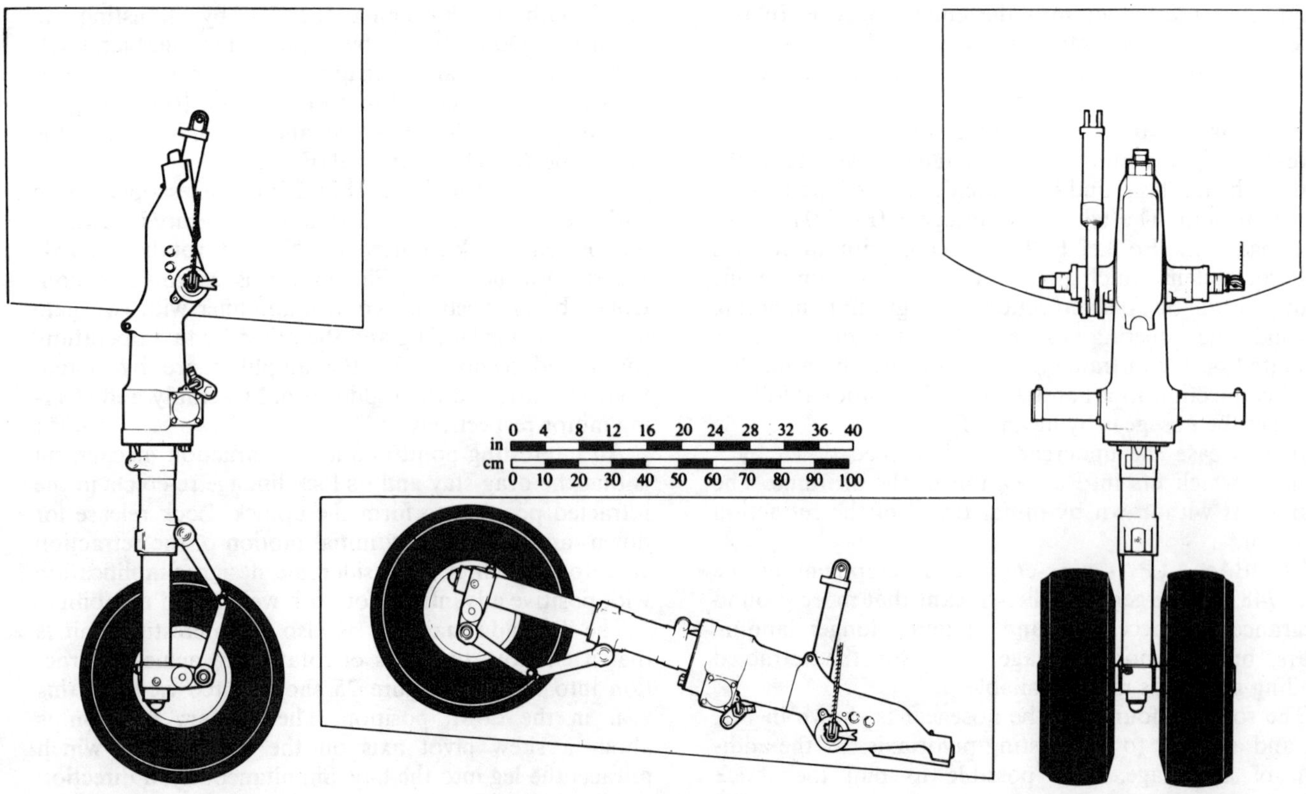

Fig. 21 F100 nose landing gear

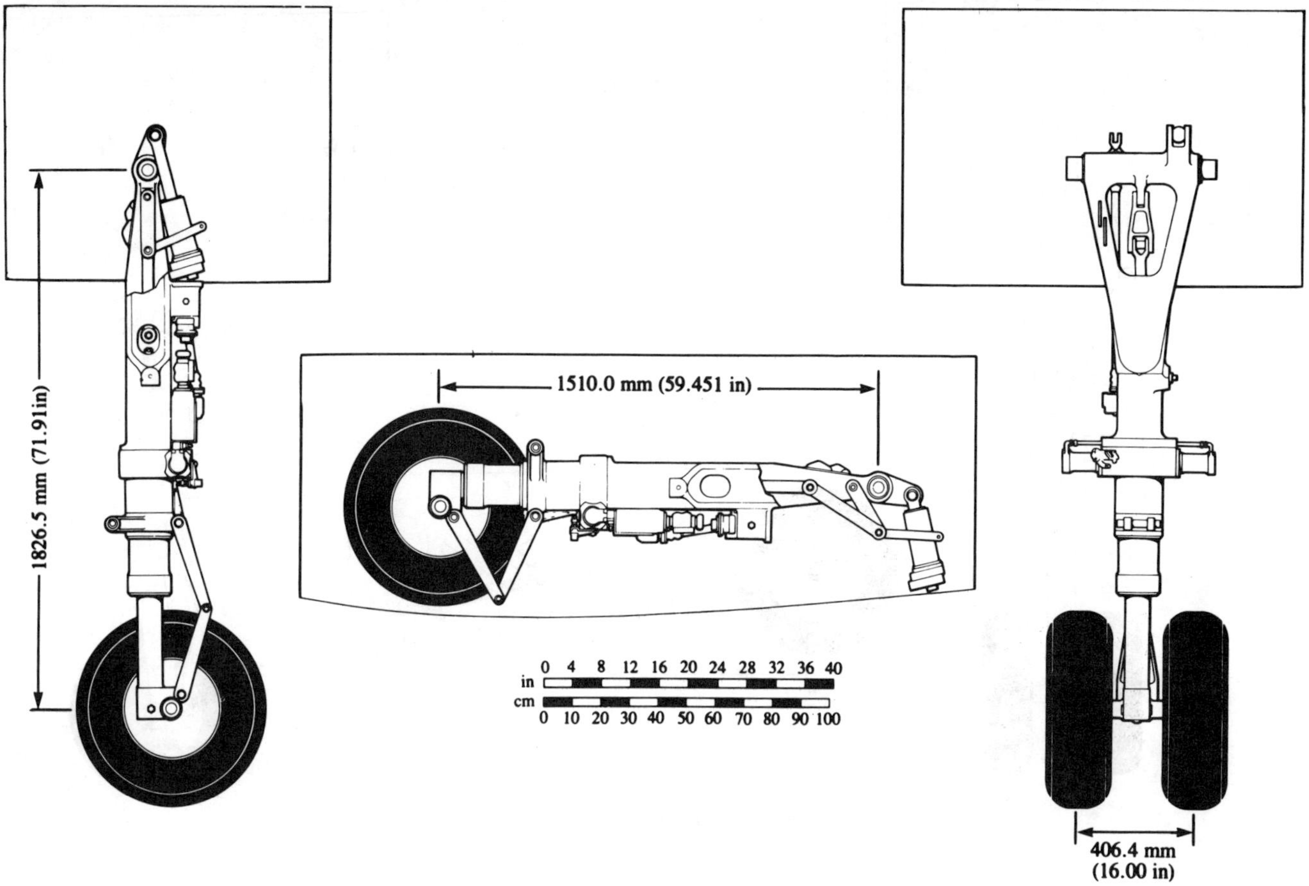

Fig. 22 BAe ATP nose landing gear

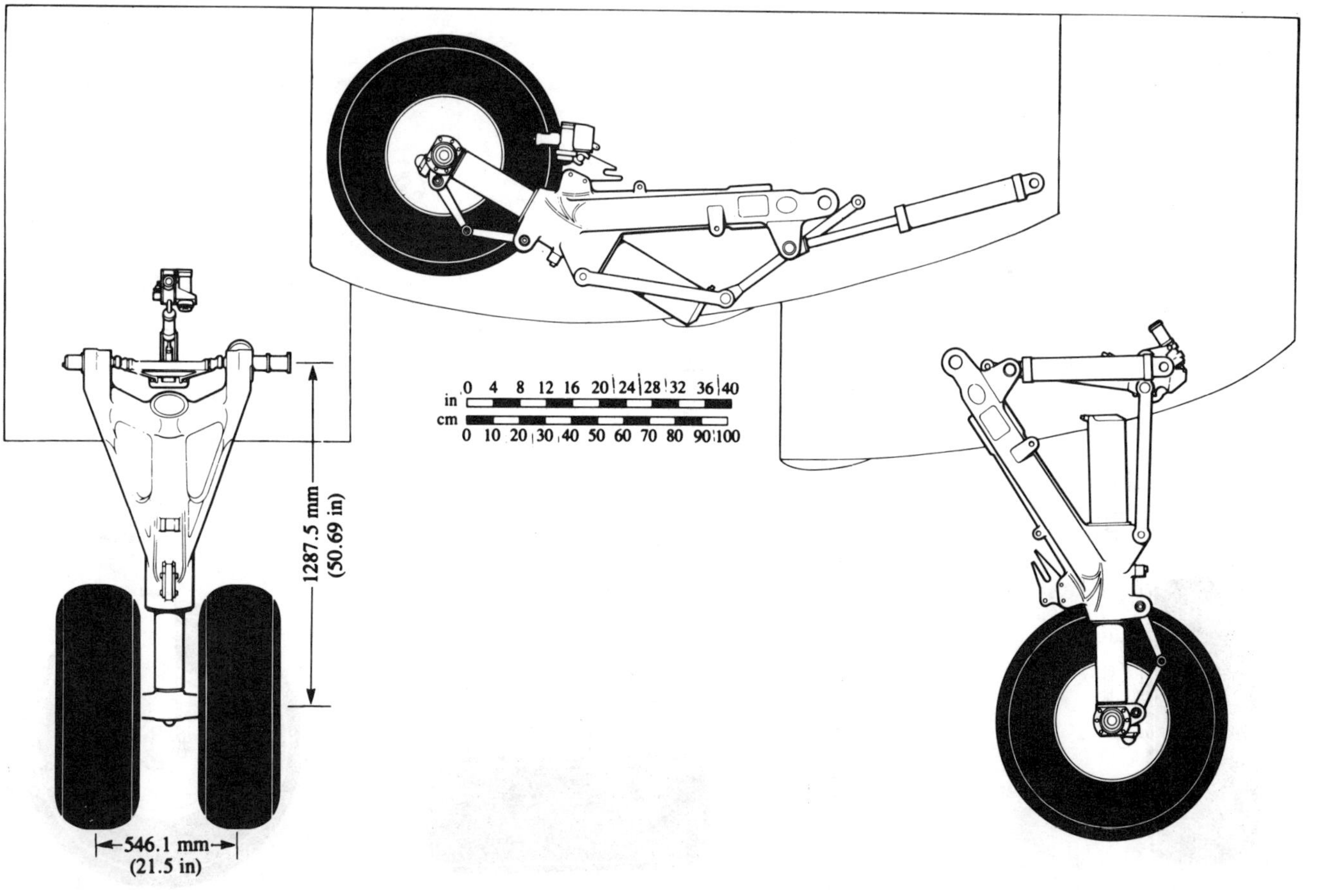

Fig. 23 BAe ATP main landing gear

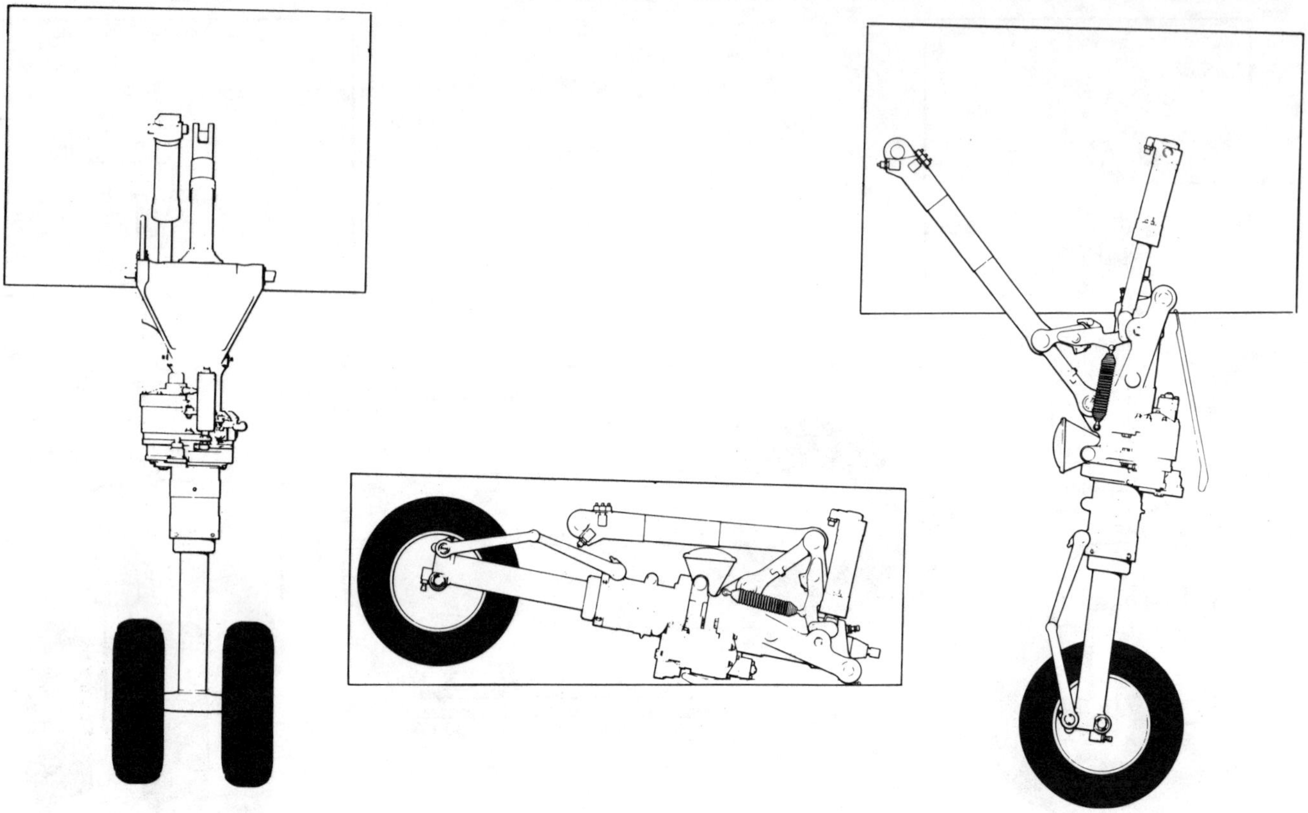

Fig. 24 Tornado nose landing gear

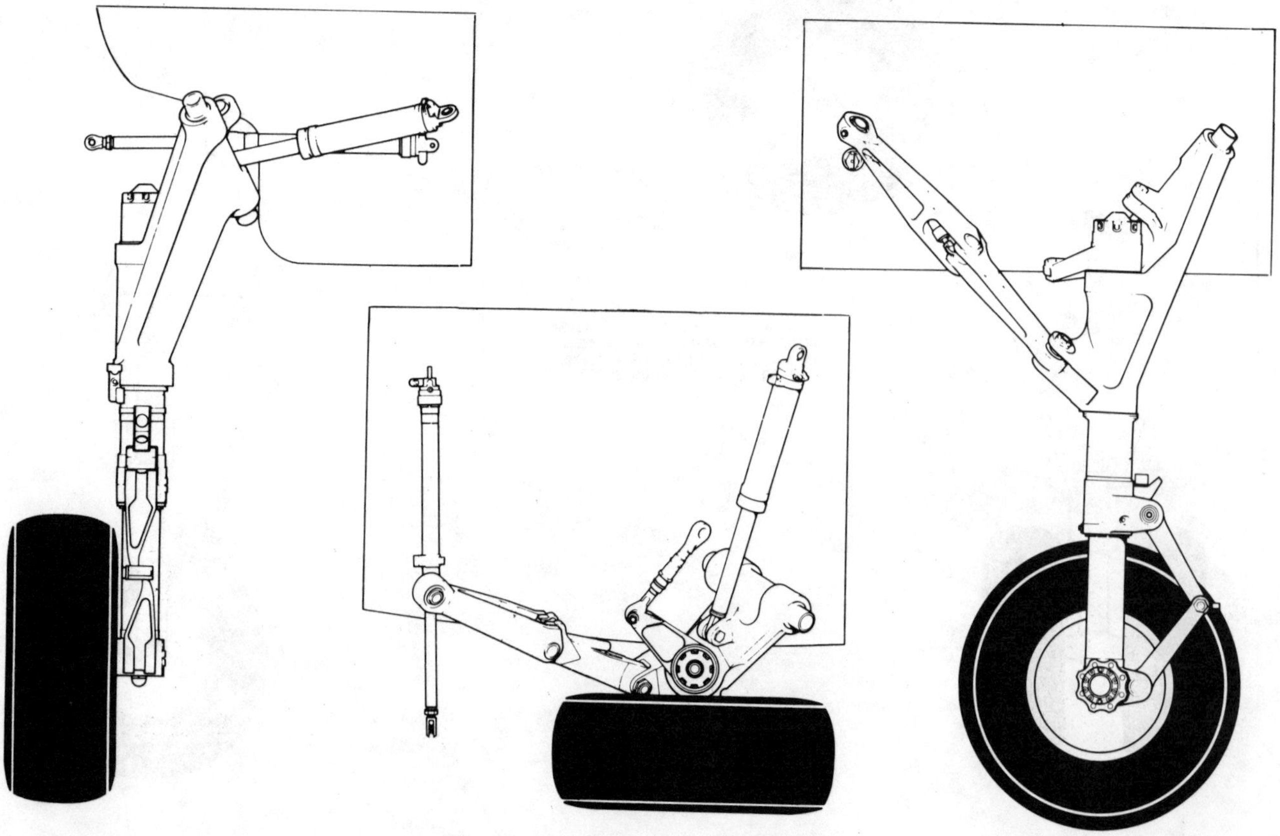

Fig. 25 Tornado main landing gear

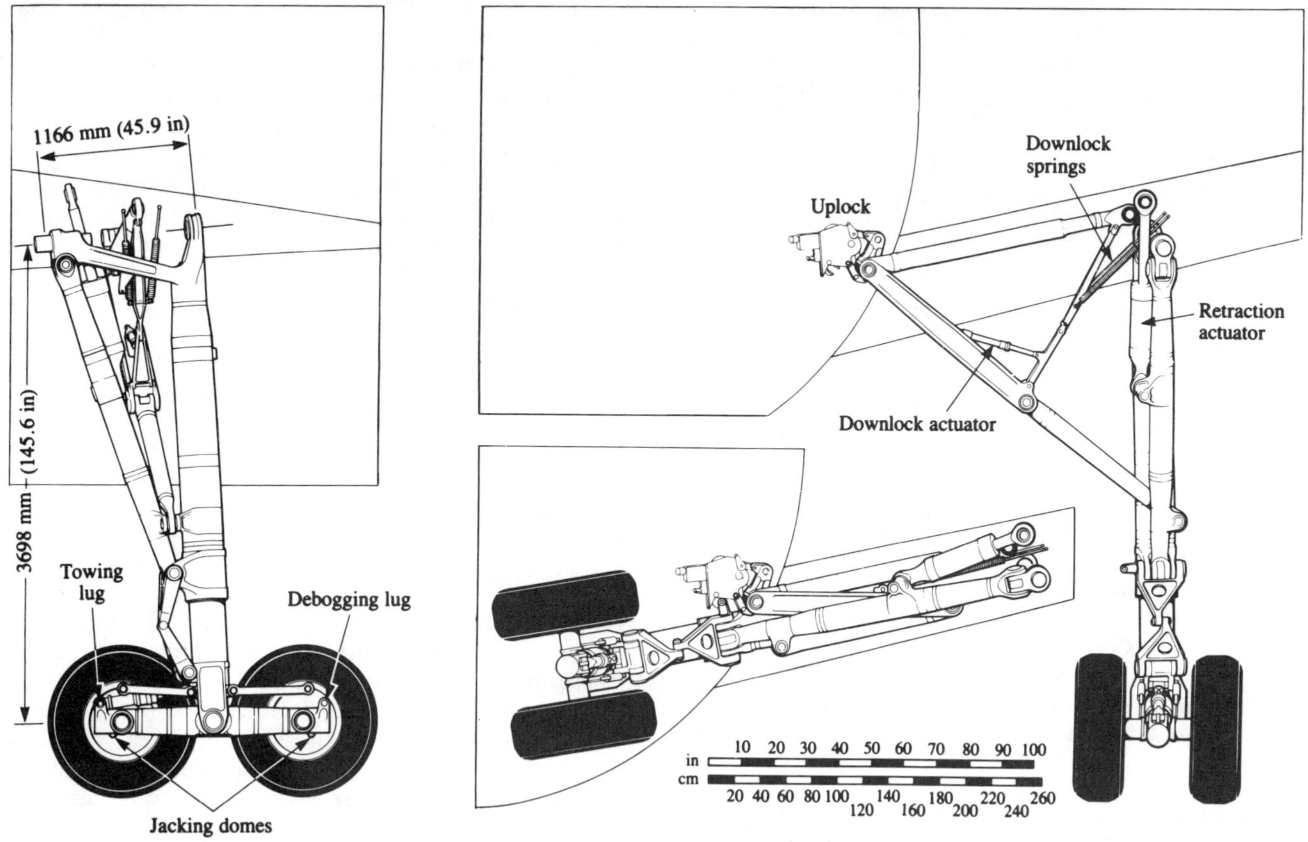

Fig. 26 A310 main landing gear

position the wheel in a substantially horizontal attitude in the aircraft so that the assembly occupies the minimum volume possible.

In this case, the folding drag-stay is a hydraulically operated self-breaking and self-locking type using a latch in the shape of a 'T' which engages with double abutments to retain it in the locked down condition.

Figure 26 illustrates the bogie undercarriage fitted to the A310 aircraft. The overall responsibility for this equipment is with Messier-Hispano-Bugatti in France but 40 per cent of the design and manufacture of the maingear is by Dowty Rotol. The construction is mainly of ultra-high tensile steel with the side-stay upper member of aluminium alloy.

The brake reaction bars and the pitch damper are located above the bogie beam which hinges in a fork on the end of the sliding member. The side-stay is a folding type stabilized by over-centre lock links. In order to minimize the side loads to be reacted by the fuselage, a reaction bar connects the side-stay to the structure near the pivot axis and the loads are fed circumferentially into the fuselage frame by a short link. This arrangement also minimizes the effects of structural deflections on the landing gear geometry.

For most landing gears, an uplock is needed, and Fig. 27 shows a tumbler type which may also be released manually for emergency operation. This type of lock can accommodate airframe deflections and tolerances and the use of a roller sear, which engages with the tumbler when locked, minimizes release loads. Emer-

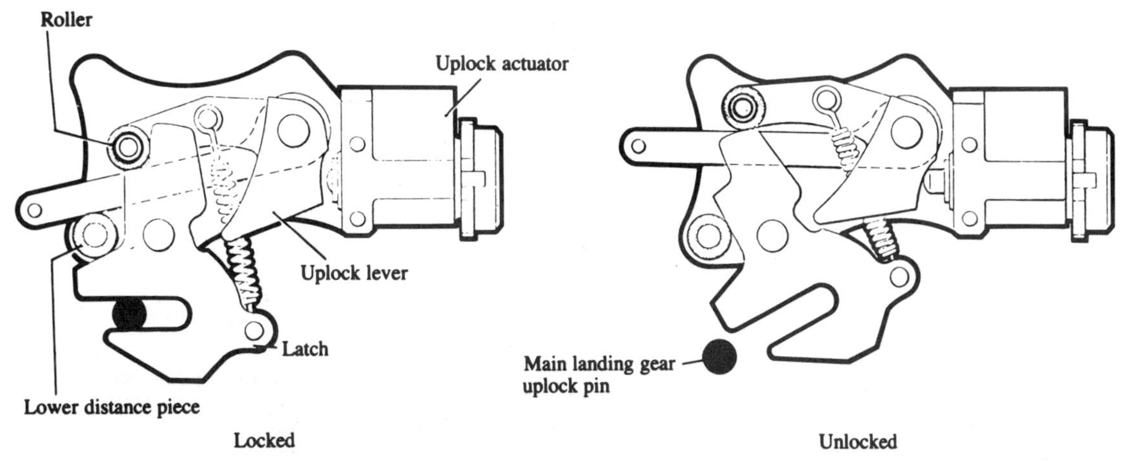

Fig. 27 Landing gear uplock

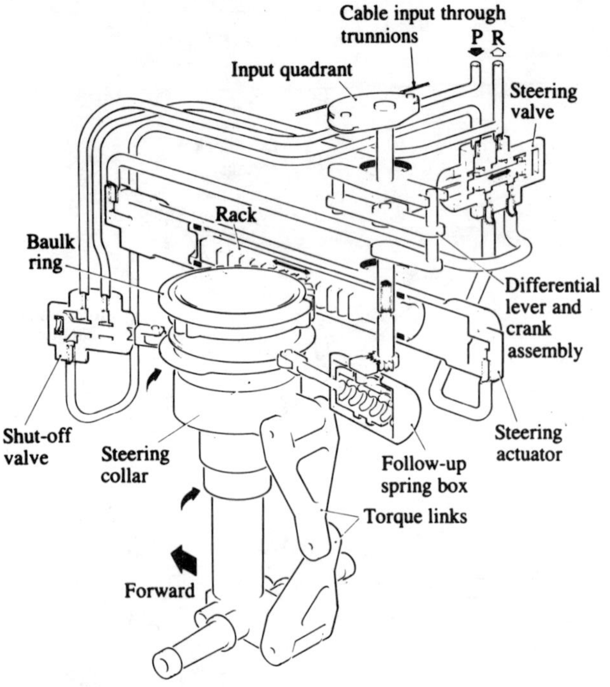

Fig. 28 Operational diagram of steering mechanism for BAe 146 nosegear

gency release can also be by a second hydraulic system supplying pressure to a separate piston in the release actuator.

4.4 Steering

For larger aircraft, power steering is a necessity, and although there are some electrically powered steering systems, the majority are hydraulic boosters. Early nosegears had limited steering angles because of the mechanical constraints of a single actuator. The need for larger steering angles for manoeuvring in restricted places resulted in twin actuators, particularly in the USA, and the now increasingly common rack and pinion steering mechanism. This latter device has been developed to provide constant steering torque with large steering angles, with automatic disconnection to allow manoeuvring and towing using large nosegear rotation angles without the need to disconnect the steering manually.

Figure 28 illustrates a typical rack and pinion steering system with mechanical control which is fitted to the BAe 146. Signals to the valve are provided by cable movements which operate a differential linkage. Position feedback with gain variable with steering angle is provided by a feedback cam mounted eccentrically to the leg centre-line. On this aircraft an option is available which provides $\pm 7\frac{1}{2}°$ rudder pedal controlled steering in addition to full handwheel control.

Steer-by-wire systems are used on many more aircraft now. These have advantages with respect to low operating forces and low mechanical backlash and tend to provide a lower weight installation. The ease of controlling the aircraft by electronic control using a servo-valve is liked by pilots, and installation in the airframe is eased as only wiring has to be routed.

4.5 Materials

In recent years, Dowty Rotol have standardized on a range of materials, with increasing in-service usage. These are summarized in Table 2. The materials now used by Dowty Rotol result from gradual improvements over the years rather than significant steps. In some ways, we have not moved far, as steel of 85 tonnes ultimate strength was an Air Board requirement for axles during World War I! From the strength–density ratios (specific strength) of ultra-high tensile steel and aluminium alloy, one could conclude that high strength light alloy is no longer weight effective. However, for fittings which are complex and from which it is not feasible to remove all ineffective material, its use can produce a weight and cost effective solution.

Ultra-high strength steels are used solely to reduce weight. Service experience over the last fifteen years has been good, but there are adverse factors, including substantially increased manufacturing costs, and greater material notch sensitivity.

4.6 Weight

For the aircraft designer, weight is always a matter of considerable concern. It is not possible to report a steady improvement of landing gear structure weight

Table 2 Principal landing gear materials used by Dowty Rotol

Material specification	0.2% proof MPa	UTS MPa	E GPa	Density kg/m³	Remarks
Steels					
S99	1080	1230	205	7890	35NCD16 and S153 similar
DTD5212	1070	1800	186	8000	to maraging 18% nickel, stable, expensive
35NCD16THQ	1450	1800	202	7890	Stable, air or gas hardening, 75°C heat treat after hardening
S155	1500	1900	207	7833	Considerable worldwide usage as 300M
Titanium					
IMI551	1095	1250	114	4623	Similar to 6Al–6V–2Sn
Aluminium alloys					
L161	385	450	72	2800	Similar to 7075T73 excellent stress corrosion resistance
7010T736	430	500	72	2820	Good stress corrosion resistance
AZ74	450	510	72	2820	0.25–0.4% silver, expensive

Table 3 Main landing gear weight

Aircraft type	Landing gear configuration	% take-off weight
Low wing with jet engines under wing	Bogie wing-mounted	
	Diabolo wing-mounted	
Low wing with fuselage or wing-mounted engines	Diabolo wing-mounted	
	Bogie wing-mounted	
High wing	Diabolo wing-mounted	
	Diabolo fuselage-mounted	

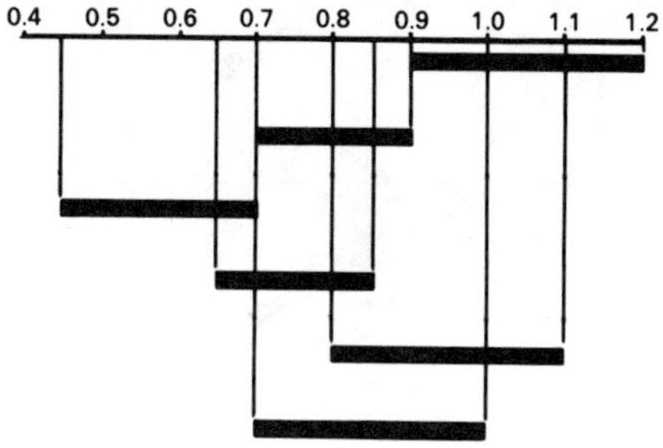

which reflects improvements of technology over the years since the War, as a firm baseline for judging weight effectiveness is difficult to set.

Table 3 indicates the wide range of percentile weight for main landing gears of different configurations. This range is due to many factors, driven by the basic specification for the landing gears and their local geometry and includes the effect of more severe fatigue spectra and longer service life requirements. It is, however, worth noting that landing gears of current design tend to be at the lower end of their appropriate range.

5 DEVELOPMENTS FOR FUTURE LANDING GEARS

5.1 Materials

The development of any material to a stage where the risk of using it on an aircraft is small is an extended process. Table 4 gives the characteristics of a few materials in the pipeline compared with current reference materials.

Developments in the USA for steels are aimed at developing the 300M type steel from 1900 MPa to 2160 MPa minimum ultimate tensile strength. Their efforts have not been wholly successful with respect to consistent quality, but doubtless they will achieve their aim in due course. Whether the material will offer enough advantages for general adoption remains to be seen.

Improvements of aluminium alloys focus on aluminium–lithium alloys which are now becoming available in plate and sheet form, but high strength forging material, which holds the most interest for the landing gear designer, is unlikely to be available to industry for evaluation before the end of 1986. Testing of specimens, when this material becomes available, may result in its introduction for parts where its greater than 10 per cent weight saving potential for the individual components can be realized, and the increased cost justified.

Powder metallurgy materials have high purity which may eventually result in high strength with good fatigue characteristics. Work carried out in the USA indicates that the powder metallurgy versions of some of the 7000 series alloys have superior fatigue characteristics to their normally produced equivalents. The use of these materials for castings, together with the hot isostatic pressing process, offers the possibility of precision castings with consistent properties approaching those of wrought materials. This advance offers the possibility of considerable cost and weight savings, and structurally more efficient designs are possible once the constraints imposed by forging and machining are removed. In order to realize weight savings, it is essential that consistency of materials and technique are demonstrated so that the strength super factors imposed on castings can be removed.

What of composite materials, in particular carbon fibres? For landing gears, the main problems emerge from the need for joints, the many design cases and the directional properties of the material. Unfortunately, most landing gear components have a multiplicity of joints and the variety of load cases generates many different stress paths within components. Pin-jointed struts are the most likely first use, but even here the efforts of

Table 4 Comparison of current materials with future landing gear materials

Material	0.20% Proof MPa	UTS MPa	E GPa	Density specific gravity	Weight* relative to S99	Volume* relative to S99	Flexibility* relative to S99
Reference materials							
Steel S99	1080	1230	205	7.89	1.00	1.00	1.00
Aluminium alloy 7010T736	430	500	72	1.82	0.88	2.46	1.15
Future landing gear materials							
Steel HP310	1870	2160	202	7.83	0.56	0.57	1.78
Aluminium–lithium alloy, Alcoa goal D	427	503	80	2.55	0.79	2.44	1.05
Uni-directional carbon fibre 0°	−1000†	1370	130	1.65	0.19	1.88	1.75
XAS/epoxy 90°	− 200†	42	9	1.65	6.13	19.27	0.77

* Based on ultimate tensile properties.
† Ultimate compressive strength.

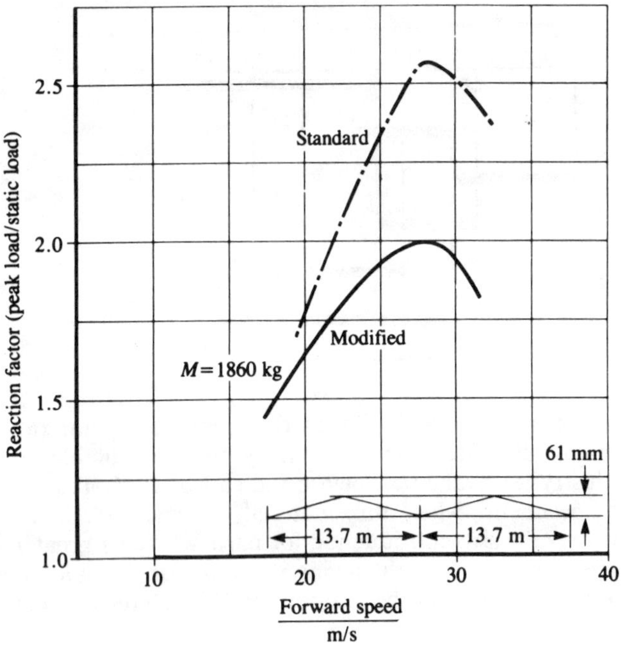

Fig. 29 Nose undercarriage reaction factor for two-ramp input: comparison of standard and modified shock absorbers

various landing gear manufacturers have not been crowned with complete success.

The design of metal–composite joints is difficult due to the effect of temperature as well as the structural aspects. For landing gear use, the effect of impact damage on the compressive strength of parts also needs to be considered. Much development needs to be done before the 20 per cent weight saving possible for simple parts, such as struts, can be realized.

Metal matrix composites, particularly an aluminium matrix with silicon carbide, appears to provide the best prospects for weight saving, but it will be several years before we are in a position to evaluate this type of material for real applications.

5.2 Rough ground capability

The need for both civil and military aircraft to operate from rough airfields has long been recognized. The ability to do so extends the operational scenarios for both types with the prospects of more sales for civil aircraft and improved operational readiness for military types.

Dowty Rotol have research programmes aimed at developing passive and adaptive (or semi-active) devices for shock absorbers to improve rough ground performance.

Figure 29 illustrates the degree of load reduction achieved using a lever suspension optimized for landing performance, compared with one containing a passive damping control valve which can provide suitable ground ride characteristics as well as the requisite landing performance.

The passive device developed for this purpose will also provide lower fatigue loads for taxiing, so modifying the fatigue spectrum to give significant improvements of landing gear (and perhaps also the airframe) life. This is potentially an important development when fatigue lives for landing gears of 60 000 landings are now commonplace with some manufacturers aiming at safe lives of 90 000–100 000 landings.

5.3 Steer-by-wire

Steer-by-wire is likely to be generally adopted in the future. The ease of installation, lower total weight, and the absence of mechanical backlash are in its favour as is its flexibility with respect to secondary inputs.

It is possible to adjust the gain of the system with increasing speed and harmonize with the aircraft's aerodynamic directional control and the system may be cross-connected to differential braking to provide neutral steering characteristics. For the fundamentally unstable aircraft now being developed, integration of steering control with the flight control system seems to offer many advantages with respect to harmonizing both steering and aerodynamic control to eliminate unstable situations, particularly during landing.

6 CONCLUSIONS

Each new aircraft brings a new challenge for the landing gear designer and manufacturer.

Engineering technology improvements and new engineering design tools reduce the risks by providing an increasing depth of information on the design. We must take care to have a sound baseline on which to judge this information, but the background of experience of dedicated landing gear designers is helpful in preventing over-reaction in this respect.

I am sure that landing gear manufacturers will remain responsive to the needs of the aircraft industry and will continue to develop new methods, use new materials, and devise new devices to produce hardware which will meet the future needs of the aircraft industry and its customers.

ACKNOWLEDGEMENTS

The author would like to acknowledge the facilities provided by Dowty Rotol Limited for preparing the paper, together with the advice freely given by Mr D. G. M. Davis, Mr S. W. H. Wood and Mr D. B. Hughes, and many others during its preparation.

Locking Actuators Today and Beyond

James D. Helm and Walter G. Gellerson
Dowty Decoto, Inc.

* Paper 881434 presented at the Aerospace Technology Conference and Exposition, Anaheim, California, October, 1988.

ABSTRACT

Hydraulic actuators with internal mechanical locks are attractive when safe repeatable positioning is required along with insensitivity to load and hydraulic pressure variations. Internal locking designs generally offer a lighter, smaller, better protected solution than external locks for applications such as: landing gear, doors, highlift devices, inlet/exhaust geometry control, and armament positioning. The evolution of higher pressure hydraulic systems imposes design challenges because of smaller areas, higher stresses, and high release loads. This paper looks at the experience of one supplier of locking actuators, Dowty Decoto, Inc. (DDI), and at the technology that is, and will be, required to provide the industry with functional, reliable locking actuators.

ITEM DEFINITION

THE TERM "LOCKING ACTUATOR" generally refers to a linear, hydraulic actuator that has an internal mechanism to lock the piston or ram in one or more positions, most commonly at either end of the stroke, or both. This lock mechanism is usually configured so that it automatically engages at the desired piston position, remains engaged with the absence of hydraulic pressure, and disengages only when hydraulically or mechanically signaled. While there are rotary hydraulic actuators and linear and rotary mechanical actuators that are, indeed, locking actuators, this paper deals primarily with the more common linear, hydraulic type.

There are several mechanisms in use today to provide actuator locking, but they all perform the same basic function; that of locking the moving part of the actuator to the non-moving grounded part of the actuator so that there can be no relative motion between the parts. The locking mechanism must be controllable so that actuator locking or unlocking occurs only when desired, and the locks must be configured to meet the performance, load carrying, and envelope requirements of the specific application.

ADVANTAGES OF LOCKING ACTUATORS

There are many things that make locking actuators attractive for use in mechanism design. First and foremost are weight and space savings. Any actuated mechanism must have an actuator for it to function. This actuator supplies the hydraulic muscle to move the mechanism within its defined geometry. Now, if it becomes desirable to lock the mechanism in one or more positions, what better, more direct way to do it than to locate the lock mechanism internal to the actuator? This means that the lock mechanism merely becomes an internal addition to an already existing piece of equipment and does not create the need to add heavy, envelope consuming links and over-center mechanisms to the design. The lock within the actuator can be configured to meet most any requirement with the addition of only a few small, minimum complexity actuator parts. These lock parts benefit from their installation in this protected, oil-wetted environment and, thus, typically exhibit excellent longevity and reliability with little or no maintenance. The parts can be configured to accommodate other actuator functions such as snubbing and rate control, and to be tolerant of extremes in load, pressure, and temperature. Lock condition (locked or unlocked) can be readily indicated externally, both electrically and visually, and emergency or manual unlock provisions can be added with minimal complexity impact.

Another big advantage to locking actuators is the complexity reduction and resultant reliability increase that occurs through their use. The positive locking of a mechanism requires that one or more of the primary moving parts be restrained from motion. If the actuator itself is not used for this purpose, it is necessary to add some sort of linkage or cam device that will operate at the proper moment to secure the mechanism in the desired position. This usually requires the addition of a separate actuator to lock/unlock the linkage, necessitates strengthening of the linkage-to-mechanism attach points, and mandates the addition of numerous lugs, bushings, pins, and links. While this method can provide an effective lock, its complexity, parts count, and exposed location can contribute directly to reliability reductions. Reduced reliability means increased scheduled and unscheduled maintenance and higher mechanism costs.

When examining the cost aspects of the "locking issue", it is easy to see that locking actuators have an advantage. In a locking actuator, the lock becomes a part of an already existing piece of equipment, and it is unnecessary to add duplicative actuation, linkages, external indication mechanisms, tubing and wiring runs. This, typically, means a lower cost can be realized through the use of a locking actuator.

In general, locking actuators offer the lowest weight, minimum envelope consumption, highest safety and reliabiity, and maximum cost effectiveness for mechanism design. They are versatile, rugged pieces of equipment that can offer excellent solutions to tough design problems.

TYPICAL APPLICATIONS

Locking actuators have been successfully employed in a wide variety of aircraft and ground vehicle applications. Figure 1 is indicative of the experience of just one major actuator supplier. Among the more common applications are:

LANDING GEAR - Everyone recognizes the importance of aircraft landing gear and the abuse it takes. A locking actuator in this application generally provides the extend and retract action for the gear, as well as the downlock dragbrace, and uplock functions. This is probably the single most demanding application for a locking actuator because of the severe environment, high loads, multiple functions, and extreme reliability demands! These actuators are often equipped with all the "bells and whistles", including lock indication, rate control, snubbing, emergency extend systems, area balancing stand-pipes, and strong, fatigue resistant structural properties.

AIRCRAFT LANDING GEAR DOORS - These doors must be opened/closed in sequence with landing gear operation and usually are locked in the closed position. Locking actuators for this application typically contain snubbing mechanisms to lessen door bottoming effects, lock indication mechanisms to annunciate door closed and locked condition, and a secondary, emergency unlock/door opening system. They are, of course, a very high usage device and must exhibit exceptional reliability!

LIFT CONTROL DEVICES - These devices are critical to the low speed performance of modern aircraft. They include such things as the ground spoilers, leading edge slats, and leading and trailing edge flaps. They usually require locking in the fully deployed or fully stowed positions, and the integrity of these locked positions is flight critical. Typically these actuators may include servo control, lock indication, blocking valves and other "add-ons" depending upon the application.

LIFT CONTROL DEVICES - These devices are critical to the low speed performance of modern aircraft. They include such things as the ground spoilers, leading edge slats, and leading and trailing edge flaps. They usually require locking in the fully deployed or fully stowed positions, and the integrity of these locked positions is flight critical. Typically these actuators may include servo control, lock indication, blocking valves and other "add-ons" depending upon the application.

HOOK/PROBE DEPLOYMENT - Aircraft tail hooks and inflight refueling probe mechanisms often use locking actuators for extension/retraction and to lock the device in the stowed (tail hooks) or deployed (refueling probe) positions. These actuators are usually rather basic but sometimes include lock indication and damping provisions.

SPEEDBRAKES - Aircraft speedbrakes often require locking in the stowed positon to ensure against inadvertent inflight deployment. Locking actuators are a natural for this application and are most often used with an indication device that annunciates the stowed and locked position.

ARMAMENT POSITIONING - Aircraft weapons pylons, external guns, and missile launchers often use locking actuators for weapon positioning (aiming) and for stowing and locking the device in stowed position during non-combat flight. The devices range in complexity from very simple to fully-optioned, servo-controlled, position feedback, semi-smart actuators. Often, these units must have tailored dynamics to help meet extremely demanding stability and accuracy requirements typical of fire control systems.

Locking actuators have also been used for jet engine inlet/exhaust geometry control, aircraft and marine vehicle (hydro-foil) flight controls, land vehicle suspension lockout, and a host of other special and unique applications; but the majority of the devices fall into those categories listed above.

DESIGN REQUIREMENTS FOR SUCCESS

If the weight, envelope, reliability, and cost advantages of internal locking actuators are to be realized, specific criteria must be addressed by the systems designer and the actuator supplier who is eventually to satisfy his demands. Figure 1 is indicative of the scope of experience in internal locking actuators that is available from specialists in this field. Suppliers of this specialty hardware are eager to participate early in programs to assure that specifications are realistic and that the resulting internal locking actuator will be successful in its chosen application.

Consulting with the internal locking actuator supplier community, the systems designer should determine early if, in fact, internal locking actuators are the optimum solution for his system. Most important are criteria such as available envelope, actuator holding loads, actuator unlock loads, geometry and force amplifications, hydraulic pressure and fluid type, environmental conditions, and desired options such as annunciation of locked condition,, emergency release, and any special system interfaces.

To his relationship with the prime system designer, the internal locking actuator design specialist will bring the following design emphasis:

ACHIEVING FIDELITY AND RELIABILITY OF THE LOCK MECHANISM - This is the first and foremost design consideration. The lock mechanism must hold and release within specification tolerance and work reliably over the design lifetime.

CONTROLLING STRESS WITH LOAD RELEASES - This is the key to lock element wear control. As the lock releases, it would appear at first glance that bearing areas may approach zero, causing stresses to approach infinity. Sophisticated designs treat this paradox by controlling and managing strain energy, lock loading at the moment of release, and quick response of lock parts to avoid heavy loading during periods of reduced lock engagement.

AVOIDING GENERATION OF CONTAMINATION - This goes hand in hand with bearing stress control for all lock group part interfaces. A whole litany of compatible materials is available from the internal locking actuator community to match lock part stresses over the number of lock releases/actuator cycles to be encountered in the actuator lifetime.

REACHING SPECIFIED ENDURANCE LIFE - The lifetime of an internal locking actuator will be determined by a combination of lock load at release, stroking endurance cycles, and pressure impulse cycles. Realism by the system's designer in specifying the criteria early in the design cycle will assure that an optimum solution is obtained by the actuator supplier in his analytical and empirical approach.

The empirical approach to internal locking actuator design is vital and cannot be avoided. Even with modern computer aided design (CAD) systems with greatly enhanced analytical and graphical tools for exploring the intricate geometries involved in high energy load releases, it has still been found necessary to build and test lock hardware to validate the design process. For all of the actuators indicated in Figure 1, Dowty Decoto has found that actual hardware testing is a very important part of the design process.

ADDED CHALLENGES OF HIGHER PRESSURES

With today's aircraft hydraulic systems moving toward higher pressures, the demands on internal locking actuators are becoming more severe. The push to higher pressure is being driven primarily by the need to provide smaller, lighter hydraulic components to fit in the limited envelopes of today's modern fighter aircraft. Most current locking actuator technology has been developed around the 3000 psi hydraulic system, and the higher pressure systems, usually 5000-8000 psi, present some new challenges.

Of particular concern to the actuator designer are the reduced cylinder bore diameters that occur through the use of higher system pressures. It is common practice in locking actuators to have the locking elements engage a lock seat that is part of the cylinder to prevent piston motion. As this seat becomes smaller with bore diameter reductions, lock element contact area is reduced, thus increasing stress levels. Stress level control is very important in controlling lock wear and performance, and compensation for this area loss becomes a challenge. Use of higher strength materials for lock elements together with increased radial stroke for the lock elements can compensate for some of the circumferential loss of area.

Another big challenge presented by the smaller bores is the fundamental task of fitting the lock itself within the actuator. The physical size limitations present a whole set of challenges all by themselves. One observation, however, is that while the higher system pressures allow adequate force output with reduced diameters, it often becomes necessary to increase the diameter again to meet column loading and other structural requirements. Thus, the challenge of limited space for the lock is relieved. Figure 2 is a photo of the 5000 psi Boeing Helicopter V-22 Main Landing Gear Drag Strut Actuator which embodies compact locks at both the extend and retract positions. The actuator also contains an area balancing standpipe, rate control, snubbing, visual and electrical lock indication, and manual unlock mechanisms that further reduce space available for the locks.

Another more generic concern is with the seals and sealing technology that must be used with the higher pressure systems and lower lubricity fluids such as CTFE. This concern affects all hydraulic components, not just locking actuators, and has been extensively investigated by the seal suppliers. Seals have been developed that give excellent sealing and friction characteristics in lock actuators at higher pressures.

LOCKING TECHNIQUES IN USE

Over the years, many different mechanisms have been designed to internally lock hydraulic actuators, from simple blockage of the hydraulic ports to sophisticated mechanical systems.

On the simple end of the spectrum are the hydraulic locks which rely on a valve to prevent passage of fluid and thereby block movement of the piston. While simple, this type of lock has no external indication that the lock is secure, and it is dependent upon no leakage past seals and valve elements.

Mechanical locks generally operate on a common principle; a shear element between the cylinder and the piston. The types of locks in use today use different configurations of lock elements and methods to move the elements. Simplified diagrams of some of these configurations are shown in Figure 3.

"Ball Locks" are made of a series of balls usually contained in radial holes in the piston head. When locked, these balls rest partially in a radiused circumferential groove in the cylinder,thereby locking the piston to the cylinder. This type of lock is inexpensive but has a low load carrying capability.

"Finger Locks" are made up of a set of long finger-like projections with the locking elements formed on the ends. The fingers may be part of a common base and require an elastic deformation to engage the cylinder or piston, or they may be hinged at the base end. This lock design usually has several drawbacks. First, it takes up more axial envelope and second, it is usually designed with small radial engagement. Therefore, the load carrying capabiity is limited.

The "Ring Lock" uses as the lock element a circular ring cut at one point on its circumference. The ring is radially displaced into a groove in the cylinder which locks the actuator. Since almost the entire circumference of the ring is carrying load, an enormous amount of lock area can be obtained for a small radial deflection. Since Ring Locks are flexural devices, they must be designed for a specified fatigue life.

Finally, there is the "Segment Lock". The segment type of lock element is seen in various shapes; round, square, and rectangular. This type of lock differs mainly in the type of mechanism used to move the lock elements; conical pistons, flattened piston, or pistons driving toggle links. Segment locks are the predominant type in use today. They require a minimum of axial envelope and are relatively inexpensive to manufacture.

All of these types of "locks" have been used in various applications throughout the aircraft industry.

SPECIAL CASE: THE REPEATABLE RELEASE HOLDBACK BAR

Locking actuator technology is applied to other devices as well. One such device with which DDI has had extensive experience is the Repeatable Release Holdback Bar (RRHB).

The RRHB, shown schematically in Figure 4, is a device used during catapult launching of aircraft. It attaches to the aircraft nose gear and functions to restrain the aircraft against engine thrust and shipboard motion until the catapult is fired and sufficient thrust is developed to cause the RRHB to release. Release loads are in the 20,000 to 100,000 lb. range depending upon the individual aircraft.

DDI became involved with these high energy release systems in 1969. The company's high performance actuator locking mechanisms were adapted for the original RRHB which releases at 73,864 to 78,296 pounds with loading rates to 1,200,000 pounds/sec., reliably and repeatably for a total of 2,000 cycles. The concept chosen for the RRHB used the strain associated with the loading of a structural member to control the release of an internal locking mechanism.

It was learned early in RRHB development that the area of most difficulty was in the lock mechanism. The load magnitudes and strain energies involved were too severe for then existing lock element technology. Early lock mechanism configurations, though appearing satisfactory in theory, would either fail or become erratic after only a few cycles. The solution was a two-stage mechanism with a small, lightly loaded secondary lock element used to trigger a larger, main load carrying primary lock element. The lower loads on the secondary elements allowed much more precise control. The primary elements, now free of triggering requirements, could be designed with low mass and proper contact angles for much greater tolerance to high level/rate load release. The design worked so well that DDI applied for, and was granted, a patent for it.

The technology developed for the "severe usage" RRHB has, subsequently, been put back into locking actuators. DDI has been very successful in using variations of the RRHB lock mechanism in locking actuators where extremely heavy loads must be released, yet absolute reliability and release point control are vital.

TWO-STAGE LOCKS

The lock mechanism that DDI developed for the repeatable release holdback bar is called the Two-Stage Lock. It is shown schematically in Figure 5. This type of lock uses two sets of lock elements, a primary or "main load carrying" set and a secondary or "triggering" set. DDI uses lock ring or lock segment configurations in these applications, and it is common for the primary elements to be considerably larger than the secondary elements.

This type of lock is especially useful in applications where very heavy loads are present on the lock mechanism at the time of unlock or where it is desirable to be able to prove analytically that the lock does not rely on friction to maintain locked condition. In this mechanism the primary lock elements are held in position by a primary lock piston. This primary lock piston is designed to be of a low mass, quick reaction configuration and is held in position by the secondary lock elements. The secondary lock elements are then, in turn, held in position by the secondary lock piston. It is the secondary lock piston that senses hydraulic pressure or mechanical displacement and initiates the unlock sequence. There are two advantages here.

The first is that the load on the secondary lock piston is much lower than that on the primary lock piston which means less frictional drag and more precise control of unlock pressure. It is this light loading that allows the use of a flat (not inclined) interface between the secondary lock piston and secondary lock segments. This is the feature that makes the two-stage lock independent of friction to maintain locked condition. It should be noted that the load on the secondary lock piston is easily manipulated by altering the angular interfaces of the parts in the lock group. This adds to the "controllability" of the lock mechanism.

The second advantage is that this mechanism places the extra mass that is necessary to house the unlock control seals onto the secondary lock piston. This means that the primary lock piston is freed of this requirement and can thus be designed in a minimum mass configuration. Minimum mass is essential in the primary lock piston and segments to achieve the rapid accelerations that are necessary during unlock to avoid high stress and subsequent wear on the lock parts.

LOCK SEGMENT DESIGN

Of all the components in a locking actuator, the lock element is probably the most critical and represents the greatest design challenge. Segment type locks and the basics for design of that type of lock element are discussed here.

The first consideration is the selection of the number of segments to be used. As loading goes up, it is imperative to maximize the lock engagement area to minimize stress on both the lock seat interface and lock piston interface areas of the segment. Concurrently, piston head integrity must be examined as the number and size of segments used has a direct effect on the amount of load reacting material that is left between the segment slots.

If the number of segments selected is small, and the cylinder bore diameter is large, the circumferential span of each segment increases. In this case it is important to configure the segments so that their outside radius coincides closely with that of the bore when the unit is unlocked and out of lock position. This helps to distribute radial relock forces against the bore more uniformly over the segment circumference, thus lowering the stresses present at this interface and minimizing bore wear. This is especially important when extremely hard materials are used for the lock segments.

Radial lock engagement in the lock seat and circumferential lock seat contact distance are the two controllable factors that establish lock segment bearing areas and thus control stresses. Besides the number of segments selected, their actual shape and their radial "throw" caused by the lock piston design are the features most easily manipulated. Of course, nothing is "square" in lock segment design, and actual achieved lock bearing area is affected by angular interfaces between the segments and their mating parts, radii used to smooth unlock/relock motion, and tolerance build-ups in all lock group parts. All geometry must be thoroughly analyzed, and manufacturing techniques must provide required tolerances if performance predictions are to be accurate. Testing is a necessary adjunct to the segment design process.

Once basic segment configuration is established, the remaining choices are materials, processes, and finishes. The selections made here are based upon the severity of the application and the outcome of the geometry creation discussed above. Figure 1 illustrates the wide range of segment counts and materials that have been used in DDI segment lock design.

ASSURING LOCK/RELEASE RELIABILITY

Reliability criteria for internal locking actuators center around the locked condition, the unlocked condition, and the life expectancy of the actuator. Each of these conditions depends upon the internal design of the lock itself but may also be enhanced by external means. For example, a landing gear drag brace that is required to remain locked during ground operations may have a ground lock pin added to ensure that it meets this requirement. Electrical and visual indication also add to the assurance that an internal lock is functioning properly.

There are often requirements for redundant unlock systems for locking actuators. Normally unlock is achieved by the application of hydraulic power to the lock mechanism. In the event of a loss of hydraulic power, alternate means of unlocking the actuator may be used such as an alternate hydraulic or pneumatic source directed through a shuttle valve, or a manual release mechanism. All of these mechanisms may be added to a locking actuator to ensure that the desired result is achieved even in the event of a failure in the primary mechanism.

When looking at the basic reliability issue, however, it is very important that reliability is a critical part of the locking actuator program from the onset. For instance, lock mechanism configuration decisions must be approached early in the design process. A thorough analysis of the loading spectrum indicates whether the lock is to be designed for a high release load, a high static load, or both. This one point dictates the type of lock to be used. For a high static load but low release load, a single-stage lock would be used. For a high release load requirement, a two-stage lock would be used. The spring rate of the supporting structure also enters into the picture at this point. A structure with a low spring rate (high strain energy) relative to the release load may require a two-stage lock even if the unlock load is relatively low. This lock configuration decision is very basic to achieving reliability goals.

Equally important decisions must be made on material selection and processing, part finishes, tolerances, and lock element geometry. Special attention must be given to load sharing between lock elements and "manufacturability" of all parts. All of these considerations, along with the myriad of minute details that accompany locking actuator design, are the building blocks that are used to form the final product, and as such, are essential to final product performance.

Additionally, the designers (and the specifying agency) should always remember that the actuator is a part of a system and, in order to achieve maximum reliability, must be recognized and designed as a part of that system, not just a single element. This will help to assure not only actuator reliability, but system reliability as well.

FUTURE DEVELOPMENT FOR HIGH PRESSURE/LOADS

As with any technical area in our industry today, the locking actuator is undergoing constant change and refinement. Presently being investigated are several mechanisms for the future which are aimed at higher pressure, higher release load capability.

Efforts are now underway to research the capabilities of an enhanced "ring lock" design. The "ring lock" has been in use for many years, but its full capability has never been explored. Also in work right now is a design to completely divorce radial movement from the lock mechanism. The design will totally eliminate the load from the segments at the time of unlock initiation. This would mean that the lock load is removed from the segments at the same time that the engagement area is at its maximum.

A third area of lock development is that of segment acceleration control. Instead of relying on the motion of the loaded elements to move the mechanism, the release portion of the device will be mechanically linked to the entire lock.

On-going research and development programs are being conducted in design techniques and materials for improving performance and reliability of internal locking actuators to meet the higher pressure/load demands of the future.

Actuator	Lock Type: Single Stage	Lock Type: Two Stage	Lock Type: BE-CU Lock	Lock Type: Steel Ring	Lock Type: BE-CU Lock	Lock Type: Steel Segments	Dynamic Unlock Operating Load (LBS): 0-2,000	2,000-5,000	5,000-10,000	10,000-20,000	> 20,000	Static Lock Operating Load (LBS): 0-2,000	2,000-5,000	5,000-10,000	10,000-20,000	> 20,000
CITATION MLG ACTUATOR	✓		✓				✓								✓	
CITATION NLG ACTUATOR	✓		✓				✓								✓	
727 LEADING EDGE SLAT ACTUATOR	✓		✓				✓							✓		
F-14 RRHB		✓		✓							✓					✓
CITATION SPEEDBRAKE ACTUATOR	✓		✓				✓						✓			
707 THRUST REVERSER ACTUATOR	✓				✓				✓					✓		
COBRA TML ACTUATOR	✓		✓					✓						✓		
COBRA ARMAMENT PYLON ACTUATOR	✓				✓			✓						✓		
F/A-18 IFR PROBE ACTUATOR	✓					✓		✓						✓		
F/A-18 RRHB		✓				✓					✓					✓
757 NLG DOOR ACTUATOR		✓				✓			✓					✓		
757 MLG DOOR ACTUATOR		✓			✓	✓			✓					✓		
T-46A MLG ACTUATOR	✓				✓		✓					✓				
T-46A NLG ACTUATOR	✓				✓		✓					✓				
APACHE GUN TURRET ACTUATOR	✓				✓			✓					✓			
XV-15 FLAP CTRL ACTUATOR	✓				✓		✓						✓			
XV-15 LAT. CYCLIC ACTUATOR	✓				✓		✓						✓			
T-45 RRHB		✓				✓					✓					✓
V-22 DRAG STRUT ACTUATOR	✓				✓	✓	✓	✓						✓		
EDO POWER MODULE		✓				✓				✓					✓	
EDO C&R MECH.	✓					✓	✓					✓				
AV-8B WING GEAR ACTUATOR	✓				✓			✓					✓			

Actuator	Lock Position: Retracted	Mid-Stroke	Extended	Unlock Modes: Pressure	Manual	Emerg. Pneu.	Lock Ind.: Visual	Electrical	Snub.: Retract	Extend
CITATION MLG ACTUATOR			✓	✓		✓	✓	✓	✓	
CITATION NLG ACTUATOR	✓			✓		✓		✓		
727 LEADING EDGE SLAT ACTUATOR	✓		✓	✓				✓		
F-14 RRHB	✓				✓		✓			✓
CITATION SPEEDBRAKE ACTUATOR	✓			✓				✓	✓	✓
707 THRUST REVERSER ACTUATOR	✓			✓	✓			✓		
COBRA TML ACTUATOR		✓		✓						
COBRA ARMAMENT PYLON ACTUATOR		✓		✓						
F/A-18 IFR PROBE ACTUATOR	✓			✓				✓		
F/A-18 RRHB	✓				✓		✓			✓
757 NLG DOOR ACTUATOR	✓			✓	✓				✓	✓
757 MLG DOOR ACTUATOR	✓			✓	✓				✓	✓
T-46A MLG ACTUATOR			✓	✓			✓	✓		
T-46A NLG ACTUATOR	✓			✓			✓	✓		
APACHE GUN TURRET ACTUATOR	✓			✓	✓				✓	✓
XV-15 FLAP CTRL ACTUATOR		✓		✓						
XV-15 LAT. CYCLIC ACTUATOR		✓		✓						
T-45 RRHB	✓				✓		✓			✓
V-22 DRAG STRUT ACTUATOR	✓		✓	✓	✓	✓	✓	✓		✓
EDO POWER MODULE	✓				✓					✓
EDO C&R MECH.	✓				✓					
AV-8B WING GEAR ACTUATOR	✓		✓	✓		✓		✓		

Actuator	System Pressure (PSI): 1500	3000	5000	5000-8000	Endurance Test Cycles: 1,000-5,000	5,000-20,000	20,000-100,000	100,000-500,000	> 1,000,000	Impulse Cycles: 1,000-20,000	20,000-100,000	> 100,000
CITATION MLG ACTUATOR	✓					✓						
CITATION NLG ACTUATOR	✓					✓						
727 LEADING EDGE SLAT ACTUATOR		✓						✓				✓
F-14 RRHB					✓							
CITATION SPEEDBRAKE ACTUATOR	✓					✓						
707 THRUST REVERSER ACTUATOR			✓				✓					
COBRA TML ACTUATOR	✓								✓			
COBRA ARMAMENT PYLON ACTUATOR		✓							✓			
F/A-18 IFR PROBE ACTUATOR		✓				✓						✓
F/A-18 RRHB					✓							
757 NLG DOOR ACTUATOR		✓						✓				
757 MLG DOOR ACTUATOR		✓						✓				
T-46A MLG ACTUATOR		✓						✓				✓
T-46A NLG ACTUATOR		✓						✓				✓
APACHE GUN TURRET ACTUATOR		✓							✓			✓
XV-15 FLAP CTRL ACTUATOR		✓							✓			✓
XV-15 LAT. CYCLIC ACTUATOR		✓							✓			✓
T-45 RRHB						✓						
V-22 DRAG STRUT ACTUATOR			✓				✓				✓	
EDO POWER MODULE				✓	✓					✓		
EDO C&R MECH.				✓	✓					✓		
AV-8B WING GEAR ACTUATOR		✓				✓						✓

FIGURE 1

DDI INTERNAL LOCKING ACTUATORS

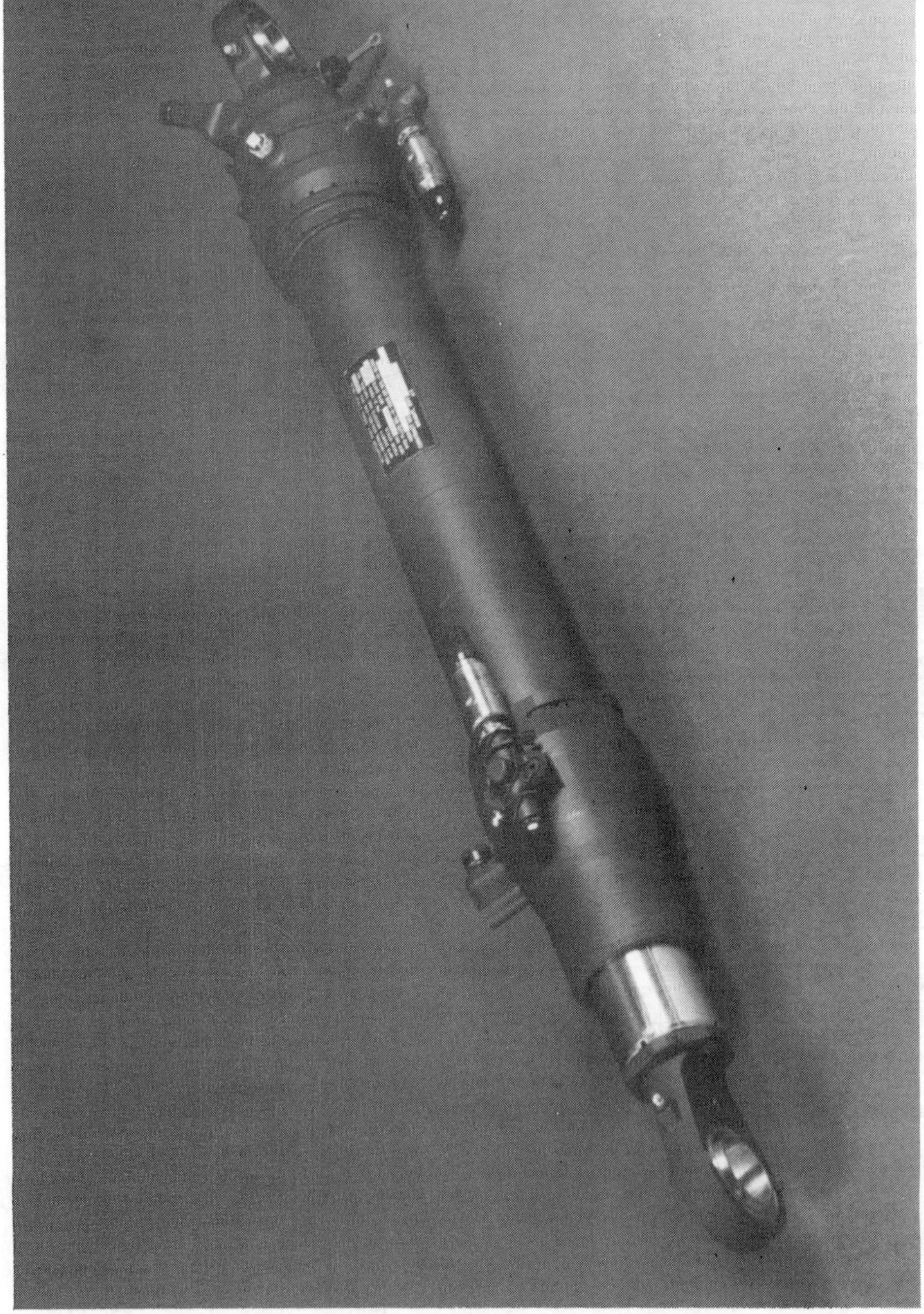

FIGURE 2

DRAG STRUT/RETRACTION ACTUATOR
V-22 MAIN LANDING GEAR

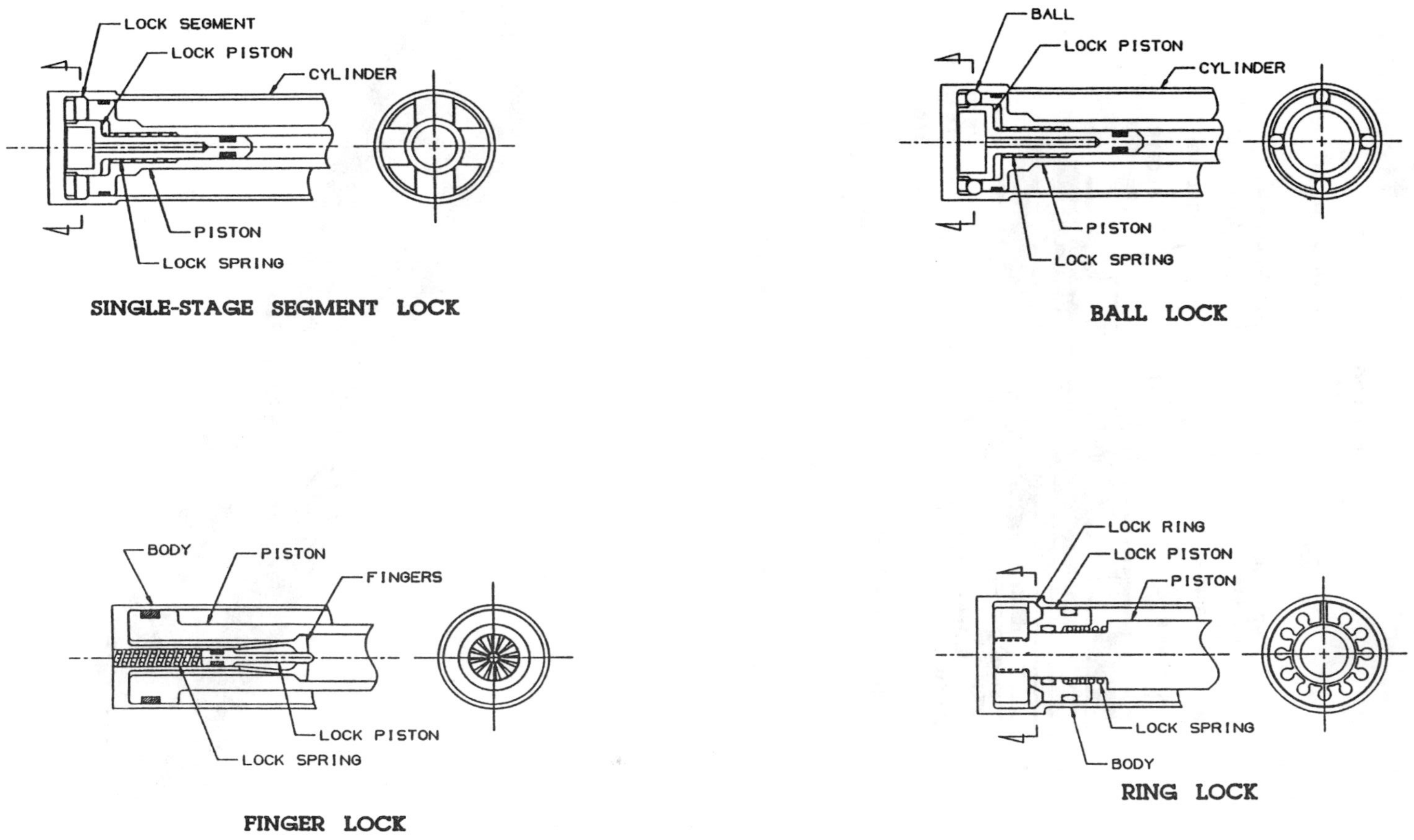

FIGURE 3

LOCKING TECHNIQUES

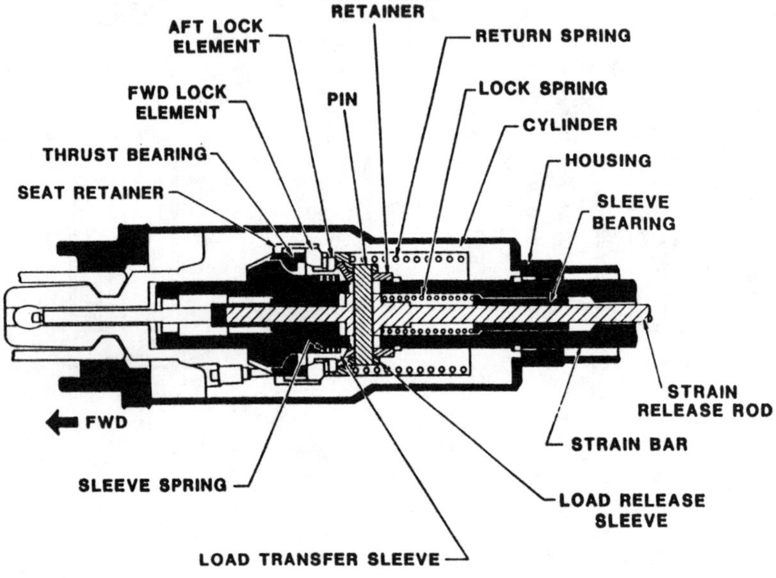

Repeatable Release Holdback Assembly (Top View)

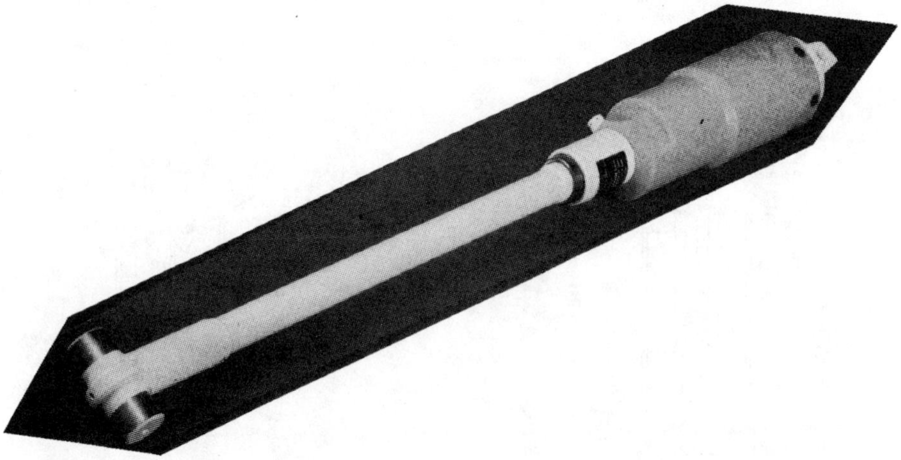

FIGURE 4

REPEATABLE RELEASE HOLDBACK ASSEMBLY

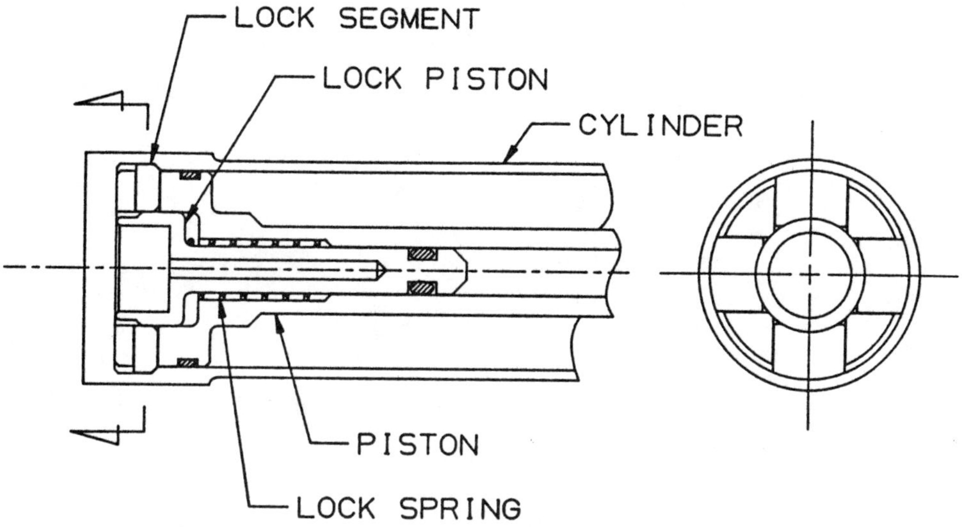

SINGLE-STAGE SEGMENT LOCK

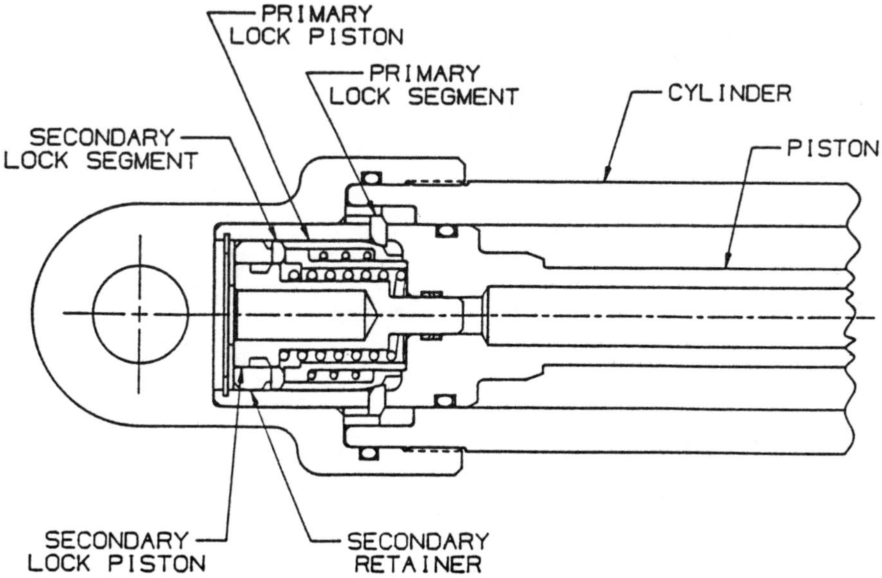

TWO-STAGE SEGMENT LOCK

FIGURE 5

SIMPLIFIED DIAGRAM
OF
SINGLE/TWO STAGE SEGMENT LOCK

Improved Steel for Landing Gear Design

William W. Macy, Mark A. Shea, Rigoberto Perez, and Robert E. Newcomer
McDonnell Aircraft Co.

* Paper 892335 presented at the Aerospace Technology Conference and Exposition, Anaheim, California, September, 1989.

ABSTRACT

An improved high-strength, high-toughness steel has been developed which shows considerable promise for landing gear applications. Previous materials provided high strength or high toughness, but not both. The improved material is a modified chemistry of AF1410 steel with increased carbon and an altered heat treat process. Tensile ultimate strengths (F_{tu}) of over 260 ksi (1.79 GPa) have been achieved while maintaining plane strain fracture toughness (K_{Ic}) in excess of 100 ksi$\sqrt{}$inch (110 MPa$\sqrt{}$m). In addition, the material has low sensitivity to environmental factors such as hydrogen embrittlement and stress corrosion cracking.

INTRODUCTION

Since the introduction of contemporary landing gear designs in aircraft, advancements in their structural materials have been a constant part of aircraft development. The volume and weight restraints put on the landing gear by the vehicle configuration have generally driven this material development to continually higher strength alloys. 300M steel, with an ultimate tensile strength in excess of 280 ksi (1.93 GPa), has been the accepted standard material for Navy, Air Force, and commercial aircraft landing gear for the past 20 years. The high specific strength performance of 300M, though, is a result of sacrificing other key properties especially important to Navy landing gear. The sacrificed properties include fracture toughness and resistance to stress corrosion cracking and hydrogen embrittlement.

In the F/A-18 aircraft, fracture toughness became a material selection criterion for certain "fracture critical" components in the launch train of the nose landing gear and the arresting system. HP9-4-.30 steel with F_{tu} of 220 ksi (1.52 GPa) and K_{Ic} of 90 ksi$\sqrt{}$inch (99 MPa$\sqrt{}$m) was originally used. However, it was replaced on the arresting gear by AF1410 which provided both higher strength, F_{tu} of 235 ksi (1.62 GPa), and higher fracture toughness, K_{Ic} of 130 ksi$\sqrt{}$inch (143 MPa$\sqrt{}$m). While the AF1410 allowed some weight savings relative to the HP9-4-.30, it could not compete with the minimal weight and volume allowed by the use of 300M throughout the remaining areas of the gear.

While its high strength allows efficient landing gear design, the sensitivity of 300M to various environmental factors such as hydrogen embrittlement and stress corrosion cracking, along with its low fracture toughness, results in delayed failures in service. Often the failures are traceable to a hard landing, or to some other high peak load condition, which caused local yielding of a 300M part in compression. Upon returning to normal loading, springback of the unyielded sections of the part sets up residual tensile stresses within the locally yielded areas. The residual tensile stresses, probably assisted by migration of hydrogen inherent from processing, initially accelerate fatigue crack initiation in the part. Subsequently, the same hydrogen, or environmentally induced stress corrosion cracking, increases crack growth until the critical flaw size for rapid fracture is reached. The salt water environment and potential for hard landings make this problem especially critical for carrier based aircraft. For these reasons, any steel alloy used on new Navy landing gear was to have a minimum fracture toughness of 100 ksi$\sqrt{}$inch (110

MPa$\sqrt{m}$). New Navy aircraft now being designed, would have to use lower strength alloys to meet the toughness requirement. As a result, those landing gear systems would suffer a significant weight penalty reducing aircraft performance and increasing operational costs.

Consequently, McDonnell Aircraft Co. (MCAIR) began evaluating new materials as possible replacements. The material properties sought were higher fracture toughness, reduced sensitivity to hydrogen embrittlement and stress corrosion cracking, and a tensile strength equal, or near equal, to that of 300M.

MATERIAL DEVELOPMENT

Work performed by USS's Monroeville Research Facility in the mid 1970's (Reference 1) demonstrated that high purity in chemistry formulations significantly increased fracture toughness of both 10Ni-modified (AF1410) and 18Ni maraging steels at a strength level around 250 ksi (1.72 GPa). In addition, high purity also significantly reduced the environmental-cracking susceptibility of the 10 Ni grade. During the 1980's, MCAIR has made the arresting gear for the F/A-18 Hornet from AF1410 with the nominal composition of 10Ni-14Co-2Cr-1Mo-.16C.

During this period, fracture toughness of the material, as measured by K_{Ic} tests, has significantly improved. Figure 1 illustrates how the average strength and toughness have changed, in terms of consecutive heats produced, for one of the three major materials suppliers. This increase in toughness with minimal change in strength is believed to be due to several factors. The first is improvements in melting practice, including ladle additions to tie up residual sulphur (Reference 2), which allow current material to approach the properties of the "high purity" material of Reference 1. The second is improved mill reduction practices. The third is modification of heat treat practice which increased austenitizing temperature, slightly lowered F_{ty}, and improved ductility and fracture toughness.

As the fracture toughness of production heat lots progressively increased, it became apparent that some of this increased fracture toughness would be sacrificed to obtain higher strength levels by modifying the aging cycle. Under Reference 3, it was determined that AF1410, aged at 900°F (482°C) to a tensile strength of approximately 260 ksi (1.79 GPa), was significantly more resistant to SCC than 300M. Later

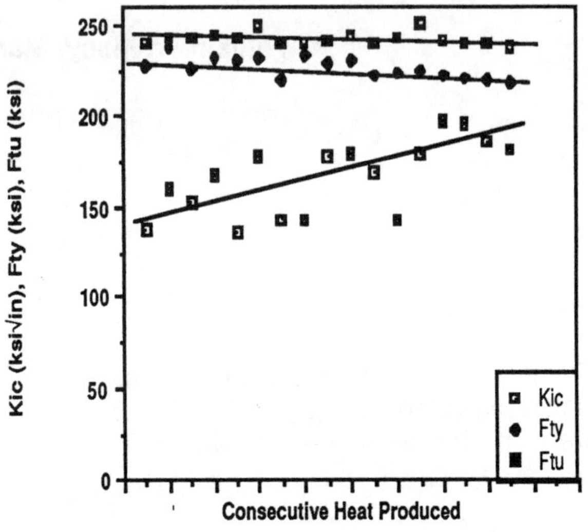

Figure 1. AF1410 Properties Improvements

work indicated that while thin sections of AF1410 could be heat treated to the 260-280 ksi (1.79 - 1.93 GPa) range, heavy sections would probably have reduced strength.

Carpenter Technology Corporation (CARTECH) is a supplier of AF1410 to MCAIR for the F/A-18 arresting gear. In late 1987, CARTECH had some success in modifying the alloy composition and increasing the strength level to over 300 ksi (2.07 GPa). In this range, however, the toughness and stress corrosion cracking (SCC) resistance had been compromised too much for airframe structural applications, such as landing and arresting gear. To improve the balance of strength, toughness, and environmental resistance in an alloy based on AF1410, a cooperative development program was initiated between CARTECH and MCAIR. CARTECH used their capabilities to melt alloys and fabricate them into products suitable for testing. MCAIR set up and tested the products to assess their suitability for aircraft landing and arresting gear applications.

A 300 lb. (136kg) laboratory heat of AF1410 was made with a nominal carbon content of 0.20, instead of the normal chemistry of 0.16. It was called modified AF1410 or 0.2C AF1410. Initial evaluations were so successful that a specification was written around the material (Reference 4). Production scale heats were made and some production applications planned. Figure 2 shows how F_{tu} and K_{Ic} have varied with aging temperature for production scale heats of material. While tensile strengths in excess of 280 ksi (1.93 GPa) can be obtained by aging at 900°F (482°C), K_{Ic} is relatively low. However, aging at 925°F

(496°C) provides a better balance of properties with F_{tu} in excess of 260 ksi (1.79 GPa) together with K_{Ic} over 100 ksi√inch (110 MPa√m).

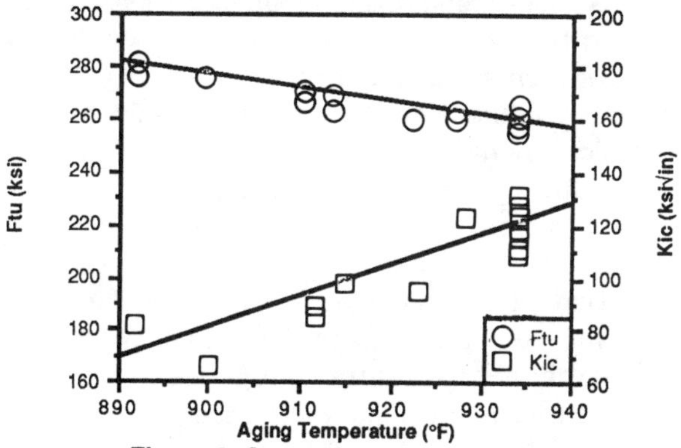

Figure 2. Strength and Toughness vs. Aging Temperature .20 C - 10 Ni 14 Co Steel

TEST RESULTS

Tests by MCAIR in 1988 and 1989 revealed the superior properties of modified AF1410 when compared to 300M (Figure 3). High carbon AF1410 has twice the fracture toughness of 300M and the same yield strength, with only slightly lower ultimate strength. For landing gear applications, the high fracture toughness allows modified AF1410 to sustain larger cracks before fracture occurs. Larger cracks are more likely to be detected during inspections.

MATERIAL	Ftu ksi (GPa)	Fty ksi (GPa)	KIc ksi√in (MPa√m)	RESISTANCE SCC	RESISTANCE HYD
300M - MIN	280 (1.93)	230 (1.65)	50 (60.5)	LOW	LOW
AF1410 - MIN	235 (1.62)	215 (1.48)	130 (143)	HIGHEST	HIGH
.2 C Af14110	274 (1.89)	246 (1.70)	113 (124)	HIGH	HIGH

Figure 3. Material Properties Test Results

Crack initiation tests of the modified AF1410 under constant amplitude loads have also been performed, and the fatigue lives compared with 300M data. The comparison in Figure 4 shows that both materials have similar fatigue lives in laboratory air.

Cracking in salt water environment is another area where the new alloys are superior to 300M.

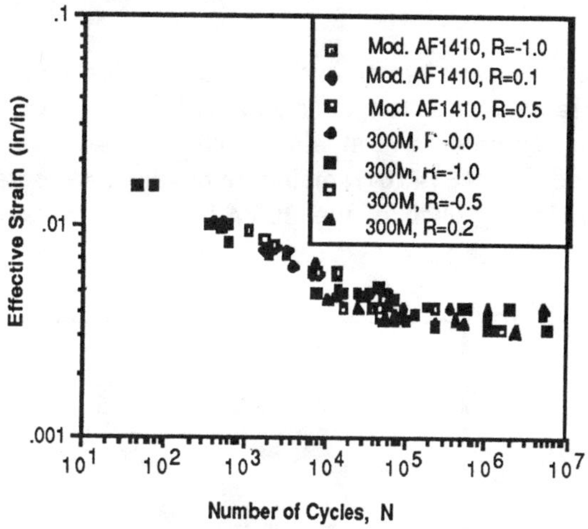

Figure 4. Fatigue Life; 300M vs. 0.2 C AF1410

Figure 5 compares crack growth rate as a function of stress intensity factor for AF1410, high carbon AF1410, and 300M in laboratory air. Although the three materials fall along the same band at low stress intensities, 300M deviates from the curve and shows a faster crack growth rate near a stress intensity factor equal to 50 ksi√inch (55 MPa√m). This is caused by the low fracture toughness of 300M.

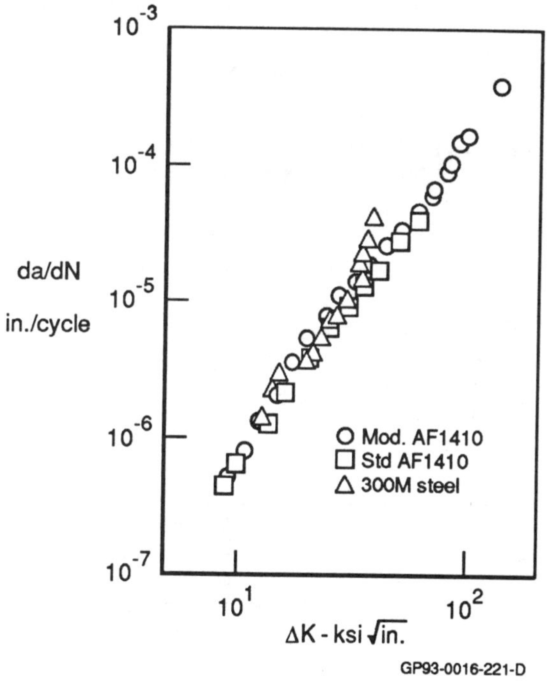

Figure 5. Crack Growth In Laboratory Air; 300M vs Mod AF1410

Figure 6 summarizes a similar set of data for specimens tested in a salt water environment at different cyclic load frequencies. Although the 300M salt water curve at 10 Hz is the same as the air data in Figure 5, the 1.0 Hz and 0.1 Hz salt water data show a tremendous increase in crack growth rate. Slower loads allow the salt water and corrosion effects to act in the crack for a longer period of time. In an aggressive environment at slower cycle rates, da/dN of the modified AF1410 remains relatively unaffected; however, crack growth of the 300M becomes markedly faster.

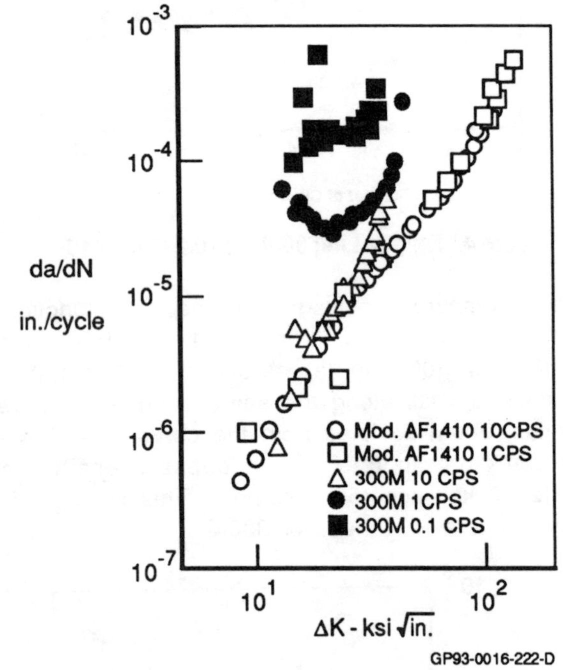

Figure 6. Crack Growth in Salt Water; 300M vs Mod AF1410

Modified AF1410 material was also supplied to the Naval Air Development Center for long term K_{Iscc} testing. These results are shown in Figure 7. The data for the modified AF1410 shows essentially the same properties as conventional AF1410. It is on the low side of the bandwidth because it is a higher strength material.

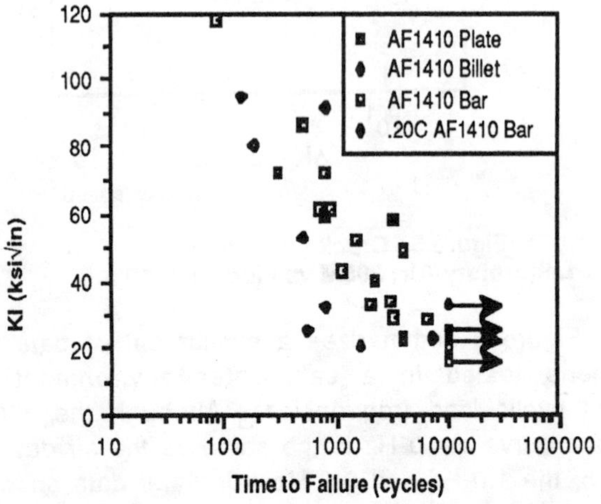

Figure 7. Stress Corrosion of Landing Gear Steel

SUMMARY AND CONCLUSIONS

A high-strength, high-toughness steel has been developed based on the chemistry of AF1410 but having the carbon content increased from 0.16 to 0.20. It will be used on future Navy programs to avoid the landing gear weight penalties which would ensue from using existing lower-strength, high-toughness materials. The benefits from the material are as follows:

- Use of the selected materials in landing gear will result in increased reliability due to its excellent fracture toughness and stress corrosion cracking resistance.
- With a structural efficiency nearly equal to 300M, it will allow designs with minimum envelope and will avoid the weight penalty associated with using lower strength steels.
- An increase in life cycle costs due to the weight penalty would be avoided.
- The reliability improvements and associated reductions in operating costs will result in lower life cycle costs.

REFERENCES

1. Novak, S. R., "Effect of Purity on Reliability Characteristics of High Strength Steels," et al. (U.S. Steel), AFML-TR-78-89, 1977.
2. Garrison, W. M., Jr., "Micromechanism of Ductile Fracture and Design of Ultra High Strength Steels," proceedings of the Earl R. Parker Symposium on Structure Property Relationships, TMS-AIME Warrendale, PA 1985.
3. Garland, K.C., "SCC Threshold for 300M and Other Alloy Steels," MCAIR TWD LMA03.11-042R, TR 3NA-667.ED.
4. McDonnell Material Specification 216, Bars and Forgings, Modified 10Ni-14Co Steel, December 1988.

Titanium Matrix Composite Landing Gear Development

William W. Macy and Mark A. Shea
McDonnell Douglas Corp.
David L. Morris
WRDC/FIVMA

* Paper 892337 presented at the Aerospace Technology Conference and Exposition, Anaheim, California, September, 1989.

ABSTRACT

Design and fabrication of a titanium matrix composite (TMC) F-15 nose landing gear (NLG) outer cylinder is discussed. Results of a field experience survey examining landing gear (LG) operations are also discussed. Weight, supportabilty and cost benefits are summarized for this component and projected for production applications of the material.

BACKGROUND

The landing gear system is tasked with repeatedly and reliably performing in a severe environment of loads, contamination, and temperature. The extreme nature of these requirements is reflected in the continued presence of landing gear systems near the top of the list of aircraft systems with regard to failure rates and total maintenance actions. With the landng gear strut in particular, design constraints of minimum weight and volume allowance coupled with the high loading have resulted in almost exclusive use of high strength steel and aluminum materials. However, these materials have essentially reached their limit of structural efficiency. Even higher efficiency materials are needed to keep pace with the weight reductions in composite airframes. Also, the conventional metals have a major problem with corrosion which results in considerable maintenance and overhaul tasking. The use of protective surface treatments for these materials such as anodizing, plating, and painting is insufficient to eliminate this corrosion and results in additional overhaul work to remove these coatings for inspection and refurbishment with subsequent reapplication of the coatings.

The application of advanced materials to landing gear struts began over 20 years ago with the design, fabrication and testing of a boron epoxy outer cylinder, piston and side brace for the A-37 MLG. These components were only able to withstand approximately limit load conditions. The promise of greater strength and fabricability led to the design, fabrication and testing of carbon epoxy side braces and torque arms for the A-37 in 1971. While the side brace designs were successfully tested, the torque arms could not meet the strength requirements within the given volume envelope. A similar program involving the design, fabrication and testing of an A-37 outer cylinder & trunnion assembly in 1973 also identified landing gear stowage volume as the limiting factor for this application. Likewise, a 1975 program involving an F-15 main landing gear upper drag brace reached the same conclusions as the two previous programs. For a new design with well defined loading paths and a relatively simple shape, carbon epoxy may serve as an excellent material for landing gear component applications.

Titanium has been successfully utilized as a landing gear material in the past with both main and nose gears of the SR-71, A-7 nose gear, F-111 main gear upper cylinder and F-18 MLG piston head as the principal examples. Other applications that were tested include a P-2V NLG, C-130 MLG outer cylinder and C-141 MLG bogie beam. Several titanium braces are currently utilized on production aircraft including the F-15 A/B and 747.

With the advent of Superplastically Formed Diffusion Bonded (SPF/DB) fabrication technology for titanium, a six inch long segment of the F-100 MLG outer cylinder was successfully designed, built and tested using SPF/DB to create a truss core sandwich structure. A follow-on to this program resulted in the first TMC landing gear application. As in the previous program, the same six inch section truss core sandwich was fabricated. But this time, the Ti-6Al-4V was reinforced with boron fiber resulting in further performance improvements over the previous design. A current TMC application involves fabrication of an arresting hook shank for an F/A-18 aircraft using Ti-15-3Al-3Cr-3Sn reinforced with silicon carbide fibers.

This program is intended to develop TMC material as a viable alternative to conventional materials for landing gear structure. Key advantages of the TMC are very high levels of specific strength and corrosion resistance, both of which are essential for landing gear applications. Additionally, longer critical crack lengths compared to high strength steels and minimum protective coating requirements will simplify inspection procedures and reduce maintenance and overhaul costs. This combination of high strength and reduced maintenance should result in lighter, more structurally efficient landing gear components with significantly reduced life cycle costs.

TECHNICAL PROGRESS

TMC is a continuous fiber reinforced composite offering very high strength-to-weight ratios. It consists of a silicon carbide fiber embedded in a titanium matrix. The fiber is a patented product of Textron Specialty Materials.

The fabrication process of the TMC barrel begins with the production of SCS-6 fiber. As shown in Figure 1, the fiber is made

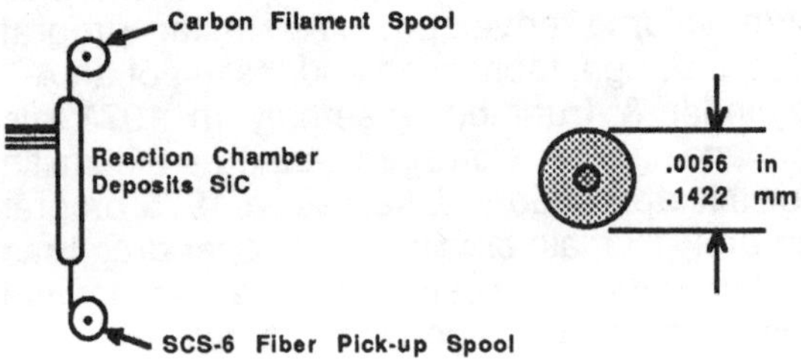

Figure 1. SCS-6 Fiber Production

by depositing silicon carbide onto a carbon filament. The outer coating is a carbon rich treatment to prevent any detrimental interaction between the fiber and the titanium. The fibers are woven into a mat with .050 (1.27 mm) wide molybdenum foil ribbon. The mat is then laid down over a sheet of Ti-15-3 foil. The mat is sprayed with acrylic to help hold the mat in place during layup. One large piece of the fabric was used for the 0° ply set while the 30° plies required several smaller pieces. The layup for the F-15 cylinder began with several layers of titanium foil to provide a machining surface for the inside bore of the finished part. The foil is wrapped around the steel mandrel and inner end collars followed by the fabric for the longitudinal plies. Each of the crossply pieces of fabric were cut so that their fibers were oriented at the proper angle. They were then laid up individually. The composite layup is then placed inside the steel outer tool and the outer titanium collars are inserted. Steel end caps are welded to both ends of the outer tool and inner mandrel sealing off the composite. The layup was evacuated through ports in the end caps and then subjected to a debulk cycle to remove the adhesive and any contaminants present in the layup. A diagram of this process is shown in Figure 2.

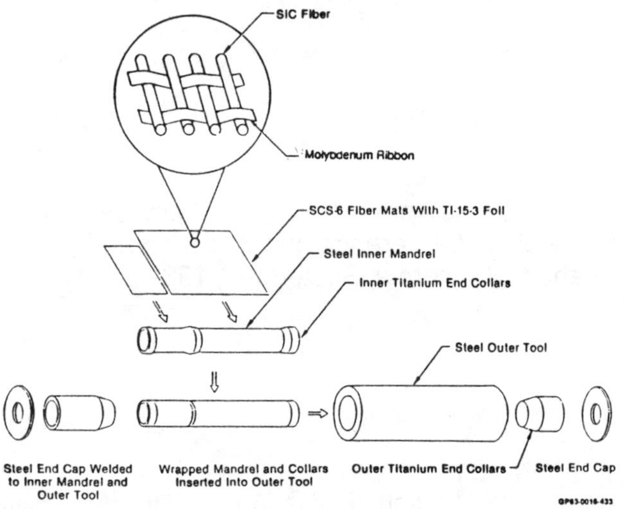

Figure 2. Weaving and Layup Sequence

Consolidation of the composite takes place in a hot isostatic press (HIP) at a maximum temperature of 1800°F (982°C) and pressure of 15000 psi (103422 kPa). Under these conditions, the inner mandrel expands forcing the composite and titanium end collars out against the outer tool. The TMC layup and titanium collars completely diffusion bond to form the cylinder blank assembly. Because of the expansion in the HIP cycle, the crossplies must be located on the outside of the laminate. The increased strain on the inside would be too high for off-axis fibers and their failure would result in severe stress concentrations and loss of strength in the final part.

Among the first tasks in the program was the selection of the test component which was driven by the following criteria. First, the part had to be a major landing gear component. This precluded use of braces and links. Secondly, the part should be one with complex loading conditions such as pressure and concentrated normal loads as well as bending and axial loads. Also, the part design should be easily adaptable to using TMC. At this point in the technology development, this means a long, simple section with a minimum of miscellaneous lug and arm attachments. In conjunction with the size and simplicity, the current part should be made from aluminum in order to avoid envelope problems in titanium portions of the part.

The components examined closely for this program included the pistons and outer cylinders from the main and nose gears of both the F-4 and F-15 aircraft. From the criteria discussed above, selection was narrowed to the F-4 and F-15 nose gear outer cylinders. The F-15 component, seen in Figure 3, was eventually chosen because it was applicable to a production aircraft, documentation for its analysis was readily available, and a test fixture and test gear would be much easier to acquire.

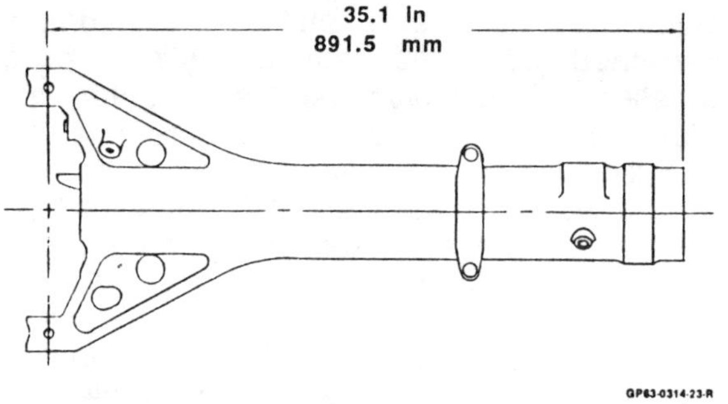

Figure 3. F-15 NLG Outer Cylinder

An element test cylinder was the first component to be attempted as part of this program. It was a generic article designed and built to verify the material capabilities and several design aspects which were to be used on the F-15 cylinder. The element cylinder was made with the same 30 ply laminate used in the F-15 part. Ti-6Al-4V collars were substituted for Ti-15-3 because of availability. Also, a large collar for a drag brace lug was incorporated along the center of the length to test a highly loaded attachment in the TMC. The element test cylinder fabrication sequence is shown in Figure 4.

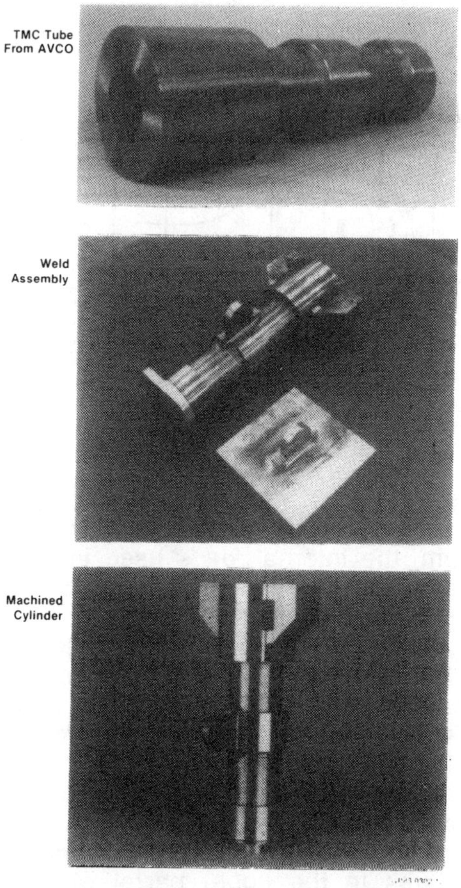

Figure 4. Element Test Cylinder

The fabrication of this test cylinder, however, was not successful. The TMC did not completely consolidate and evidence strongly suggests that this was caused by an incorrect HIP cycle and thermal problems in the tooling. Figure 5 shows photos of properly consolidated TMC (on right) and a cross section of material from the element test cylinder. The incomplete consolidation of the element resulted in cancellation of its structural test. The part was still assembled and inspected to verify the remaining fabrication processes. Changes to the F-15 part as a result of the element include an improved thermal analysis of the tooling, alteration of the collar design, establishing more rigid HIP parameters, and eliminating a chemical milling operation.

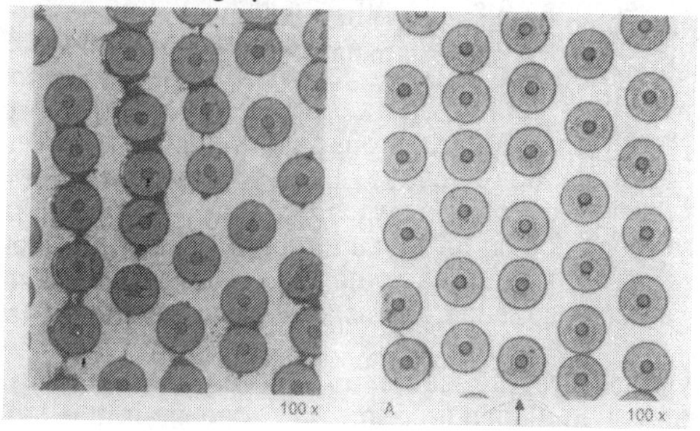

Figure 5. TMC Consolidation Comparison

The design of the composite in the F-15 NLG cylinder has progressed through several stages. The original configuration of the TMC laminate is shown in Figure 6. A single 30 ply laminate extended over a

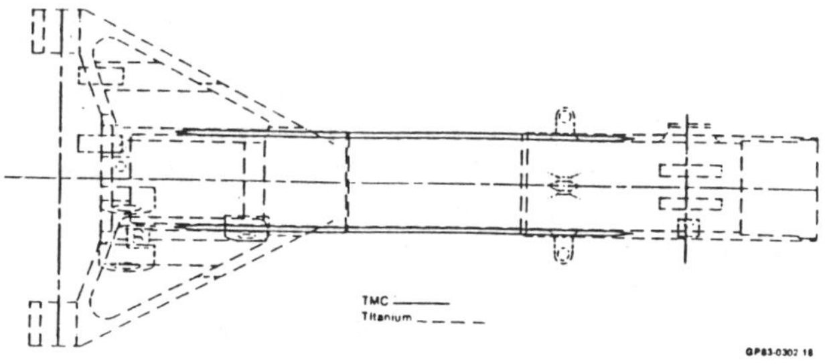

Figure 6. Original TMC Concept

large portion of the barrel including the section above the trunnion arm attachments. However, closer examination of the cylinder loading conditions showed that the barrel above the trunnion arms would be over-designed with the thick wall. The design was changed to incorporate a second, smaller laminate for the upper portion of the barrel. This section could have been fabricated as a separate tube which would be butt welded to the main barrel. This approach was permanently changed to the configuration shown in Figure 7: a single tube containing both TMC sections.

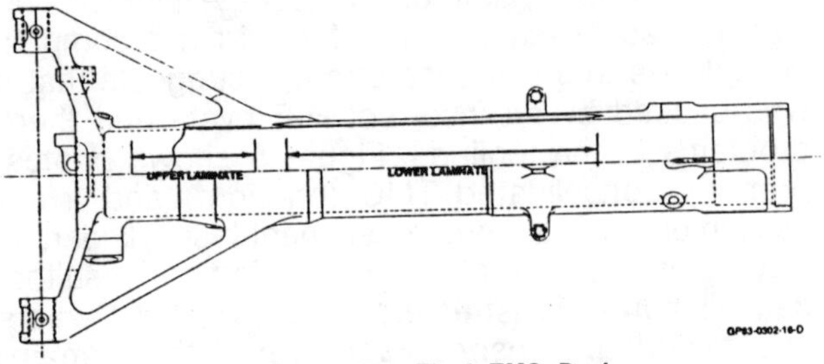

Figure 7. Final TMC Design

Although slightly more ambitious, it had the advantage of improving the ability to hold tolerances between the two sections. A single laminate with ply drop-offs at or below the trunnion arms was also considered but the risk would have been too high for the present fabrication technology. At the lower end of the cylinder, the TMC runs out several inches above the drag brace lugs because of holes through the wall for the weight-on-wheels switch and the oil fill port. Since the technologies for these through-the-laminate holes are still under development, the lower end of the cylinder was made from monolithic titanium.

The other solid titanium portions of the cylinder are the same as, or very similar to, the production aluminum part. An exception to this is the trunnion area where the webs were removed and holes were put in the upper trunnion arms to accommodate the welding thickness limitation.

The selection of the laminates used was facilitated by the use of a computer program developed at McDonnell Aircraft Co, (MCAIR). It was a quick and relatively easy way to iterate the most efficient laminate design based on the loading conditions for a given section of the cylinder. A more detailed analysis then provided information on more specific criteria such as matrix cracking and ply by ply failure. The program does indicate that over a 30% reduction in wall thickness is possible for the lower barrel with the implementation of a few key technical developments.

The strength analysis of the F-15 cylinder was not typical of landing gear structure since considerable material development was taking place. At the start of the final design, only the internal loads had been sufficiently defined. The available TMC allowables accounted for only the orthotropic tensile and compressive behavior of the original 30 ply laminate. Shear strengths and moduli were not available for this fiber system. Also, methods for analyzing a TMC cylinder had not been substantiated in the element test cylinder.

Except for the lack of shear allowables, the data would have been sufficient to analyze the cylinder if the 30 ply laminate had been used throughout. However, as discussed above, the upper barrel wall was reduced to 8 plies to save weight. A progressive ply failure method was used to predict the strengths of the baseline upper laminate and the missing shear allowables. Different strengths from changes in titanium thickness and laminate iterations were determined from the laminate sizing program discussed above which used a weighted rule of mixtures approach. Available laminae properties in conjunction with values extrapolated from the laminate test data were used to estimate "non-linear" stress-strain plots. These tension and compression plots provided the yield and ultimate stresses, strains, and initial moduli required for analysis. The final allowables were then fitted to quadratic failure criterion. These contours, shown in Figures 8 and 9, were used to size the TMC laminate in the regions of the cylinder subjected to combined internal loads.

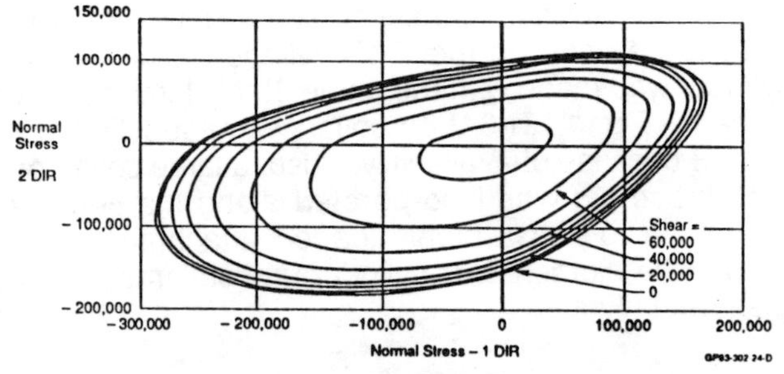

Figure 8. Quadratic Allowables Plot for Upper Laminate

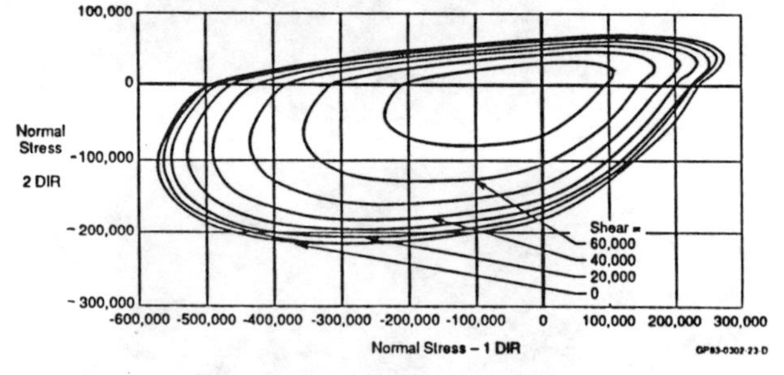

Figure 9. Quadratic Allowables Plot for Lower Laminate

Because TMC has much higher stiffness than aluminum, the internal loads used in the strength analysis of the original F-15 NLG had to be recalculated to account for the decrease in beam-column effects. A non-linear NASTRAN model was created to generate design loads. The model consisted of beam truss and rigid bar elements with the applied loads from the F-15 analysis. Separate models were made for each of the four critical stroke positions.

To size the barrel, three critical sections were selected for analysis from shear and bending moment diagrams plotted relative to the cylinder axis. One section was in the upper barrel where rebound pressures cause the largest transverse stresses. The other two locations included the trunnion arm intersection and at the bottom of the cylinder. Both places have concentrated side loads. At these locations, two criteria are met. First, matrix-fiber interface damage is not permitted at limit load anywhere in the TMC laminate. Such damage is assumed related to the "knees" in the stress-strain curves. These knees are produced by the separation of the fiber from the matrix . Because of the limited plastic behavior of the TMC, a second requirement mandated that the monolithic material be sufficient to resist ultimate load while the TMC is considered ineffective.

Margins of safety in the upper barrel were calculated from a limit Tsai-Hill stress contour. At the other TMC locations, a lamination program was used to determine "first ply damage". A modified Tsai-Hill criterion was used for the margin. A straightforward plastic analysis was used in the monolithic areas using the criteria that no yielding can occur at limit load. A summary of the critical margins are listed in Figure 10.

Element	Critical Condition	M.S.	
TMC Cylinder (Lower Barrel)	Rebound	+0.120	
Trunnion Arm Intersection	Cosine Dip	Cracking @ Limit	+0.103
		Ultimate After Cracking	+1.040
Cylinder End	Towing, 0° + D	+0.021 (Plastic)	
Drag Brace Load Introduction	Towing, 0° + D	+0.250 (Plastic)	
Jury Link Lug/ Upper Trunnion Arm	Pressurized Actuator	+0.100 (Plastic)	
Lower Trunnion Arm	Towing, 0° + D	+0.381 (Plastic)	

GP63-0302-8-T

Figure 10. Critical Margins of Safety

The first HIP attempt of the F-15 cylinder was not successful because a tooling weld had failed eliminating the pressure delta across the laminate. Although the reason for the failure was never made certain, it was presumed to be due to exposing an extension of the mandrel to the high radiant temperatures of the HIP furnace. A second HIP was successful after wrapping the mandrel with a thermal blanket. After consolidation, the outer tool and inner mandrel were removed by machining and acid etching.

The consolidated TMC tube, shown in Figure 11, was inspected by several non-destructive testing (NDT) methods including dimensional checks, ultrasound and eddy current.

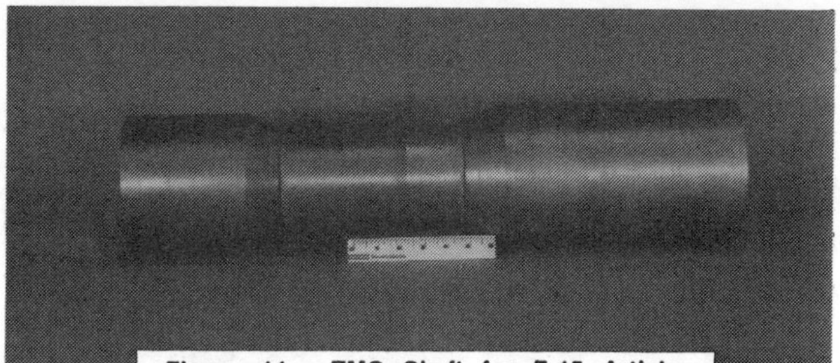

Figure 11. TMC Shaft for F-15 Article

Figure 12 contains a photograph of the TMC shaft in the water tank of the ultrasonic test machine. The c-scan plot, shown in Figure 13, indicates several small areas of possible disbonds but no appreciable flaws affecting the structural integrity of the part. These areas were supposedly caused by using several small pieces of 30° material to complete the layup.

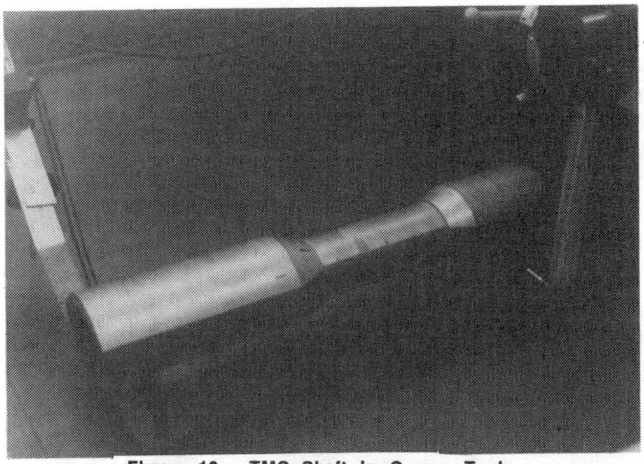

Figure 12. TMC Shaft In C-scan Tank

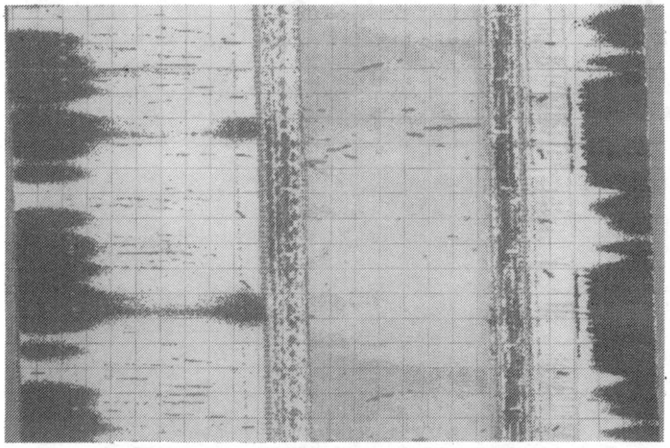

Figure 13. C-scan Plot of TMC Laminate

The machining and welding stages of fabrication are illustrated in Figure 14. Following NDT,

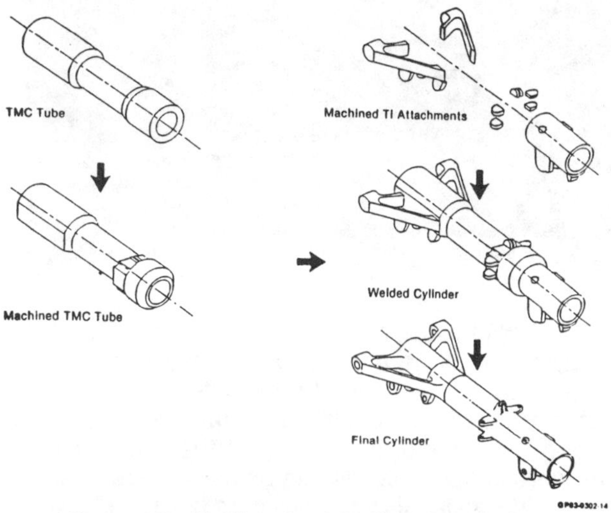

Figure 14. Machining and Welding Sequence

the titanium collars on the cylinder blank were machined to provide the necessary welding surfaces for attachment of the titanium details. These parts, including the lower barrel, trunnion arms, lugs and fittings were machined separately from Ti-15-3 bar stock. The photos of Figure 15 show the TMC tube and the trunnion arms, and the various other attachments such as the lugs and lower barrel.

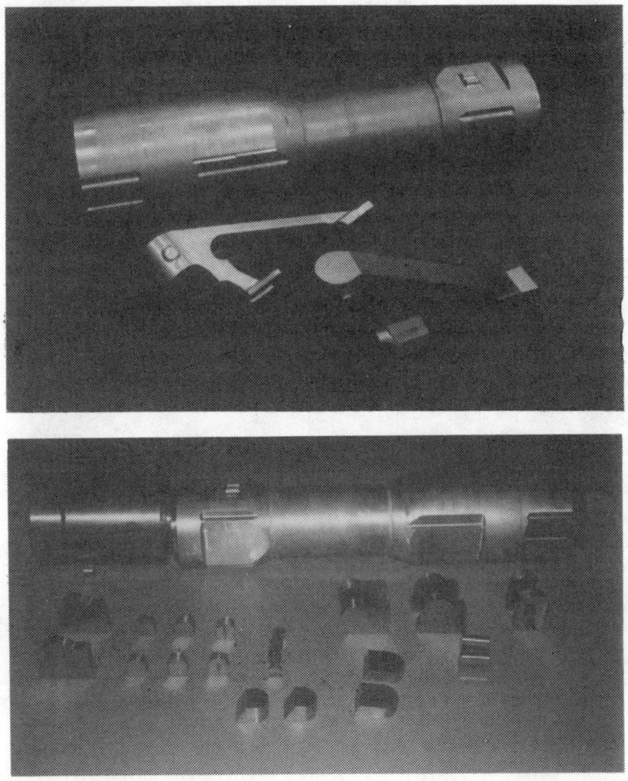

Figure 15. Machined TMC Shaft and Titanium Attachments

Assembly of the cylinder took place in MCAIR's electron beam weld shop. A total of 22 welds were required. All the attachments were slightly oversized to allow for any warpage that might have ocurred in welding or stress relief and to provide stock material for weld clean up and finish machining. Two of the welds were of relatively high risk. These were the continuous butt welds for the end cap on the top of the cylinder and lower barrel. The upper end cap weld was done first before any other parts were added because the cylinder had to be turned to make the weld. Figure 16 shows the welds at the top of the cylinder which include those for the trunnion arms, the jury brace lugs, and the steering actuator lugs. Note that the inboard jury link lug was welded in a circular arc. This was necessary to prevent the radius of the jacking lock lug from interfering with the weld. Instead, the radius blends into the attached part away from the weld path.

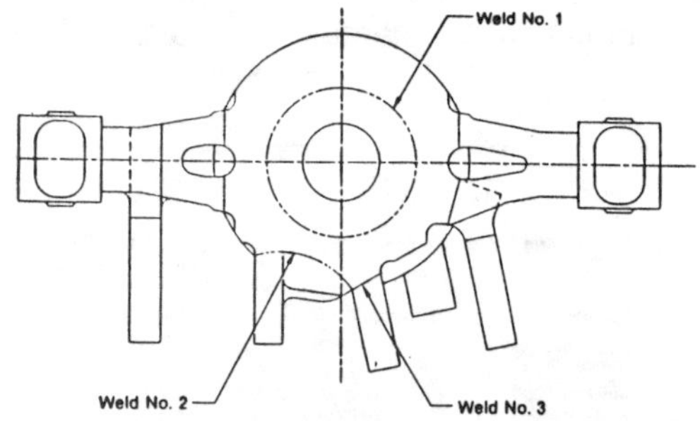

Figure 16. Welds In Upper End of Cylinder

Several welding limitations including length, thickness and accessibility restricted the design of the trunnion arms. Three changes were required: 1) the trunnion web was removed, 2) the high and low pressure service ports were removed, and 3) holes were cut in the upper trunnion arms to make two welds of acceptable thickness. Figure 17 contains a photograph of the completed weld assembly.

Figure 17. Weld Assembly of F-15 NLG Cylinder

No NDT was performed on the welds because the surface discontinuities precluded the use of ultrasonic and radiographic inspections. Visual inspection performed on the element test cylinder proved to be adequate for assessing the weld quality. The element test cylinder also demonstrated that warpage of the welds was quite small. Therefore, a minimum amount of stock was provided for finsihing the lug faces. No stress relief fixture was used but a simple 4340 steel plate, bolted to both trunnion lugs, limited warpage of the arms.

Following welding, the cylinder was taken to the machine shop where small bores were made on both ends to establish part centerline for the final machining sequence. This step is crucial to prevent cutting into the fibers on the final bore.

The weld assembly was sent to Scarrott Metallurgical Co. for the heat treatment. It was recommended that the facilities used be "cleaner" than the normal steel heat treat furnace to minimize contamination of the part surface. The cycle was a 4 hour soak at 1000°F (538°C) followed by an air cool. No serious warpage was detected after the stress relief.

The next stage of fabrication is the final machining which is similar to that required by a rough forging. Although not completed at the time of this writing, the final machining is being done by Cleveland Pneumatic Co. (CPC), the subcontractor for the production F-15 landing gear. The final cylinder configuration will be fully compatible with the rest of the gear assembly and interchangeable with other F-15 C/D NLG cylinders.

A couple of potential functional problems, associated with using titanium in the cylinder, were solved for this program. The first is the possibility of galvanic corrosion from cadmium plating on the lug bushings. Unplated bushings will be used in the test to avoid any problems. Secondly, titanium's tendancy to gall in bearing applications was a serious problem in past production applications. An electroless nickel plating was applied to the cylinder bore for the strut bearings. Past MCAIR studies have found that the nickel and TFE liner system is superior for protection of the titanium. Finally, the machining of the cylinder itself was, at first, very difficult because of the characteristics of the titanium. Ti-15-3 is an alloy suited for cold forming and not machining. Previous programs in combination with the experience from the cylinder attachments dictated a very slow speed and limited cutter depth to be successful.

The final cylinder will be part of a test gear assembly. The other parts for the gear such as the piston, strut internals and jury link are being procured. The drag brace from a static test gear will be used while the Air Force will supply a spare wheel and tire.

The testing will be done in two stages. The first is a structural test at CPC in an F-15 NLG fixture. The cylinder strain gages will be calibrated to known load levels the same way as in the production tests. CPC will use the real wheel and tire to apply the static limit loads as shown in Figure 18. A total of 40 channels will be used during the structural test excluding those required for an acoustic emission test which was recently added to the test plan. Acoustic emission is expected to increase the knowledge base for the behavior of the TMC under load.

Condition	Strut Stroke, in (mm)	Load, kips (N) V	D	S	Moments, in lb (N m) MV	MD	MS
Towing	13.0 (330)	10.7 (47600)	12.3 (54710)	0 (0)	0 (0)	0 (0)	12.3 (54710)
1-cos Dip	16.0 (406)	29.1 (129437)	0 (0)	0 (0)	0 (0)	0 (0)	0 (0)
Jacking	13.0 (330)	12.9 (57340)	-3.8 (-16902)	3.8 (16902)	-33.3 (-148k)	-117.5 (-522k)	-4.4 (-19570)

Figure 18. Structural Test Matrix

After the structural test, the instrumented test gear assembly and fixture will be shipped to the WRDC facility at WPAFB for functional testing. The functional test matrix is shown in Figure 19. The test runs will include static load-stroke, drops from 3 to 10 fps, rebound snubbing at various strokes, and gear life tests.

Condition	A/C Wt, kips (N)	Strut Stroke, in (mm)	Sink Rate, fps (mps)	No. of Tests	Comments
Static Load/ Stroke	N/A	0-15-0 (0-381)	N/A	1	
Rebound Snubbing	N/A	5,10,13,16.5 (127,254,330, 419)	N/A	4	Retract Load RAM @ Max Rate
Jig Drops	35 (155680)	N/A	3-10 (.92-3)	6	Wheel Spin w/ A/C Lift
Life Tests					
Jig Drops	35 (155680)	N/A	5 (1.5)	100	5 min Wait
Cycling	35 (155680)	Static	N/A	50000 cycles	1.5 Hz, 2 in P-P (51 mm)

Figure 19. Functional Test Matrix

BENEFITS

The first part of the program involved researching operations data from five current Air Force aircraft. The key supportability drivers for landing gear structure in terms of both cause (How-mal) and Work Unit Code were documented. The data was gathered from the 66-1 database and from the Ogden Air Logistics Center at Hill AFB. The aircraft included the F-4E, F-111, A-10, F-15 and F-16. The study indicated that TMC could result in a potential 30% increase in reliability because of its corrosion resistance and improved fatigue life. Depot level maintenance would be improved at least as much. Because the eventual goal is to put TMC in service without paint, this assumption can be justified through the elimination of paint and corrosion removal. A reduction in embrittlement relief bakes is also expected.

Although the supportability benefits of TMC are substantial, weight reduction will probably be the principal driver in selecting TMC for landing gear structure. The weight studies performed for this program have verified that component weight savings are proportional to the part size. Other factors, although to a lesser extent, include load levels and the simplicity of design. Long, highly loaded, simple parts such as braces and beams can realize savings approaching 50%. Since TMC has such remarkable compressive strength, long trunnion arms can be reduced up to 65% over the applicable length. Navy arresting hook shanks, though they are certainly perfect from a form and fit standpoint, will not reach their weight saving potential because the TMC needs protection from the severe cable slap.

When the amount of TMC in the part is limited to a small percentage of the total, the benefits will be correspondingly small. For example, short braces and very complex components will have weights that offer little or no savings over monolithic titanium alloys. Of course, these conclusions are drawn from early fabrication technology. The inevitable expansion of TMC usage into more complex areas of the parts will bring the weights down to competitive levels.

Upon completion of the design of the F-15 NLG cylinder, a weight study was made which indicated the TMC cylinder would weigh approximately 33.2 lb (147 N). This compares to 25.96 lb (115 N) for the aluminum production cylinder. The high weight is a result of two major drivers: 1) the titanium portions were slightly over designed to assure that the TMC barrel was the critical part of the structure, and 2) extensive optimization was excluded because of cost, yet undeveloped technologies, and the interchangeability requirement. No weight savings was originally expected in this first material demonstration part. The aluminum cylinder was selected to insure adequate envelope. MCAIR estimates that by the time TMC is implemented in production, most of the undeveloped technologies will be available. With these, the weight savings possible on the F-15 would average 21% on the major components.

A life cycle cost estimate was made of the F-15 landing gear system based on a conversion to TMC with the benefits described above. The life cycle cost study accounted for changes in weight, aircraft growth, replacement costs, overhauls, aircraft resizing, fuel consumption and RDT&E to create a cost model from MCAIR's Advanced Concepts Cost Model (ACCM). The ground rules for the cost model included number of aircraft, utilization rate, labor rates, and aircraft life. The resulting life cycle cost savings was estimated to be $ 26k per aircraft.

CONCLUSIONS

TMC has been identified as a possible landing gear material of the future offering large benefits in aircraft performance and cost. Using the F-15 NLG outer cylinder, this program has successfully demonstrated the ability to design, build, and test a full-scale TMC landing gear component. The program experience provides a better understanding of the material and its effective implementation into production.

Ground Operational Problems

Aircraft Flotation Analysis - Current Methods and Perspective

Norman S. Currey
Lockheed-Georgia Co.

* Paper 851936 presented at the Aerospace Technology Conference and Exposition, Long Beach, California, October, 1985.

ABSTRACT

The objective of this paper is to inform the aircraft flotation analyst of the currently acceptable methods of calculating flotation, to show how those methods relate to each other, and to discuss the flotation classification (ACN-PCN) that is on the threshold of acceptance.

Worldwide, some sixteen methods of calculating flotation are currently in use. This creates confusion for the procuring agencies, aircraft companies, and airport authorities. The International Civil Aviation Organization has addressed this problem by advocating universal use of the LCN method, later the LCG method, and now the ACN-PCN classification. However, many analyses are still being made using methods such as PCA and S-77-1.

This paper defines flotation terminology and flotation parameters such as single wheel load, radius of relative stiffness, surface type, subgrade and surface fatigue. It summarizes the analysis methods (both old and new), and compares the LCN and LCG methods. Recommended analysis methods are listed, categorizing them for commercial and military usage, by military service to be satisfied, and by the type of runway. Obsolete methods are listed, with appropriate discussion of superseding documents. Finally, the correct usage and limitations of flotation analysis are reviewed-emphasizing that the calculated values should be considered as relative numbers rather than absolute values.

AN AIRCRAFT SPENDS MORE OF ITS LIFE ON THE GROUND than in the air, and yet it seems that prior to the Boeing 747 and Lockheed C-5 an aircraft's compatibility with the ground was often ignored! High performance, better payload-range and improved economy were emphasized without realizing that airfield compatibility is just as important as high performance. Unless the aircraft has appropriate flotation, the number of fields it can operate from is restricted, or payload/range must be degraded, and this affects route structure and economics in the commercial aircraft, and mission capability plus payload-range on a military aircraft. Boeing and Lockheed did not use a multitude of tires on their 747 and C-5 in order to make the tire, wheel and brake vendors happy; they paid the associated weight penalty so that the aircraft could operate at their high weights from a wide variety of airfields all over the world--and in the case of the C-5, from bare soil.

Prior to 1960 very little thought was given to flotation, apart from spasmodic usage of Unit Construction Index. Runway traffic was generally lower than it is today, tire pressures tended to be lower, and aircraft weights were lower. The Boeing 707-320C, for instance, weighs only 43 percent as much as the 747B; and the C-141A weighs 42 percent as much as the C-5A.

Flotation analysis depends on the type of pavement design, and diverse design methods have been devised by various countries, military and civil authorities, and by government agencies. Consequently, there has been a profusion of methods for calculating flotation--and this profusion has led to confusion! For rigid, flexible and unsurfaced airfields, there are sixteen methods of calculating flotation, and AIR 1780 discusses most of them. To some extent, this can be attributed to different countries using different rules, but the fact remains that in the U.S.A. alone eight different methods are used for analysis of rigid pavements.

The International Civil Aviation Organization (ICAO) tried to resolve some of these differences when it published "Aerodrome Design Manual Part 3, Pavements" (First Edition, 1977), in which they advocated universal adoption of the Load Classification Number (LCN) method. This method was devised in Great Britian to replace their World War II system that classified pavements in broad categories such as "Fighter," "Heavy Bomber," and "Very Heavy Bomber." There is a more detailed discussion of the LCN system later in this paper.

The British went on to develop the Load Classification Group (LCG) method, and this also was adopted and promoted by ICAO (in 1974). Then, in an attempt to further simplify and generalize the usage of flotation--and in particular the aircraft/airport relationship--ICAO is now advocating Aircraft Classification Number-Pavement Classification Number (ACN-PCN) classification. This is also detailed later in this paper.

In the United States there is a willingness to promote an ICAO-sponsored worldwide system, but there are also many legacies of the past in which flotation capabilities have been evaluated by Portland Cement Association, Asphalt Institute, New York Port Authority, FAA, and military methods.

This country's development of methods for calculating flotation on bare soil has been particularly useful, and procedures are now well established. The side effects of operating on bare soil have also been researched in the United Stated for more than 20 years--items such as braking, rolling drag, and roughness.

DEFINITIONS AND PARAMETERS

FLOTATION - Flotation is a measure of an aircraft's ability to operate on an airfield surface of defined strength. These runways and taxiways may be paved or unpaved. The aircraft parameters that influence flotation are wheel load, tire contact area or pressure, and tire footprint spacing.

EQUIVALENT SINGLE WHEEL LOAD (ESWL) - On multiwheel landing gears the ESWL concept is used. In simple terms, it is the load on a single wheel that has the same effect on a particular pavement as that of a multiple wheel arrangement. These arrangements are generally characterized as dual, tandem, twin tandem, tri-twin tandem, dual twin, dual twin tandem, twin tricycle, and triple. ESWL determination varies with the particular flotation method and surface type; i.e., the U.S. Army Corps of Engineers, Waterways Experiment Station, the LCN, and the unsurfaced airfield methods all use different methods, for instance, to determine ESWL. This is necessary because of the difference in the way that tire load is distributed into the subgrade--the rigid, flexible and unsurfaced fields have completely different distribution systems.

RIGID PAVEMENT - These are characterized as being made of Portland Cement concrete. Their thickness is considered to be the thickness of concrete, commonly 8 to 14 inches (20 to 36 cm). Three kinds of loading are considered: interior, edge, and corner. Interior loading is that which is applied away from the edges, and most of the earlier methods used this as the basis for design, the philosophy being that load transfer devices at the edges were adequate to make the paved area act as a single slab. These edge restraints are frequently used to relieve the bending moment on the slab edges, but for heavy aircraft the edge conditions are still critical in many cases--hence the current trend to use this as a design basis for such aircraft. In the United States, corner loading is rarely considered.

Rigid pavement design is based on Westergaard's theories (References 1 and 2), and these in turn use the Radius Of Relative Stiffness as a primary parameter in determining ESWL. Its physical meaning is illustrated in Figure 1, and the numerical value of "ℓ" is a function of the concrete modulus of elasticity, concrete thickness, Poisson's ratio and modulus of subgrade reaction. It is defined as follows:

$$\ell = \sqrt[4]{\frac{Eh^3}{12\ (1 - \mu^2)k}}$$

where E = Young's Modulus of Concrete (psi)

h = Slab Thickness (inches)

μ = Poisson's Ratio of Concrete

k = Subgrade Modulus (lb/cu in.)

Typical values that may be used for the above are as follows:

E = 4,000,000 psi (280,000 kg/cm^3)

h = Variable

μ = 0.15

k = 300 lb/cu in.^{3} (8.3 kg/cm^3)

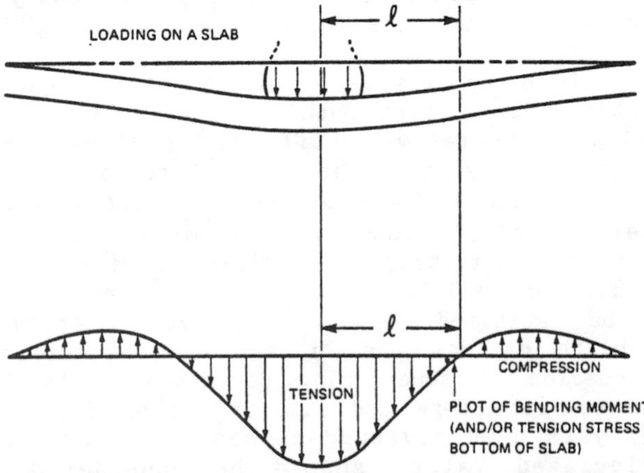

Figure 1. Physical Meaning of Radius of Relative Stiffness,

Figure 2 shows a typical flotation chart based on the Portland Cement Association method. Concrete can be expected to have a 90-day modulus of rupture (strength) of 700 psi (50 kg/cm^2), and to allow for a 1.7 (approximately) safety factor, 400 psi (28 kg/cm^2) is a good value to use in flotation evaluation. Safety factors can vary to some extent between runway ends and runway interiors.

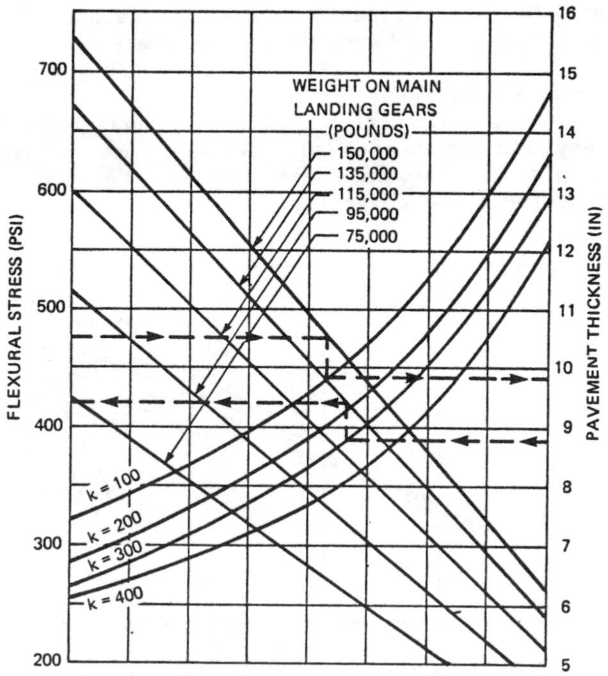

Figure 2. Rigid Pavement Design Chart for L-100-30, PCA Method

FLEXIBLE PAVEMENT - Unlike the rigid pavement, the thickness of flexible pavement includes various layers of material that are compacted beneath the surface course. Figure 3 illustrates this type of construction, and total thickness generally ranges from 8 to 60 inches (20 to 150 cm).

OTHER PAVEMENT - Apart from the conventional concrete and asphalt pavements, there are several other types, depending to some extent upon local economics, materials availability, and local topography. The Port Authority of New York and New Jersey method evaluates pavement made of lime, cement, and fly ash (L-C-F) mixed with sand. Soil stabilization provides another type of pavement; it is obtained by mixing the soil with materials such as cement, bitumens, L-C-F, and so on.

SUBGRADE - Subgrade strength is usually measured in terms of Modulus of Subgrade (or soil) Reaction (k), or CBR (California Bearing Ratio) for flexible pavements. The modulus is the applied pressure on the ground, divided by soil deflection of a rigid plate--thus k is measured in lb/in.^{3}, and typical values vary from 50 to 500 (13.6 to 135.7 MN/M^3). As noted previously, it is a parameter used in evaluation of rigid pavement, and typical values range from 200 to 300. CBR is essentially the ratio of the bearing strength of a given soil sample to that of crushed limestone gravel; it is measured as a percentage of the limestone figure, so CBR 10 is ten percent of the strength of crushed aggregate, and CBR 4 is the lowest strength upon which heavy airfield construction equipment can operate effectively. Typical CBR values of 10 to 20 can be expected on typical commercial airfields subgrades, and CBR 6 to 9 is the range commonly referred to as a soft field when an unpaved field is being considered. The procedures for measuring k and CBR are given in Reference 3, and an approximate relationship between the two is depicted in Figure 4. If CBR is being expediently measured by using a cone penetrometer, then the curve in Figure 5 may be used to relate the penetrometer readings to a CBR values. It should be realized that CBR varies widely with soil moisture content, and the readings obtained may not represent the strength available when the subgrade has stabilized.

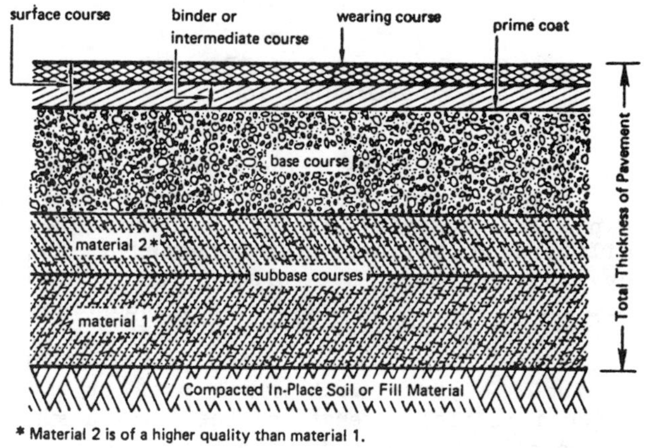

* Material 2 is of a higher quality than material 1.

PAVEMENT	Combination of subbase, base, and surface constructed on subgrade
SURFACE COURSE	A hot mixed bituminous concrete designed as a structural member with weather and abrasion resisting properties. May consist of wearing and intermediate courses.
PRIME COAT	Application of a low viscosity liquid bitumen to the surface of the base course. The prime penetrates into the base and helps bind it to the overlying bituminous course.
SEAL COAT	A thin bituminous surface treatment containing aggregate used to waterproof and improve the texture of the surface course.
COMPACTED SUBGRADE	Upper part of the subgrade which is compacted to a density greater than the soil below
TACK COAT	A light application of liquid or emulsified bitumen on an existing paved surface to provide a bond with the superimposed bituminous course
SUBGRADE	Natural in-place soil, or fill material

Figure 3. Typical Flexible Pavement and Terminology, U.S. Military Method

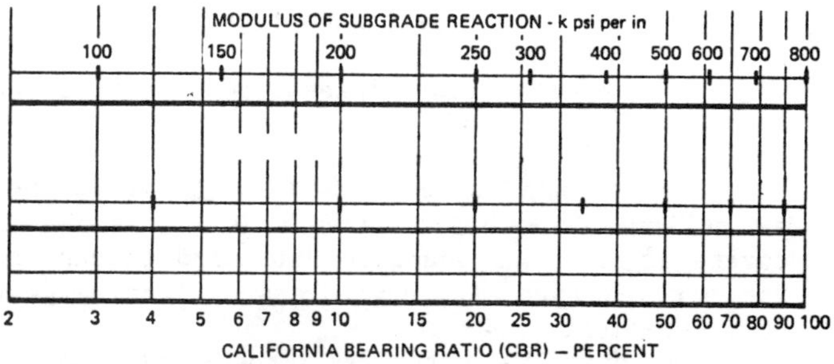

Figure 4. Approximate Interrelationship of Bearing Values

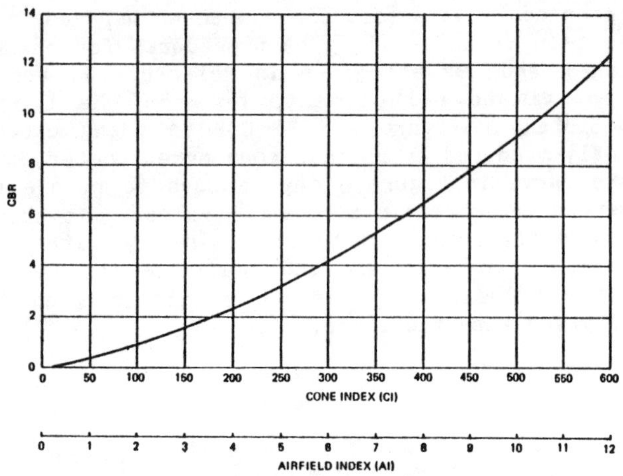

Figure 5. Correlation of CBR, CI, and AI, Fine-Grained Soil

EXPEDIENT SURFACE - There are a surprisingly large number of operations from unpaved and expedient (matted) airfields. Not only are military organizations interested in it, for obvious reasons, but also large commercial operators in undeveloped countries frequently operate on such fields. These landing areas may be natural (e.g., bare soil), gravel, turf, or surfaces covered with metal mats or plastic membranes. The strength of the in-situ surfaces is usually measured in terms of CBR, and the parameters affecting flotation are tire pressure, wheel load, and wheel spacing.

EFFECT OF LIMITED OPERATION

In paved surface analysis, the capability of an aircraft to land on a given runway is based on essentially unlimited operation. However, there are many instances in which restricted usage is desired. For instance, the aircraft might appear to be too heavy for a given runway, but it would be acceptable for a limited number of operations.

Instruction Report S-77-1 (Reference 4) provides details of the U.S. method for recognizing limited usage on flexible pavements. It uses the principle that

$$t = \alpha T$$

where t = thickness for specific number of operations (design thickness)

α = load repetitions factor

T = standard thickness for a particular aircraft

Prior to the issue of S-77-1, α was defined as follows: = 0.23 log c + 0.15 where c = coverages and was equal to "standard thickness" for a given aircraft, regardless of landing gear configuration.

Work done by Waterways Experiment Station showed that α was dependent upon both coverages and the number of wheels in the gear assembly, and S-77-1 includes a graph to relate α to coverages as a function of the number of wheels (Figure 6). This replaces the previously-used formula shown above.

For a two-wheel assembly, it shows that if a very limited number of coverages are required, such as 10, the thickness need only be about half that for 5,000 coverages.

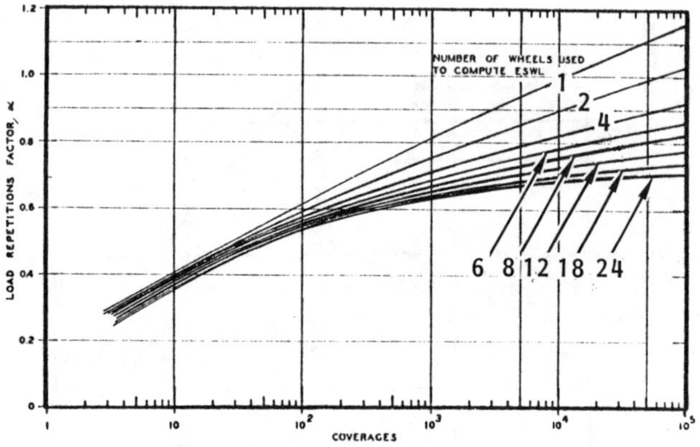

Figure 6. Load Repetitions Factor versus Coverages for Various Landing Gear Types

The British-devised LCN system also recognizes limited usage, and it uses a tabular method illustrated in Figure 7. It also recognizes the differences between channelized traffic (e.g., taxiways) and nonchannelized traffic (e.g., runways). It does this by changing the passes-to-coverage ratio, where a pass in this instance equals a takeoff and a landing. Figure 8 depicts the values associated with this system.

Ratio of Aircraft LCN / Pavement LCN	Movements	Remarks
Up to 1.1	Unlimited	
From 1.10 to 1.25	3,000	Entails acceptance of some minor failures. *
From 1.25 to 1.50	300	Some cracking may occur in concrete and possibly local failure in flexible surfaces.
From 1.5 to 2.0	Very Limited	Permission given only after examination of pavement and test data.
Greater than 2.0	Emergency	

Figure 7. LCN for Limited Pavement Use

GEAR CONFIGURATION	PASS TO COVERAGE RATIO	
	CHANNELIZED	NON-CHANNELIZED
LARGE AIRCRAFT, e.g., C-5 AND 747	2.00	2.75
DUAL TANDEM GEAR	2.25	4.00
DUAL GEARS	5.00	10.00
SINGLE WHEEL GEARS	10.00	20.00

IN THEIR USAGE, A PASS IS A TAKEOFF AND A LANDING

Figure 8. Channelized and Non-Channelized P/C in the United Kingdom

FLOTATION EVALUATION METHODS - PAVED SURFACES

In discussing flotation evaluation, it is important to recognize the difference between analyses required for pavement design, the capability of an aircraft to operate on that pavement, and the reporting of that capability.

Rigid and flexible pavements are designed by using methods such as the Portland Cement Association, FAA, or the U.S. Army Corps of Engineers' CBR method (S-77-1). These relate the strengths of subgrade and pavement, the pavement thickness, number of operations and the loading/dimensional characteristics of the landing gears expected on that pavement.

The analysis of an aircraft's capability to operate on a given pavement is essentially an inversion of the pavement design process. It uses the same methods and parameters to calculate allowable loads on a given pavement, knowing the landing gear geometry and loads of the particular aircraft involved.

Finally, there is the reporting of that capability, and until more recent years it merely involved the publishing of charts such as those from the Portland Cement Association. The ACN/PCN method, discussed later in this paper, is not used for pavement design or for detailed evaluation of aircraft capability on a known surface. It is, instead, a convenient method for quickly checking whether or not an aircraft can operate from a particular runway, both of which have been categorized in broad terms.

PORTLAND CEMENT ASSOCIATION (PCA) METHOD - The result of one of these analyses was shown previously in Figure 2. Over the years, this has been a very popular method for rigid (concrete) pavement analysis. When PCA published their manual, "Design of Concrete Pavements," in 1955 it involved a laborious "counting of blocks" on the influence chart depicted in Figure 9. The publication was revised in 1973 (Reference 5) and now includes computer methods devised by Robert H. Packard.

Prior to the development of a computer program, it was necessary to draw the tire footprint pattern on transparent paper to a scale defined by the scale of the chart being used, and the properties of the slab and its supporting subgrade. The drawing was placed on the chart in various positions to determine which one gave the greatest loading--done by counting blocks. The bending moment in the concrete can then be determined by a formula that relates the number of blocks, pavement rigidity, subgrade rigidity, and intensity of loading. From this moment the stress can be found by dividing the moment by the section modulus.

This tedious method has now been replaced by a PCA computer program (Reference 6).

The Aerospace Industries Association (AIA) has published a National Aerospace Standard NAS 3601 (Reference 7), and in the rigid pavement flotation section of this document the PCA computer program (PDILB) is specified as the method to be used.

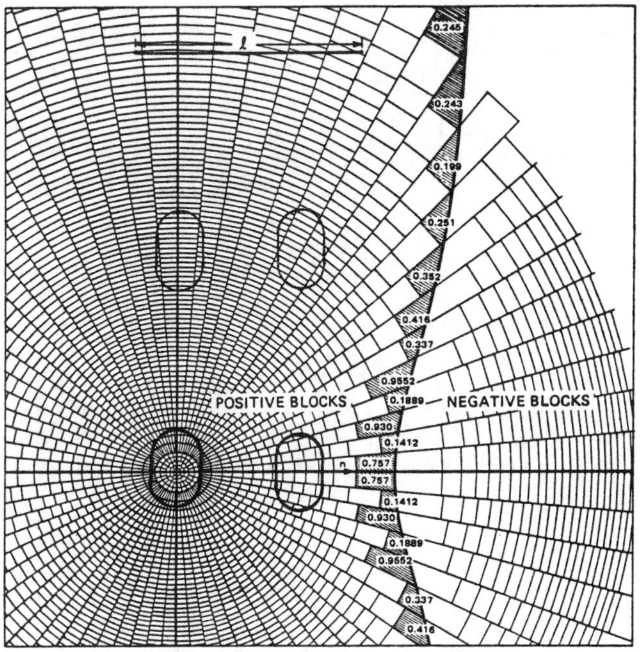

Figure 9. Influence Chart for Interior - Loaded Slab

PORT AUTHORITY OF NEW YORK AND NEW JERSEY (PANYNJ) METHOD - The PANYNJ wanted to use locally-available materials and therefore developed its own pavement design method. The pavement material used is a mixture of lime, cement, and fly ash (L-C-F) with sand. Crushed stone is added when a high strength base is required. Features of this method are a unique method of determining equivalent single wheel load, the use of theoretical expressions for soil stress and deflection, and the interrelationship between pavement roughness, aircraft, response, and pavement stress.

FEDERAL AVIATION ADMINISTRATION (FAA) METHOD - Advisory Circular AC 150/5320-6C (Reference 8) details the FAA method. On rigid pavements it is based on the Westergaard analysis of an edge-loaded slab, whereas previous FAA methods considered interior loading. The parameters used in determining pavement thickness requirements for a given aircraft are the 90-day concrete flexural strength, subgrade modulus (k), aircraft weight, and annual departures.

On flexible pavements the approach is often referred to as the CBR method. The parameters are subgrade CBR, subbase CBR, aircraft weight, annual departures, and tire contact area.

The Advisory Circular includes a figure that relates subgrade modulus to other aspects of soil classification, and also a series of graphs, such as Figures 10 and 11, that show pavement requirements for a given gear arrangement in a simple, easy-to-use manner. These graphs assume that tire pressure and wheel spacing increase with gross weight. For noncritical areas, the pavement thicknesses can be 10 to 30 percent less than those shown on the charts.

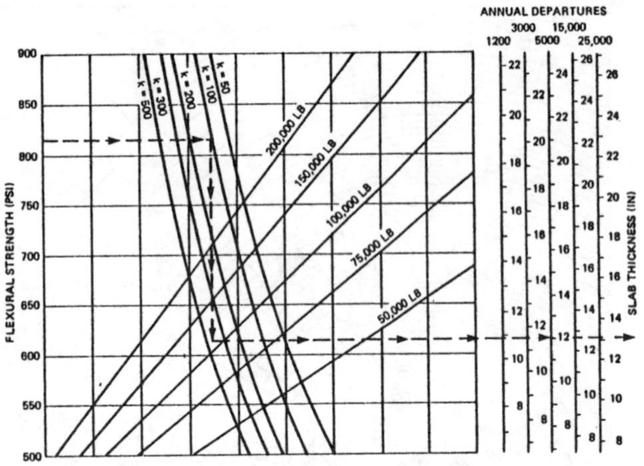

Figure 10. FAA Rigid Pavement Design Curves - Dual Wheel Gear

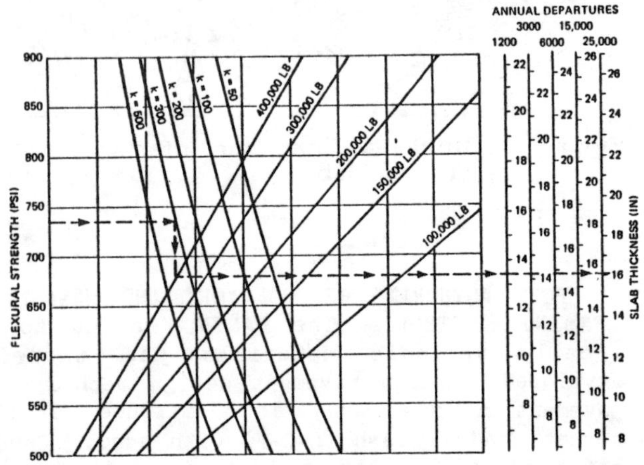

Figure 11. FAA Rigid Pavement Design Curves - Dual Tandem Gear

U.S. AIR FORCE METHOD FOR PAVED SURFACES - The method quoted in ASD-TR-70-43 (Reference 9.) is still valid and is applicable to both rigid and flexible surfaces. For rigid pavement, the USAF method is an adaptation of the computerized PCA method, but uses the U.S. Air Force/U.S. Army definitions of Type A, B, C, and D traffic areas. For flexible pavement, the method is an adaptation of the CBR method and it results in a graph of subgrade CBR versus total pavement thickness for specific numbers of passes as depicted in Figure 12.

TRI-SERVICE MANUAL METHODS (U.S. ARMY-NAVY/ MARINES-AIR FORCE - The U.S. Department of Navy, Army, and Air Force have jointly published documents that reflect their pavement design criteria. For rigid pavements the USAF criteria are given in AFM 88-6 Chapter 3 (Reference 10). Pavement design charts define thickness requirements for various traffic areas, based on modified Westergaard theory to recognize edge loading. The U.S. Army method is detailed in TM5-823-3 (Reference 11) and is almost identical to the Air Force method. Manual DM-21 (Reference 12) describes the U.S. Navy method. It is based on interior pavement loading, but is also very similar to the others. The résulting pavement thickness is almost the same, no matter which of the three methods is used.

In Reference 13 the tri-service criteria are given for flexible pavements. This comprehensive manual includes subgrade evaluation and stabilization as well as requirements for all levels of the pavement including overlays. Pavement thickness is determined by the U.S. Army Corps of Engineers' CBR method (Instruction Report S-77-1, Reference 9).

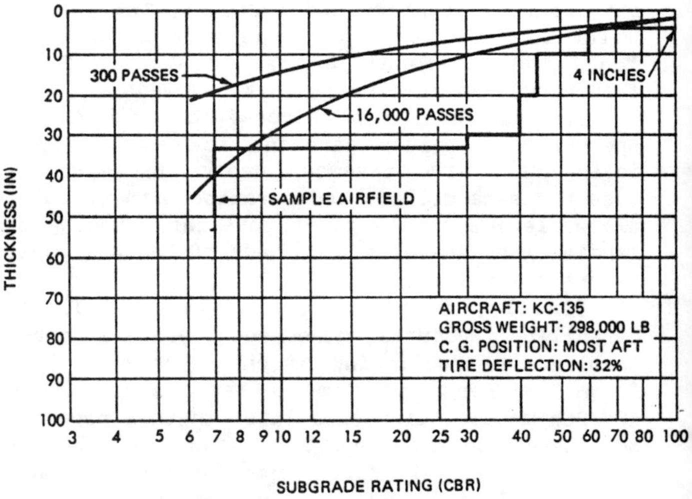

Figure 12. Asphalt Pavement Requirements, KC-135

CANADIAN METHOD - This is sometimes **referred** to as the Mcleod Method, and it applies to flexible pavements. It was developed from various studies and extensive testing conducted by the Canadian Department of Transportation (Reference 14), and uses the equation $t = k \log P/S$, where t = pavement thickness, k is defined by a curve shown in the referenced report and in Reference 15 from ICAO; P and S are the loads carried at the top of the base course and top of the subgrade, respectively, when using a loading test palet of a given size. The Canadian Method for flexible pavements does not recognize specific wheel patterns and spacings. On rigid pavements, the values for P and S are used to obtain the modulus of subgrade reaction (k). From that point onwards, the method uses the Westergaard equation for center-loading conditions.

LOAD CLASSIFICATION NUMBER (LCN) METHOD - This method originated in Great Britian immediately after World War II, and ICAO Aerodrome Design Manual Part 3 (Reference 15) provides an excellent account of its early history and subsequent development. The method was adopted by ICAO in 1956 and used worldwide; it is quite straightforward and involves the following steps: Determine the ESWL and then apply that load to the standard curve (Figure 13) that shows LCN as a function of ESWL, tire pressure, and tire contact area.

For rigid pavements, ESWL is based on Westergaard's theory, and in order to simplify its use, a series of graphs have been prepared to allow rapid determination of ESWL for gears having conventional wheel patterns such as dual wheel and dual tandem. If the radius of relative stiffness (ℓ) of the runway is known, then use it; otherwise, calculate ESWL for a range of values, say 30, 40 and 50 inches (76, 102 and 127 cm).

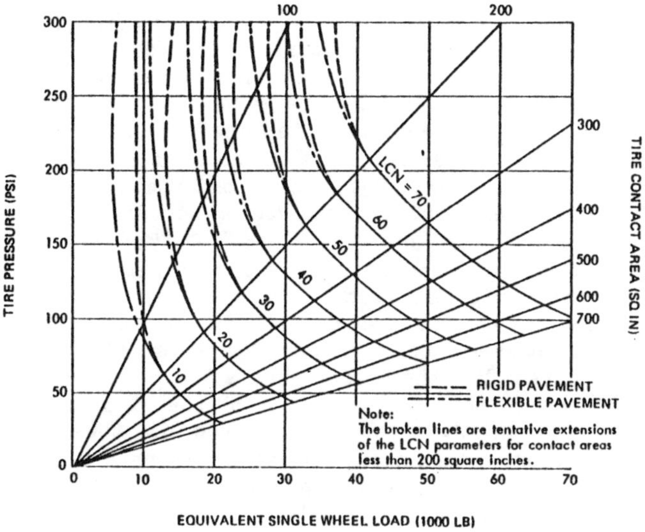

Figure 13. Load Classification Number Chart

For flexible pavements the same system is used; i.e., calculate ESWL and then use the same standard curve to determine LCN. However, the ESWL calculation is different. Recognizing that flexible pavements are more complex, being built up from several layers of different materials, it is assumed that the critical area is that where the pavement meets the subgrade. The ESWL is the load which produces the same stress in the subgrade as that produced by the whole assembly. The calculation method of this ESWL is detailed in the ICAO Aerodrome Design Manual, Part 3 (Reference 15).

Determination of the runway LCN, as opposed to the aircraft LCN described above, is ideally derived from plate bearing tests of corner loadings and thickness measurements of flexible pavement. In some cases, however, the LCN values have been established by considering the LCNs of the heaviest aircraft known to have operated on that surface. LCN values are now available for most runways in British- and ICAO-influenced countries.

The values obtained by this method permit unlimited operation. For restricted operation the table shown previously in Figure 7 may be used to show permissible LCNs.

LOAD CLASSIFICATION GROUP (LCG) METHOD - The LCG method was developed by the British in the early 1970s and adopted by ICAO in 1974; it is a modification of the LCN method and is generally considered to have replaced that method in the United Kingdom. Aircraft and pavement LCN values are now categorized into groups.

This method recognizes the fact that flotation is not an exact science; both the surface materials and subgrade have properties which vary according to time, weather and location along the runway. It is, therefore, more appropriate to define flotation by group rather than a specific number; and the ICAO Aerodrome Design Manual, Part 3, shows the LCG as a function of LCN ranges. For example, LCGIII includes LCN 51 through 75.

When applied to the LCG method, there is one significant change. The LCN contact area versus failure load average curve had the following equation:

$$\frac{W_1}{W_2} = \left(\frac{A_1}{A_2}\right)^{0.44}$$

where W_1 and W_2 were the failure loads on contact areas A_1 and A_2. Since that was formulated, analyses have indicated that the 0.44 power should be changed to 0.27, and that is the value used in determining thee curves used for establishing LCN when applied to LCG. One of the main improvements in this revised LCN calculation is that it is less sensitive to tire pressure. It is important that the latest graphs be used, therefore, for establishing LCN when considering LCG. These "new" LCNs will, in most cases, be lower than those obtained by the original system.

AIRCRAFT CLASSIFICATION NUMBER/PAVEMENT CLASSIFICATION NUMBER (ACN/PCN) - This is the latest method of evaluating or categorizing flotation, and it has been adopted by ICAO for worldwide usage. It is intended only for reporting aircraft weight bearing limits on pavements where these limits have been separately determined. It is NOT used for design and evaluation of pavements; it is, instead, a simple method for showing a relationship between aircraft and pavement. Unlike some previous methods, the ACN/PCN does not attempt to recognize effects of small variations in tire pressure or subgrade strengths--it wisely uses categories such as "high" and "low" strengths and pressures. The system also recognizes that, in the past, there has been difficulty in obtaining adequate data on airfield capability due to insufficient data on the pavement characteristics. With the ACN/PCN system, approximate pavement capability is obtainable by merely knowing the subgrade strength and the aircraft that have used the runway, although it is obviously preferable to use analysis methods.

To calculate ACN, there are subroutines available for the rigid and flexible computer programs (Reference Appendix 3 to ICAO Annex 14). To do it manually, use the existing (NAS 3601) pavement requirements charts to obtain the thickness required at the aircraft weight being considered, and then use ICAO conversion charts to translate this "reference thickness" to "Derived Single Wheel Load" (DSWL) and ACN. The ACN is the DSWL (in kilograms) divided by 500. The pavement requirements charts should have been made by using the PCA (PDILB) or the S-77-1 (CBR computer methods. On rigid pavement the thicknesses are determined for subgrade strengths (K) of 75,150,300 and 550 pci (20, 40, 80, 150 MN/M^3), with 400 psi (2.75 MPa) concrete working stress. On flexible pavement, thicknesses are obtained for subgrade CBRs of 3, 6, 10, and 15 for 20,000 coverages.

Sometimes, ACN values are required for weights and tire pressures below the maximums used in the standard ACN evaluation. When this is needed, ICAO has provided a chart to show the appropriate correction factors.

ACN values for current aircraft are listed in Attachment B of the ICAO Annex 14, and Figure 14 illustrates an example. PCN values may be obtained by any method deemed appropriate by the airport authority, including those which are primarily an evaluation of aircraft that have used the runway. If it is known that a given aircraft, at a certain weight, represents the limit for that pavement for unlimited operation, then by working backwards an ACN is determined, and this becomes the PCN. Similarly, the runway capability can be calculated by any of the normal methods for this aircraft and the ACN determined. The PCN is the same as this ACN.

When these have been determined, the ACN/PCN values are quoted in general categories, and typical ACN values are depicted in Figure 15. A PCN 80 RBWT, for instance, represents a pavement with a PCN of 80, on a rigid pavement with medium strength subgrade, can withstand high tire pressure, and has been evaluated by technical analysis rather than relating it to a limiting aircraft. This PCN establishes the allowable ACN for virtually any type of aircraft considered for use on that pavement.

FLOTATION METHODS - UNPAVED SURFACES

The method used for calculating flotation on unpaved surfaces is defined in Report ASD-TR-68-34 (Reference 18). It does not include surfaces covered by mats or membranes--so-called "expedient surfaces." The parameters involved are tire inflation pressure, tire dimensions, wheel load, wheel spacing, and overall dimensional relationship between nose and main gears. Tire pressure is one of the most important factors and in some countries this is used as the only factor in determining the ability of an aircraft to use a certain field. Both nose gear and main gear flotation are calculated, and the results are combined to show total aircraft flotation. This is necessitated because of the nature of unpaved surfaces--the damage caused by the nose wheels could be exacerbated by the main wheels such that the combined effect is worse than the individual effects. Another unusual feature is the ESWL calculation. It is determined by graphic means, using wheel spacing and contact area as parameters. Nose gear loads include those caused by braking at 10 feet per second per second (3 meters per sec per sec) deceleration with forward center of gravity; aft center of gravity is used for determination of main gear flotation.

					RIGID SUBGRADES				FLEXIBLE SUBGRADES			
AIRCRAFT TYPE	AIRCRAFT MASS (kg)	AIRCRAFT WEIGHT (LB)	LOAD EACH MAIN GEAR (PERCENT)	TIRE PRESSURE (psi)	HIGH 150 (MN/m^3)	MEDIUM 80 (MN/m^3)	LOW 40 (MN/m^3)	ULTRA-LOW 20 (MN/m^3)	HIGH 15 (PERCENT)	MEDIUM 10 (PERCENT)	LOW 6 (PERCENT)	ULTRA-LOW 3 (PERCENT)
B727-100	73,028*	161,000	46.7	158	40	43	46	48	37	38	44	49
	41,322**	91,100	46.7	158	20	22	23	25	19	19	21	25
B727-200 (STANDARD)	78,471	173,000	46.4	167	46	48	51	53	41	43	49	54
	44,293	97,650	46.4	167	23	25	26	27	21	22	24	28
B727-200 (ADVANCED)	84,006	185,200	47.8	150	50	53	56	59	46	48	55	60
	44,298	97,661	47.8	150	23	25	26	28	21	22	24	29
B737-100	45,722	100,800	46.3	148	24	26	28	29	22	23	26	30
	25,941	57,190	46.3	148	12	13	14	15	12	12	13	15
B737-200	50,349	111,000	46.35	148	27	29	31	32	25	26	29	33
	27,005	59,535	46.35	148	13	14	15	16	12	12	14	16
B747-100	334,752	738,000	23.125	225	44	51	60	69	46	50	60	81
	162,704	358,700	23.125	225	18	20	23	26	19	20	22	28
B747-200 B, C, F	373,307	823,000	23.075	210	49	58	68	78	52	58	71	93
	168,873	372,300	23.075	210	18	20	23	27	20	21	23	30
CARAVELLE SERIES 10	52,000	114,640	46.1	109	15	17	20	22	16	17	19	23
	29,034	64,009	46.1	109	7	8	9	10	7	7	9	11
CARAVELLE SERIES 12	55,960	123,371	46.0	128	16	19	22	25	17	19	21	26
	31,800	70,107	46.0	128	8	9	10	12	8	9	10	12
CONCORDE	185,06[illegible]	408,000	48.0	183	61	71	82	91	65	72	81	98
	78,698	173,500	48.0	183	21	22	25	29	21	22	26	32

*FIRST ROW FOR EACH AIRCRAFT PERTAINS TO MAXIMUM TAKEOFF GROSS WEIGHT

**SECOND ROW FOR EACH AIRCRAFT PERTAINS TO MAXIMUM OPERATORS EMPTY WEIGHT

Figure 14. Typical ACN Values

PCN	Pavement Classification Number
( )	(Bearing strength for unrestricted operations)

Code	Pavement Type
R	Rigid
F	Flexible

Code	Subgrade Category
A	High (k = 550 psi/in or CBR = 15%)
B	Medium (k = 300 psi/in or CBR = 10%)
C	Low (k = 150 psi/in or CBR = 6%)
D	Ultra Low (k = 75 psi/in or CBR = 3%)

Code	Tire Pressure Category
W	High (No limit)
X	Medium (Limited To 218 psi)
Y	Low (Limited To 145 psi)
Z	Very Low (Limited To 73 psi)

Code	Evaluation Method
T	Technical
U	Using Aircraft

Figure 15. Pavement Classification Number Category Definitions, ACN-PCN Method

In calculating flotation, the ESWL is determined, and from this the CBR required for one coverage is read from a graph. The number of coverages permissible on the desired surface is then obtained from the formula

$$C = \left(\frac{\text{CBR of surface}}{\text{CBR for 1 coverage}}\right)^6$$

Passes per coverage are determined by landing gear geometry, and from this the number of passes can be calculated. This is done for the main gear and then the nose gear, after which an equation is used to determine total airplane passes. Passes in this case refers to the movement of an aircraft past one point on the surface; i.e., one landing or one takeoff.

FLOTATION CAPABILITY - EXPEDIENT SURFACES

LANDING MATS - There are three mat categories; heavy duty, medium duty and light duty. Capabilities of heavy and medium duty mats are as follows for 1,000 coverages on a surface of CBR4, with 250 psi (1.72 NM/m^2):

Type	Single Wheel Load	Tire Contact Area
Heavy	50,000 lb (222 KN)	200 in^2 (0.129 in^2)
Medium	25,000 lb (111 KN)	100 in^2 (0.065 in^2)

The types and usage of mats are generally described in TM 5-330 and TM 5-337 (References 19 and 20).

MEMBRANES. Membranes are often placed beneath landing mats as a water-proofing device and for dust control, but have been used as expedient surfacing of landing strips to maintain in situ strengths despite weather effects. Heavy, medium and light duty membranes have been developed. These can withstand wheel loads of aircraft as heavy as the C-130H (155,000 lb, 70,310 kg gross weight) for up to 1 year.

Details of membranes and their use are given in TM 5-330 and TM 5-337 (References 19 and 20).

RECOMMENDED AND OBSOLETE PRACTICES SUMMARY

CIVIL APPLICATIONS - As noted earlier in this paper, it is important to recognize the difference between flotation evaluation methods and the aircraft loading/runway capability reporting methods. Among the former, the PCA evaluation method, using the PDILB computer program, is still valid for rigid pavements, as is the FAA AC 150/5320-6C method. The U.S. Army Corps of Engineers' S-77-1 method is applicable for flexible pavements and replaces SEFL 165A. Airports controlled by the Port of New York and New Jersey Authority use the PANYNJ method. The original LCN method is still widely used, but the British developers consider it to be generally out of date. Its successor, LCG is used on both rigid and flexible pavements in the United Kingdom, and is essentially a reporting method, although its updated LCN evaluation is now replacing the original LCN as an evaluation tool. ACN/PCN is now adopted by ICAO as a capability reporting method, and will gradually become the preferred method for relating aircraft and pavement capabilities.

MILITARY APPLICATIONS - The Defense Mapping Agency still widely quotes LCN values for military applications, but LCG is becoming more acceptable. These will soon be replaced by ACN/PCN for paved airfields. The U.S. Navy, however, still uses DM-21 for rigid surfaces and DM-21.3 for flexible surface evaluation. Use ASD-TR-68-34 for unsurfaced fields, and on fields covered with mats or membranes use TM-5-330 and TM 5-337.

OBSOLETE METHODS - It is anticipated that several of the methods that are currently in use will soon become obsolete when ACN/PCN is widely used. Also, FAA AC 150/5320-6C replaces the -6 and -6A methods. However, there are certain methods that are completely outdated, and these are listed below:

- Unit Construction Index, Technical Memorandum WCLS 53-13
- Construction Index, Miscellaneous Report 4-100
- Developing a set of CBR Curves, WES Instruction Report 4
- SEFL Report 167 for unsurfaced airfields
- SEG-TR-67-52 flotation on paved airfields
- SEFL 165A flotation on flexible pavements

PERSPECTIVE

Unlike airframe structures, aerodynamics, mechanisms and systems--where there is wide acceptance of the absolute results from using certain designs and materials--flotation analysis cannot provide such results, and there has been difficulty in obtaining worldwide usage of any given method. If it is calculated, for instance, that an aircraft can make 100 passes on a CBR 6 unprepared field, it assumes that all points on that landing area are at least CBR 6, and that there are no load excursions due to roughness--a wild assumption to say the least. Unlike a structural material, the soil strengths may change from day to day and from one area of the runway to another.

Still on the subject of unprepared fields, the surface is said to have failed when 2- to 3-inch (5.1 to 7.6 cm) ruts appear. The reasons for initially selecting the 2- to 3-inch failure criteria was that experience showed that fields deteriorated very rapidly after that rut depth was reached. For transport aircraft having larger-size tires, such ruts would be of little consequence, and the inability to take off in the length available would be a more meaningful criteria for failure. One aircraft tested recently, for instance, at close to its maximum gross weight, was able to maneuver in loose sand while making 6- to 8-inch (15.2 to 20.3 cm) ruts, and maneuvered, loaded, offloaded, and exited a CBR 7 to 10 test area with 8-inch (20.3 cm) ruts. Then there is the matter of airfield roughness on these surfaces--traversing bumps causes higher-than-static loads, and these could cause premature failure. Clearly, there is a need for trained ground personnel to evaluate a prospective landing site for military aircraft rather than using arbitrary failure criteria.

It is sometimes thought that a thin layer of pavement could prolong the usability of an otherwise unprepared field. This is not necessarily the case. While it undoubtedly removes load excursions due to bumps (for a time), the pavement may soon break up and prevent further operations. An unpaved surface, allowed to rut (particularly to the point of takeoff distance problems), may well last longer than the broken pavement.

Also, at a given CBR, one surface may disintegrate rapidly under a certain series of loads, whereas another may compact and permit far more landings.

In conclusion, it should be emphasized that flotation analysis is extremely useful when comparing one aircraft with another, but the calculated ability of an airplane to operate on a given surface is not usually precise enough, for the reasons just stated, to conduct mission analyses that depend heavily on the number of operations before surface failure. (The answers would usually be very conservative). However, if the flotation capabilities of various aircraft are calculated by the same method, then their relative abilities to use specified airfields can be determined. This approach can be used successfully in operational evaluation, and also in selecting one competing design against another in aircraft procurement.

ACKNOWLEDGEMENT

The investigation of current and obsolete methods of analyzing aircraft flotation was conducted by the SAE A-5 Aerospace Landing Gear Systems Committee, and in particular by a panel spearheaded by John A. Osterman. The work of this panel is documented in Report AIR 1780, "Aircraft Flotation Analysis," issued November 30, 1981; and this resulted in the publication of Aerospace Recommended Practice ARP 1821, "Aircraft Flotation Analysis Methods," dated September 30, 1982.

REFERENCES

1. Westergaard, H. M., "Stresses in Concrete Pavements Computed by Theoretical Analyses," Public Roads, Vol. 7, No. 2, April 1926.

2. Westergaard, H. M., "New Formulas for Stresses in Concrete Pavements of Air Fields," Transactions of the American Society of Civil Engineers, Vol. 73, No. 5, New York, NY 10017, May 1947.

3. MIL-STD-621A, "Test Method for Pavement Subgrade, Subbase, and Base-Course Material," Naval Publication and Forms Center, Philadelphia, PA 19120, 22 December 1964.

4. Tereira, A. Taboza, "Procedures for Development of CBR Design Curves," Instruction Report S-77-1, U.S. Army Engineer Waterways Experiment Station, Vicksburg, MS 39180, June 1977.

5. "Design of Concrete Airport Pavement," Portland Cement Association, Chicago, IL 60619, 1973.

6. Packard, Robert G., "Computer Program for Concrete Airport Pavement Design, (Program PDILB)," Portland Cement Association, Chicago, IL 60610, 1968.

7. NAS 3601, "Recommended Standard Data Format of Transport Airplane Characteristics," National Aerospace Standards, Aerospace Industries Association of America, Inc., Washington, DC 20036, June 1968.

8. AC 150/5320-6C, "Airport Pavement Design and Evaluation," Advisory Circulars, Federal Aviation Administration, Washington, DC, 7 December 1978.

9. Creech, Dale E., and Donald H. Gray, ASD TR-70-43, "Aircraft Ground Flotation Analysis Procedures - Paved Airfields," Aeronautical Systems Division, AFSC, Wright-Patterson AFB, OH 45433, January 1971.

10. AFM 88-6, Chapter 3 (TM5-824-3), "Rigid Pavements for Airfields Other Than Army," 7 December 1970.

11. TM 5-823-3 "Army Airfield and Helicopter Rigid and Overlay Pavement Design," Department of the Army, October 1968.

12. NAVFAC DM-21, "Airfield Pavements," Naval Facilities Engineering Command, Alexandria, VA, 22332, June 1973.

13. Navy DM 21.3, Army TM 5-825.2, and Air Force AFM 88-6 Chapter 2, "Flexible Pavement Design for Airfields," Department of the Navy, the Army and the Air Force, August 1978.

14. McLeod, N. M., "Airport Runway Evaluation in Canada," Highway Research Board, Research Report 4-B, 1947 and 1948.

15. DOC 9157-AN/901, Part 3, "Aerodrome Design Manual, Pavements," International Civil Aviation Organization, Montreal, Quebec, Canada, H3A2R2, First Ediction 1977 and Second Edition 1983.

16. "Aircraft Loading on Airport Pavements," Aerospace Industries Association of America, Inc., Washington, DC 20036, March 1983.

17. "International Standard and Recommended Practices, Aerodromes," Annex 14 to Amendment 37, International Civil Aviation Organization, Montreal, Quebec, Canada, H3A2R2, March 1983.

18. Gray, Donald H., and Donald E. Williams, "Evaluation of aircraft Landing Gear Ground Flotation Characteristics for Operation from Unsurfaced Fields,: ASD TR-68-34, Aeronautical Systems Division, AFSC, Wright-Patterson AFB, OH 45433, September 1968.

19. TM 5-330 (AFM 86-3, Vol. II) "Planning and Design of Roads, Airbases, and Heliports in the Theatre of Operations," Department of the Army, September 1968.

20. TM 5-337, "Paving and Surfacing Operations," Department of the Army, February 1966.

The Generation of Tire Cornering Forces in Aircraft with a Free-Swiveling Nose Gear

Robert H. Daugherty and Sandy M. Stubbs
NASA Langley Research Center

** Paper 851939 presented at the Aerospace Technology Conference and Exposition, Long Beach California, October, 1985.

ABSTRACT

Various conditions can cause an aircraft to assume a roll or tilt angle on the runway, causing the nose tire(s) to produce significant uncommanded cornering forces if the nose gear is free to swivel. An experimental investigation was conducted using a unique towing system to measure the cornering forces generated by a tilted aircraft tire. The effects of various parameters on these cornering forces including tilt angle, trail, rake angle, tire inflation pressure, vertical load, and twin-tire configuration were evaluated. Corotating twin-tires produced the most severe cornering forces due to tilt angle. A discussion of certain design and operational considerations is included.

IN 1966, landing tests were conducted on a model of the HL-10 manned lifting entry vehicle (1)* to determine viable steering methods. The vehicle had a conventional nose gear but had skids on the main gear struts. It was learned that asymmetrical main gear strut deflection, which produced a tilt angle on all three gears, caused cornering forces to be developed by the nose gear which was free to swivel. It was determined that this phenomenon could be used as a steering method, although the cause of the cornering forces was not investigated.

More recently, excessive differential braking has been required on some Space Shuttle orbiter landings to keep the orbiter aligned with the runway center-line. Crosswinds and runway crown cause the orbiter to assume a tilt angle and since the nose gear is operated in a free-swivel mode, unwanted cornering forces are developed by the nose gear tires which must be reacted using differential braking.

Several military aircraft accidents can be attributed in part to this phenomenon. Most military fighter aircraft have a small main-gear spacing, therefore a blown tire during the landing roll-out can cause a substantial tilt angle that creates large cornering forces at the nose gear which can cause a loss of directional control by the pilot.

The purpose of this paper is to present the results of tests conducted at the NASA Langley Research Center to determine the effects of various parameters on the magnitude of cornering forces produced by tires on a free-swiveling nose gear. The parameters studied included tilt angle, trail, rake angle, tire inflation pressure, vertical load, and for a twin-tire configuration, whether or not the tires corotate. A discussion of the cause of these cornering forces is included along with operational and design considerations.

APPARATUS

The vehicle used in this investigation was an airboat modified with a retractable tricycle landing gear. The vehicle weighed 2007 kg (4425 lbm) and the vertical load on the nose gear was typically 2135 N (480 lbf). A more detailed description of the vehicle can be found in reference 2. A photograph of the vehicle in a 10-degree tilted attitude is presented in figure 1. The tires used on the nose gear were 6.00x6 TT 8-ply Type III aircraft tires with a

* Numbers in parentheses designate references at end of paper.

rated load of 10.5 kN (2350 lbf).

Figure 1.- Test vehicle.

A schematic of the towing system used in this investigation is shown in figure 2. The lead or tow tug towed both the test vehicle and the instrumentation tug, while both tugs followed straight expansion joints on a flat concrete surface. A force transducer in the cable between the vehicle nose gear and the instrumentation tug accurately measured the lateral force produced at the nose gear.

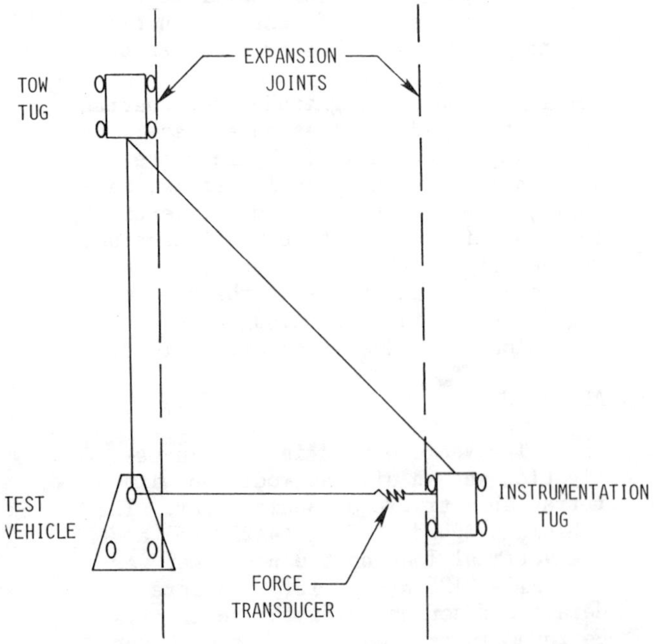

Figure 2.- Towing system.

For each test, the vehicle was tilted away from the instrumentation tug to the desired tilt angle using the retractable gear, while maintaining the proper rake angle which was normally set to zero degrees. Each tow test was conducted at a speed of approximately 1.5 m/sec (5 ft/sec) and covered a distance of approximately 61 m (200 ft).

RESULTS AND DISCUSSION

During this investigation, the effects of various parameters on the cornering force generated by an aircraft tire installed on a free-swiveling nose gear were evaluated. These parameters included tilt angle, trail, rake angle, tire inflation pressure, vertical load, and whether or not a twin-tire configuration corotates.

THE ORIGIN OF CORNERING FORCES DUE TO TILT ANGLE - Several tests were conducted to identify the mechanism that produces cornering forces when a tire is tilted. The test that most clearly defined this mechanism made use of a piece of graph paper lightly coated with grease, with a string laid straight across it. The test vehicle was tilted 10 degrees to the left and toward forward so that one of the main-gear tires (not free to swivel) rolled across the paper perpendicular to the string. The results of this test are shown in figure 3.

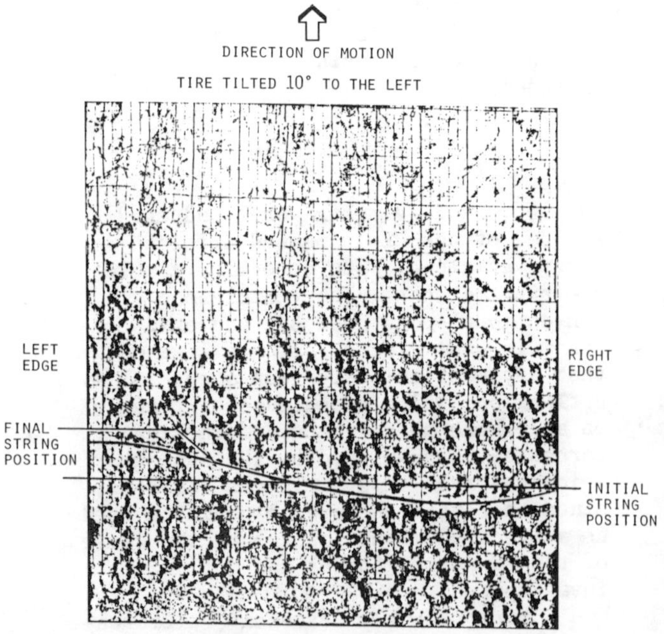

Figure 3.- Indication of differential slip in the tire footprint.

Both the initial and final positions of the string are indicated. The purpose of the grease was to keep the string from moving after the tire rolled off the paper. The movement of the string on the left edge of the footprint was in the direction of motion, indicating this side of the footprint was in a slipping or braking condition. The right side of the footprint moved the string rearwards, indicating a spinning or driving condition occurred there. This differential slip in the tire footprint is caused by the tilt angle of the system, which causes one side of the tire to have a smaller rolling radius than the other. Since the angular velocity of both edges of the wheel must be equal, one side of the tire footprint must slip while the other side actually must spin to some degree. This phenomenon is usually associated with twin-tire systems that corotate. This differential slip creates a torque which tends to steer the tire in the direction of tilt. As yaw angle develops in the tire footprint, cornering force increases until the torque created by the differential slip is balanced by the torque produced by the cornering force acting behind the steering axis. It should be noted that differential slipping due to tilt angle occurs for both non- and free-swiveling nose gears. However, significant cornering forces cannot be developed by fixed or non-swiveling nose gears since the moment produced by differential slip is balanced by the strut and vehicle structure rather than by the development of cornering forces.

EFFECTS OF VARIOUS PARAMETERS

TILT ANGLE - A table summarizing all of the test conditions and results can be found in reference 3. The effect of tilt angle on the cornering force coefficient produced by the tire is presented in figure 4. The cornering force coefficient is defined as the force perpendicular to the direction of motion produced by the tire during a tow test divided by the normal load on the tire. The cornering force coefficient increased linearly with tilt angle, from 0 at no tilt angle to approximately 0.65 at a tilt angle of 10 degrees. Most aircraft are not capable of attaining a 10 degree tilt angle on the runway, however testing to 10 degrees improves the confidence in results at the lower angles. The tire was operating at 20 percent of its rated load for these tests.

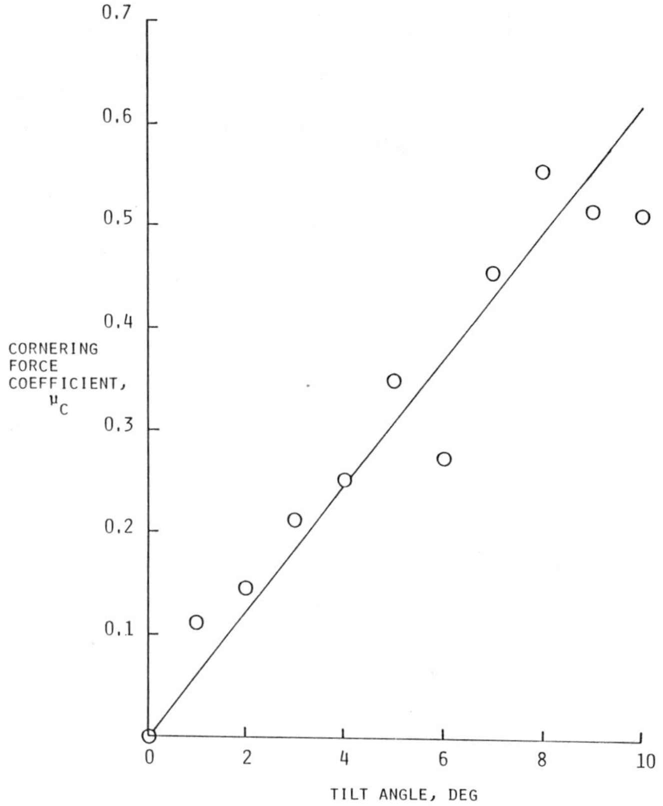

Figure 4.- Effect of tilt angle on cornering force coefficient.

TRAIL - Trail is defined as the distance between the wheel axle and the rotational axis of the strut. Two inserts for the nose gear piston were fabricated to increase the trail from 0 cm (0 in.), its normal value, to 3.8 cm (1.5 in.) and 25.4 cm (10 in.). Figure 5 presents a plot of cornering force coefficient as a function of tilt angle for three different trail configurations. The figure shows a non-linear decrease in cornering force coefficient for increasing trail. This trend can be attributed to a moment balance. For a fixed tilt angle, if the torque due to differential slipping is independent of trail, then increasing trail increases the moment arm through which smaller cornering forces are required to balance the torque.

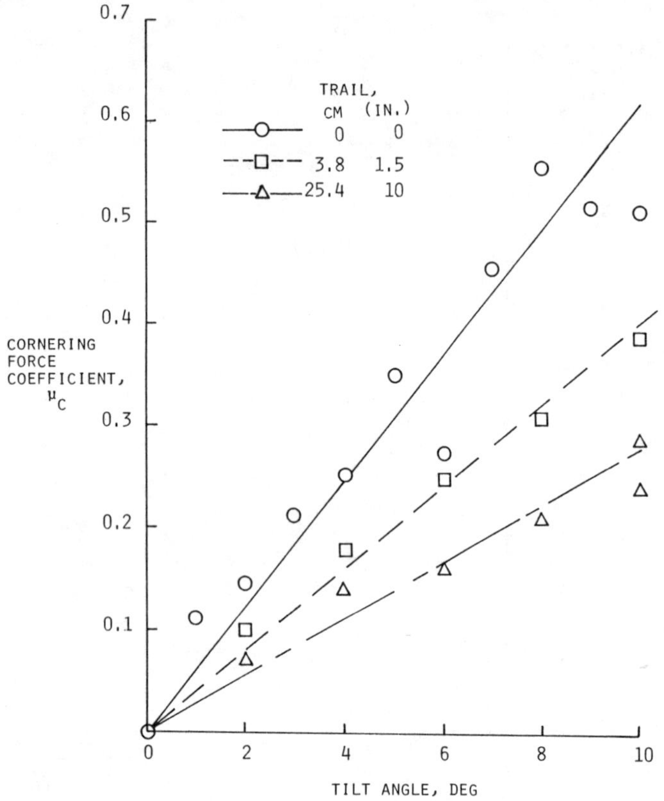

Figure 5.- Effect of trail on cornering force coefficient.

RAKE ANGLE - Rake angle is defined as the angle between the rotational axis of the strut and a line perpendicular to the runway surface. An insert for the nose gear drag link was fabricated and installed producing a forward rake of 10 degrees on the nose gear strut, so that the wheel axle was forward of the nose gear upper attachment points. Figure 6 shows the cornering force coefficient as a function of tilt angle for both 0 and 10 degrees of rake. Both sets of tests were conducted using 3.8 cm (1.5 in.) of trail. The plot shows that forward rake has the effect of reducing the cornering force coefficient by a small amount for a given tilt angle.

TIRE INFLATION PRESSURE - Figure 7 presents the cornering force coefficient as a function of tilt angle for various inflation pressures and trails. The figure shows virtually no effect of pressure on the coefficient for the zero trail configuration, even though the higher pressure is almost three times the lower pressure. The cornering force coefficients for the 3.8 cm (1.5 in.) trail configuration appear to be reduced when the inflation pressure is decreased from 165 (24) to 83 (12) kPa (psi).

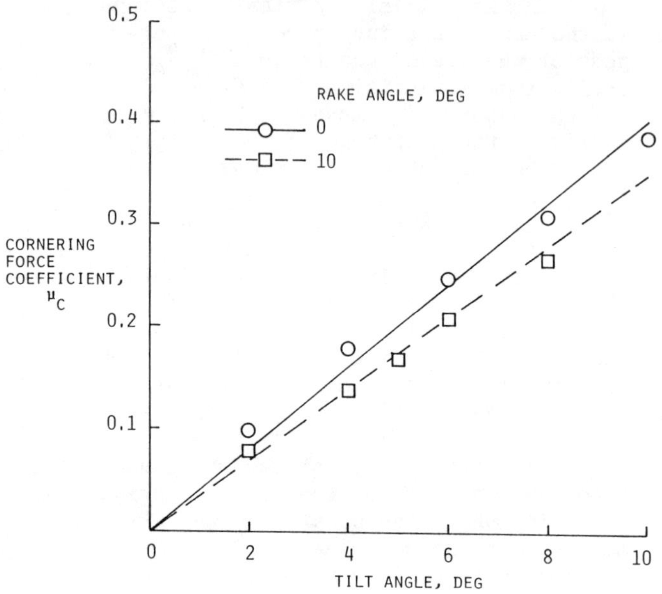

Figure 6.- Effect of rake angle on cornering force coefficient.

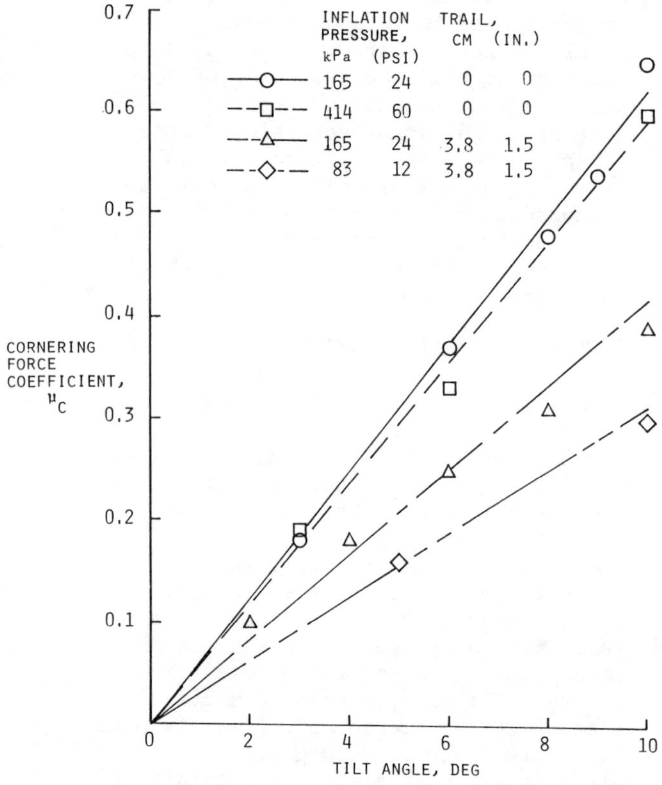

Figure 7.- Effect of tire inflation pressure on cornering force coefficient.

This is most likely due, however, to the tire being extremely underinflated for the 2135 N (480 lbf) load it was supporting, causing deflection past the design limit of approximately 35 percent. The insensitivity of cornering force coefficient to tire inflation pressure can also be attributed to a moment balance. For a fixed tilt angle and trail, increasing or decreasing the width of the tire footprint with inflation pressure causes an increase or decrease in the torque due to differential slip. However this also causes an increase or decrease in the length of the cornering force arm, requiring approximately the same force to balance torques.

VERTICAL LOAD - The effect of vertical load on the cornering force coefficient is presented in figure 8. Vertical load is normalized and presented as a percentage of the tire rated load. These tests were conducted using 3.8 cm (1.5 in.) of trail. As the vertical load on the tire is increased, the slope of the cornering force coefficient versus tilt angle curve is decreased from 0.041/degree at 20 percent rated load to 0.022/degree at 75 percent rated load. This indicates the tire becomes less efficient at producing cornering force as the vertical load is increased. This behavior is very similar to that seen in references 4 and 5, where the tire becomes less efficient at producing side force at a fixed yaw angle as the vertical load increases.

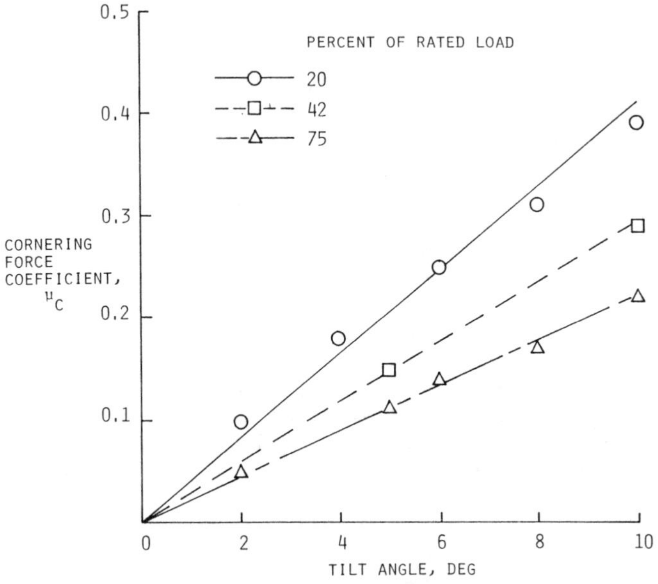

Figure 8.- Effect of vertical load on cornering force coefficient.

INDEPENDENTLY ROTATING TWIN-TIRES - An assembly was installed on the nose gear piston to allow two wheels to be mounted on the same axle and rotate independently. The total load on the nose gear for these tests was 2135 N (480 lbf), representing 10 percent of the rated load for each of the twin tires compared to 20 percent of the rated load for a single tire. The tests were conducted at both 3.8 (1.5) and 25.4 (10) cm (in.) of trail. Figure 9 presents the cornering force coefficient as a function of tilt angle, comparing twin-tire data with previously presented single tire data. No significant changes in the coefficient occur for either trail case.

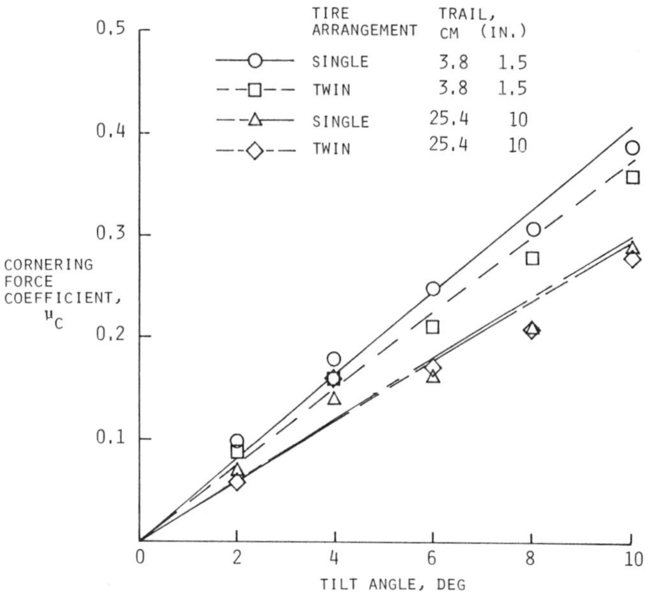

Figure 9.- Effect of independently rotating twin-tires on cornering force coefficient.

COROTATING TWIN-TIRES - The independent twin-tire system was made to corotate by bolting the split-rims of each wheel together. The spacing of the tires was 20.3 cm (8 in.) center to center. Figure 10 presents the cornering force coefficient as a function of tilt angle, comparing corotating twin-tire data with independently rotating twin-tire data for both 3.8 (1.5) and 25.4 (10) cm (in.) of trail. The plot shows a significant increase in sensitivity of the cornering force coefficient to tilt angle, with a factor of almost 5 appearing for the 3.8 cm (1.5 in.) trail case. For both trail cases, as the tilt angle reached and passed about five degrees, one tire began to leave the ground causing the system to tend to act like a single tire. At a tilt angle of about eight degrees, the tire was completely free of the ground and duplicated single-tire data.

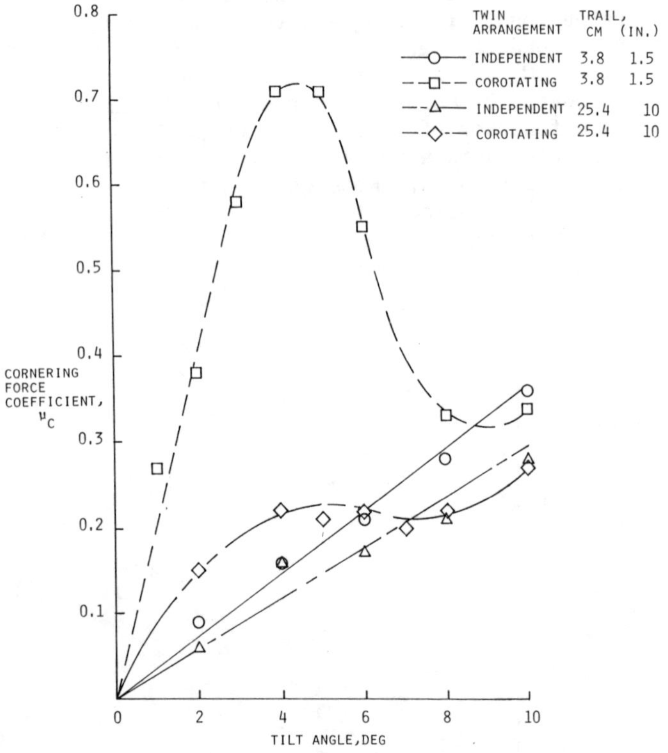

Figure 10.- Effect of corotating twin-tires on cornering force coefficient.

DESIGN AND OPERATIONS

The design of aircraft landing gears is influenced by a myriad of parameters, including structural, dynamic and operational characteristics. To keep the magnitude of cornering force due to tilt angle to a minimum, several design considerations can be made, as long as they do not compromise landing gear or aircraft stability. Trail should be as large as feasible, and rake angle should help to reduce cornering forces. The main gear should be spaced as far apart as practical to reduce the tilt angle that can be achieved should the tires fail or should strut pressure be lost. Also, if a twin-tire nose gear is used, the tires should be free to rotate independently if possible.

Operationally, several things can be done to reduce the magnitude of cornering forces developed by tilted tires. The most effective reduction of these forces is accomplished by engaging nose gear steering if it is available. The torque produced by differential slipping in the footprint does not manifest itself as cornering force if the nose gear is not free to swivel. Indeed, if the nose gear is locked straight ahead, cornering forces due to tilt angle will be essentially eliminated. Once a tilt angle has been established on the runway, whether it is due to blown tires, crosswind, etc., the tilt angle can be reduced using aerodynamic controls and by steering to the side of the runway opposite the tilt angle, to take advantage of the effect of the crown to cause a tilt in the opposite direction. Increased load on the main gear through the use of aerodynamic control can also help to reduce the cornering forces developed by the nose gear.

CONCLUDING REMARKS

An experimental investigation was conducted to examine the effects of various parameters on the cornering forces produced by aircraft tires installed on a free-swiveling nose gear. Parameters studied included tilt angle, trail, rake angle, tire inflation pressure, vertical load, and whether or not a twin-tire system corotates. Tow tests were conducted on an airboat modified to accept a conventional tricycle landing gear.

The mechanism responsible for generating the cornering forces was determined to be differential slipping in the tire footprint that occurs as a result of tilt angle. Significant cornering forces are produced only when the nose gear is free to swivel. Cornering force coefficients increased linearly with tilt angle. Increasing trail decreased the magnitude of cornering force coefficients observed. Forward rake angle decreased the cornering force coefficients slightly. Tire inflation pressure had essentially no effect on these coefficients. A nose gear with independently rotating twin-tires produced the same cornering force coefficients as a single-tire gear. When the tires were made to corotate, however, the cornering force coefficients were greatly magnified for a given tilt angle.

REFERENCES

1. Stubbs, Sandy M.: Landing Characteristics of a Dynamic Model of the HL-10 Manned Lifting Entry Vehicle. NASA TN D-3570, November, 1966.

2. Daugherty, Robert H.: Braking and Cornering Studies on an Air Cushion Landing System. NASA TP-2196, September, 1983.

3. Daugherty, Robert H.; and Stubbs, Sandy M.: A Study of the Cornering Forces Generated by Aircraft Tires on a Tilted, Free-Swiveling Nose Gear. NASA TP-2481.

4. Vogler, William A.; and Tanner, John A.: Cornering Characteristics of the Nose-Gear Tire of the Space Shuttle Orbiter. NASA TP-1917, October, 1981.

5. Tanner, John A.; Stubbs, Sandy M.; and McCarty, John L.: Static and Yawed-Rolling Mechanical Properties of Two Type VII Aircraft Tires. NASA TP-1863, May, 1981.

Flow Rate and Trajectory of Water Spray Produced by an Aircraft Tire

Robert H. Daugherty and Sandy M. Stubbs
NASA Langley Research Center

* Paper 861626 presented at the Aerospace Technology Conference and Exposition, Long Beach, California, October, 1986.

ABSTRACT

One of the risks associated with wet runway aircraft operation is the ingestion of water spray produced by an aircraft's tires into its engines. This problem can be especially dangerous at or near rotation speed on the take-off roll. An experimental investigation was conducted in the NASA Langley Research Center Hydrodynamics Research Facility to measure the flow rate and trajectory of water spray produced by an aircraft nose tire operating on a flooded runway. The effects of various parameters on the spray patterns including distance aft of nosewheel, speed, load, and water depth were evaluated. Variations in the spray pattern caused by the airflow about primary structure such as the fuselage and wing are discussed. A discussion of events in and near the tire footprint concerning spray generation is included.

THE EFFECTS of wet runway operation of aircraft such as reduced braking and cornering have long been known. On flooded runways, a reduction in take-off acceleration and hydroplaning are common dangers. The advent of multi-engine aircraft, particularly those with aft-mounted turbojets brought with it an increased potential for ingesting the water spray thrown up by the tires. Large amounts of water can cause an engine to flame out, a situation that can be especially dangerous near rotation speed.

Commercial aircraft certification requires the airframe manufacturer to demonstrate the ability to operate on a runway with approximately 1.3 cm (0.5 in.) standing water without experiencing any spray ingestion problems. Some aircraft have geometries that make spray ingestion a problem, so these are the ones typically fitted with spray suppression devices such as chined tires or nosewheel spray deflectors. References 1 and 2 describe some military aircraft that have experienced spray ingestion problems and show some geometries highly susceptible to the phenomenon. One of the considerations in the design of the aircraft geometry should be the spray ingestion potential. Thus, it is desirable to be able to estimate the path of the water spray.

The purpose of this paper is to present results of tests conducted at the NASA Langley Research Center to determine the effects of various parameters on flow rate and trajectory of aircraft tire-generated water spray. The parameters studied include distance aft of nosewheel, speed, tire load, and water depth. A discussion of aerodynamic structure effects as well as characteristics of the spray near the tire is included.

APPARATUS

The tire-generated spray investigation was conducted in the Hydrodynamics Research Facility, an 884 m (2900 ft) long towing tank with an electrically driven carriage that rides on rails 6 m (20 ft) apart. A sketch of the carriage in the tank is shown in figure 1. A feedback control system allows the carriage operator to select and maintain a test speed within ± 0.2 m/sec (±0.5 ft/sec). A more detailed description of the facility can be found in reference 3. A 1.2 m (4 ft) high runway was installed in the bottom of the tank, with rubber dams and aluminum pans added to provide a 0.9 m (3 ft) wide water trough. Adjustable dams were added for water depth control, allowing a maximum depth of 1.6 cm (0.625 in.). The runway itself was 15.2 m (50 ft) long and had a 6 m (20 ft) ramp on each end to load and unload the test tire smoothly.

As shown in Figures 2 and 3 the test aircraft, nose strut, and nosewheel used for most of this investigation were those of a general aviation twin-engine aircraft in the 26.7 kN (6000 lbf) class. The tire used in this investigation was a 6.00x6 TT 8-ply type III aircraft tire with a rated load of 10.5 kN (2350 lbf) and was typically inflated to 241 kPa (35 psi). The first tests used the nose gear and tire alone (figure 2) mounted on an "I" beam truss and attached to the underside of the

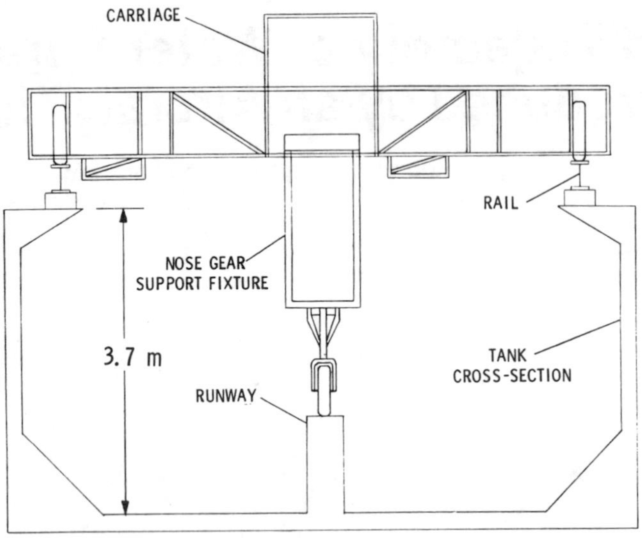

Figure 1.- Schematic of test facility.

carriage. Varying the nitrogen gas charge in the nose strut allowed different vertical loads to be tested without moving the upper portion of the strut or carriage mounting hardware. Additional tests were conducted using a fuselage alone with the nose gear installed, as well as with a partial wing attached to the fuselage in two fore-and-aft locations. A photograph of the wing and fuselage installed is shown in figure 3.

Shown in both figures 2 and 3 are water collection devices developed for this investigation. Two different size collectors were used, but both consisted of plexiglass tubes with a 4.1 cm (1.625 in.) inside diameter. The tubes were mounted in a square array so that their centers were spaced at 7.6 cm (3 in.) intervals. The purpose of the array was to collect a portion of the spray generated by the

Figure 2.- Original test fixture and small water collector.

Figure 3.- Fuselage and wing installed with large water collector.

aircraft tire and retain the volume of water using a rubber stopper placed in the aft end of each tube. After a test run, the volume of water was measured and, based on the speed of the test, a flow rate for that position area was calculated. The position of these discrete flow rates was also noted so that the shape of the spray pattern was known. One should note that the tubes were tilted with their aft-ends low so that carriage deceleration after a test did not cause the captured water to be lost out of their forward ends. The collector face was also perpendicular to the ground. The small collector shown in figure 2 consisted of an 8 by 8 tube array, while the large one shown in figure 3 had a 22 by 22 tube array. The collectors were movable in three dimensions, and normally one complete plane aft of the tire was surveyed before changing fore-and-aft collector locations. The small collector was used for collecting data within about 2 m (6 ft) aft of the tire; the large collector was used to collect data 5 m (16.6 ft) aft of the tire.

RESULTS AND DISCUSSION

In the spray investigation, the effects of various parameters such as distance aft of nosewheel, speed, load, and water depth on the flow rate and trajectory of the water spray generated by an aircraft tire operating on a flooded runway were determined. Before discussing the effects of these parameters on spray, an understanding of the events in and near the tire footprint is needed. Extensive high speed film was taken to observe the characteristics of the spray in the vicinity of the tire. The classic "bow wave" and "rooster tail", the fore-and-aft components of the spray, were determined to be low in density and not able to contribute significantly to the body of water that typically finds its way to an engine inlet. The bow wave acquires a forward ground speed higher than that of the tire and atomizes very quickly. The major contributor to the volume of water available for engine ingestion is the water which is ejected laterally from the tire footprint, or the "side plume". As water is expelled laterally from the tire footprint, it encounters a wall of water next to it which absorbs some of its lateral energy. This action causes the original unit of water to change direction and be thrown upwards, while the next unit of water on the surface, having had lateral energy imparted to it, undergoes the same process and is thrown upwards but with less initial velocity. This effect is shown and sketched in figure 4, and is identified as the wake-generated spray similar to the wake of a boat as it passes through the water. The wake contains enough energy to eject perhaps 5 to 10 times the amount of water that is in the direct path of the tire footprint.

a. Photograph of wake-generated water spray.

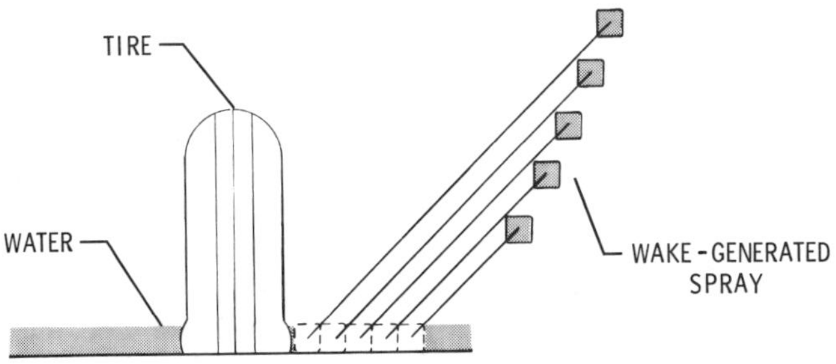

b. Sketch of wake-generated water spray.

Figure 4.- The wake phenomenon.

High-speed film was also used to define the initial trajectory angles of the side plume, using a short, narrow water trough so that wake effects were not present. Combining film taken from two directions allowed a three-dimensional trajectory to be computed. The initial angles for the trajectory were 43 degrees up from horizontal in the lateral plane, and 11 degrees forward in the ground plane. Ranges of speeds, loads, and water depths were tested and filmed, yet no significant difference in the trajectory angles was noticed. A trajectory for a 12 m/sec (40 ft/sec) test is shown in figure 5. The vertical lines represent equal time increments of 0.1 seconds.

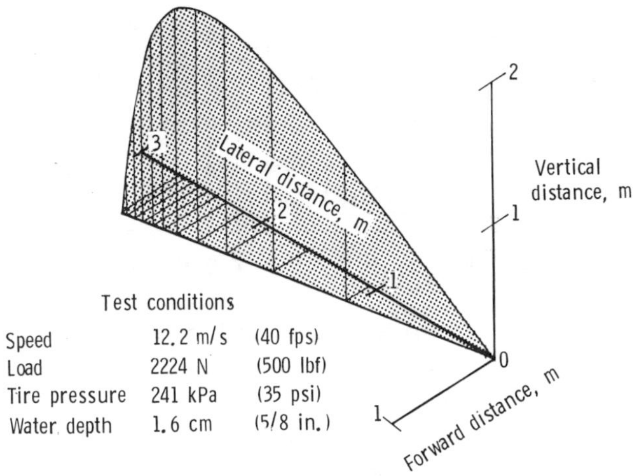

Figure 5.- Typical spray trajectory relative to point of origin.

A table summarizing all of the test conditions and a complete set of data for the tire-generated spray investigation can be found in reference 4. Figure 6 shows a set of typical data obtained for a single test. The x, y, and z coordinates represent the longitudinal, lateral, and vertical distances from the center of the tire footprint to the lower inboard collector tube face. The figure is oriented so the reader is aft of the collector looking forward. The number in each square represents the flow rate in milliliters per second averaged over the individual tube area at that location. Only about 25% of the area at the collection plane was actually comprised of collector tubes, so the remaining 75% of the spray was not collected. Figure 7 presents a contour map generated using the data shown in figure 6. The map shows a typical spray pattern, with higher flow concentration in the lowest region identified. The angle of the majority of the spray is much higher than the 43 degree angle identified for small units of water coming out of the tire footprint. The reason is that the wake phenomenon shown in figure 4 dominates the pattern. This is discussed in more detail in the next section.

											1	2	4	4	6	10	12	10	11	8	8
							1		1	1	2	4	5	6	10	11	12	10	10	6	6
							1		1	2	4	5	7	11	12	12	14	12	8	6	5
						1	1	1	2	4	5	6	8	13	14	13	11	8	5	5	2
						1	1	2	4	4	6	6	12	14	14	16	13	6	4	2	2
				1	1	1	2	4	5	7	12	18	19	17	16	13	5	4	4	4	1
					1	1	2	4	6	8	14	13	18	19	12	12	5	2	1	1	
					1	1	4	5	7	12	17	23	22	19	13	7	5	2			
				1	1	4	5	7	11	14	17	24	25	22	16	10	4	1	1		
			1	1	2	4	6	8	11	19	24	26	30	20	12	6	1	1			
				1	1	2	5	7	12	22	28	35	23	16	4	1					
				1	1	2	5	8	12	20	34	37	32	13	1						
					1	2	5	7	12	24	32	36	25	8	1						
					1	4	4	8	16	24	38	31	18	6							
						2	4	7	24	32	35	24	8	2							
						2	6	10	24	34	31	16	6								
						4	7	13	31	37	30	13	5	2							
					1	4	6	20	32	38	25	10	4								
						4	12	24	55	42	20	8	6								
				1	4	6	23	42	58	34	16	7	4								
					4	16	34	65	55	24	8	4									
				4	10	23	59	83	42	10											

x = 5.1 m y = 0 m z = 0.4 m

Figure 6.- Typical spray test data.

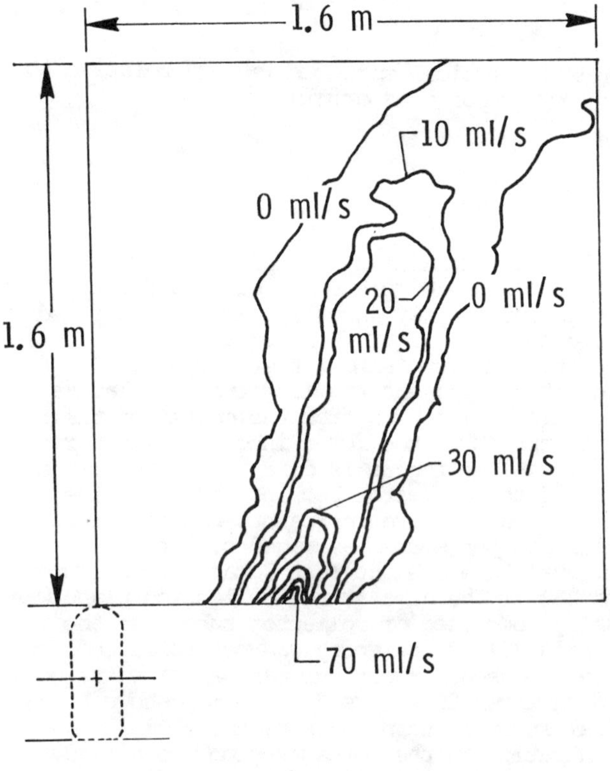

Figure 7.- Computer-generated spray contour map.

EFFECTS OF VARIOUS PARAMETERS

DISTANCE AFT OF NOSEWHEEL- Three planes aft of the nose tire were surveyed to determine the spray flow rate and trajectory generated by the nose tire. The planes were 0.8 m (2.6 ft), 1.9 m (6.3 ft), and 5.1 m (16.6 ft) aft of the tire respectively. Figure 8 presents a position plot of the lines of maximum flow concentration for each plane. Each line represents the position of the most intense localized flow and the values at each end of the lines denote the maximum flow rates in ml/sec at the bottom and top of the collected sample. The center of the tire footprint in the lateral and vertical plane is at the origin for all plots of this type. The flow concentration line for the plane closest to the tire shows a slope very similar to the 43 degree trajectory component identified by movie film.

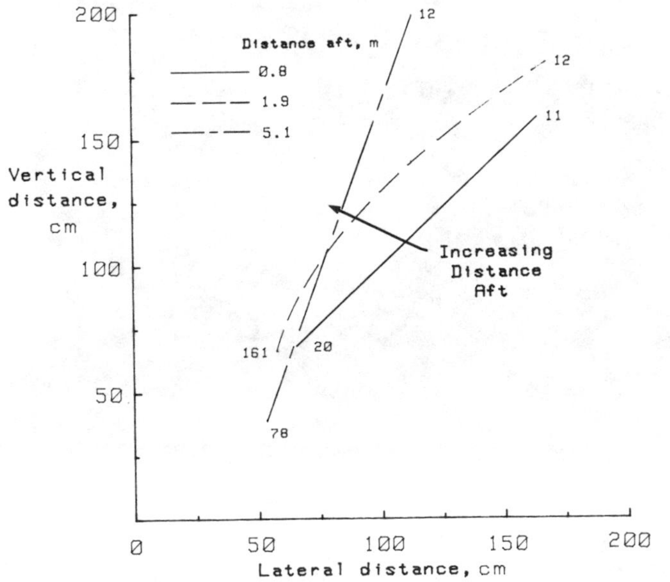

Figure 8.- Effect of distance aft of nosewheel on spray pattern.

This flow is comprised of water actually leaving the tire footprint, since wake-generated spray does not have enough time to reach the collector in this position. At 1.9 m (6.3 ft) aft of the tire, the flow concentration line retains the low slope in its upper region, identifying again the water leaving the footprint. But the slope of the line in its lower region is increased, identifying more water which is actually ejected by the wake. At the most aft plane, the spray pattern is steeper yet, indicating that the wake-generated spray dominates the pattern.

SPEED- Generally, three speeds were tested in this investigation: 12.2 m/sec (40 ft/sec), 18.3 m/sec (60 ft/sec), and 24.4 m/sec (80 ft/sec). Figure 9 presents a plot of the lines of maximum flow concentration at the plane 5.1 m (16.6 ft) aft of the nose tire. The water depth for these runs was 1.6 cm (0.625 in.) and the vertical load was 11.1 kN (2500 lb). The figure shows an increase in the spray intensity as speed is increased, which is due to the increased available energy to eject water upward and to impart greater lateral energy to the wake. The flow concentration lines also move inboard slightly as speed is increased. This may suggest a non-linear relationship between the acquired water velocity and the tire velocity.

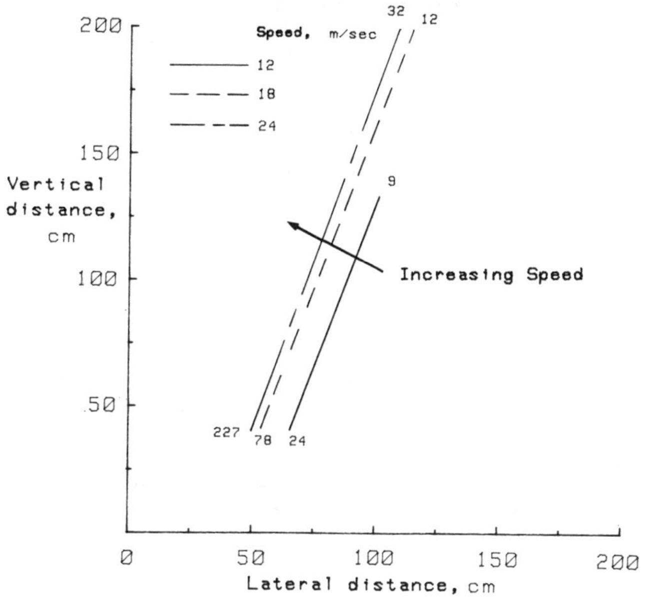

Figure 9.- Effect of speed on spray pattern.

LOAD- The loads tested in this investigation ranged from 2.2 kN (500 lb) to 11.1 kN (2500 lb) which represent an underload as well as an overload condition for the tire. Figure 10 presents a plot of the lines of maximum flow concentration for the two loads at the aft plane, with a water depth of 1.6 cm (0.625 in.) and at a speed of 18.3 m/sec (60 ft/sec). The lines originate at the base of the collector in the same area, yet in the upper region of the spray, the higher load condition moves the flow inboard and local flow rates decrease due to the pattern spreading out more. The water leaving the footprint, which has been shown to dominate the spray upper region, must acquire a higher lateral speed to leave the wider footprint in the same amount of time as that water leaving the more lightly loaded footprint, causing the upper region of the spray to show the speed effect discussed earlier.

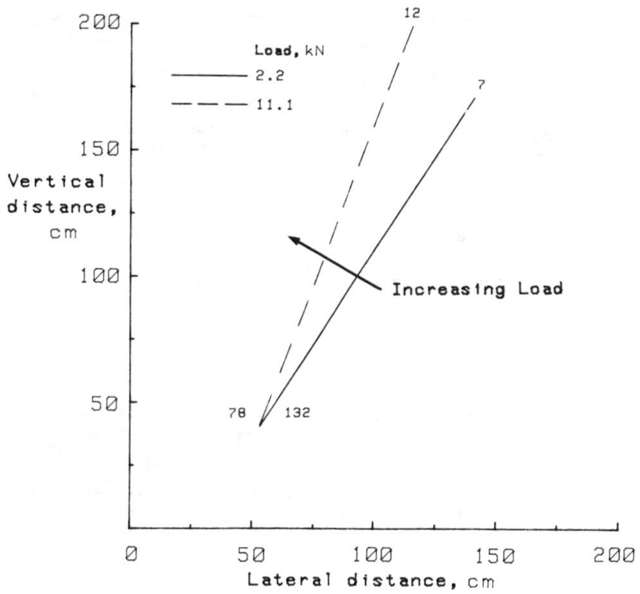

Figure 10.- Effect of load on spray pattern.

WATER DEPTH- Water depths of 1.3 cm (.5 in.) and 1.6 cm (0.625 in.) were tested in this investigation. Figure 11 shows the variation in the spray patterns due to the two water depths at a plane 1.9 m (6.3 ft) aft of the tire, at a speed of 18.3 m/sec (60 ft/sec) and at a load of 11.1 kN (2500 lb). The lines of maximum flow concentration show similar positioning, with the higher water depth condition slightly inboard. The magnitudes of the flow rates are significantly increased for the higher water depth test. Variations due to water depth appear to be affected by distance aft; a similar comparison at the aft-most plane shows almost identical position of the lines and very similar flow rates.

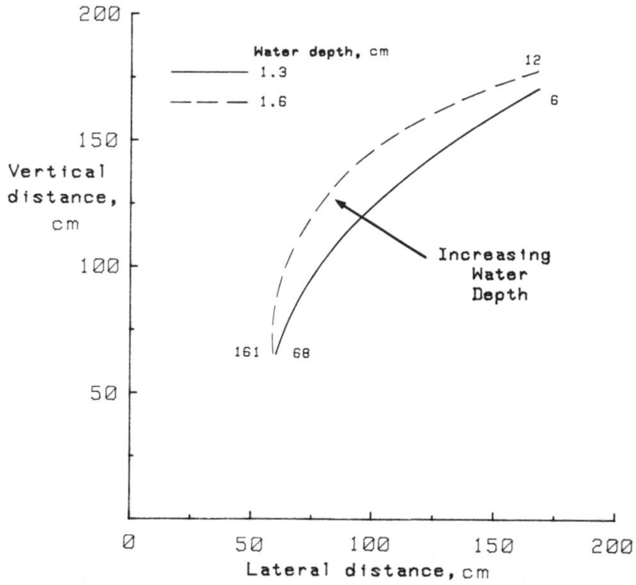

Figure 11.- Effect of water depth on spray pattern.

AERODYNAMIC STRUCTURE EFFECTS- A twin-engine aircraft fuselage was installed on the test carriage (see figure 3) without the wing, to study the effects of fuselage aerodynamics on the tire spray patterns. Figure 12 shows a comparison of the spray pattern with and without the fuselage. The data indicate that the presence of the fuselage causes the slope of the line of maximum flow concentration to be slightly lower than when the fuselage is not present. The flow rates for the two test conditions are virtually identical however. Figure 13 is a planform of the fuselage and wing stub illustrating the two wing locations evaluated in this investigation. The data presented in figure 14(a) indicate that adding a wing stub to the fuselage in the forward position caused nearly all of the spray to be deflected downward by the underside of the wing and to be concentrated in the lowest regions of the collector. The fuselage and wing are silhouetted on the data array. This configuration would not create an ingestion hazard. Figure 14(b) illustrates the effect of moving the wing 1.4 m (4.5 ft) aft. The data indicate that the low pressure region above the wing actually caused the spray to be more heavily concentrated in the upper region of the collector than was observed in tests conducted with no wing (see figure 6, for example). The change in data with wing position, indicates that high-lift devices can possibly increase spray ingestion potential.

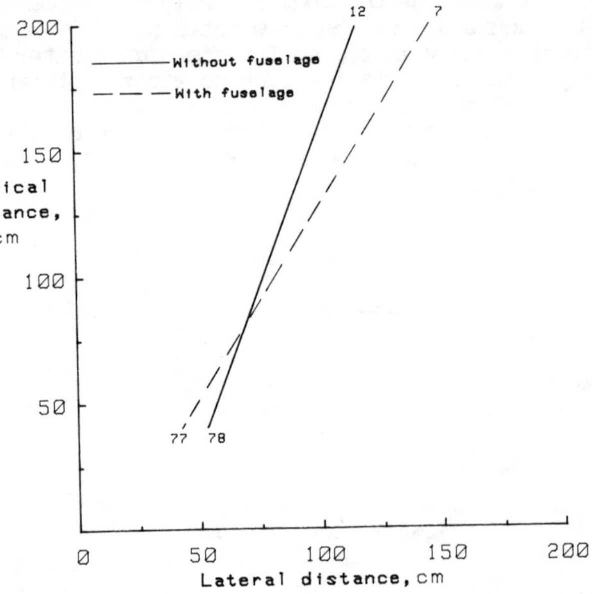

Figure 12.- Effect of fuselage on spray pattern.

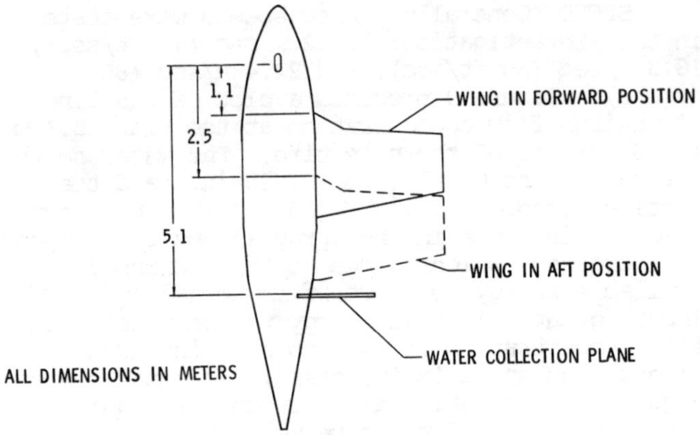

Figure 13.- Planform showing fuselage, wing, and water collection plane.

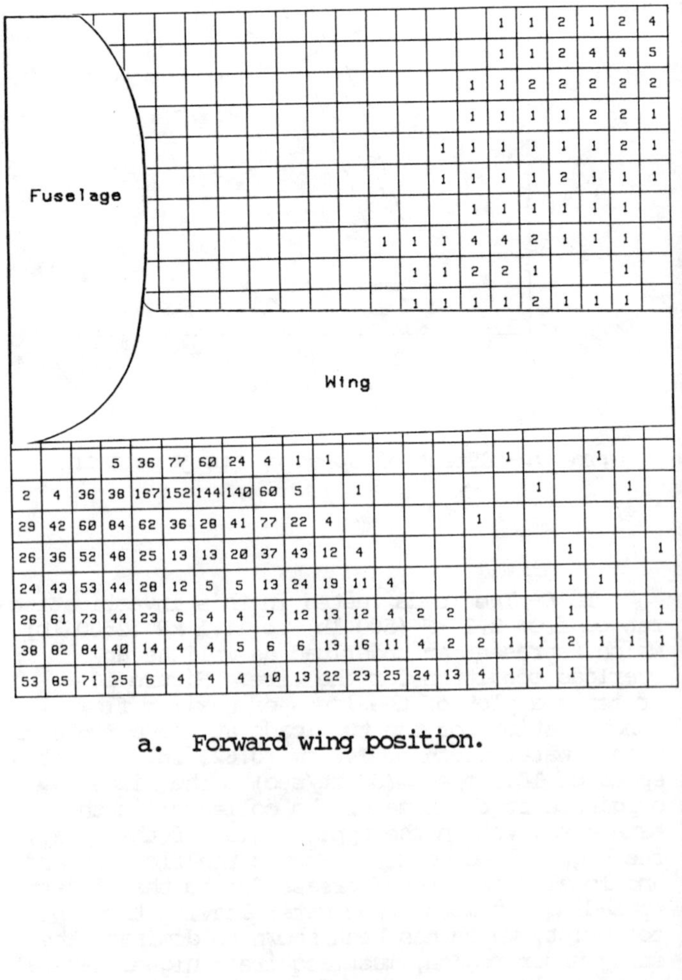

a. Forward wing position.

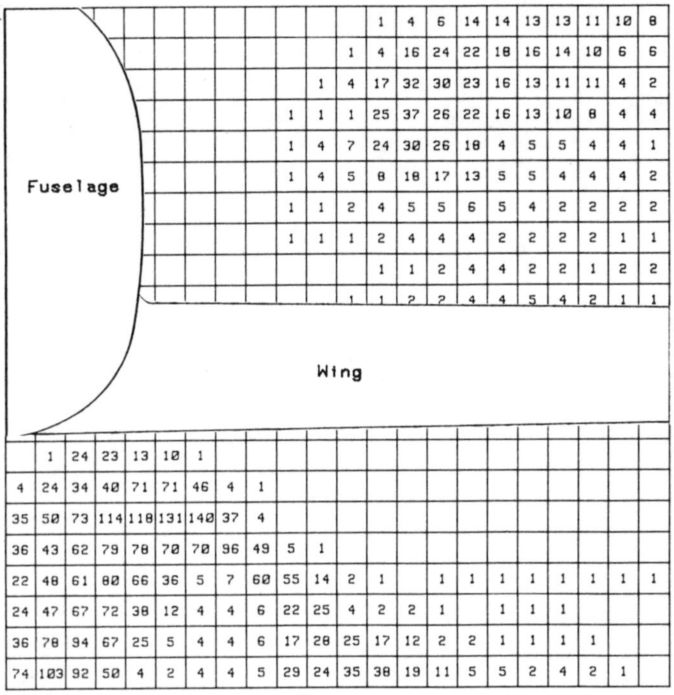

b. Aft wing position.

Figure 14.- Effect of wing position on spray pattern.

CONCLUDING REMARKS

An experimental investigation was conducted to examine the effects of various parameters on flow rates and trajectory of water spray produced by an aircraft tire operation on a flooded runway. Parameters include distance aft of nosewheel, speed, load, and water depth. Tests were conducted using an electrically-operated carriage to which aircraft hardware and a water collection device were mounted.

It was determined that the side plume of water produced by the aircraft tire is the most significant of the various classical plumes relative to spray ingestion problems.

Trajectory angles of water leaving the tire footprint are insensitive to variations in any of the parameters tested, although water particle speed and volume were affected. The wake produced on the surface by the tire contributes greatly to the volume of water ejected upward and forward, and actually dominates the spray pattern at planes far aft of the tire. Increasing the speed moves the bulk of the spray pattern inboard. Increasing load causes the upper regions of the spray pattern to move inboard. Increasing water depth causes increases in localized flow rates, though the further aft, the less significant water depth variations become. The effects of aerodynamic structures such as the fuselage and wing were also investigated. The fuselage tends to depress the apparent spray angle slightly. The wing caused all spray to be deflected downward when mounted in a forward position, yet when mounted further aft the water spray was actually concentrated above the wing slightly.

REFERENCES

1. MacGregor, C. A.; and Bremer, R. J.: An Analytical Investigation of Water Ingestion in The B-1 Inlets. NA-73-181, B-1 Division Rockwell International, Los Angeles, CA, June 15, 1973.

2. Jaeger, Frederick J.: F/FB-111 Nosewheel Tire Tests. AF FTC-TR-74-5, ASD/YBT Wright-Patterson AFB, OH, March, 1974.

3. Olson, Roland E.; and Brownell, William F.: Facilities and Research Capabilities. High Speed Phenomena Division, David Taylor Model Basin, Langley Field, VA, David Taylor Model Basin Report 1809, April, 1964.

4. Daugherty, R. H.; and Stubbs, S. M.: Measurements of Flow Rate and Trajectory of Aircraft Tire-Generated Water Spray. NASA TP , December, 1986.

Alternate Launch and Recovery Surface Traction Characteristics

Thomas J. Carter and David H. Treanor
Flight Dynamics Laboratory
Martin D. Lewis
Air Force Engineering Services Center

* Paper 861627 presented at the Aerospace Technology Conference and Exposition, Long Beach, California, October, 1986.

ABSTRACT

The rapid repair of bomb-damaged runways is of increasing concern to the U.S. Air Force, therefore, expedient repair concepts are being developed. Aircraft performance effects imposed by the repair treatments include: tire flotation, aircraft weight, landing dynamics, and the forces generated at the tire/runway surface interface. This study focuses on tire/runway surface interface forces and was initiated to evaluate several surfaces with respect to their relative tractive and lateral force potential. Three damage repair surface materials, a baseline concrete surface, and a ceramic aluminized marking strip were tested. Quasi-static tests were run at seven tire yaw angles, with and without braking under dry, wet, and icy conditions.

BACKGROUND

SINCE 1979, THE U.S. Air Force has been conducting an intensive research and development program aimed at improving their capability to rapidly repair bomb-damaged airfield pavements. At that time, the existing technology, dating from the 1950's, was limited to aluminum matting composed of 2 ft. × 12 ft. × 1.5 in. panels manually assembled to create a 40 ft. × 70 ft. patch. This patch was anchored with bolts set in the pavement using hot liquid sulphur. Increasing demands of new generation fighter aircraft, heavy transport aircraft, and expanding threats made this method of repair unsuitable to meet Air Force mission requirements.

Therefore, the U.S. Air Force, at the Air Force Engineering and Services Center (AFESC) has been developing interim and future techniques to rapidly repair airfield pavements. Their research has centered on alternative pavement surfaces and techniques to temporarily repair bomb craters. Alternative pavement surface research has focused on the ability of low-cost pavement construction to support a relatively low number of passes before failure. These alternate surfaces range from unsurfaced soil or sod to thin asphalt layers over cement-treated aggregate bases.

The basic concept in bomb crater repair is to back fill the crater with the ejected debris to within approximately a foot of the runway surface; select fill is then placed and compacted to complete the filling of the crater. Since loose particles of fill remain on the surface of the repair, various surface treatment concepts are being studied to address the foreign object damage (FOD) potential due to crater repairs. To alleviate the problem, surface stabilization and/or installation of a FOD cover over the compacted fill are the primary techniques utilized to date. In addition, rapid-setting concrete-like materials are being investigated as a one-step repair procedure.

To better understand the effects on aircraft performance imposed by the various surfaces, and vice versa, several factors must be evaluated. These factors include but are not limited to: tire flotation, aircraft weight, landing dynamics, and forces generated at the tire/runway surface interface. This study was conducted to evaluate several surfaces with respect to tractive and lateral force potential. Laboratory tests were conducted under controlled surface and environmental conditions at the Landing Gear Development Facility (LGDF) of the Air Force Wright Aeronautical Laboratories at Wright-Patterson Air Force Base, Ohio.

SCOPE

Testing was accomplished under quasi-static conditions using the Tire Force Machine (TFM) (see Figure 1) at the LGDF to control yawing and loading of the tire, and to measure the reactions developed between the tire and surface. After load data was gathered from all of the tests, the surfaces were quantitatively compared in terms of tractive forces developed for the various conditions. Seven rapid runway repair (RRR) surfaces were included in the original test plan, five of which were completed (the other two surfaces were deleted due to test set up problems). The five surfaces tested were:

a) Portland Cement Concrete (PCC)
b) Fiberglass Mat (FM)
c) Polyurethane Polymer Concrete (PPC)
d) Asphaltic Concrete (AC)
e) Ceramic Aluminized Strips (CAS)

The CAS surface is not considered a RRR type surface since it is being evaluated for use as an alternative to painted runway centerlines, i.e., a maintenance improvement proposal. The CAS surface data is presented separately and discussed on its own merits.

Fig. 1 - Tire Force Machine

DESCRIPTION OF SURFACES

Each test surface had its peculiarities in terms of similarity to the operational environment. Any differences between the laboratory and operational environments were kept to a minimum, realizing that the bottom line of these tests was to acquire data about the friction developed between the tire and the surface.

Each of the surfaces, except for the FM surface, was mounted in test forms so that concrete foundations (or rock matrix in the case of the PPC tests) could be poured and the surface applied to this foundation. These test forms were then secured to the moving test table on the TFM. This modular surface setup allowed for easy installation and removal, as well as constructing one surface as another was being tested on the TFM.

Portland Cement Concrete (PCC)–The specifications for the preparation of this surface called for a light brushing as the concrete was setting up to enhance the friction qualities. This surface was considered to be the baseline for the other surfaces since it represents the operational environment and is equivalent to surfaces used in current day-to-day operations.

Fiberglass Mat (FM)–This surface utilizes a polyurethane impregnated fiberglass mat, typically 2-3 ply and 3/8-1/2 inch thick, which is anchored to the pavement using rock bolts and specially designed bushings. The crater is filled with 18-24 inches of well-compacted crushed stone graded level with surrounding pavement.

The original test setup concept prescribed securing the FM over a gravel subsurface compacted in the test forms. However, as a result of subsurface instability observed during preliminary check-out runs; the FM was bolted directly to the TFM test table to provide a better simulation of the mat installed over a well-compacted subsurface.

Polyurethane Polymer Concrete (PPC)–This allweather polymer concrete uses a polyurethane resin as a binder. This catalyzed two-component polyurethane resin is mixed and poured over the crater which has been prepared with a level 10-inch layer of open-graded aggregate. The resin percolates through the aggregate and sets in 20-30 seconds. In 30 minutes it cures to flexural strengths twice that of traditional portland cement concrete.

This laboratory surface was constructed by filling test forms with rounded river rock (0.75 inch minimum to 1.5 inch maximum diameter) and pouring the polyurethane polymer concrete mix directly over the rock matrix (the water-like consistency of the mixture enabled it to penetrate the rock matrix to the full depth of the forms). The rough surface created by protruding rocks provided a natural enhancement for friction.

Asphaltic Concrete (AC)–Like the PCC surface, the AC surface was constructed to simulate existing airfield pavement surfaces. This surface simply consisted of 2 inches of road grade asphalt applied over a 4 inch concrete base.

Ceramic Aluminized Strip (CAS)–The CAS test surface was constructed by gluing 1/16-inch thick ceramic aluminized steel plates to the surface of PCC test sections. The plates were bonded to the PCC surface as recommended for operational installations. The ceramic coating (applied to one side only) was installed upward to contact with the tire.

CAS is being proposed for use as runway centerlines to reduce the maintenance required for painted lines. Therefore, CAS is seen as an airfield maintenance enhancement rather than a RRR surface. For operational use, the plates are to be installed flush with the runway surface, either by providing a recess during new construction or by grinding a shallow recess in existing concrete. The effects of a PCC to CAS transition were examined by installing PCC test sections on the first half of the test surface and CAS test sections on the remainder.

OBJECTIVE

The test objective was to acquire data describing the tire/surface interface in terms of the forces and moments developed. This information was used to determine the tire/surface coefficient of friction. Primarily, the data of interest consisted of the forces developed in the direction the tire was yawed, perpencidular to the direction of yaw, and in the direction the load was being applied, as illustrated in Figure 2.

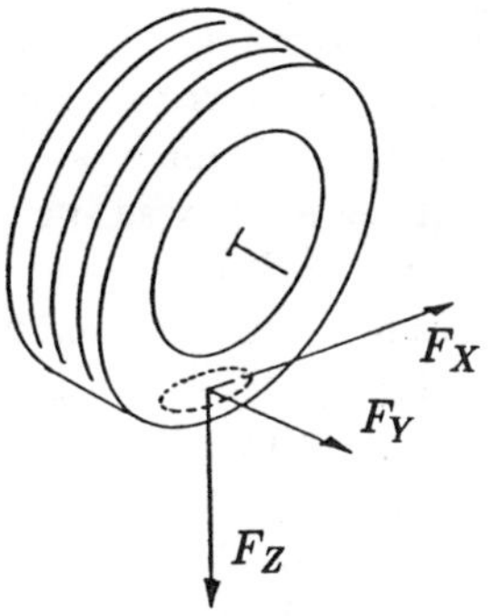

Fig. 2 - Forces Sensed On TFM

By knowing the yaw angle, the forces in line with and perpendicular to the tire, as well as the vertical load force, the "effective" coefficient of friction (μ_{Eff}) could be calculated by dividing the vertical load into the effective drag (see Figure 3). With the exception of speed effects (all testing was accomplished at low speed, i.e., quasi-static), μ_{Eff} is the coefficient of friction an airplane would develop in an operational situation, and hence is used as the basis of comparison throughout the report. The component forces are useful for the component contribution for a particular μ_{Eff} value.

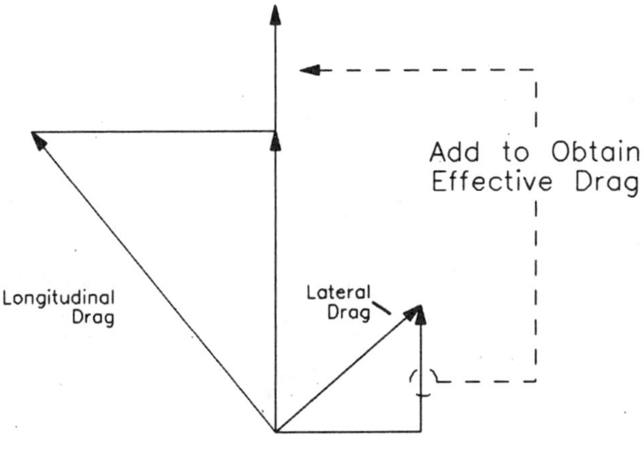

Fig. 3 - Effective Drag Force

Each test surface was affixed to the TFM moving table and an F-4 main gear tire mounted in the stationary yoke. The surface condition was simulated as dry, wet or icy, and the amount of tire yaw was set. Then a vertical force approximating the aircraft weight supported by the landing gear was applied to the tire and the tire began to roll as the table mounted surface moved underneath it at a velocity of 1 inch per sec (in./s). If braking was included, the brake was gradually applied to a preset limit after the tire began to rotate. As the tire rotated, the load cell (see Figure 4) outputs were sampled at 20 hz. This data was then converted from analog to digital form, recorded on magnetic tape and converted to engineering units to be averaged and plotted. Since the distances between load cells are known, the loads data was used to compute the moments about the various axes.

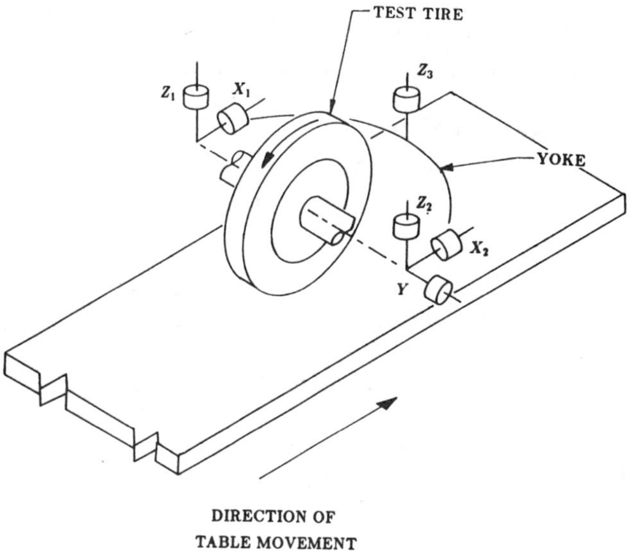

Fig. 4 - TFM Load Cells

TEST CONDITIONS

Three parameters were investigated which affect the development of friction between the tire and the proposed surfaces. These parameters included: (1) Environmental conditions (dry, wet, or icy); (2) braking conditions (braked or unbraked); and (3) yaw angle with respect to the direction of travel. All testing was accomplished under quasi-static condition (table velocity of 1 in./s).

Environmental Conditions–Each test surface was evaluated under dry, wet, and icy conditions. For the wet condition a small water retention dam was constructed to surround the length of the test section. The section was flooded with enough water to create standing puddles up to 1/8-inch deep, and the test performed on the wet surface. For the icy surfaces, liquid nitrogen was pumped directly on the surface about 6 inches in front of the location where the tire made contact with the test surface. To create the initial ice layer, spray bottles were used to apply a fine water mist which froze as soon as it came in contact with the liquid nitrogen cooled surface. This method was useful for controlling the ice formation in a uniform manner over the length of the test section.

Braked and Unbraked Conditions–The tire was operated in braked and unbraked modes to simulate braking after landing and the free rolling takeoff condition. A disk brake integral to the TFM was used to apply up to a maximum of 10 percent braking. This maximum was reduced if the tire tended to lock up and slip without rolling under icy or wet conditions.

Yaw Variations–Yaw variations were imposed for each condition to consider aircraft landing in adverse yaw, tire forces developed during turning, as well as indicating any unsymmetrical tire friction properties to be expected at either negative or positive yaw angles. Yaw angles of 0°, ±1°, ±6° , ±12° (negative values indicate yaw to right of direction of travel, positive to the left) were set for each surface and brake condition.

TEST SEQUENCE

For each surface, 126 runs were performed to satisfy the test matrix. The 126 runs per surface consisted of the following parameters:

7 yaw angles	(0°, ±1°, ±6° , ±12°)
with 2 braked conditions	(unbraked, 10% brake)
and 3 environments	(dry, wet, icy)
and 3 runs per combination	(quality assurance).

Approximately 500 test runs were conducted during this program. Each surface was prepared and mounted on the TFM, the yaw angle would be set and three runs would be executed in the unbraked condition. With the same yaw angle setting three runs would then be executed in the braked condition. After all the dry runs were accomplished, the same procedure was repeated for the wet and icy conditions.

DATA REDUCTION

Each run took about 180 seconds to complete; at a sampling rate of 20 hertz, a total of 3600 data points were acquired for each parameter (longitudinal drag, for example) per run, therefore approximately 75,000 data points were obtained for each condition.

To compare all of the combinations as efficiently as possible, the following averaging scheme was used. First, the raw data for each run was stored in a computer file containing the seven parameters sampled. Each run file was then reviewed and an overall average value for each parameter was computed (see Figure 5). Symbolically, this is represented in Equation 1:

$$\frac{\sum_{i=1}^{N} X_i}{N} = A \qquad (1)$$

where A is the overall average and N the number of data points for each parameter. Once an overall average was computed, a second pass of the data was made to determine the upper and lower average values. Symbolically, this is represented in Equations 2:

$$\frac{\sum_{i=1}^{N} X_i > A}{U} = A_U \qquad \frac{\sum_{i=1}^{N} X_i < A}{L} = A_L \qquad (2)$$

where U and L are the number of data points greater than and (Upper) less than (Lower) the overall average. The upper and lower averages are represented by the variables A_U and A_L. The vertical forces were averaged and a single representative value was recorded since this parameter varied little. A file was then produced holding the yaw angle, vertical force, and the lower, overall, and upper average values for each parameter for all seven yaw angles. This process was performed for each of the three runs of the similar combinations of conditions, resulting in three files of average data.

Next, for each given set of conditions, the closest two overall average values of the three data sets were determined. These two overall average values and the and the corresponding upper and lower averages were averaged again to yield single values representing the upper, overall, and lower values for each condition. This sequence was repeated for all of the yaw angles, braking conditions, and environmental conditions so that the data gathered for each combination could be represented by these three average values.

The parameters investigated in this study were: (1) the longitudinal drag force; (2) the lateral drag force; (3) the "effective" drag force; (4) the aligning moment; and (5) "effective" coefficient of friction (μ_{Eff}). Illustrated

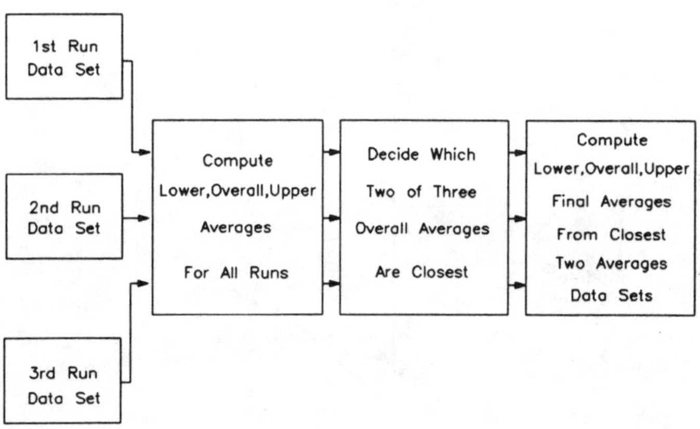

Fig. 5 - Data Reduction Flow Chart

in Figure 6 are some of the main tire/surface traction force components present as a tire rolls over a surface. Two of these components, the lateral drag (perpendicular to the plane of the tire) and the longitudinal drag (in line with the plane of the tire) combine to produce a total "effective" drag which acts in the direction of travel of the tire (denoted by the off-axis arrow). The aligning moment is the restoring torque exerted on the tire about the vertical axis (perpendicular to the surface along which the tire rolls, i.e., about the z-axis) as it reacts to the surface. The yaw angle, θ, describes the orientation of the direction of travel to the longitudinal plane of the tire. Finally, μ_{Eff} was computed by taking effective drag values and dividing by the vertical force. In the unbraked situation, μ_{Eff} is a function of the yaw angle, increasing in value as the absolute value of the yaw angle increases. The effective coefficient of friction is the parameter presented to compare the surfaces and conditions in a nondimensional manner and is the most important factor when discussing control and braking of aircraft in landing and taxi.

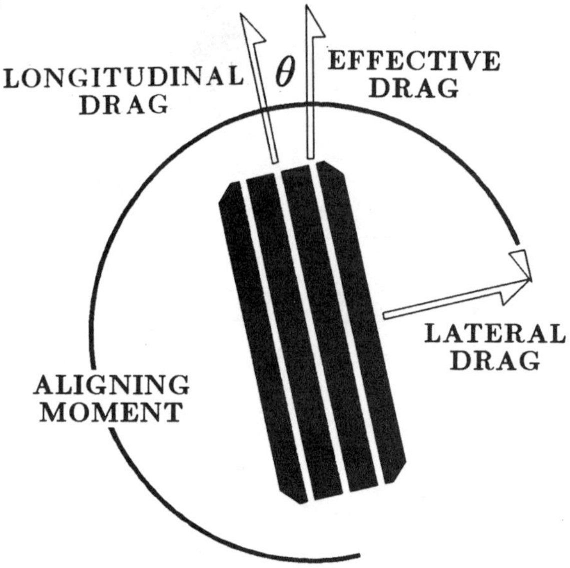

Figure 6. Forces and Moment at Tire Interface.

RESULTS

The reduced data was plotted in bar graph form to clearly illustrate the range of μ_{Eff} for each combination of conditions. Each bar represents the upper and lower average values, and the overall average. Each constant environment plot (dry or wet or icy) has eight bars for each yaw angle, corresponding to the baseline and three RRR surfaces in the braked and unbraked conditions.

When reviewing the μ_{Eff} plots, note that a μ_{Eff} value of 0.01 is roughly equivalent to 0.01 × 25,000 pounds (the aircraft weight supported by each main wheel) or 250 pounds effective drag. Comparison on this basis will provide for a better understanding of the differences in the numbers from surface to surface.

Dry Surfaces–The μ_{Eff} values versus yaw angle for the three RRR surfaces are plotted in Figure 7 along with the baseline PCC surface for dry (braked and unbraked) conditions. Two trends are immediately apparent. First, the unbraked μ_{Eff} tends to increase as the absolute value of the yaw angle increases for all of the surfaces and second, the braked μ_{Eff} remains at a fairly constant level across the range of yaw angles.

Wet Surfaces–The μ_{Eff} versus yaw angle for the three RRR surfaces are plotted in Figure 8 along with the baseline PCC surface for wet (braked and unbraked) conditions. The general trends observed for the unbraked wheel on the dry surfaces still exists.

Icy Surfaces–The μ_{Eff} versus yaw angle for the three RRR surfaces are plotted in Figure 9 along with the baseline PCC surface for icy (braked and unbraked) conditions. A definite drop in the values for braked conditions is observed. An unexpected observation is seen in the baseline PCC surface having a markedly decreased μ_{Eff} level. This was probably caused by the application of a thicker ice layer and the tendency of the PCC to retain this layer.

Constant Environment–A summary of the constant environment comparisons for the braked wheel is shown in Table 1. This table was developed by taking the baseline PCC to be a 100 percent effective surface in all conditions and comparing the other surfaces in the various environmental conditions to the baseline by dividing the PCC μ_M into the RRR environment μ_M.

TABLE 1. SURFACE EFFECTIVENESS.

Surface (Braked Wheel)	**Percent Effectiveness** Dry μ_M	Wet μ_M	Icy μ_M
PCC	100 %	100 %	100 %
FM	82 %	94 %	129 %
PPC	96 %	93 %	142 %
AC	102 %	115 %	145 %

For the dry surfaces, the only significant reduction in performance occurs with the fiberglass mat (-18 percent), the AC and PPC levels remaining well within 10 percent of the baseline effectiveness. In wet environments the asphaltic concrete surface turns out to be especially effective (+15 percent) and the FM and PPC surfaces are both within 10 percent of baseline effectiveness. An improved capability in icy environments is clearly shown by all RRR surfaces over the PCC. The probable cause for these results is the tendency of the PCC to retain a good coating of ice versus the tendency of the ice to crack and break up on the RRR surfaces.

A summary of the effect of transition from one environment to another on the baseline PCC surface and the three RRR surfaces (FM, PPC, AC) is shown in Table 2. This was accomplished by determining the percent change in μ_{Eff} developed from the dry to the wet, the wet to the icy, and the dry to the icy environment conditions for all surfaces.

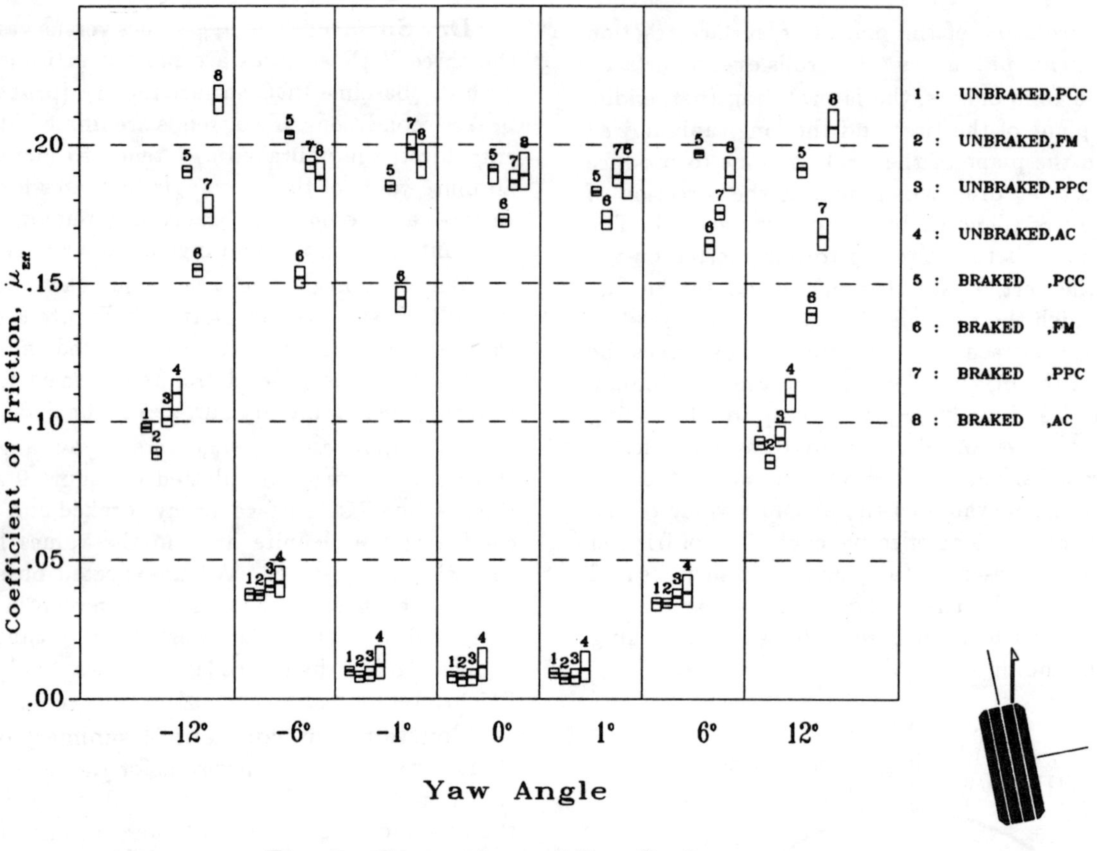

Fig. 7 - Comparison of Dry Surfaces

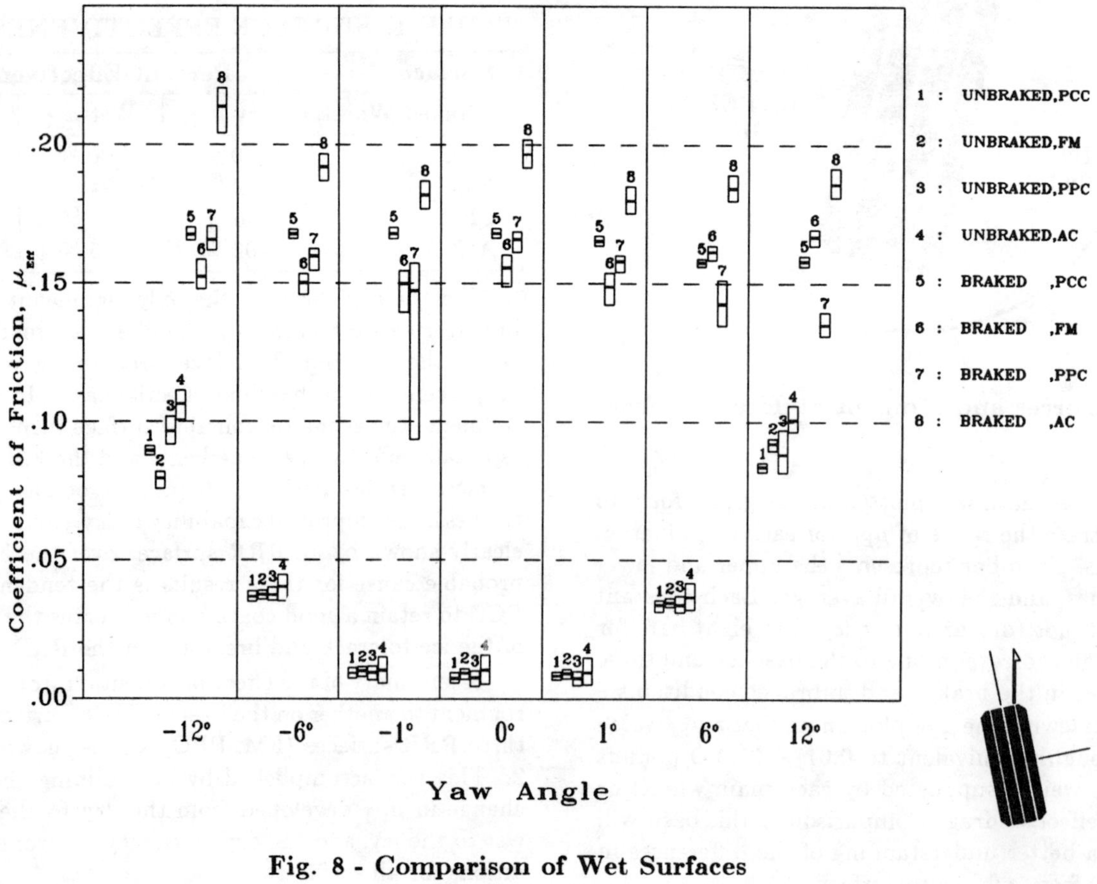

Fig. 8 - Comparison of Wet Surfaces

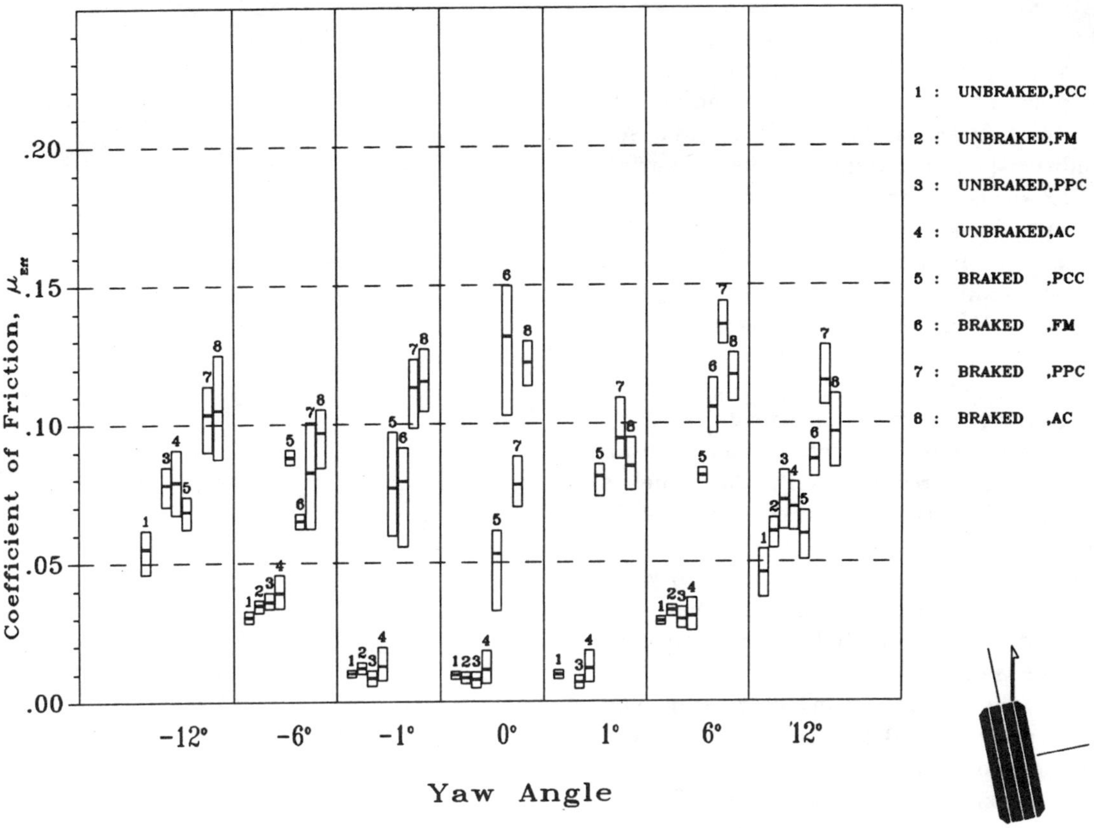

Fig. 9 - Comparison of Icy Surfaces

TABLE 2. CHANGE IN PERFORMANCE.

Transition from :	PCC	FM	PPC	AC
Dry to Wet	-14 %	- 1 %	-16 %	- 3 %
Wet to Icy	-56 %	-40 %	-33 %	-45 %
Dry to Icy	-62 %	-40 %	-44 %	-46 %

Note the small percent change in performance for the transition from Dry to Wet environment conditions for the FM and AC surfaces. This contrasts with the fairly large percent changes for PCC and PPC under the same transition. Even though the Wet to Icy and Dry to Icy transitions for the RRR surfaces show relatively large reduction, this reduction is still not as large as that for the PCC baseline surface.

Ceramic Aluminized Strip–This surface was analyzed on its own merits and not in comparison to any RRR-type surface. This analysis examined the effect of environment on performance for the PCC to CAS transition. These tests were characterized by a buildup of the drag forces while the tire traveled on the PCC half. This was followed by a decrease of forces on the CAS section for the wet and icy environments. A summary of the average drag decrease is shown in Table 3.

TABLE 3. PCC TO CAS DRAG DECREASE.

Yaw Angle	ENVIRONMENT		
	Dry	Wet	Icy
0°	—*	300 lbs	1500 lbs
1°	—*	270 lbs	360 lbs
6°	—*	600 lbs	550 lbs
12°	—*	2000 lbs	1000 lbs

* Difference not significant.

This test was particularly difficult to obtain consistent data during the wet and icy environment tests. This was because the tire would tend to roll normally for a part of the run and then slip for a portion. The brake setting would be reset to a lower value to enable tire roll. At some value less than the 10 percent braking value (that all the other tests were run) the test would be repeated, only to observe the same slipping phenomena. This observation points to the conclusion that the CAS surface exhibits a great variablity in performance in wet and icy environment conditions.

CONCLUSIONS

Of the three RRR surfaces tested, the asphaltic concrete maintained superior friction characteristics, the polyurethane polymer concrete ranking second and the fiberglass mat third. Based on these comparisons of surfaces and conditions, it is apparent that the AC surface is the best RRR surface (from a μ_{Eff} point of view) as it consistently provided for the highest μ_M regardless of the environment. Likewise, though the difference is small in the wet conditions, the PPC yielded μ_{Eff} values second to AC with the FM surface falling in at third choice.

The PPC surface caused extensive gouging of the tire surface. However, the extent of damage under operational conditions would be decreased since exposure would be limited to repaired areas only. Nevertheless, the damage potential exists and must be considered for future operational use.

A major problem area with the FM laboratory test setup was the anchoring methodology and mat flexibility. Instability of the simulated subgrade required mounting of the FM on the TFM test table.

Installation of the CAS results in a decrease of available traction; no painted line data was provided for comparison, therefore, no direct evaluation of CAS versus current runway marking methodology was possible.

Shuttle Orbiter Arrestment System Studies

Pamela A. Davis and Sandy M. Stubbs
NASA Langley Research Center

* Paper 881361 presented at the Aerospace Technology Conference and Exposition, Anaheim, California, October, 1988.

ABSTRACT

Scale model studies of the Shuttle Orbiter Arrestment System have been completed. The system was tested with a 1/27.5 scale model at the NASA Langley Research Center and a 1/8 scale model at All American Engineering Company. The purpose of these studies was to determine the proper net arrestment system configuration to bring the Orbiter to a safe stop in the event of a runway overrun with minimal damage. Tests were conducted for centerline engagements and off-center engagements at simulated speeds up to 95 knots full scale. The results of these studies defined the net-orbiter interaction, corrections to prevent underwing engagements, corrections necessary to prevent net entanglement in the main gear, the dynamics of off-centerline engagements, and the maximum number of vertical straps that might become entangled with the nose gear.

THERE ARE A NUMBER OF ABORT LANDING sites designated for use by the Space Shuttle Orbiter. However, there is the possibility of landing abnormalities during aborts that could lead to hazardous runway overrun incidents on these runways. A runway overrun has the potential of significantly damaging the Orbiter and the possibility of injury or loss of crew. Because of these possibilities, a properly designed net arrestment system has been suggested as a means of bringing the orbiter to a safe stop with a minimum amount of damage in the event of a runway overrun. Scale model tests have been conducted to develop an effective net arrestment system using a 1/27.5 scale model at the NASA Langley Aircraft Landing Dynamics Facility and a 1/8 scale model at the All American Engineering Company.

The purpose of this paper is to present the data from tests conducted with the 1/27.5 scale model and to analyze this data with respect to the net-orbiter interaction. These tests were conducted at simulated speeds up to 95 knots full scale using five nets of different geometries. The data from the 1/8 scale model tests are also presented. The areas of interest were the effect of various net geometries on net engagement of the nose gear and main gear, whether or not the top bundle contacts the crew cabin window and where the bundle comes to rest on the payload bay doors, under wing engagements, and the dynamics of off-center engagements. All data in this report are presented in model scale values unless otherwise indicated.

APPARATUS AND TEST PROCEDURES

1/27.5 SCALE MODEL STUDIES -

Model - Figure 1 shows the 1/27.5 scale model of the Space Shuttle Orbiter used in the NASA Langley tests. The model was made of fiberglass and adjusted for the appropriate scaled weight. The nose and main gear tires used for the model were solid rubber hobby-model type tires with the nose gear tires cut down to obtain the proper diameter. All landing gear struts and braces were made of steel and properly dimensioned. In order to simulate the strut failure load in the drag direction, the gears were free to pivot but held by soft wire that would stretch at the properly scaled force simulating the failure load for the gear. Copper wire was used in parallel with the main gear drag braces and nickel wire was used on the nose gear. A nose gear failure indication system was used to determine if and when the nose gear failed during arrestments. It consisted of a switch inside the model that was used to trigger a flash bulb mounted on the tail of the model so that the time of the nose gear failure could be determined from camera film. The nose gear was designed to be either free castoring or fixed to simulate nose gear steering engaged or disengaged. Black tape was used to mark the payload bay doors and the crew cabin window in order to determine where the top

Figure 1 - 1/27.5 scale model of the Space Shuttle Orbiter.

bundle rested and whether it contacted the pilot's window. Scaling factors for the model and apparatus used in the 1/27.5 scale tests are given in table I.

Net - A multiple element net consisting of individual elements with an upper and lower horizontal member connected by vertical members that are tied into place was used (figure 2). Groups of elements were bundled together to form the entire net assembly with each element acting independently to apply force on the arrested vehicle and at the same time minimizing damage to the vehicle by minimizing localized loading.

A 30 element net was used for the first three net configurations tested. Nylon chording of 0.036 inch (0.09 cm) in diameter was used to construct the net. Scale strength verticals were used in a 6 inch (15.24 cm) center portion of the net to accurately determine if the nose gear failed. The horizontal members were approximately 12 ft (3.7 m) long and the height and horizontal spacing of the vertical members varied for the different net designs. Figure 3 shows the net window geometry for the first three nets tested. The vertical members were double knotted to the horizontal members and then sealed with superglue. Six groups of five elements in the 30 element net were painted yellow, red, orange, green, blue, and black in order to distinguish the various net elements. The nets were supported with a suspension chord and the upper horizontal bundle was attached to breakaways with a breaking force of 0.5 lb (2.22 N).

The fourth and fifth nets were constructed using the net geometry All American Engineering Company (AAE) proposed to NASA as shown in figure 4. These nets had a total of 36 elements divided into three different groups of elements (A, B, C) with two sets of each group and six elements in each group. There were two different net window widths for the elements in each group. The fourth net was constructed of the nylon parachute chord and the fifth net of scale strength nylon string that had a breaking strength of 1.2 lb (5.34 N). The same construction technique was used as for the previous nets and the groups of elements were also painted. The fourth and fifth nets were supported in a similar manner as the previous nets as well. Figure 5 shows a completely assembled 1/27.5 scale multiple element net.

Energy Absorbers - A water turbine system is to be used as the energy absorber system for the full scale vehicle arrestment. This type of system is difficult to model at the 1/27.5 scale and it was determined that chains of different weight working through a pulley would adequately model the energy absorber system as shown in figure 6. Nylon parachute chording was attached to the net horizontal bundles and then to the small, medium, and large link chains. The purpose of the nylon chording and the graduated chain links was to allow a gradual application of loads as the net encompassed the model and to obtain a peak arrestment load gradually with little or no damage to

TABLE I - SCALING FACTORS

1/27.5 Scale Model (λ=1/27.5)

FULL SCALE	MODEL SCALE	DEFINITION
A	λ^2 A	Area
F	$\lambda^{2.85}$ F	Force
I	$\lambda^{4.85}$ I	Inertia
L	λ L	Length
V	λ^3 V	Volume
a	1 a	Acceleration
m	$\lambda^{2.85}$ m	Mass
t	$\sqrt{\lambda}$ t	Time
v	$\sqrt{\lambda}$ v	Velocity
w	$\lambda^{2.85}$ w	Weight

1/8 Scale Model (λ=1/8)

FULL SCALE	MODEL SCALE	DEFINITION
A	λ^2 A	Area
F	λ^3 F	Force
I	λ^5 I	Inertia
L	λ L	Length
V	λ^3 V	Volume
a	1 a	Acceleration
m	λ^3 m	Mass
t	$\sqrt{\lambda}$ t	Time
v	$\sqrt{\lambda}$ v	Velocity
w	λ^3 w	Weight

the model.

Runway - The 83 ft (25.3 m) runway was made of 0.75 inch (1.9 cm) plywood stacked two sheets high and secured to the concrete floor. The spacing between the plywood sheet edges was filled with dental plaster and then sanded down to provide a smooth surface. A 0.25 inch (0.635 cm) groove was cut into the plywood along the net line in order to recess the bottom bundle of the net during a run if desired. A mass-pulley system was used to catapult the model into the net. The launch system had an electronic trigger that, when released, allowed the model to move forward, accelerated by a falling mass. By varying the mass, the proper net engagement speed could be obtained. To keep the model on centerline, a pulley was attached under the fuselage aft of the nose gear which allowed the model to be guided between the two aluminum rails spaced 2.3 inches (5.9 cm) apart and 24 ft (7.32 cm) long to net engagement. Figure 7 shows the test set up for the 1/27.5 scale model studies

Instrumentation/Photo Coverage - The tests were recorded using one panning video camera for quick look purposes and four 16 mm cameras located at various positions to obtain film coverage of each test from different angles. Photos were taken of the model after each test. The speed of the model was recorded by two miniature magnetic pick ups one foot (0.305 m) apart and mounted at the end of the catapult section of the runway as shown in figure 8. Four load links were used to measure the arresting load of the top and bottom horizontal bundles on each side of the model.

Test Procedure - For the 1/27.5 scale model, the arresting system was tested at three different speeds 3.82, 11.56 and 19.10 kts (20, 60, and 100 kts full scale). Prelaunch preparations included hanging the net, checking its height and lateral position, and repositioning the energy absorber system. The model was prepared and loaded into the launch system and locked into place. After each test, the following information was recorded: model runout distance, model speed, nose gear offset from centerline, position of top bundle over payload bay doors, number of verticals enveloping the wing, net entanglement of main and/or nose gear, whether nose or main gear failure occurred, number of strings broken, arresting system force, and still pictures and close-up videos were taken of the model after arrestment.

1/8 SCALE MODEL STUDIES -

Model - The 1/8 scale orbiter model shown in figure 9 was made of plywood, rigid styrofoam and fiberglass and was scaled to the proper weight and geometry. The landing gear was made of steel, the wheels were fabricated from phenolic resin castors and the tires were solid rubber. The center of gravity of the model was properly positioned and the mass moments of inertia were also scaled. Scaling factors used in the 1/8 scale tests are given in table I.

Net - Nylon thread was used to construct four nets. In order to achieve a scaled weight for the 1/8 scale net, two pieces of nylon rope were used to ballast the upper and lower horizontal bundles. Primary holddowns using a single loop of polyester thread were used to anchor the lower horizontal bundle. Tearaway straps, shown in figure 10, were used on the upper horizontal bundle to maintain tension until the net completely enveloped the wings. The tearaway straps were made of nylon straps sewn together in such a way that they acted as a zipper when the model applied tension to the net.

Energy Absorber System - AAE Model 14 energy absorbers were used to simulate the loads produced by the dual Model 44B-SOAS energy absorbers to be used on the full-scale arrestment system (figure 11).

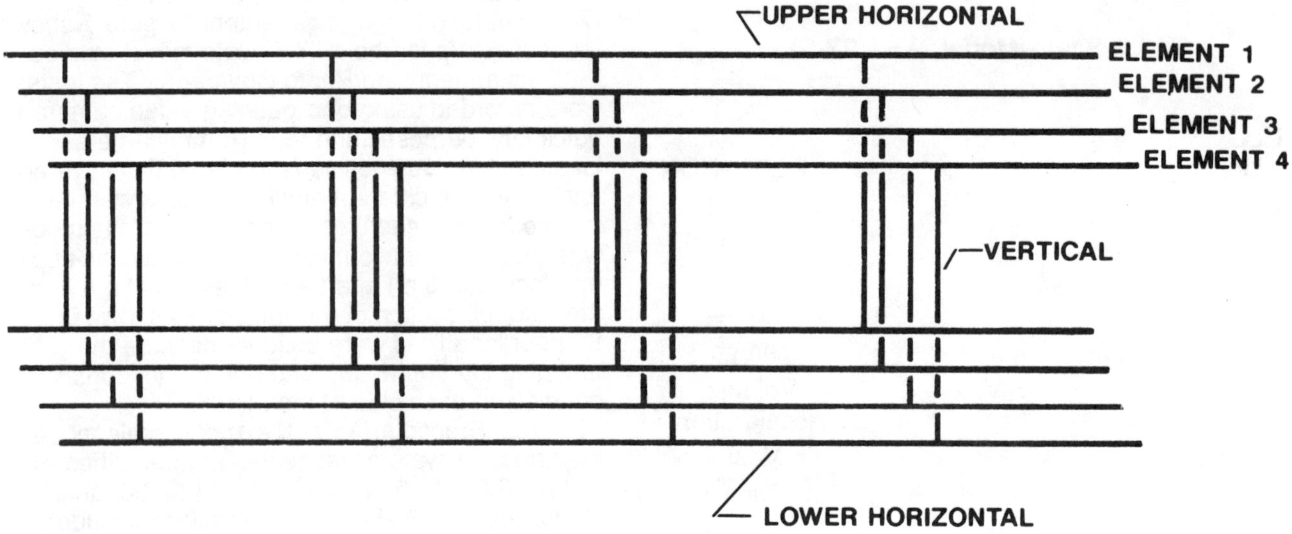

Figure 2 - Multiple element net.

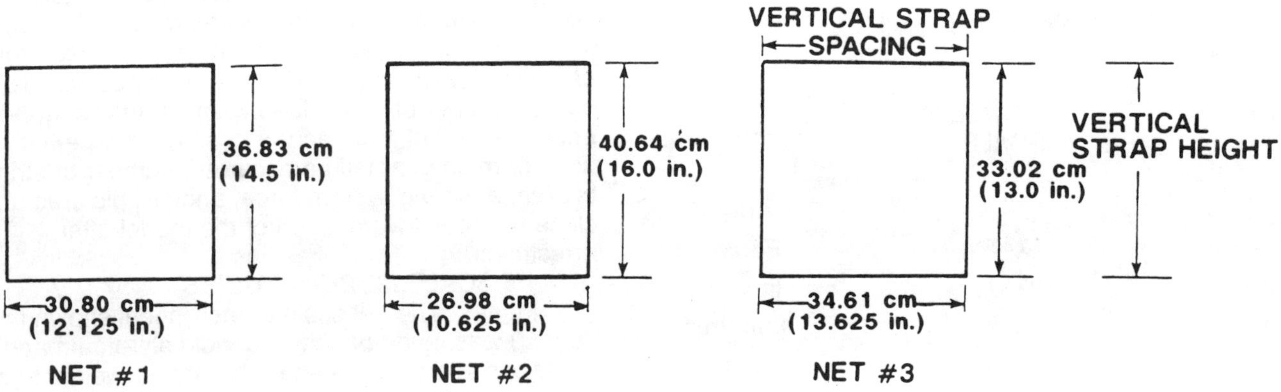

Figure 3 - Window geometries for nets 1 through 3.

Runway - The 1/8 scale tests were conducted in a warehouse where the concrete floor simulated the runway. The model was launched pneumatically into the net using guide rails to maintain directional control of the model catapult stroke. For off-center engagements, the launch system was laterally displaced from the runway centerline. For yawed tests, the net was adjusted to the proper angle. Figure 12 shows the 1/8 scale model test set up.

Instrumentation/Photo Coverage -Photographic coverage of the tests utilized stationary and panning video cameras. The launch velocity was obtained from elapsed time across 1 ft (0.305 m) of travel of the model. The net engagement velocity was determined by a pair of infrared light sources and sensors outboard of the runway edges. The engagement velocity was recorded as the model penetrated the net.

RESULTS AND DISCUSSIONS

1/27.5 SCALE MODEL STUDIES -

Net #1 - There was a total of 13 centerline engagements and four off-centerline engagements at average speeds of 3.73, 11.55 and 18.0 kts. The average rollout distances for these speeds were 6.61, 17.93, and 27.40 ft (2.02, 5.46 and 8.35 m).

The net usually tangled around the right main gear axle more than the left main gear axle causing the model to veer to the right of centerline for centerline engagements. The nose gear remained untangled for all tests except for one run when one vertical caught on the nose gear axle. An average of 8 scale strength verticals were broken for the six engagements this was recorded and the top bundle of the net usually came to rest on the fourth payload bay door.

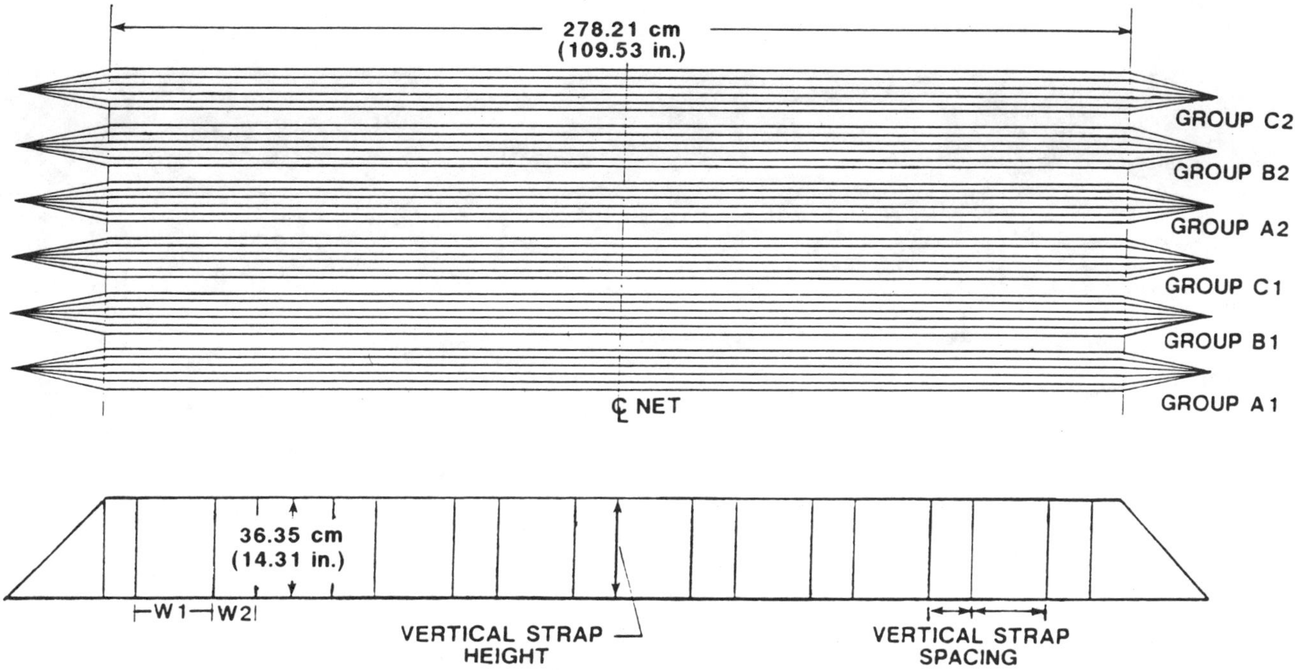

Figure 4 - Nets 4 and 5 geometry.

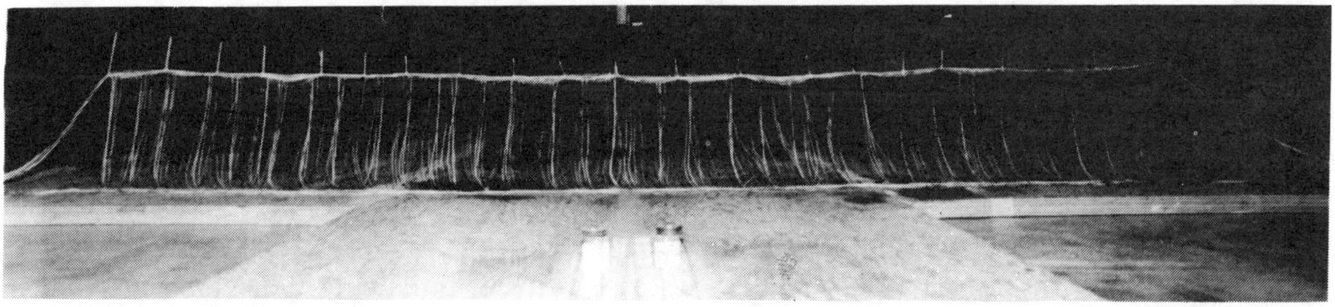

Figure 5 - 1/27.5 scale multiple element net.

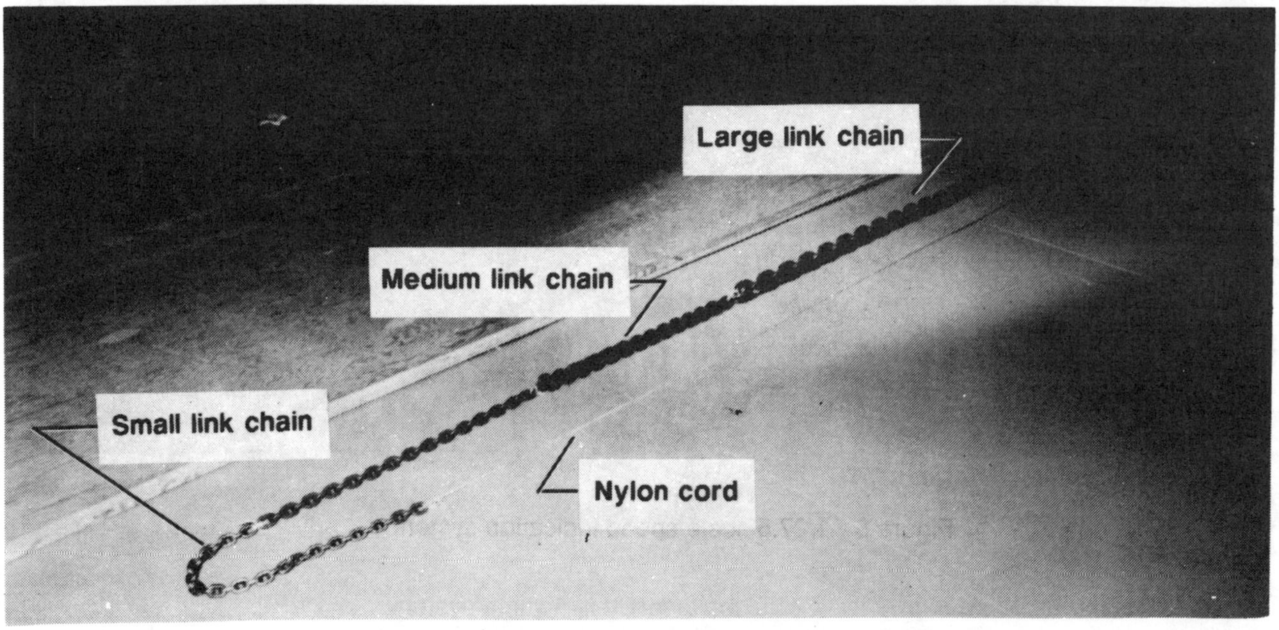

Figure 6 - 1/27.5 scale energy absorber system.

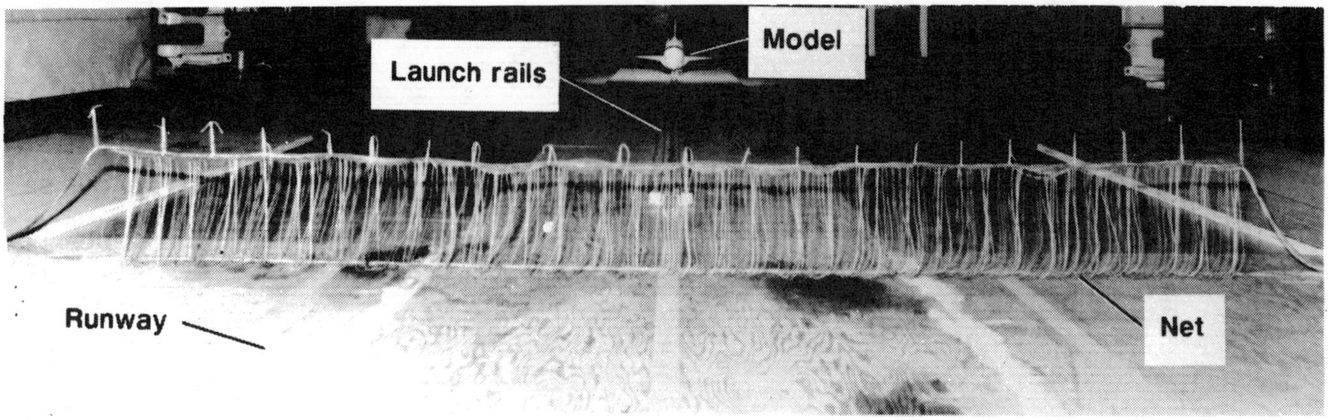

Figure 7 - 1/27.5 scale test set up.

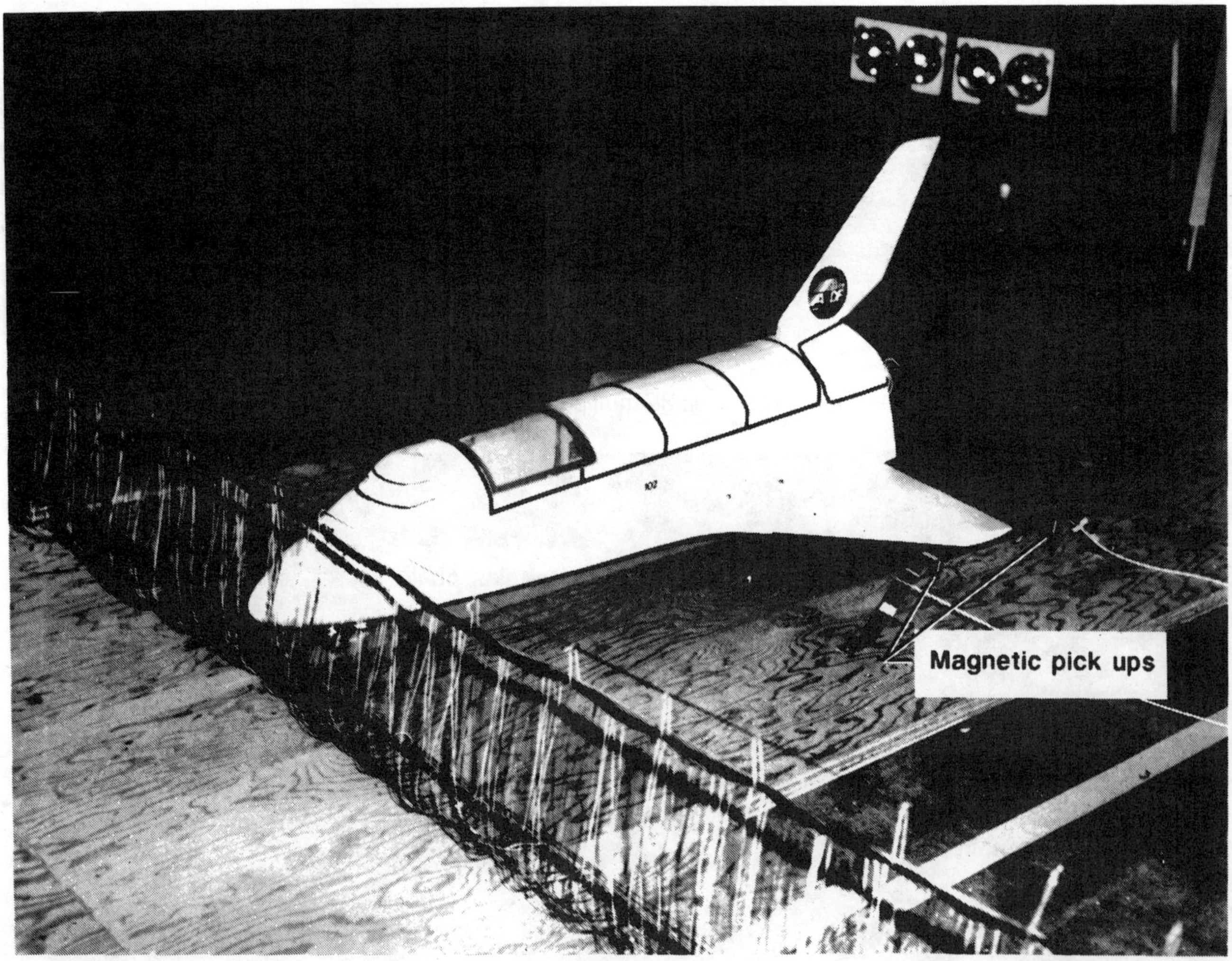

Figure 8 - 1/27.5 scale speed indication system.

Figure 9 - 1/8 scale model of the Space Shuttle Orbiter.

The off-center engagements to the right of centerline were conducted with verticals repositioned to allow a free nose passage in the net to evaluate the off-center arrestment without nose gear entanglements. Only the left main gear axle was tangled with the net. The top bundle came to rest on the third payload bay door. The off-centerline engagements were conducted in the over-strength part of the net thus, no scale strength verticals were broken. Table II gives the centerline and off-center test results obtained with net #1.

Net #2 - There were a total of 13 engagements with an average net engagement speed of 4.26, 11.56, and 17.90 kts. The runout distances for these speeds averaged 8.72, 18.92, and 25.35 ft (2.66, 5.76 and 7.73 m). The main gear was tangled with the net on most of the runs and the nose gear was not tangled at all. For the centerline engagements, the nose gear failed on 23 percent of the runs. The left and right main gears had a failure rate of 54 percent. Some scale strength verticals were broken on 8 of 10 centerline runs and the top bundle came to rest on the third or fourth payload bay door area, sometimes in both areas.

There were a total of three off-center engagements. The nose gear failed on one of three runs and and the main gear failed on two of three runs. The net was tangled in both main gear on all three tests but the nose gear remained untangled. The top bundle again came to rest in the third payload bay area. The large number of runs with the main gear entangled could be attributed to the longer vertical strap height. Table III gives the test data obtained for net #2.

Net #3 - A total of 19 runs with the third net geometry were performed including fifteen centerline engagements and four off-centerline engagements. The average speeds were 4.29, 11.31, and 18.26 kts with runout distances of 7.49, 18.5, and 27.15 ft. (2.28, 5.64, and 8.28 m) respectively.

The nose gear and/or main gear failed 7 of 15 runs for the centerline engagements and there were no landing gear failures with the off-center engagements. The net was successfully engaged through all the runs with one entanglement in the nose gear. The top horizontal bundle came to rest in the third and fourth payload bay door areas. On average, ten scale strength verticals were broken for 10 arrestments. Table IV gives the data obtained for the third net configuration.

Net #4 - Thirty-three pull-thru tests were conducted with the 36 element AAE net configuration in order to determine if any vertical members caught on the nose gear. For the first 20 tests, the nose

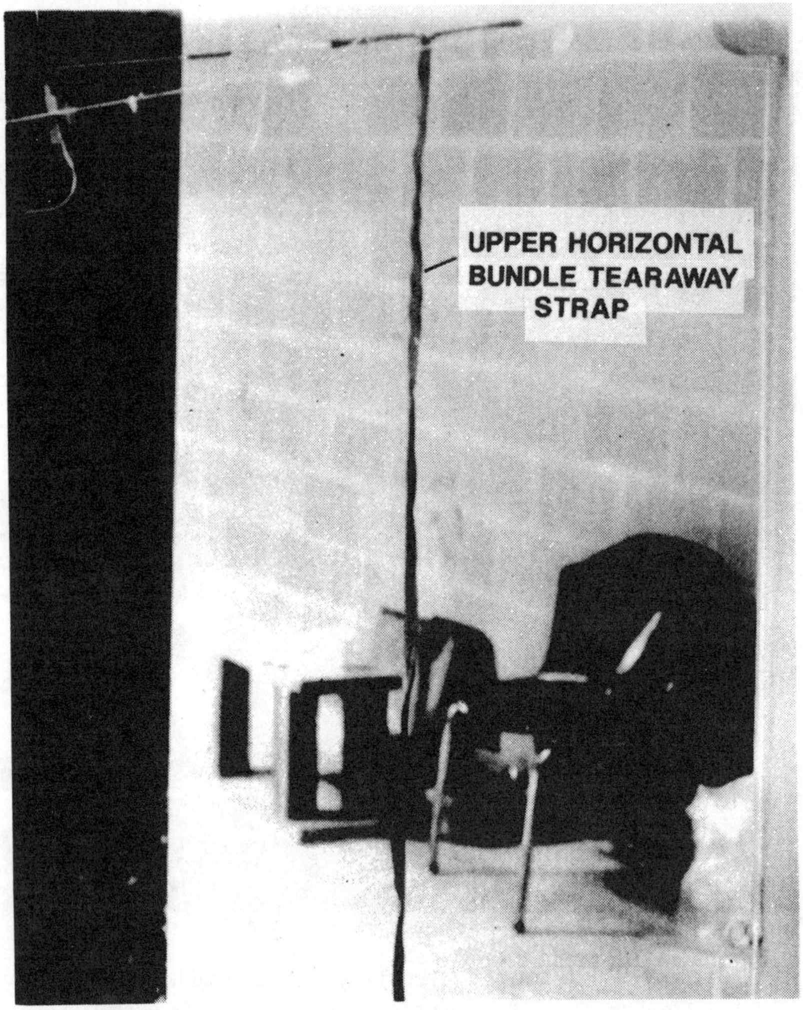

Figure 10 - 1/8 scale upper horizontal bundle tearaway strap.

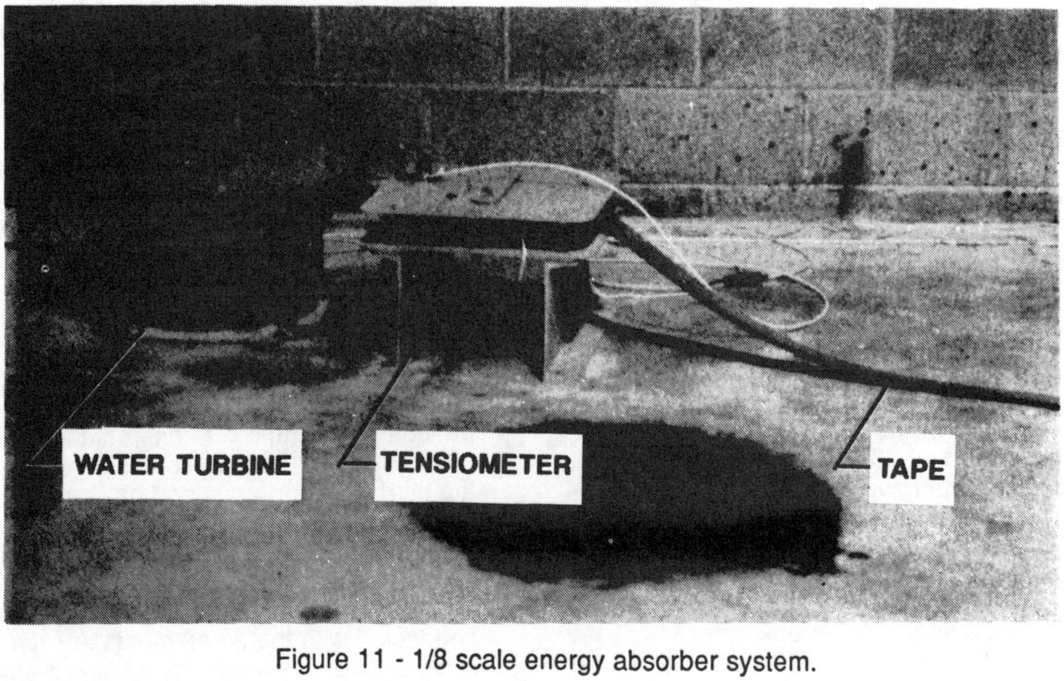

Figure 11 - 1/8 scale energy absorber system.

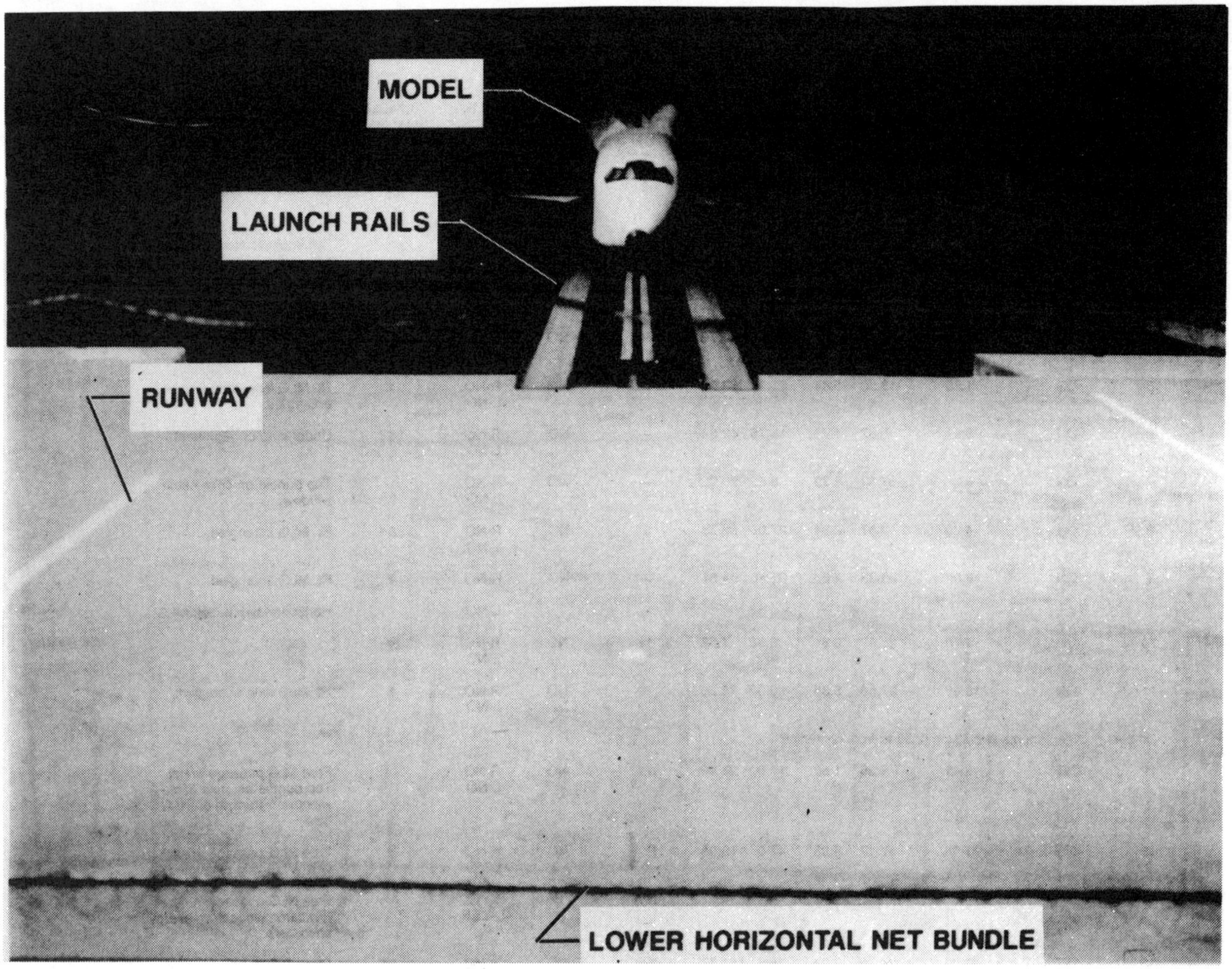

Figure 12 - 1/8 scale test set up.

gear was free swiveling . Thirty-five percent of these tests resulted in one or two verticals caught on the model's nose, one or two verticals caught between the nose gear tires for 45 percent of the runs and one vertical caught on the nose gear strut.

The nose gear was locked straight ahead (not free swivelling) for the next set of 10 tests. One vertical caught on the nose on one test and one vertical caught between the nose gear tires during 3 tests. No verticals caught on the nose gear strut.

The final three runs were specialized with an anchored net or an open window that allowed free nose gear passage. One vertical caught between the nose gear tires.

Table V gives the data for the catapult tests with the fourth net. Tests 1 through 18 had average speeds of 11.78 and 18.62 kts with average runout distances of 21.9 and 32.19 ft (6.77 and 9.81 m) respectively. These 18 runs were made to see how many vertical members caught on the nose gear, the initial loading of the energy absorber system on the model, and the distribution of the top bundle on the payload bay doors.

Of the 18 catapult tests completed, either the left or right main gear failed 4 of 18 tests. The nose gear was overstrength and never failed. The nose gear was tangled 8 of 18 runs and the left and right main gear was tangled 9 of 18 runs and 7 of 18 runs respectively. When the verticals caught on the nose gear, the top bundle was prevented from spreading back over the payload bay doors properly as shown in figure 13. When vertical members did not catch on the nose gear, or when the vertical members were broken by the nose gear, the top bundle spread appropriately over the third or fourth payload bay door as shown in figure 14.

Additional tests, 19 through 27, with net #4 were conducted to determine at what speed the top bundle goes under the wing. The tests were all centerline engagements and started at low speeds and gradually went up to higher speeds. Table V gives the test data for these additional tests. The top bundle tended to go over the left wing but stayed in front of the right wing as shown in figure 15, for the

TABLE II - 1/27.5 SCALE MODEL NET #1 TEST DATA

RUN NO.	AIM POINT ON/OFF CENTERLINE	ENGAGEMENT VELOCITY (KTS)	ROLLOUT (FT)	ROLLOUT (M)	DEVIATION FROM LAUNCH CENTERLINE (IN)	DEVIATION FROM LAUNCH CENTERLINE (CM)	NUMBER SCALE STRENGTH STRAPS BROKEN	NLG FAILURE	MLG FAILURE R-RIGHT L-LEFT	TOP BUNDLE PAYLOAD BAY DOOR AREA	REMARKS
1	ON	11.50	20.67	6.30	18.5	47.00	6	NO	R-NO L-NO	4	Rt. MLG axle entangled.
2	ON	---	---	---	---	---	---	---	---	---	Cable interefered with pulley. No data.
3	ON	11.45	13.58	4.14	1.75	4.44	5	NO	R-YES L-YES	4	MLG entangled & locked up. Breakaways by-passed.
4	ON	11.50	17.96	5.47	7.00	17.78	9	NO	R-NO L-NO	4	Rt. MLG axle entangled.
5	ON	11.50	18.50	5.64	25.0	63.50	10	NO	R-NO L-NO	4	Rt. MLG locked-up & axle entangled.
6	ON	3.61	5.67	1.73	16.75	42.54	---	NO	R-NO L-NO	1&2	Underwing engagement.
7	ON	3.67	7.83	2.39	8.63	21.92	---	NO	R-NO L-NO	---	Top bundle on crew cabin window.
8	ON	4.20	7.67	2.34	27.00	68.58	2	NO	R-NO L-NO	3&4	Rt. MLG entangled.
9	ON	18.06	28.33	8.63	21.50	54.61	---	NO	R-NO L-NO	4	Rt. MLG entangled. Horizontal behind left MLG.
10	ON	17.76	28.92	8.81	3.00	7.62	16	NO	R-NO L-NO	4	
11	ON	18.06	27.50	8.38	30.00	76.20	---	NO	R-NO L-NO	4	Rt. MLG axle entangled.
Off-center engagements at 0.79 cm (2.58 ft) from centerline.											
12	OFF	3.43	5.25	1.60	10.00	25.40	N/A	NO	R-NO L-NO	---	Free NLG passage in net. Top bundle on crew cabin window. 1 vertical on NLG axle.
13	OFF	11.56	17.17	5.23	47.0	119.38	N/A	NO	R-NO L-NO	3	Free NLG passage in net. Left MLG axle entangled.
14	OFF	---	---	---	---	---	N/A	YES	R-YES L-YES	---	Free NLG passage in net. Net came loose from energy absorbers.
15	OFF	18.06	24.75	7.54	75.5	191.77	N/A	NO	R-NO L-NO	---	Free NLG passage in net. Model tracked off runway to right.
16	ON	18.06	29.00	8.84	15.00	38.10	---	NO	R-NO L-NO	4	---
17	ON	11.78	19.67	6.00	34.50	87.63	---	NO	R-NO L-NO	4	---

N/A - Not applicable

first four tests at a speed range of 4.29 to 7.82 kts. In an effort to correct this problem, the sheaves in line with the net's bottom bundle were moved upstream 7.27 ft (2.22 m) as shown in figure 16. This simulated moving the full scale water turbine system 200 ft (61 m) upstream.

Even with the arresting gear sheaves upstream, there was still a problem with the top bundle going under the model wings. Another correction to the arrestment system was made using simulated extended tearaways on the top bundle on both sides of the net. This proved to be an effective correction that resulted in the top bundle going over the wings at slow speeds of 5.33 kts and at higher speeds of 11.66 kts (figure 17).

Net #5 - A scale strength net was constructed as close to the AAE net specifications as possible. One test was run at 19.66 kts with a runout distance of 39.5 ft (12.04 m). The top bundle came to rest in the latter part of the third payload bay door area as shown in figure 18. One horizontal member wrapped twice around the right main gear axle. Three verticals were under the main gear axle or caught on the axle; one was behind the main gear but not caught. Two verticals and one horizontal broke. Neither the nose gear nor the main gear failed.

1/8 SCALE MODEL STUDIES - Dynamic 1/8 scale barrier engagement tests for the Shuttle Orbiter Arresting System were conducted by All American Engineering Company in Glenn Riddle, Pennsylvania. The purpose of these tests was to verify that the net configuration was appropriate for complete and proper penetration and wing envelopment by the net. Low speed, centerline, off-centerline, and yawed engagements were conducted. There were a total of 29 test runs with locked nose gear. Table VI gives the test data and conditions for these tests.

TABLE III - 1/27.5 SCALE MODEL NET #2 TEST DATA

RUN NO.	AIM POINT ON/OFF CENTERLINE	ENGAGEMENT VELOCITY (KTS)	ROLLOUT (FT)	ROLLOUT (M)	DEVIATION FROM LAUNCH CENTERLINE (IN)	DEVIATION FROM LAUNCH CENTERLINE (CM)	NUMBER SCALE STRENGTH STRAPS BROKEN	NLG FAILURE	MLG FAILURE R-RIGHT L-LEFT	TOP BUNDLE PAYLOAD BAY DOOR AREA	REMARKS
1	ON	11.54	18.50	5.64	11.00	27.94	7	YES	R-YES L-NO	4	NLG failed after Right MLG failed. No net around NLG when failed.
2	ON	11.66	17.58	5.36	31.25	79.38	---	NO	R-NO L-YES	3 & 4	Left MLG failed at end of rollout. Net wrapped around left MLG axle. NLG yawed to right - tilt steering.
3	ON	11.60	19.0	5.79	3.25	8.26	8	NO	R-NO L-NO	3 & 4	---
4	ON	---	8.21	2.50	1.50	3.81	1	NO	R-NO L-NO	3	Left MLG entangled.
5	ON	4.14	8.46	2.58	4.75	12.07	5	NO	R-NO L-NO	3	Both MLG entangled.
6	ON	4.26	8.5	2.59	10.50	26.67	2	NO	R-YES L-NO	3	NLG - tilt steering
7	ON	18.00	25.33	7.72	72.00	182.88	18	YES	R-YES L-YES	4	Energy absorbers disconnected on left side. Both left MLG wheels came off. NLG failed after MLG. Both MLG entangled. Model yawed to rt.. and ended perpendicular to runway.
8	ON	18.00	29.25	8.91	1.75	4.44	18	NO	R-NO L-NO	4	---
9	ON	18.12	24.5	7.47	6.50	16.51	17	NO	R-YES L-YES	3	Both MLG entangled.
10	ON	11.72	23.58	7.19	0.50	1.27	---	NO	R-YES L-YES	4	---
Off-center engagements at 27.94 cm (11 inches) from centerline.											
11	OFF	SLOW	9.21	2.81	7.00	17.78	N/A	NO	R-NO L-NO	3	Strings in front & behind MLG struts.
12	OFF	11.27	15.92	4.85	23.00	58.42	N/A	YES	R-YES L-YES	3	Both MLG entangled. NLG broke when right wing hit runway, then left MLG broke.
13	OFF	17.46	22.33	6.81	16.50	41.91	N/A	NO	R-YES L-YES	3	Both MLG entangled.

N/A - Not applicable

Tests 1 through 3 were run at 11 kts. (31 kts full scale) with the model successfully arrested although the lower horizontal bundle was tangled in the main gear. The problem was recognized as a lack of tension in the lower horizontal bundle and was resolved by using tearaway straps between the lower bundle and auxiliary ground anchors. These tearaways, shown in figure 19, maintained tension in the lower bundle and allowed the kink wave generated at engagement to run until the net and tape assembly became taut with the energy absorbers. The fourth test was at 14.1 kts with the tearaway straps in the lower horizontal bundle. Two vertical straps were broken on the nose gear but the net completely enveloped the wing with no entanglement in the main gear (figures 20 and 21). Tests 5 through 9 were run from 17.8 to 33.8 kts and where all successful arrestments.

Runs 10 through 12 were at 5° yawed engagements at speeds from 14.8 to 25.2 kts. All were successful engagements with the model stopping to the right of center.

Off-center engagements with an offset of 5.44 ft (1.66 m) were tested at 0° and 5° yaw configurations. The 5° yawed engagements were made at 12.0 and 19.7 kts. For these two runs, 13 and 14, the model veered left immediately after leaving the end of the launcher. The tracking problem was thought to be caused by a slight bow in the nose gear rails of the launcher. On both runs, the engagement was successful and no vertical or horizontal members were broken. The off-center engagements with a 0° yaw were conducted at speeds of 12.2 and 19.4 kts. The model tracked straight from the launcher to the net and then drifted to the left edge of the runway during arrestment (figure 22). Two additional tests at 15.2 and 24.6

TABLE IV - 1/27.5 SCALE MODEL NET #3 TEST DATA

RUN NO.	AIM POINT ON/OFF CENTERLINE	ENGAGEMENT VELOCITY (KTS)	ROLLOUT (FT)	ROLLOUT (M)	DEVIATION FROM LAUNCH CENTERLINE (IN)	DEVIATION FROM LAUNCH CENTERLINE (CM)	NUMBER SCALE STRENGTH STRAPS BROKEN	NLG FAILURE	MLG FAILURE R-RIGHT L-LEFT	TOP BUNDLE PAYLOAD BAY DOOR AREA	REMARKS
1	ON	---	8.46	2.58	9.25	23.50	3	NO	R-NO L-NO	---	Rope to mass came off pulley.
2	ON	11.22	20.67	6.30	8.50	21.59	10	NO	R-NO L-NO	4	---
3	ON	11.13	16.68	5.08	5.75	14.60	10	NO	R-YES L-YES	3	Bottom horizontal bundle broke MLG
4	ON	---	16.25	4.95	15.75	40.00	18	YES	R-YES L-YES	4	---
5	ON	11.07	17.38	5.30	28.50	72.39	11	NO	R-YES L-NO	4	Nosewheel steering. Model veered to edge of runway.
6	ON	SLOW	6.67	2.03	37.75	95.89	3	NO	R-NO L-NO	3	Model veered to right off runway.
7	ON	SLOW	8.33	2.54	19.50	49.53	4	NO	R-NO L-NO	3	Model veered to right.
8	ON	4.29	8.21	2.50	26.75	67.94	1	NO	R-NO L-NO	3	Vehicle veered to right close to edge of runway.
9	ON	18.77	23.33	7.11	2.25	5.72	16	NO	R-YES L-YES	4	---
10	ON	17.94	28.83	8.79	5.50	13.97	---	NO	R-NO L-NO	4	---
11	ON	18.47	27.83	8.48	18.00	45.72	16	NO	R-NO L-NO	4	Nosewheel steering. Left MLG entangled.
Bottom horizontal bundle bunched to see if NLG fails when hits bundle; for next two runs.											
12	ON	18.00	27.0	8.23	21.50	54.61	---	YES	R-NO L-YES	4	SS verticals not repaired for this run.
13	ON	18.06	28.25	8.61	14.50	36.83	N/A	YES	R-NO L-NO	4	NLG failed when hit bottom bundle.
14	OFF	SLOW	6.75	2.06	2.25	5.72	N/A	NO	R-NO L-NO	---	Top bundle on crew cabin window. 2 verticals on NLG.
15	OFF	11.49	17.83	5.43	42.00	106.68	N/A	NO	R-NO L-NO	4	Right MLG entangled. Model veered to right.
16	OFF	18.35	27.50	8.38	32.50	82.55	N/A	NO	R-NO L-NO	4	Model veered to right
17	OFF	18.24	25.88	7.89	8.00	20.32	N/A	NO	R-NO L-NO	4	Left MLG axle entangled. Model veered to rt. and pulled back to left.
18	ON	18.24	28.58	8.71	2.25	5.72	---	NO	R-NO L-NO	4	SS verticals not repaired.
19	ON	11.66	19.92	6.07	21.50	54.61	---	YES	R-NO L-NO	4	Bottom bundle bunched; model veered to right. NLG failed after NLG was over bottom bundle.

N/A - Not applicable

TABLE V - 1/27.5 SCALE MODEL NET #4 TEST DATA

RUN NO.	AIM POINT ON/OFF CENTERLINE	ENGAGEMENT VELOCITY (KTS)	ROLLOUT (FT)	ROLLOUT (M)	DEVIATION FROM LAUNCH CENTERLINE (IN)	DEVIATION FROM LAUNCH CENTERLINE (CM)	NLG FAILURE	MLG FAILURE R-RIGHT L-LEFT	TOP BUNDLE PAYLOAD BAY DOOR AREA	REMARKS
Used pulleys for lead-off sheaves, parachute cord to chains, 0.61 m (2 ft) small, 0.61 m (2 ft) medium, heavy chain, overstrength net. NLG fixed and overstrength. Bottom bundle not buried.										
1	ON	11.58	16.00	4.88	6.50	16.51	NO	R-YES L-YES	2&3	MLG broke at end of rollout. Rt. MLG axle entangled.
2	ON	12.08	26.96	8.22	3.50	8.89	NO	R-NO L-NO	3	Open nose passage for NLG in net.
3	ON	11.82	17.42	5.31	15.00	38.10	NO	R-YES L-YES	1	Both MLG entangled.
4	ON	11.90	24.17	7.37	12.75	997.79	YES	R-NO L-NO	1	NLG entangled.
5	ON	---	24.33	7.42	38.50	97.79	NO	R-NO L-NO	1	---
6	ON	10.57	22.00	6.71	0.50	1.27	---	---	1	1 vertical across NLG strut. 2 verticals between NLG tires.
Two 0.61 m (2 ft) medium size chains in parallel added inbetween medium and large chains.										
7	ON	12.00	24.33	7.42	24.75	62.86	NO	R-NO L-NO	1	Left MLG entangled.
8	ON	---	16.29	4.96	55.50	140.97	NO	R-NO L-NO	3	Both MLG entangled.
9	ON	---	20.96	6.39	57.25	145.42	NO	R-NO L-NO	1	1 vertical across NLG strut. 2 verticals between NLG tires. Left MLG entangled.
10	ON	12.02	16.21	4.94	22.50	57.15	NO	R-YES L-NO	3	Both MLG entangled.
11	ON	11.84	23.58	7.19	7.50	19.05	NO	R-NO L-NO	1	2 verticals around NLG axle.
12	ON	11.78	23.96	7.30	29.00	73.66	NO	R-NO L-NO	3	Both MLG entangled.
13	ON	11.90	21.29	6.49	45.25	114.94	NO	R-NO L-NO	1	1 vertical caught between NLG tires. 1 vertical on left MLG.
14	ON	11.90	23.50	7.16	7.25	18.42	NO	R-NO L-NO	1&2	1 vertical around NLG axle. Left MLG entangled.
15	ON	11.78	23.25	7.09	9.25	23.50	NO	R-NO L-NO	1	1 vertical between NLG tires. MLG entangled.
16	ON	11.96	26.25	8.00	37.50	95.25	NO	R-NO L-NO	3	Rt. MLG entangled.
17	ON	18.65	27.96	8.52	22.00	55.88	NO	R-NO L-YES	1	NLG steering engaged. NLG axle. 3 verticals on left MLG.
18	ON	18.59	36.42	11.10	10.00	25.40	NO	R-NO L-NO	3	Open passage for NLG in net. Model stopped by instrumentation cord.
Tests to determine at what speed have underwing engagement.										
19	ON	4.29	---	---	---	---	NO	R-NO L-NO	---	Open passage for NLG in net. Top bundle infront of left wing.
20	ON	7.10	---	---	---	---	NO	R-NO L-NO	---	Top bundle on crew cabin window.
21	ON	7.10	---	---	---	---	NO	R-NO L-NO	3	Top bundle over left wing and infront of right wing. Both MLG entangled.
22	ON	---	---	---	---	---	NO	R-NO L-NO	3	Top bundle over left wing and infront of right wing. Both MLG entangled.
Moved sheaves upstream 2.22 m (7.27 ft).										
23	ON	11.72	20.42	6.22	44.50	113.03	NO	R-NO L-NO	---	2 horizontals on NLG strut. Both MLG entangled.
24	ON	11.84	22.00	6.71	27.00	68.58	NO	R-NO L-NO	3	Underwing engagement. Model veered to right.
Simulated extended tearaways installed.										
25	ON	11.49	24.54	7.48	17.50	44.45	NO	R-NO L-NO	3	Overwing engagement. Rt. MLG entangled. Model yawed to rt.
26	ON	11.66	24.25	7.39	10.00	25.40	NO	R-NO L-NO	3	Both MLG entangled. Overwing engagement.
27	ON	5.33	9.83	3.00	0.75	1.90	NO	R-NO L-NO	1	1 vertical around NLG axle. Overwing engagement.

Figure 13 - Top bundle near front of 1/27.5 scale model.

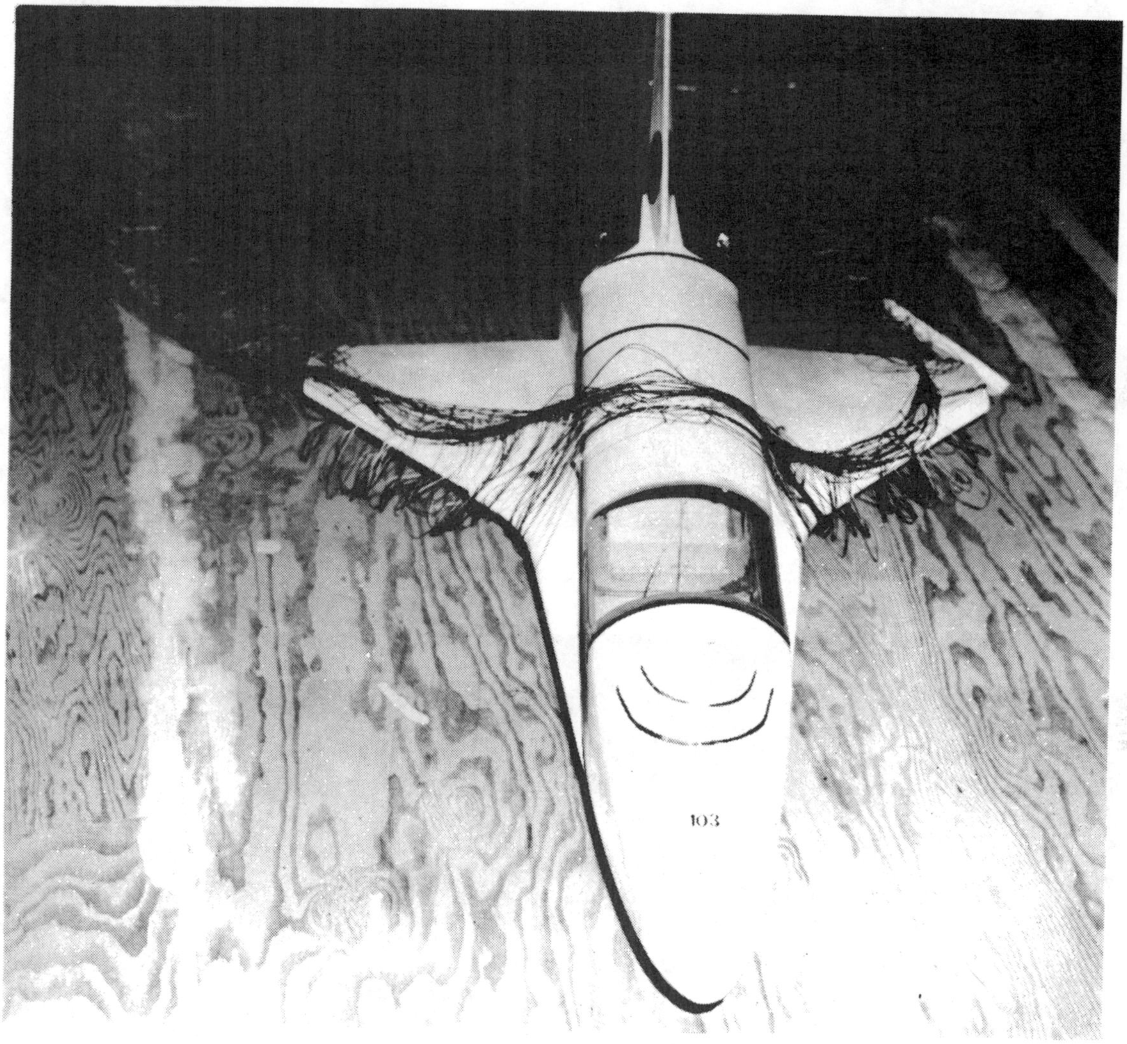

Figure 14 - Top bundle in third payload bay door area.

Figure 15 - Underwing engagement.

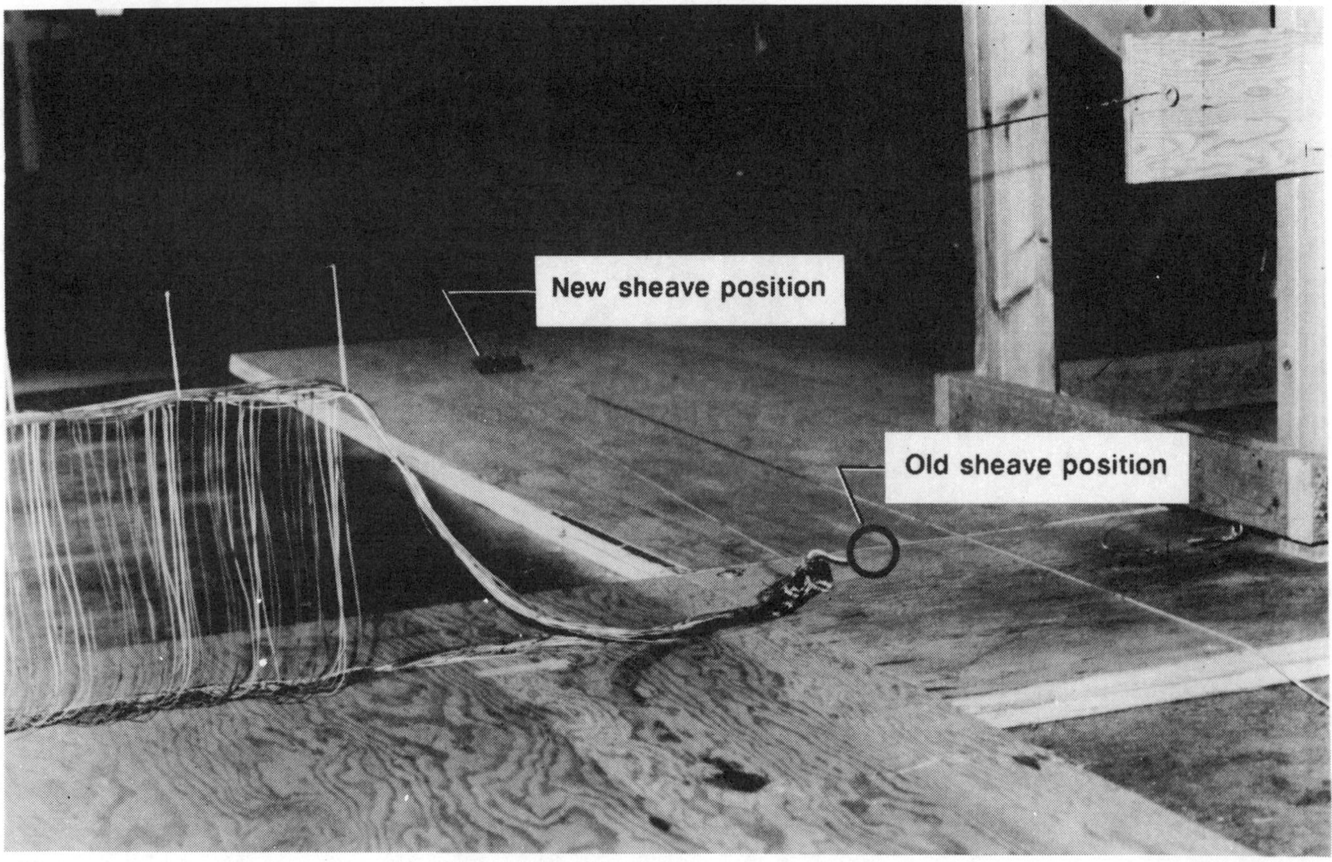

Figure 16 - Sheaves moved upstream.

Figure 17 - Overwing engagement.

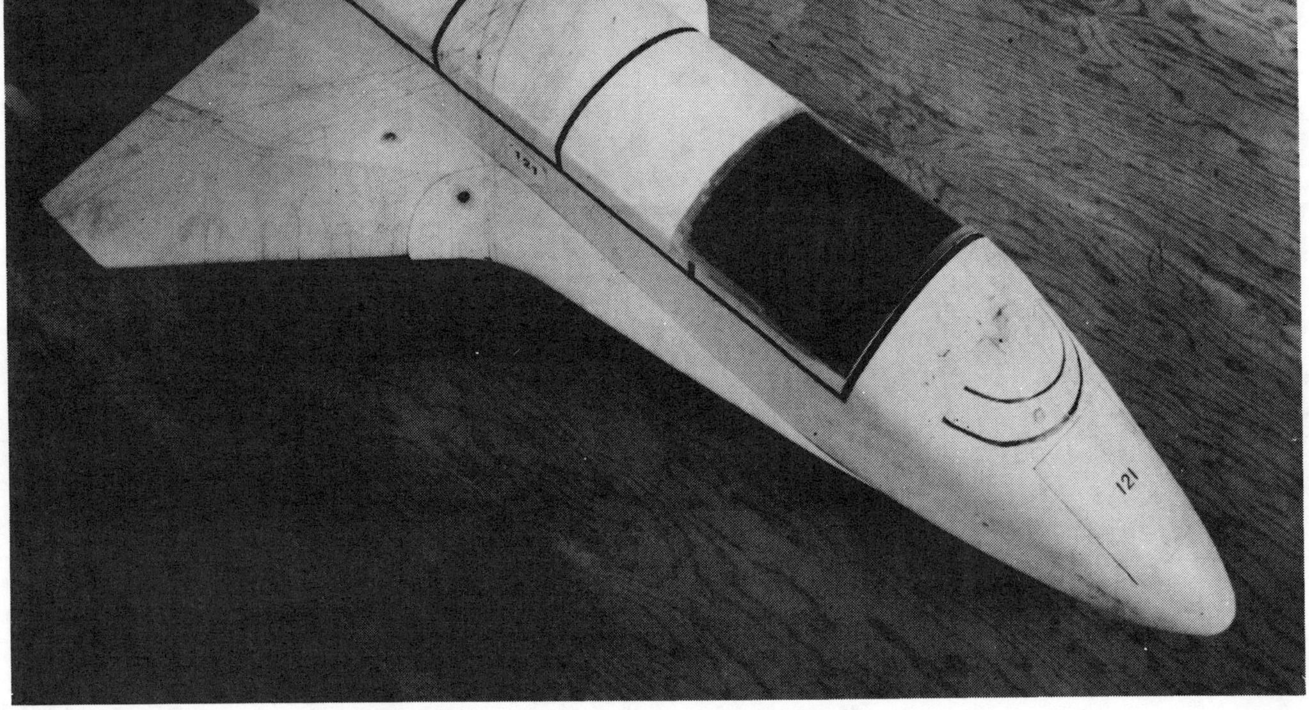

Figure 18 - 1/27.5 scale strength net arrestment.

TABLE VI - 1/8 SCALE MODEL TEST DATA

RUN NO.	AIM POINT ON/OFF CENTERLINE	ENGAGEMENT VELOCITY (KTS)	ROLLOUT (FT)	(M)	DEVIATION FROM LAUNCH CENTERLINE (IN)	(CM)	NUMBER STRAPS BROKEN	NET PLANE ENTRY (DEG)	PEAK ARRESTMENT TAPE TENSION PORT	STBD	MAXIMUM ACCELERATION (G) X	Y	REMARKS
1	ON	11.0	64.75	19.74	18.00	45.72	---	90	---	---	---	---	Lower horizontal bundle on MLG.
2	ON	10.9	63.33	19.30	16.00	40.64	---	90	---	---	---	---	Lower horizontal bundle on MLG & primary anchor
3	ON	11.1	65.33	19.91	12.00	30.48	---	90	---	---	---	---	Lower horizontal bundle on MLG.
4	ON	14.1	85.5	26.06	8.00	20.32	2 V	90	66.0	70.0	0.32	---	Added tear-aways to lower bundle.
5	ON	17.8	90.83	27.68	3.00	7.62	2 V 1 H	90	105.0	103.0	0.44	---	---
6	ON	24.2	96.25	29.34	9.00	22.86	0	90	182.0	166.0	0.72	---	Open nose passage in net.
7	ON	10.9	77.08	23.49	11.00	27.94	0	90	45.0	42.0	0.28	---	Open nose passage in net.
8	ON	33.8	99.50	30.33	13.00	33.02	0	90	---	---	---	---	Lost left NLG tire
9	ON	33.8	100.00	30.48	12.00	30.48	1 V 1 H	90	416.0	325.0	1.40	---	NLG tire seperated from hub.
Balanced energy absorber fluid levels at three quarts low.													
10	ON	14.8	90.0	27.43	127.00	322.58	3 V 1 H	85	---	57.0	0.32	---	Model tracking off centerline.
11	ON	15.0	90.25	27.51	3.00	7.62	0	85	75.0	70.0	0.34	---	Open nose passage in net.
12	ON	25.2	100.33	30.58	22.00	55.88	0	85	161.0	172.0	0.72	---	---
Installed lateral accelerometer.													
13	OFF	12.0	83.0	25.30	27.00	68.58	0	85	40.0	38.0	0.18	---	Model tracking off centerline.
14	OFF	19.7	94.67	28.85	58.00	147.32	0	85	127.0	93.0	0.48	N	Model tracking off centerline.
Realigned launcher NLG rail.													
15	OFF	12.2	81.25	24.76	10.00	25.40	1 H	90	56.0	28.0	0.16	N	---
16	OFF	19.4	93.17	28.40	74.00	187.96	2 V	90	120.0	105.0	0.46	N	---
17	OFF	15.2	89.00	27.13	12.00	30.48	0	90	73.0	58.0	0.30	N	Appeared to catch vertical-didn't break.
18	OFF	24.6	98.50	30.02	6.00	15.24	0	90	193.0	133.0	0.60	N	---
Increased arresting gear tape length by 20 ft (6.1 m).													
19	OFF	15.2	99.00	30.17	29.00	73.66	0	90	88.0	63.0	0.24	N	---
20	OFF	25.2	115.25	35.13	93.00	236.22	1 V	90	216.0	158.0	0.62	N	---
Reduced arresting gear tape length by 5 ft (1.52 m).													
21	OFF	12.2	87.92	26.80	4.00	10.16	0	90	60.0	30.0	0.20	N	---
22	OFF	14.8	96.83	29.51	33.00	83.82	0	90	85.0	32.0	0.26	N	---
23	OFF	19.2	104.58	31.87	103.00	261.62	0	90	133.0	68.0	0.40	N	Model drifted off runway.
24	OFF	24.5	108.25	32.99	145.00	368.30	1 H	90	173.0	135.0	0.64	0.24	Model drifted off runway.
25	OFF	24.5	109.50	33.37	74.00	187.96	3 V	90	166.0	141.0	0.76	0.24	---
26	OFF	33.0	113.58	34.62	134.00	340.36	12 V	90	200.0	230.0	1.34	N	Broke 12 vert. adjacent straps.
27	ON	25.2	112.00	34.14	22.00	55.88	0	90	158.0	160.0	0.68	N	---
28	ON	11.3	87.42	26.64	26.00	66.04	0	85	33.0	37.0	0.24	N	---
29	ON	33.8	117.58	35.84	53.00	134.62	0	85	218.0	217.0	0.94	N	Model corrected toward 85° C/L.

N - Negligible

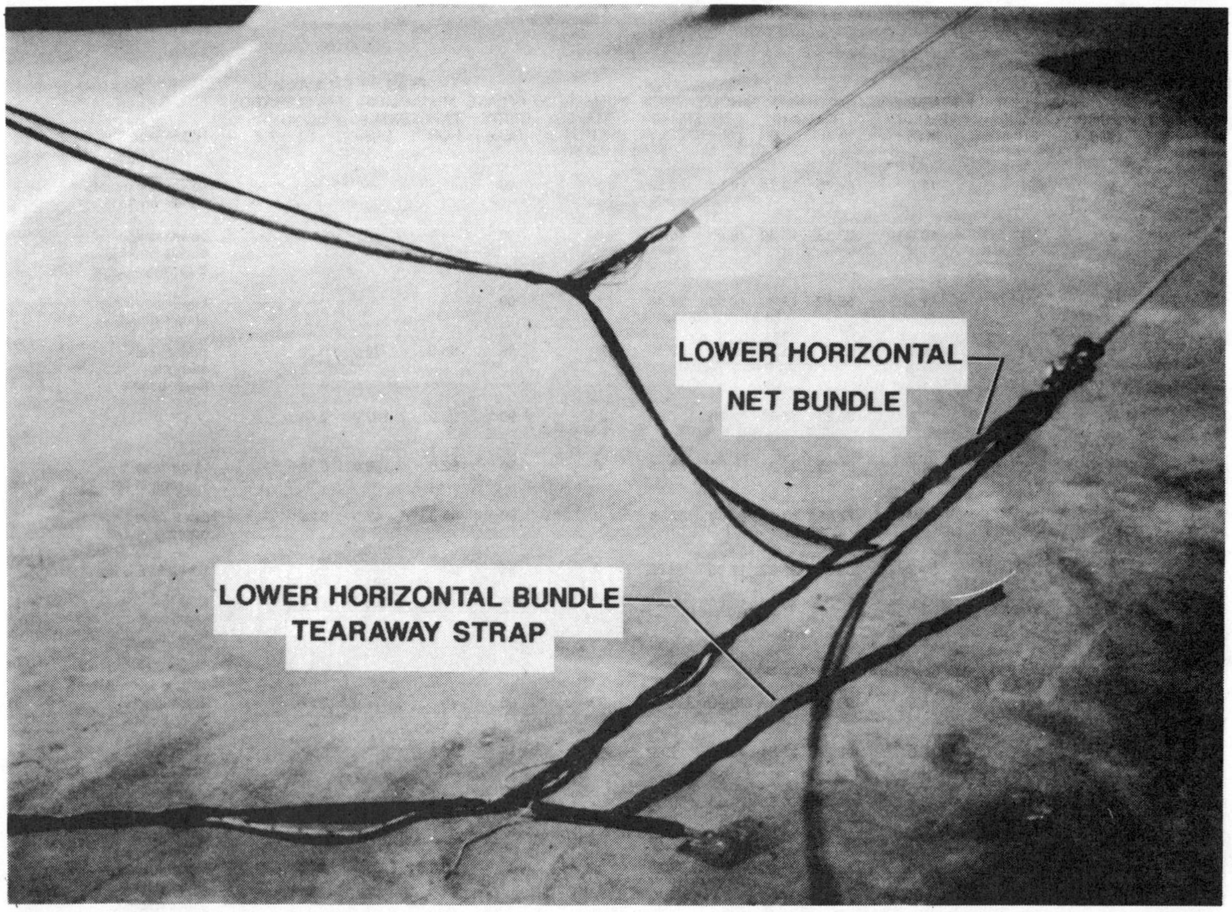

Figure 19 - Lower horizontal bundle tearaways.

kts. were conducted with a steering vector set in the nose gear to counter the drift toward the runway edge during arrestment. Engagements were successful and the model only drifted about one foot to the left of center as opposed to 5 to 6 ft (1.52 to 1.83 m) without the preset steering vector. Runs 19 and 20 were made under these same off-center conditions with the arresting gear tape length increased by 20 ft (6.09 m) but without a steering vector. These two tests were at 15.2 and 25.2 kts. The model still drifted off-center 2.42 ft (0.74 m) and 7.75 ft (2.36 m) respectively.

After the runs with the increased arresting gear tape length, the tape length was reduced 5 ft (1.52 m) from the extended length. There were six runs, 21 through 26, with the model offset 10.88 ft (3.32 m) from the centerline at 0° yaw angle. The test engagement velocities varied from 12.2 to 33.0 kts. (34.5 - 93.3 kts. full scale). Runs 21 and 22 were under this configuration and resulted in good wing envelopment with no straps broken. Runs 23 and 24 were at 19.2 and 24.5 kts and had successful wing engagements, however, the model drifted off the runway. One horizontal strap was broken during run 24. The final two off-center tests, had a steering vector applied to counter the off runway drift. Run 25 at 24.5 kts resulted in three broken vertical straps and the model coming to a stop off the runway. Run 26 was at 33.0 kts and 12 vertical members were broken. The model was again off the runway at the end of the rollout. From these off-center engagement tests, it is clear that an off-center engagement of this type will result in the vehicle tracking off the runway.

Three additional tests (27 through 29) were completed with centerline arrestment. The first test was an on centerline, 0° yaw run at 25.2 kts and was very successful. The following two runs were at 11.3 kts and 33.8 kts on centerline at a 5° yaw. Both runs were successful arrestments with no straps broken.

CONCLUDING REMARKS

Scale model studies were conducted of the Shuttle Orbiter Arresting System at the NASA Langley Research Center and at All American Engineering Company. The results of the 1/27.5 and the 1/8 scale model tests indicate the following conclusions:

(1) A net arrestment system has been properly designed to bring the orbiter to a safe stop with a minimal amount of damage.
(2) Using tearaway straps in the upper horizontal bundle and moving the energy absorber sheaves 200 ft (61 m) upstream corrected the upper bundle from going under the wings at slow speed engagements.
(3) Tearaway straps in the lower horizontal bundle will eliminate the lower horizontal bundle entangling the main gear.
(4) These model tests indicate that no more than three vertical straps will be caught by the nose gear and they will be broken with no detrimental consequences to the vehicle.
(5) For off-center engagements, the vehicle will drift towards the edge of the runway and may even depart the runway.

Figure 20 - Successful arrestment without main gear entanglement.

Figure 21 - Successful arrestment without main gear entanglement.

Figure 22 - Model tracking to edge of runway.

Orbiter Post-Tire Failure and Skid Testing Results

Robert H. Daugherty and Sandy M. Stubbs
NASA Langley Research Center

* Paper 892338 presented at the Aerospace Technology Conference and Exposition, Anaheim, California, September, 1989.

ABSTRACT

An investigation was conducted at the NASA Langley Research Center's Aircraft Landing Dynamics Facility (ALDF) to define the post-tire failure drag characteristics of the Space Shuttle Orbiter main tire and wheel assembly. Skid tests on various materials were also conducted to define their friction and wear rate characteristics under higher speed and bearing pressures than any previous tests. The skid tests were conducted to support a feasibility study of adding a skid to the orbiter strut between the main tires to protect an intact tire from failure due to overload should one of the tires fail.

Roll-on-rim tests were conducted to define the ability of a standard and a modified orbiter main wheel to roll without a tire. Results of the investigation are combined into a generic model of strut drag versus time under failure conditions for inclusion into rollout simulators used to train the shuttle astronauts.

ONE OF THE KEYS to safe and successful operation of any aircraft, including the Space Shuttle Orbiter, is realistic pilot training under normal and emergency conditions. One of the unknowns in the landing phase of orbiter operations is the response of the vehicle to tire failure and the ensuing sequential failure of other landing gear components. Landing with one tire deflated is a common baseline failure scenario, and after touchdown the other tire on the strut is almost certain to fail. A need existed to understand the failure behavior of a flat tire, whether or not it departs the wheel, and if so, what drag forces and behaviors are associated with the remaining landing gear hardware.

A proposal was made to install a skid between the two wheels on each strut to prevent overloading and failing an intact tire if one tire was deflected prior to touchdown. The proposed skid would be designed to slide faster and with higher bearing pressure than any skid ever tested.

The purpose of this paper is to present the results of tests conducted at the NASA Langley Research Center to investigate the post-tire failure drag loads and behavior of the Space Shuttle Orbiter main wheel assembly. The paper describes tests involving simulated landings with a flat tire, roll-on-rim tests to evaluate the capability of both a standard and a modified main wheel to roll without a tire, and skid tests to define the friction and wear characteristics of various candidate materials for the proposed strut overload protection skid. The results are combined into a generic model of strut friction versus time, indicating the likely sequential failure of various strut components.

APPARATUS

This investigation was conducted at the NASA Langley Research Center's Aircraft Landing Dynamics Facility (ALDF). The facility, shown in figure 1, consists of a set of rails 850 m long on which a 49 000 kg. carriage travels. The carriage is propelled at speeds up to 220 kts using a high-pressure water jet and is arrested using a set of water turbines connected across the track by nylon tapes. An assembly referred to as the drop carriage rides on vertical rails inside the main test carriage. Test fixtures can be attached to the drop carriage and hydraulically loaded onto a simulated runway surface during a test run. Typically, a dynamometer is attached to the drop carriage, but other components such as landing gear struts, etc., can be mounted for testing. A more detailed description of the facility can be found in reference 1.

Figure 1. Aircraft Landing Dynamics Facility.

The tire used in this study was a 44.5 x 16.0 - 21 bias-ply aircraft tire with a 34-ply rating. A photograph of such a tire is shown in figure 2. The tire has a 5-groove tread pattern made of natural rubber with grooves 2.54 mm deep. The tire has 16 carcass plies and the rated load and pressure is 271 kN and 2.17 MPa respectively.

Figure 2. Orbiter main wheel and tire.

All test hardware was mounted in a force measurement dynamometer pictured and sketched in figure 3. The dynamometer supports the test wheel axle using instrumented load cells so that ground - reaction forces can be measured.

The wheels used in this study were basically standard orbiter main gear wheels made of aluminum, and a cross-sectional sketch of a standard wheel is shown in figure 4. One half-wheel was tested that had one of the bead flanges full of material as shown in figure 4. This design was intended to allow for more roll-on-rim capability. During roll-on-rim testing, only the wheel inner halves were used. An outer wheel half was modified with its bead flange removed so that it

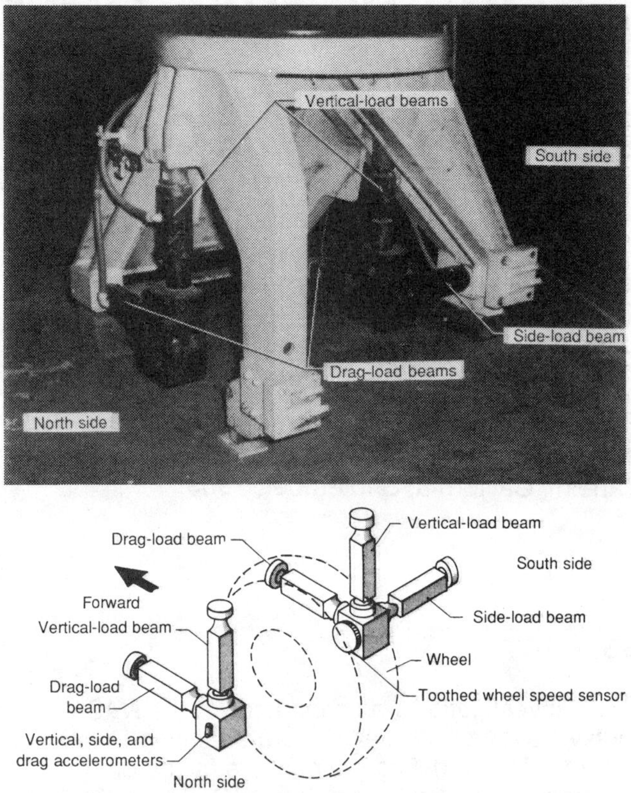

Figure 3. Force-measurement dynamometer.

could be bolted to any of the inner wheel halves tested and its axle bearing used for support. Tests were conducted on half-wheels because the maximum vertical load capability of the facility is about 312 kN and single, full-wheel loads during actual landings can reach values of over 579 kN.

Skid testing was conducted using the apparatus photographed in figure 5. This apparatus consists of a modified test wheel axle, a laterally-movable center yoke, and a pivoting shoe. Various skid specimens, shown in figure 6, included Inconnel 718, 4340 steel, Stellite, tungsten with medium or coarse carbide chips embedded in it, and 1020 steel. The specimens were mounted to the pivoting shoe at the bottom of the yoke, and the front edge of each specimen was beveled upward at 45 degrees to try to prevent runway gouging during a test. Specimen dimensions were 32.1 cm. long by 11.8 cm. wide and 3.8 cm. thick. The Stellite specimen was about 1.5 cm. thicker than the others due to the fabrication process.

Figure 7 shows a typical roll-on-rim test setup from underneath the drop carriage. The figure shows a net mounted behind the dynamometer which is designed to catch or at least bring up to speed debris after wheel disintegration to lessen the chance of test carriage damage. The runway used in this investigation was a simulation of the rough runway at Kennedy Space Center shown in figure 8. The runway has an extremely rough longitudinally-brushed texture with transverse grooves 6.4 mm. wide by 6.4 mm. deep with 29 mm. spacing.

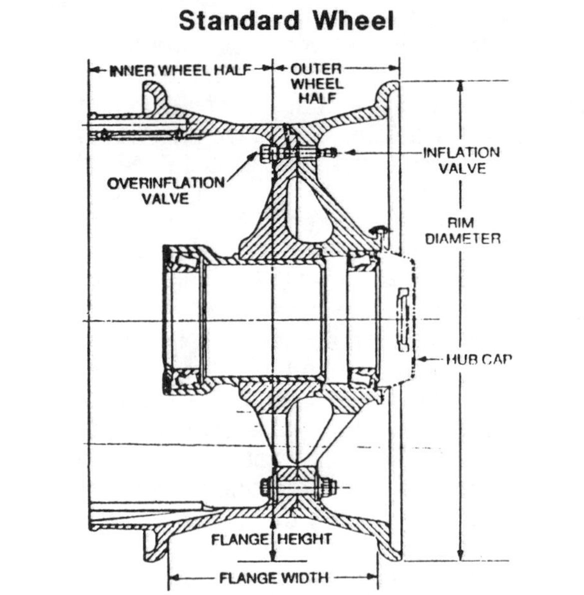

Figure 4. Standard and modified wheel designs.

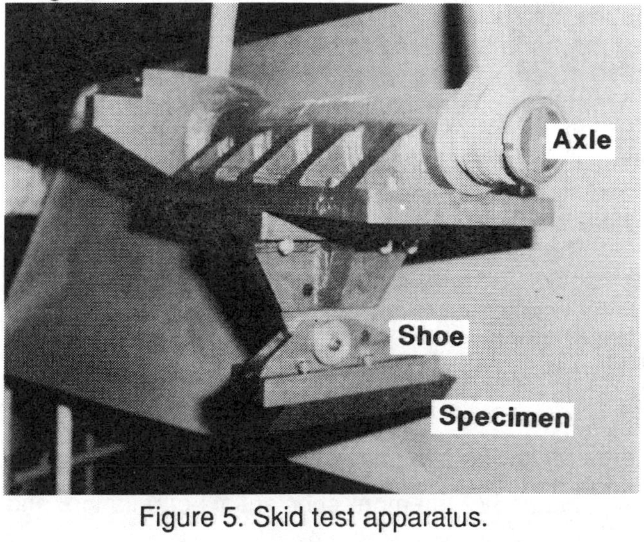

Figure 5. Skid test apparatus.

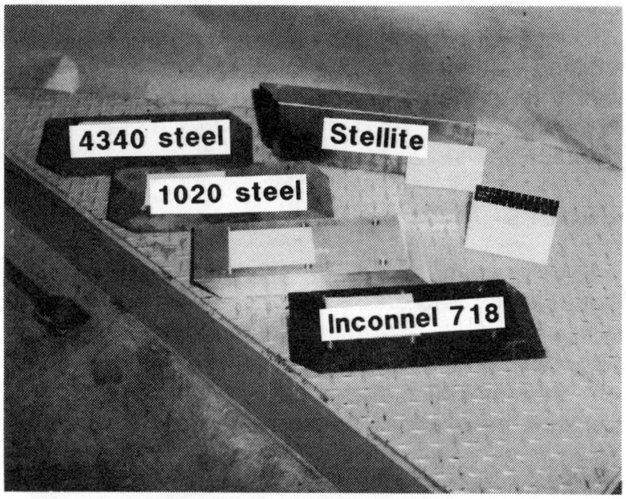

Figure 6. Skid specimens.

Figure 7. View of roll-on-rim apparatus.

Figure 8. Rough KSC runway.

Data from onboard the carriage during a run, including vertical load, drag load, speed, etc., were digitally telemetered to a receiving station where they were converted into engineering units by a desktop computer and stored.

RESULTS AND DISCUSSION

In this study, tests were conducted to examine the failure behavior and drag loads associated with landing on a flat tire, rolling on wheel rims of various design with no tire, and to examine the friction and wear characteristics of candidate materials for a proposed tire overload protection skid.

Landing on a Flat Tire

One test was conducted with the orbiter main gear tire mounted in the dynamometer but completely deflated. The test was conducted at a carriage speed of 159 kts., a nominal sink rate of about 0.75 m/sec., zero yaw angle, and at a vertical load after touchdown of about 312 kN. During the run, film taken onboard the carriage showed smoke coming from the bead area of the tire, indicating slippage between the tire bead and the wheel bead flange. After rolling about 320 m., the tire began to destroy itself due to the damage the wheel bead flanges were inflicting on the tire beads. After rolling an additional 50 m , virtually nothing was left of the tire except the two bead bundles as shown in figure 9. Rolling resistance (or drag coefficient), defined as the drag load divided by the vertical load, while rolling on the flat tire was about 0.2. With 312 kN vertical load on the wheel, the tire beads that remained were barely able to support the load as evidenced by several small contact marks on the wheel flanges and runway. The wheel rolled on the tire beads for another 245 m which ended the test. Rolling resistance while rolling on the tire beads dropped, compared to the flat-tire case, to 0.1 since the center of pressure of the footprint is closer to the fore-and-aft axle position as compared to rolling on a flat tire.

Figure 9. Remains of flat-tire landing.

Roll-On-Rim Tests

Tests were conducted to evaluate the capability of the orbiter main wheel to roll without a tire. Tests conducted on the wheel manufacturer's own test dynamometer, a large steel wheel with a smooth surface, showed that under rated load, 271 kN, the bare wheel rolled about 1200 m at about 4.5 m/sec. before the bead flange cracked. The test was stopped after sustaining a crack in the bead flange of the wheel. A similar test at the ALDF, on the flat, rough, concrete runway was conducted by rolling a standard wheel inner half loaded to about 135 kN. The wheel rolled about 200 m before the bead flange cracked, and on the next wheel revolution as the cracked area came into ground contact the wheel was completely destroyed. This test suggests that roll-on-rim capability is affected by surface roughness and steel drum dynamometer testing is likely to show unrealistically high roll-on-rim performance. For instance, the effect of a rough concrete runway in the present test was to reduce the roll-on-rim capability of the wheel by a factor of six.

Several tests were performed to try to determine the high-speed roll-on-rim capability of orbiter main wheels. Again, these tests used only inner wheel halves. During these tests, the carriage was accelerated to speed and the wheel lowered to the runway at a sink rate of about .8 m/sec. The spinup of the wheel was not part of the test, and to try to prevent wheel damage during spinup a rubber belt was bonded to the inner wheel flange. During wheel spinup, the rubber belt would rip off the wheel in an orderly fashion and wheel damage was avoided.

The first high-speed test run was with a standard orbiter inner wheel half at a test speed of 150 kts. After spinup and as vertical load was being applied in an effort to reach about 270 kN, the wheel disintegrated after rolling about 10 m. At wheel breakup the vertical load had only risen to about 135 kN. After wheel breakup, the test carriage was designed to restrict the test axle vertical travel so that the wheel center section would not be loaded during the remainder of the run. Figure 10 is a photograph of the wheel inner half debris, and analysis of the debris showed that the failure mode appeared to be failure of the bead flange in 10-13 cm. pieces which "zippered" off the wheel uniformly in one revolution. After the bending stiffness provided by the wheel bead flange was destroyed, the remaining brake housing simply could not support the bending moment imposed by the vertical load and the wheel shattered.

The next test was with the modified inner wheel half shown in figure 4. During the test at 157 kts., the wheel was spun-up, and as vertical load was passing through about 180-200 kN the wheel was completely destroyed. The wheel inner half had rolled about 36 m. Figure 11 shows the wheel inner half debris from this test, and the failure mode appeared to be the same as for the standard wheel inner half after its bead flange had been destroyed. The modified inner wheel half, because of the more substantial bead flange, and

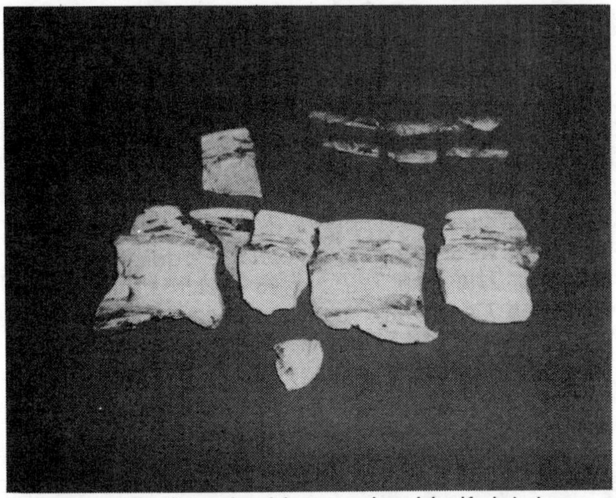

Figure 10. Standard inner wheel half debris.

Figure 11. Modified inner wheel half debris.

Figure 12. Runway damage caused by wheel failure.

hence bending stiffness, would be expected to roll further and support a higher vertical load as in fact it did. Figures 12 shows the damage to the runway incurred due to the failure of this wheel. Based on the roll-on-rim tests conducted at the ALDF, it was determined that no realistic roll-on-rim capability exists for the orbiter with either the standard or modified wheel design.

Skid Tests

Several tests were conducted to examine the friction and wear characteristics of various metal skid specimens under higher speeds and bearing pressures than every tested before. A proposal had been made to install a skid in between the tires on each main gear strut, so that if one tire was flat prior to touchdown, the skid would absorb enough vertical load during the derotation event to prevent the good tire on the strut from being overloaded and failing. Figure 13 shows a sketch of the proposed skid. Table 1 shows various orbiter conditions and strut loads with the corresponding proposed skid loads and bearing pressures.

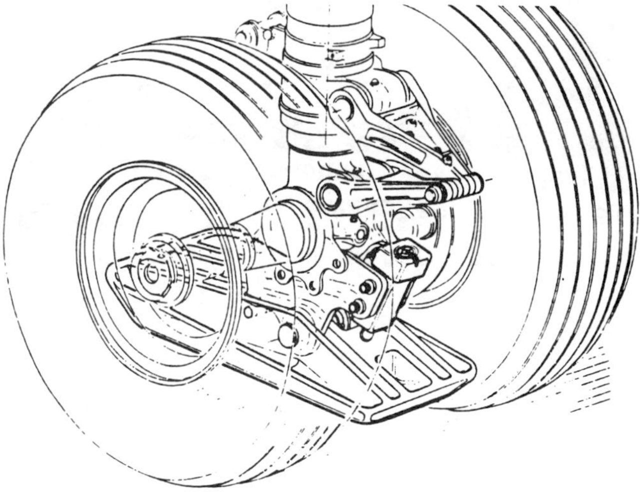

Figure 13. Proposed overload protection skid.

CONDITION	STRUT LOAD	SKID AREA	SKID LOAD	BEARING PRESSURE
DEROTATION (2 tire failure)	890 kN	1135 cm ^{2}	890 kN	7.83 MPa
STATIC (2 tire failure)	490 kN	1135 cm ^{2}	490 kN	4.31 MPa
STATIC (1 tire failure)	490 kN	1135 cm ^{2}	200 kN	1.76 MPa

ALDF TESTS

VERTICAL LOAD	SKID AREA	BEARING PRESSURE
294 kN	374 cm^2	7.85 MPa
156 kN	374 cm^2	4.16 MPa
67 kN	374 cm^2	1.79 MPa

Table1. Loads and conditions for skid tests.

ALDF testing took place with a skid approximately 1/3 the area of the proposed skid but at full-scale speeds and bearing pressures, therefore, the skid tests were only partially scaled. Five different materials were used for skid specimens including Inconnel 718, 4340 steel, Stellite, tungsten with either medium or coarse carbide chips embedded in it, and 1020 steel. These materials covered both the range of proposed skid materials and materials used on the main gear strut itself so that strut skidding after complete wheel failures without the proposed skid might be evaluated.

Table II shows the results of the skid tests with speed, load, bearing pressure, slide distance, friction coefficient, (μd), wear, and wear rate. The pivot on the yoke of the skid apparatus was behind the geometric center of the skid specimen, therefore, the skid would always wear more rapidly at the rear than at the front. Figure 14 shows the specimens after testing and the rear versus front wear on each specimen is evident.

Figure 14. Skid specimens after testing.

The Inconnel 718 specimen was tested at about 157 kts. Because of runway roughness, a large amount of "chatter" occurred which made quasi-steady loading of the specimen difficult, therefore, vertical loads and bearing pressures are given as a range of values where appropriate. Bearing pressures for the Inconnel 718 skid ranged from 3.3 to 6.9 MPa and the friction coefficient was about 0.14. The friction coefficient was apparently insensitive to bearing pressure. The wear rate of this specimen, on average, was about 7.4 cm/km. Low-speed towing tests were conducted and the friction coefficient rose to about 0.7. This extreme friction was probably caused by the destruction of the surface of the simulated KSC runway.

Tests with either the 4340 steel or the 1020 steel specimens at high speed showed friction coefficients ranging from 0.08 to 0.16. The average wear rate for these steels was about 8.8 cm/km. Low-speed tests showed a friction coefficient of 0.57. The low-speed tests turned out to be very severe on the test equipment and were discontinued at this point. The 4340 skid got extremely hot during testing and a trail of sparks and molten material could be seen behind the specimen as shown in figure 15. The 1020 steel specimen also exhibited this behavior but no other material was heated enough to trail molten material.

Tests on the stellite specimen showed a friction coefficient of 0.09 and a wear rate of about 4.9 cm/km. Specimens of tungsten with either medium or coarse carbide chips showed, on average, the lowest friction and wear characteristics, with average friction ranging from 0.05 to 0.10. The average wear for these

MATERIAL	SPEED, kts	LOAD, kN	BEARING PRESSURE MPa	DISTANCE, m	μd	WEAR,cm Front	Rear	AVERAGE WEAR cm/km
Inconnel 718	156	134-156	3.5-4.1	88	.15	.58	1.27	10.5
	158	125-263	3.3-6.9	94	.13	.43	.38	4.3
	0-5	45	1.2	NA	.70	NA	NA	NA
4340 Steel	157	165	4.3	62	.10	.28	.84	9.0
	154	89-223	2.3-5.9	65	.08-.16	.13	.86	7.6
	0-10	196	5.2	6	.57	NA	NA	NA
	100	134-156	3.5-4.1	101	.17	.56	1.32	9.3
Stellite	169	134-156	3.5-4.1	64	.09	.25	.38	4.9
Tungsten/ Medium Carbide	165	156	4.1	104	.05	.08	.10	0.9
	95	178	4.7	134	.08	.10	.25	1.3
	171	134-289	3.5-7.6	134	.05-.08	.15	.36	1.9
Tungsten/ Coarse Carbide	171	134	3.5	127	.07	.08	.10	0.7
	167	267	7.0	137	.10	.38	.51	3.2
1020 Steel	173	178	4.7	122	.07	.58	1.09	6.8
	158	134	3.5	119	.13	.89	1.78	11.2

Table 2. Results of skid tests.

specimens was about 1.6 cm/km. Tungsten with carbide chips, therefore, would likely be an appropriate skid material since it has low friction, and its low wear characteristics translate into the lowest weight for the skid.

One other test, intended to be a roll-on-rim test, involved rolling a damage wheel flange on its substantial center section to evaluate further rolling capability. The wheel actually slid during the test, though, and ground a flat spot on the wheel center section as shown in figure 16. Vertical load was 312 kN and the friction coefficient was about 0.2.

Figure 15. 4340 steel specimen during skid test.

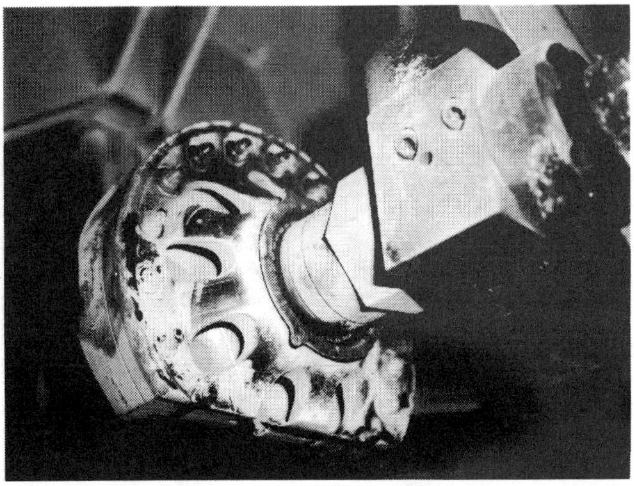

Figure 16. Results of skidding on wheel center section.

Simulator Model

Based on the tests conducted at the ALDF, a model for the strut drag coefficient was developed for landing on a concrete runway with one tire flat prior to touchdown. The model, shown in figure 17, is a function of time but can be a function of velocity or displacement as well. The model shows the drag of spinup followed by an increase of rolling drag on the inflated/flat tire combination due to derotation. Near the time of nose gear contact, the strut loads increase and the remaining good tire fails due to vertical load. At this point, both wheels are probably rolling on tire beads until maximum strut vertical load is reached and both wheels completely fail, since the tire beads can only protect each wheel at loads below 312 kN. After this, skidding on the wheels, brake stack, and axle probably will occur and based on ALDF skid tests the friction is likely to be about 0.2 until the speed becomes low and the friction may approach 0.7 as the vehicle stops. This model is only generic, and time, speed, or distance variation will affect the shape of the curve; however, the model does provide a tool for producing failure conditions for astronaut training in rollout simulations.

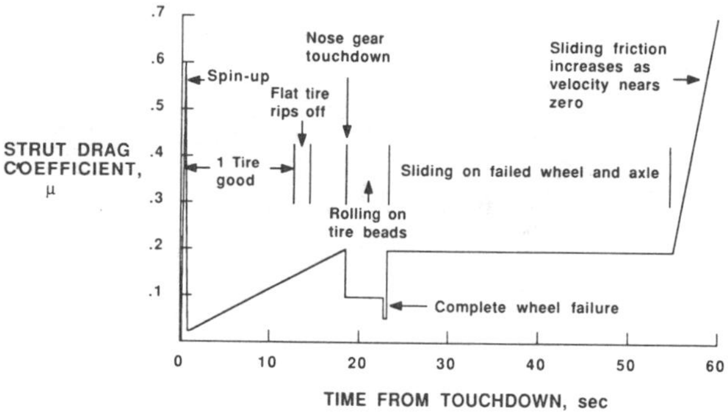

Figure 17. Model for strut drag with one tire flat prior to touchdown on concrete.

CONCLUDING REMARKS

An experimental investigation was conducted at the NASA Langley Research Center Aircraft Landing Dynamics Facility to define the post-tire failure behavior of the Space Shuttle Orbiter main wheel assembly. Tests to define friction and wear characteristics of various candidate materials for a proposed overload protection skid were also conducted.

Tests showed that landing on a flat tire causes a rolling resistance friction coefficient of about 0.2. After about 1,000 feet of rollout the tire will be cut away from the tire beads by the wheel flanges. Rolling on tire beads on an intact wheel produces a rolling resistance of 0.1.

Slow speed tests on a standard wheel showed that, without a tire, at tire rated load the wheel will fracture in about 200 m. At high speed on a rough concrete runway a standard wheel will disintegrate in about 10 m at loads over about 135 kN. Modifying the wheel flanges with increased thickness extended wheel life under the same conditions to only about 36 m. Skidding on the wheel center section at high speed produces a friction coefficient of about 0.2.

High-speed, high-pressure skid tests on metal specimens including Inconnel 718, 4340 steel, Stellite, tungsten with carbide chips, and 1020 steel showed friction coefficients ranging from 0.05 to 0.17. Wear rates of the 374 cm^2 skid specimens were lowest with harder materials with lower friction, such as tungsten with carbide chips. The softer skid materials such as Inconnel 718, 4340 steel, and 1020 steel exhibited the highest wear rates of all with values of 7-9 cm/km.

The results of the flat-tire, roll-on-rim, and skid tests were combined into a strut drag versus time model that indicates likely sequential failures for inclusion into rollout simulators to provide shuttle astronauts a more realistic model of orbiter behavior under failure conditions.

References

1. Davis, Pamela A.; Stubbs, Sandy M.; and Tanner, John A.: Langley's Aircraft Landing Dynamics Facility. NASA RP 1198, October 1987

Tire Friction, Wear, and Mechanical Properties

Cornering and Wear Behavior of the Space Shuttle Orbiter Main Gear Tire

Robert H. Daugherty and Sandy M. Stubbs
NASA Langley Research Center

* Paper 871867 presented at the Aerospace Technology Conference and Exposition, Long Beach, California, October, 1987.

ABSTRACT

One of the factors needed to describe the handling characteristics of the Space Shuttle Orbiter during the landing rollout is the response of the vehicle's tires to variations in load and yaw angle. An experimental investigation of the cornering characteristics of the Orbiter main gear tires was conducted at the NASA Langley Research Center Aircraft Landing Dynamics Facility. This investigation compliments earlier work done to define the Orbiter nose tire cornering characteristics. In the investigation, the effects of load and yaw angle were evaluated by measuring parameters such as side load and drag load, and obtaining measurements of aligning torque. Because the tire must operate on an extremely rough runway at the Shuttle Landing Facility at Kennedy Space Center (KSC), tests were also conducted to describe the wear behavior of the tire under various conditions on a simulated KSC runway surface. Mathematical models for both the cornering and the wear behavior are discussed.

THE SPACE SHUTTLE ORBITER is the first and only spacecraft to date designed to land on conventional landing gear and runways. The Orbiter is designed to land in a variety of environments and conditions, creating a need to fully understand all the factors that affect the way the vehicle handles on the runway. One of these factors is the response of the Orbiter tires to various crosswind forces and pilot steering commands. Reference 1 documents the results of tests conducted to define the cornering behavior of the Orbiter nose tires under realistic operating conditions. Although road wheel dynamometer studies have been conducted on Orbiter main tires, realistic cornering behavior is typically not obtained due to rubber contamination of the dynamometer surface.

One of the factors associated with tire cornering performance can be tire wear. Typically tire wear on aircraft tires is a concern only after dozens or sometimes hundreds of landings. With the Space Shuttle Orbiter, however, tire wear has become a concern for a single landing. The Shuttle Landing Facility runway at Kennedy Space Center has an extremely rough texture and groove pattern, and tire wear due to spin-up and rollout at this location has been excessive. Flight tire wear observed for Mission 51-D highlighted the need to understand the wear mechanisms associated with the tires.

The purpose of this paper is to present the results of tests conducted at the NASA Langley Research Center to determine the cornering characteristics and wear behavior of the Orbiter main gear tire. Cornering characteristics were evaluated by measuring parameters such as side load, drag load, and aligning torque in response to variations in vertical load and yaw angle. Wear behavior of the tire was evaluated by measuring tire wear in response to variations in vertical load, yaw angle, and tilt angle. Mathematical models used for describing cornering and two types of tire wear are presented.

APPARATUS

This investigation was conducted at the NASA Langley Research Center Aircraft Landing Dynamics Facility. The facility consists of a set of rails 850 m long on which a 49,000 kg carriage travels. The facility is shown in figure 1. The carriage is propelled at speeds up to 220 kts. using a high pressure water jet and is arrested using a set of water turbines connected by nylon tapes. A more detailed description of the facility can be found in reference 2.

The tires used in this study were 44.5 x 16.0-21 bias ply aircraft tires with a 34 ply rating. A photograph of a new tire is shown in figure 2. The tires have a 5-groove tread pattern made of natural rubber with the

Figure 1.- The Langley Research Center Aircraft Landing Dynamics Facility.

Figure 2.- Photograph of main tire mounted on wheel.

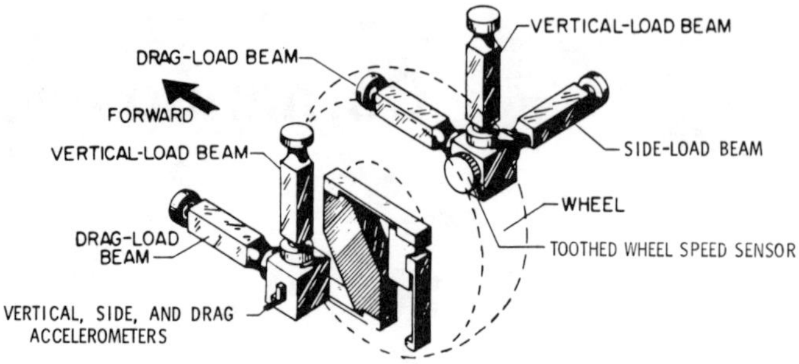

Figure 3.- Force measurement dynamometer schematic.

Figure 4.- Photograph of rough Kennedy Space Center runway.

grooves 2.54 mm deep. The tires have 16 actual carcass plies and their rated load and pressure are 271 kN and 2.17 MPa respectively. The tires were mounted on Orbiter main wheels with mass added to simulated brake rotor inertia and installed in the force measurement dynamometer sketched in figure 3. Data generated at the dynamometer were digitally telemetered to a receiving station where they were converted into engineering units by a desktop computer and stored. A 550 m long simulated Kennedy Space Center (KSC) runway was installed at the facility to conduct these tests. The KSC runway shown in figure 4 has an extremely rough longitudinally-brushed texture combined with transverse grooves 6.4 mm wide by 6.4 mm deep with 29 mm spacing. The runway was designed to retain good friction characteristics when wet which it does exceptionally well, but associated with that comes the increased tire wear.

To obtain cornering data, a test normally consisted of rotating the dynamometer and wheel assembly to the desired yaw angle, accelerating the carriage to the desired test speed, lowering the tire and applying the preselected vertical load, and recording the output from the dynamometer instrumentation. Yaw angles tested were 0, 1, 2, 4, 7, and 10 degrees. The vertical loads tested ranged from 67 kN to 543 kN. For loads above 312 kN, approximately 22,700 kg of weight was added to the carriage and due to carriage structural considerations, only tow tests (5 kt.) were conducted in this configuration. To obtain wear data, a yaw or tilt angle was preselected along with a vertical load and a full rollout at the desired speed was conducted on the entire 550 m long runway surface.

RESULTS AND DISCUSSION

In this investigation, the effects of vertical load and yaw angle on the cornering characteristics of the Orbiter main tire were measured. In addition, tests were conducted to determine the wear mechanisms responsible for the excessive flight tire wear observed for landings at the KSC runway.

CORNERING BEHAVIOR

For presenting cornering data, all parameters herein are presented as a function of yaw angle and load ratio, R, in the form of carpet plots. Load ratio is defined as the tire vertical load divided by the rated load of 271 kN. This normalization merely facilitates discussion of the effects of vertical load. Each carpet plot is the result of a least-squares bi-cubic curve fit to the test data, all of which will be published in a forthcoming NASA technical paper. The carpet plots show lines of constant load and yaw angle to aid the reader. Figure 5 shows the side force, measured perpendicular to the wheel plane, as a function of yaw angle and load ratio. The figure shows this tire to behave like most tires, in that as yaw angle is increased for a constant vertical load the side force increases, except for very low loads at high yaw angles.

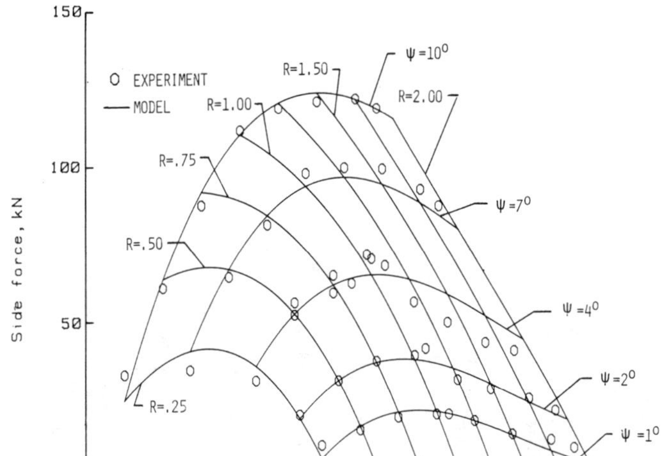

Figure 5.- Variation of side force with yaw angle and load ratio.

It is helpful to present side force data as side force coefficient, which is defined as the side force divided by the vertical load or normal force on the tire. Figure 6 presents this data which literally represents the efficiency of the tire at producing side force. The figure shows behavior very similar to that seen in reference 1, where the side force coefficient, μ_s, basically increases with increases in yaw angle, but decreases with increases in vertical load.

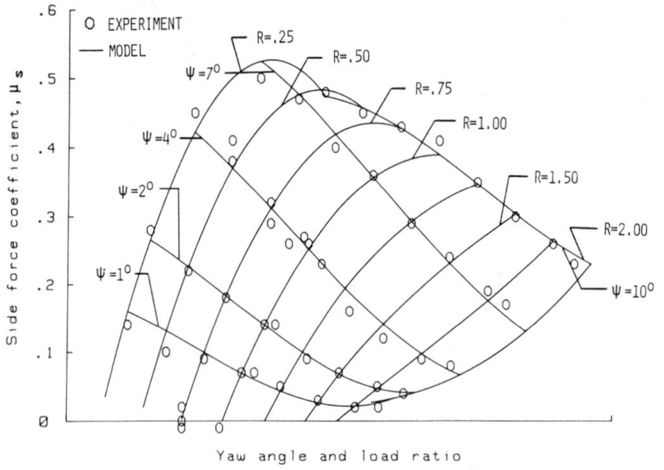

Figure 6.- Variation of side force coefficient with yaw angle and load ratio.

Figure 7 presents the drag force as a function of yaw angle and load ratio. The plot shows relatively flat behavior of the drag force for changes in yaw angle, and that increases in vertical load are mostly responsible for increases in drag force. Drag force coefficient data is not presented, though in general, a drag force coefficient of 0.02 is observed indicating this is the rolling resistance of the tire under normal conditions.

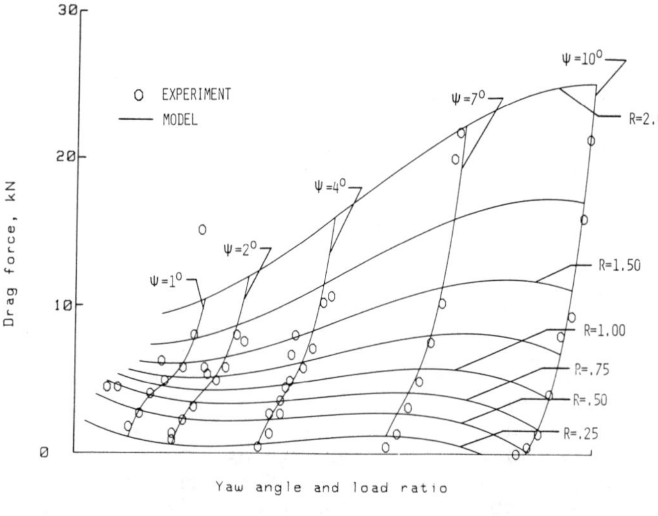

Figure 7.- Variation of drag force with yaw angle and load ratio.

Aligning torque data is presented in figure 8. Aligning torque is generated by the yawed-rolling tire to return it to the zero-yaw condition. The plot shows that for yaw angles up to about six degrees, increases in yaw angle increase the aligning torque at a given load. Further increases in yaw reduce the aligning torque indicating the tire is becoming less stable in yaw, though even at 10 degrees positive aligning torque is still observed.

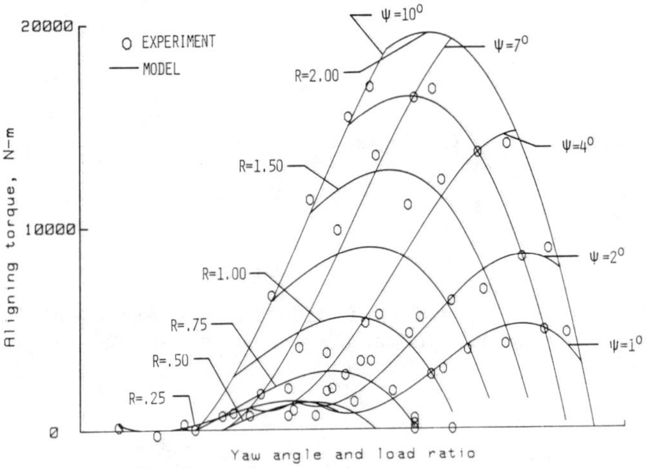

Figure 8.- Variation of aligning torque with yaw angle and load ratio.

WEAR BEHAVIOR

As stated earlier, the main gear tire has shown excessive wear during landings on the KSC runway. Parametric tests were conducted to determine the effects of vertical load, yaw angle, and tilt angle on main gear tire wear. Some flight tires have exhibited wear on the shoulder, the area of the tire normally just out of ground contact. The shoulder wear has been observed to go as deep as 3 to 4 cord layers into the tire. Based on information presented in reference 3, it was determined that this kind of wear is produced by rolling the tire in a tilted condition, i.e., the wheel plane is not perpendicular to the local surface. This causes one side of the tire to have a smaller rolling radius than the other and ultimately causes significant differential slip on that side. This action, particularly on the rough KSC surface, can cause severe tire wear. Tilt angle is contributed to by a number of factors including axle bending due both to vertical and side load, runway crown, and vehicle roll attitude. Tilt angles on Orbiter tires can be as high as 3 to 4 degrees under certain conditions. Parametric tests revealed that shoulder wear can be determined by examining a parameter referred to as tilt area, which is the area under the tilt angle versus distance curve for the tire. Figure 9 shows the results of tests conducted at various tilt angles for different distances along with a prediction line based on a least-squares curve fitting technique. Shoulder wear is expressed in absolute millimeters of an inch along with a secondary scale showing cord layers. Part of the data this curve is based on is seen in figures 10 and 11. Figure 10 shows the mission 51-D flight tires after the landing rollout along with computer predictions of the tilt angle distance history for each tire based on all available flight data. Figure 11 shows one of the tires tested at the ALDF for shoulder wear along with a plot showing its observed shoulder wear indicating that tilt area seems to be a valid parameter for determining shoulder wear.

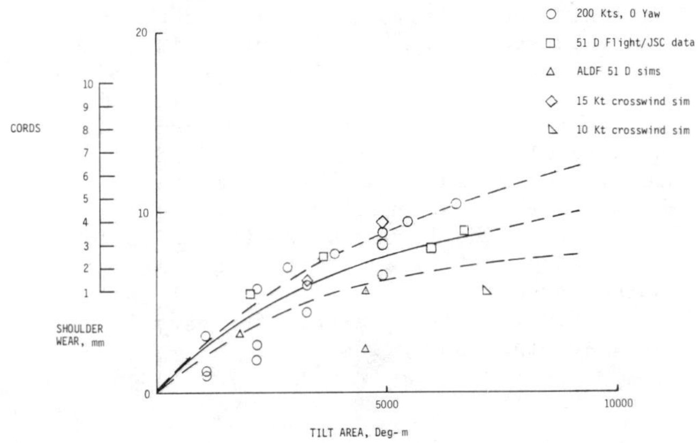

Figure 9.- Shoulder wear progression as a function of tilt area.

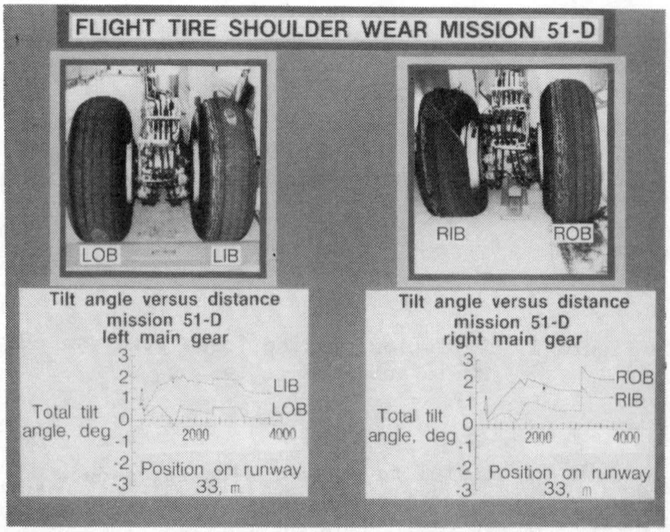

Figure 10.- Mission 51-D flight tire wear and predicted tilt histories.

Figure 11.- Results of a parametric test at the ALDF compared with flight tire wear.

The other and more important half of tire wear is that caused by yawed-rolling on top of a severe touchdown/spin-up spot. Flight experience and tests at the ALDF (fig. 12) have shown that for normal landings on the KSC runway, exposure and compromise of the tire's first cord layer is essentially unavoidable. Subsequent rollout in crosswind conditions or significant pilot steering inputs can cause this spin-up spot to increase in size and depth. The mechanism which causes this wear was determined to be the production of side force due to yaw angle and the inability of the rubber to adhere to the carcass because the texture of the surface literally cuts the rubber off. Figure 13 shows an extreme case of this type of wear condition. It was determined that the wear of the tire above and beyond the initial spin-up patch could be

Figure 12.- Result of spin-up on KSC rough surface.

Figure 13.- Example of severe wear at the ALDF.

expressed as a function of side energy, which is defined as the amount of work produced laterally by the tire during cornering. The side energy is defined as:

$$\text{Side energy} = \int_0^S F \sin \psi \, dx$$

where S = total rollout distance

F = side force

ψ = yaw angle

and both F and ψ are functions of x, the instantaneous rollout position

Some test results are plotted in fig. 14, again showing tire wear both in cord layers and millimeters. The data were obtained by running the tire at various speeds and loads under a variety of yawed conditions. Also plotted is a least-squares curve fit to the data. Note that the data all emanate from the 1 1/2 cord wear condition which, again, is what can be expected during spin-up. The prediction line is useful for determining tire wear during computer-simulated landings with various crosswinds and piloting techniques. All that is required is to keep a running sum of the instantaneous side energy produced by the tire for each computing time step. The results of simulations can provide pilots feedback which can help them modify their steering techniques to reduce tire wear to a minimum.

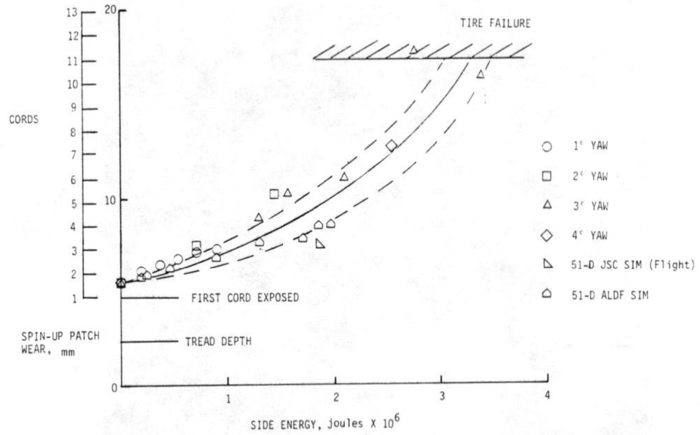

Figure 14.- Variation of initial spin-up patch wear progression with side energy.

CONCLUDING REMARKS

An experimental investigation was conducted at the NASA Langley Research Center Aircraft Landing Dynamics Facility to determine the effects of vertical load and yaw angle on the cornering characteristics of the Space Shuttle Orbiter main gear tire. Tests were also conducted to define the wear behavior of the main gear tire on the Kennedy Space Center Shuttle Landing Facility runway.

It was shown that the Orbiter main tire behaved in a manner similar to other aircraft tires relative to cornering. In general, it was found that increases in yaw angle increased the side force coefficient, a term which describes cornering efficiency, while increases in vertical load cause a reduction in side force coefficient. Like most other tires, this tire was found to be stable in yaw. Drag force measurements indicated a rolling resistance of this tire of approximately 0.02 under normal conditions which is well within the range of coefficients seen in other aircraft tires.

Excessive shoulder wear on the Orbiter main tire was found to be caused by rolling with the tire in a tilted attitude. The differential slip in the tire footprint causes one shoulder to wear more than normal. It was determined that a parameter defined as tilt area could be used to describe the shoulder wear state of the tire. Tilt area is the area under the tilt angle distance history for the tire.

Yawed-rolling was found to affect the wear progression of the spin-up patch on the tire produced at touchdown. It was found that this wear progression can be determined by examining the work produced in the lateral direction by the tire while cornering. Both side energy and tilt area can be computed in landing rollout simulations, thus providing pilots with wear information feedback.

REFERENCES

1. Vogler, William A.; Tanner, John A.: Cornering Characteristics of the Nose-Gear Tire of the Space Shuttle Orbiter. NASA TP 1917, 1981.

2. Davis, Pamela A.; Stubbs, Sandy M.; Tanner, John A.: Aircraft Landing Dynamics Facility, A Unique Facility with New Capabilities. SAE 851938. Presented at the 1985 Aerospace Technology Conference and Exposition, Long Beach, California.

3. Daugherty, Robert H.; Stubbs, Sandy M.: A Study of the Cornering Forces Generated by Aircraft Tires on a Tiltied, Free-Swiveling Nose Gear. NASA TP 2481, October 1985.

Static Mechanical Properties of 30 x 11.5-14.5, Type VII, Aircraft Tires of Bias-Ply and Radial-Belted Design

Pamela A. Davis and Mercedes C. Lopez
NASA Langley Research Center

** Paper 871868 presented at the Aerospace Technology Conference and Exposition, Long Beach, California, October, 1987.

ABSTRACT

An investigation was conducted to determine the static mechanical characteristics of 30 x 11.5 - 14.5 bias-ply and radial aircraft tires. The tires were subjected to vertical and lateral loads and mass moment of inertia tests were conducted. Static load deflection curves, spring rates, hysteresis losses, and inertia data are presented along with a discussion of the advantages and disadvantages of one tire over the other.

RADIAL TIRES have been used in the automotive industry for more than 20 years, but until about 1980 the aircraft landing gear industry felt that the mechanical properties of these radial tires would make them unacceptable for aircraft use. For example, landing gear designers were concerned about excessive lateral forces during crosswind landings (1)*, compatibility of the tire fore-and-aft stiffness characteristics and the performance of aircraft antiskid braking systems (2), and the influence of tire vertical stiffness characteristics on aircraft landing dynamics response.

In recent years, the continuing research efforts of several tire manufacturers have resulted in radial aircraft tire designs which appear to overcome many of the previously mentioned problems and have been successfully used on several different aircraft (1). Thus, it is possible that the benefits associated with radial automotive tires, i.e. longer tread life, cooler operating temperatures, and improved friction characteristics may also be realized for radial aircraft tires. The emergence of radial aircraft tire technology has also created a need to measure and document the mechanical properties of these new tires so that landing gear designers can predict the dynamic response characteristics of their landing gear systems. To aid in this endeavor, the NASA Langley Research Center in cooperation with the Federal Aviation Administration and the U.S. landing gear industry has initiated a research effort, called the Radial Tire Program, to study the mechanical properties and friction characteristics of radial and bias-ply tires under a variety of operating conditions.

The purpose of this paper is to present the results from an investigation of the mechanical characteristics of 30 x 11.5 - 14.5 bias-ply and radial-belted aircraft tires subjected to static loading. These characteristics were obtained over a range of inflation pressures from 1.7 to 2.1 MPa (245 to 310 psi), vertical loads up to 111.2 kN (25,000 lbs), and lateral loads up to 17.8 kN (4,000 lbs). The measured parameters include tire stiffness characteristics, damping, and inertia properties. Comparisons of the mechanical properties of the two tire designs are also presented. The tires used in this investigation were supplied by the U. S. Air Force.

CONSTRUCTION OF BIAS-PLY AND RADIAL-PLY TIRES

Figure 1 shows the difference in construction for the bias-ply and radial-belted tire. The bias-ply tire is a laminated construction of several rubber-textile plies that alternate at various angles from 60° near the tire bead to 30° near the crown area of the tire. The rigid shell and heavy weight of the tire are a result of the ply assembly and multiple ply casing. Because of the large number of plies, multiple bead wires on each side of the tire are required to hold these plies together. To provide tread reinforcement, additional plies or belts may be laid at a specified angle between the tire carcass and the tread (1).

*Numbers in parentheses designate references at end of paper.

The radial tire uses fewer plies of higher denier textile cords than the bias-ply tire and this results in a weight and volume reduction. The radial tire has casing plies with cords approximately oriented in the plane of the cross section and its size and load carrying requirements determine the number of plies in the tire. Generally, only one steel bead wire is needed on each side of the tire as opposed to two or more in the bias-ply tire. For the radial tires used in this investigation, a textile cord belt surrounds the casing of the tire circumferentially and a steel belt surrounds the cord belt and acts as a protector ply (1).

APPARATUS AND TEST PROCEDURE

Static loading tests and inertia measurements were conducted on bias-ply and radial-belted 30 x 11.5 - 14.5 tires with a ply rating of 24 and 26 respectively.

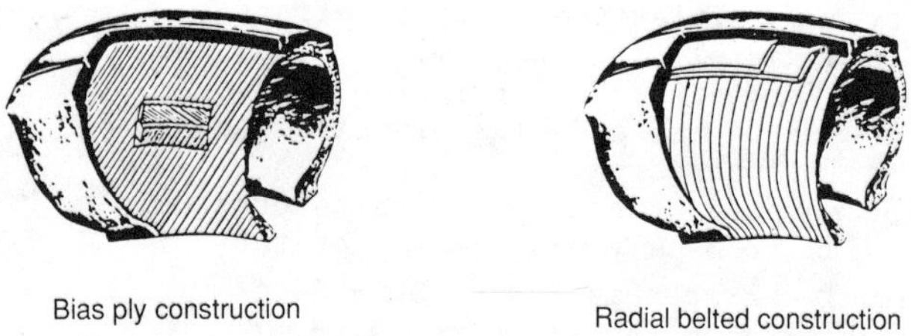

Figure 1 - Bias-ply and radial-belted tire construction.

Characteristics of the two types of tires are given in Table 1 and the tires are shown in Figure 2. The tires and wheel assemblies were supplied by the U. S. Air Force, although the bias and radial tires were from two different manufacturers. Prior to static loading tests at the NASA Langley Research Center, each tire was preconditioned at the Wright Aeronautical Labs, Vehicle Equipment Division with 3.2 km (2 miles) of taxi tests at rated load and inflation pressure at 26.1 knots (30 mph).

The test set up to measure the static mechanical properties of both the bias and radial tires is shown in Figure 3. The tire, wheel assembly, and axle are mounted on the dynamometer, which is instrumented with five strain gage beams to measure vertical load (two), lateral load (one), and drag load (two). The tire is loaded via a hydraulic cylinder onto a frictionless table. The table is instrumented with three vertical load cells mounted underneath as a standard source to measure vertical load and a horizontal load cell to measure lateral or drag load as shown in Figure 4. The tire vertical and lateral displacements are measured by displacement transducers mounted parallel to each hydraulic cylinder. The data from the various instruments were read into a computer at specific time intervals where the data were converted into engineering units and saved for further analysis.

STATIC VERTICAL-LOADING TESTS-Static vertical-loading tests were conducted to measure tire vertical load spring rate, and vertical damping characteristics. The test procedure followed was the same as that reported in Reference 3. For a static vertical loading test, the load was applied hydraulically until the maximum rated load of 111.2 kN (25,000 lbs) was reached. The load was then gradually reduced to zero. Vertical

Table 1 - Characteristics of bias-ply and radial-belted tires.

	BIAS	RADIAL
Size	30 x 11.5 - 14.5	30 x 11.5 - 14.5
Ply rating	24	26
Weight, uninflated, kg (lbs)	27.7 (61)	11.6 (25.5)
Rated vertical load, kN (lbs)	111.2 (25,000)	111.2 (25,000)
Rated inflation pressure, at 35% load deflection, MPa (psi)	1.7 (245)	2.1 (310)
at 51% load deflection (radial only), MPa (psi)	—	1.7 (245)
Outside diameter of unloaded tire, m (inches)	0.76 (30)	0.76 (30)
Max. width of unloaded tire, m (inches)	0.2 (8)	0.2 (8)
Tread description (number of circumferential grooves)	3	4

load and deflection were continously monitored during the loading and unloading cycle. This loading resulted in a load-deflection curve or hysteresis loop which gave an indication of the tire vertical-loading behavior and defines the vertical spring rate and the hysteresis loss (3). These tests were conducted at four peripheral positions around each tire.

STATIC LATERAL-LOADING TESTS-Combined static vertical and lateral-loading tests were conducted to measure tire lateral spring rate and hysteresis loss, again, according to Reference 3. For the combined loading test, the vertical load was applied hydraulically and maintained at the rated load of 111.2 kN (25,000 lbs). The lateral load was then applied hydraulically up to a load of 16% of the vertical load and then reduced to zero. The load was applied in the opposite direction up to 16% of the vertical load and again reduced to zero. A load deflection curve was obtained defining the tire's lateral-loading behavior, lateral spring rate, and hysteresis loss. Lateral tire deflection was defined as the displacement of the frictionless table perpendicular to the wheel plane and was measured by a displacement transducer. The combined static vertical-lateral loading tests were conducted at four peripheral positions around the tire.

MASS MOMENT OF INERTIA TESTS-A torsional pendulum, shown in Figure 5, was used to measure the mass moment of inertia of the bias and radial tires (4). The torsional pendulum consisted of two-61 cm (24 inch) diameter plates bolted together and suspended in parallel by three equally tensioned 5.2 m (17 ft) wire cables, attached to an overhead

Figure 2 - Bias-ply and radial-belted tires, uninflated.

support. An index in 5° increments up to $\pm$ 20° was scribbed on the top plate. A pointer was aligned with the index at 0° for a reference.

The moment of inertia of the two plates was initially determined by rotating the plate assembly through a predetermined angle and releasing it. Each plate assembly was oscillated three times at approximately 10°, 7.5^o, 5°, and 2.5^o for 20 oscillations each and the time was recorded using a standard stopwatch. The moment of inertia for the plates was then computed. The inflated tire, wheel assembly, and rotors were then suspended between the plates and similar tests were conducted. The weight of the plates and tire-wheel assembly were recorded.

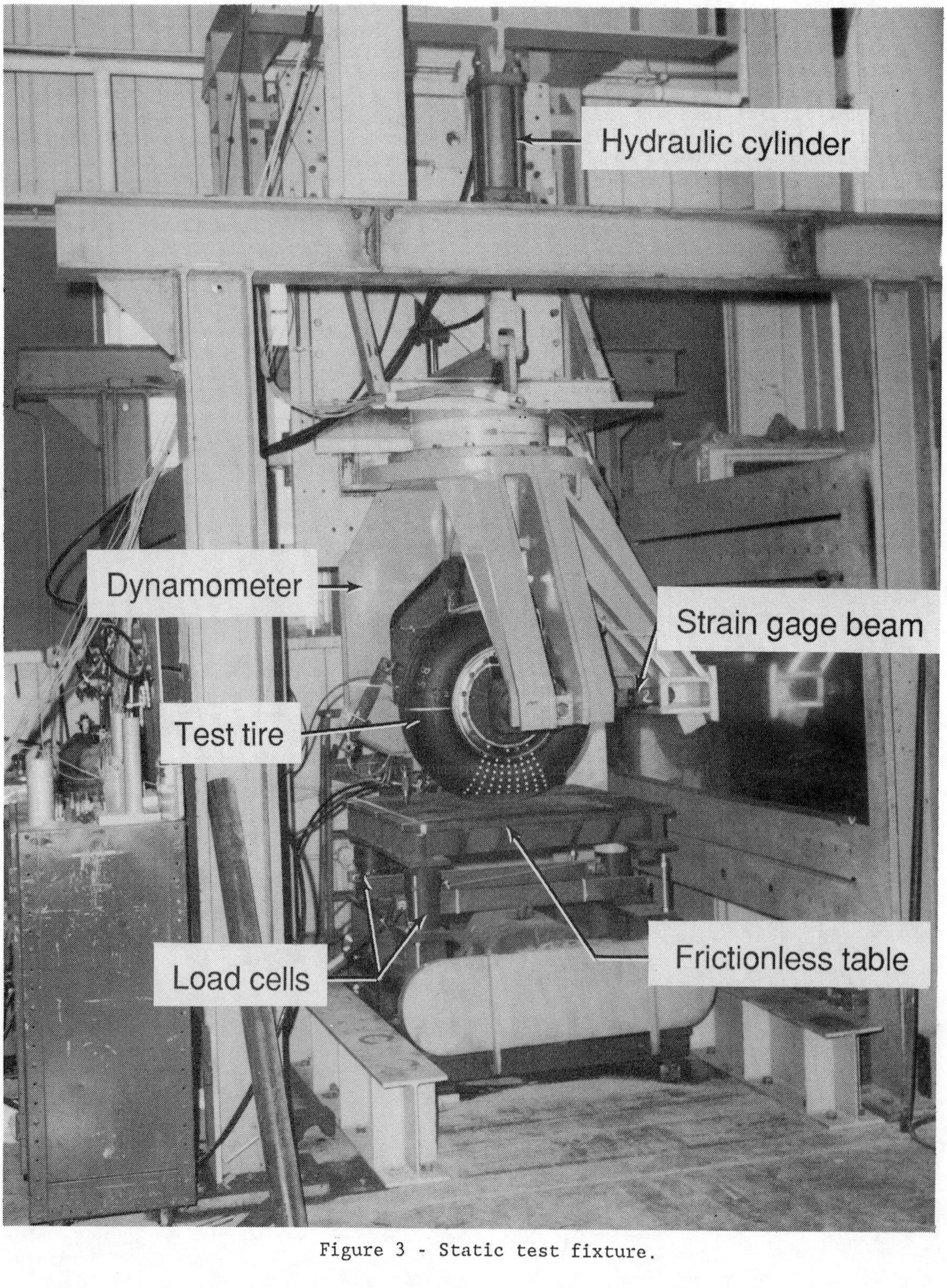

Figure 3 - Static test fixture.

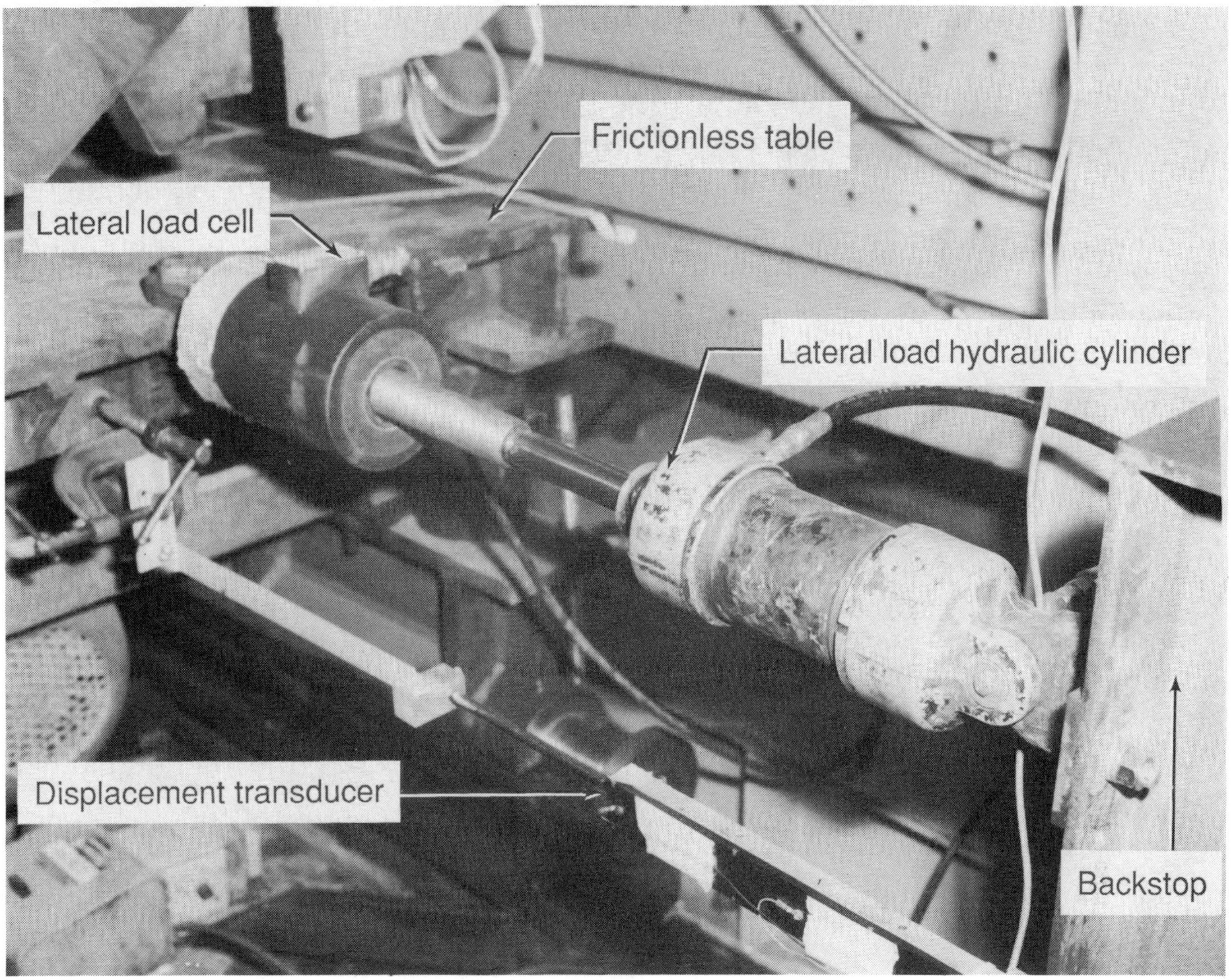

Figure 4 - Lateral load measurement set up.

ERROR ANALYSIS-The percentage of error in the data has been computed based on calibration and instrumentation accuracy. Reading errors from all nine channels of data were computed taking the voltage readings plus or minus the manufacturer's instrumentation accuracy factor as theoretical values and the voltage readings themselves as experimental values. The experimental values were subtracted from the theoretical values and then divided by the theoretical values to obtain error percentages. Maximum and minimum voltage responses from each channel were considered and analyzed. Large error percentages were expected for low voltage response due to the voltage ranges available for each item. Calibration errors were obtained from linear curve fittings that were obtained by calibrating the load cells in our test apparatus. Calibration and instrumentation percentage errors were then added to the reading errors to obtain the total error percentage.

Standard vertical load data from the load cells was shown to be 0.61% in error under both minimum and maximum voltage conditions. Vertical load measured from strain gages on the dynamometer indicated a maximum of 34% error under minimum voltage condition and 2.89% error under maximum voltage. Lateral standard load data was shown to be 0.66% and 0.61% in error for minimum and maximum voltage respectively. The dynamometer lateral load data had a 27.5% error under minimum voltage and 2.45% error under maximum voltage. Vertical and lateral displacement readings both indicated a 0.6% error under both voltage conditions.

Due to the limited amount of samples and test runs (eight for each tire), a complete statistical analysis could not be done. However, random errors such as round-off error due to data reduction techniques were considered in the data analysis.

Figure 5 - Torsional pendulum test fixture.

RESULTS AND DISCUSSION

STATIC VERTICAL LOADING-Static vertical loading tests were conducted on two 30 x 11.5 - 14.5 bias-ply and radial tires. The tires were inflated to 1.7 MPa and 2.1 MPa (245 psi and 310 psi) for the bias and radial respectively. These inflation pressures represented 35% tire deflection at the rated load of 111.2 kN (25,000 lbs). In addition, **the radial tire was tested at an inflation pressure of 1.7 MPa (245 psi)** which was equivalent to a 51% tire deflection. Therefore, the bias and radial tires were analyzed based on the same tire deflection and the same inflation pressure. Results of the vertical load deflection curves, vertical spring rate, and hysteresis losses of the tires are presented.

VERTICAL LOAD-DEFLECTION RELATIONSHIPS-The results of the static vertical loading tests are presented in this section. Four load-deflection curves were generated for each tire. A typical load-deflection curve for the bias-ply tire is shown in Figure 6. The curve for this tire shows an increasing slope with increasing load up to about 26.7 kN (6,000 lbs) thus indicating a hardening spring. The curve then becomes linear and remains constant up to the rated load of 111.2 kN (25,000 lbs). During initial load relief, the slope is much steeper than during the initial load application which suggests an even stiffer spring. A further reduction in load shows a softening spring rate until the load is returned to zero. The loading and unloading process results in a vertical load deflection curve with the area within the curve defined as the hysteresis loss.

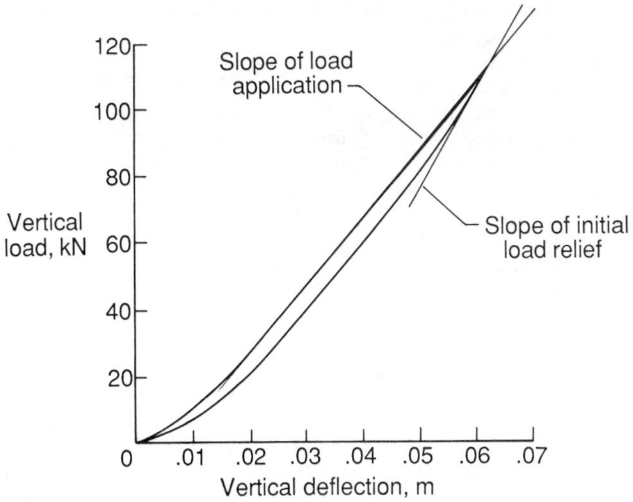

Figure 6 - Vertical load-deflection curve of the bias-ply tire at 245 psi.

Similar load-deflection curves were obtained for the radial tire at 2.1 MPa and 1.7 MPa (310 psi and 245 psi) as shown in Figures 7 and 8. As with the bias-ply tire, the radial tire has an increasing slope with increasing load but it continues to increase up to the rated load, indicating a continuously stiffening spring. At initial load relief, the slope is again much steeper than during load application. As the load is relieved, the spring softens gradually until the load is returned to zero.

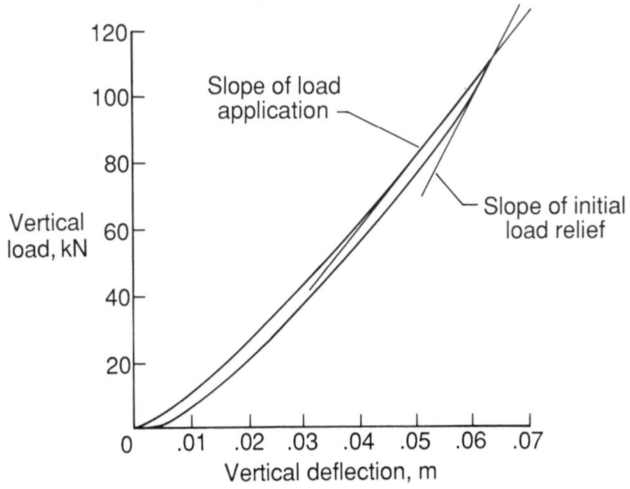

Figure 7 - Vertical load-deflection curve of the radial tire at 310 psi.

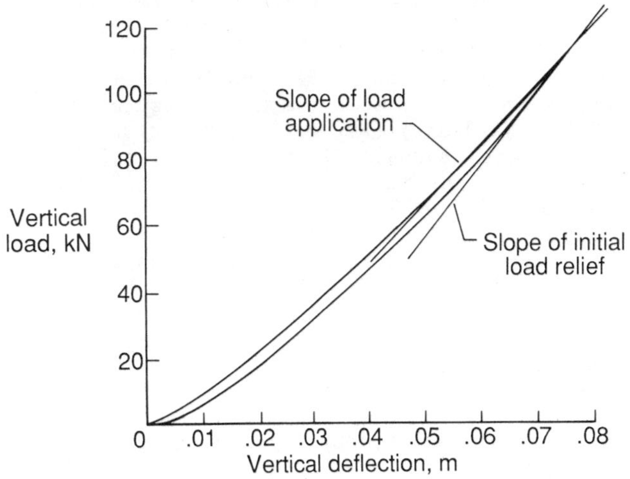

Figure 8 - Vertical load-deflection curve of the radial tire at 245 psi.

<u>Spring Rate</u> - For the three tires tested, the vertical spring rate was obtained by measuring the instantaneous slope at various points along the load application and load relief curves. The lower bound of tire vertical stiffness is represented by the load application curve and the upperbound of tire vertical stiffness is represented by initial load relief (3). The vertical spring rate for the bias-ply tire is plotted as a function of vertical deflection in Figure 9. The spring rate for the initial load application increases from about 1,312 kN/m (7,500 lbs/inch) to a maximum value of 2,012 kN/m (11,500 lbs/inch) and then remains constant for the remainder of the load application. A maximum spring constant of 2,712 kN/m (15,500 lbs/inch) is observed at initial load relief. As the load relief continues, the spring rate decreases to 2,012 kN/m (11,500 lbs/inch).

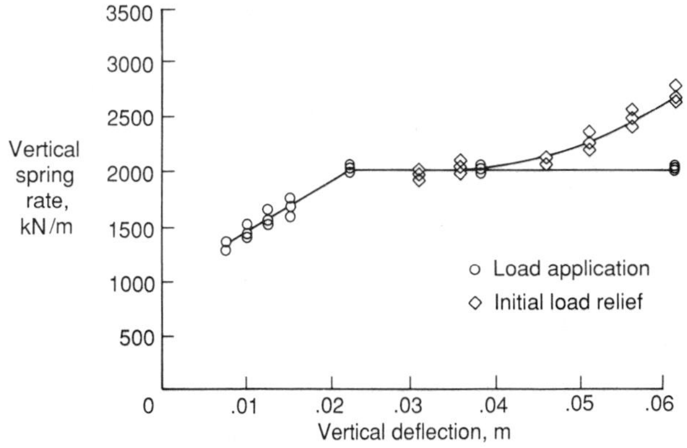

Figure 9 - Vertical spring rate of the bias-ply tire at 245 psi.

The vertical spring rates for the radial tire inflated to 2.1 MPa and 1.7 MPa (310 psi and 245 psi) are plotted as a function of vertical deflection in Figures 10 and 11 respectively. At an inflation pressure of 2.1 MPa (310 psi) (Figure 10), the tire spring rate gradually increases from 1,225 kN/m (7,000 lbs/inch) to 2,100 kN/m (12,000 lbs/inch) during load application. The maximum spring rate of 3,150 kN/m (18,000 lbs/inch) is obtained at initial load relief. It then gradually decreases to 1,925 kN/m (11,000 lbs/inch) during the unloading cycle. Similar spring rate characteristics were observed for the radial tire with an inflation pressure of 1.7 MPa (245 psi) (Figure 11). The spring rate gradually increases from 1,050 kN/m (6,000 lbs/inch) to 1,794 kN/m (10,250 lbs/inch) during load application. A maximum spring rate of 3,106 kN/m (17,750 lbs/inch) was measured during load relief and then gradually decreased to 1,488 kN/m (8,500 lbs/inch) as the load relief continued. The data in Figures 9 through 11 indicate that the vertical stiffness characteristics of the bias-ply tire and the radial tire are very similar. The implication of the result is that the landing dynamic characteristics of an aircraft equipped with this radial tire would be about the same as that aircraft equipped with the standard bias-ply tire.

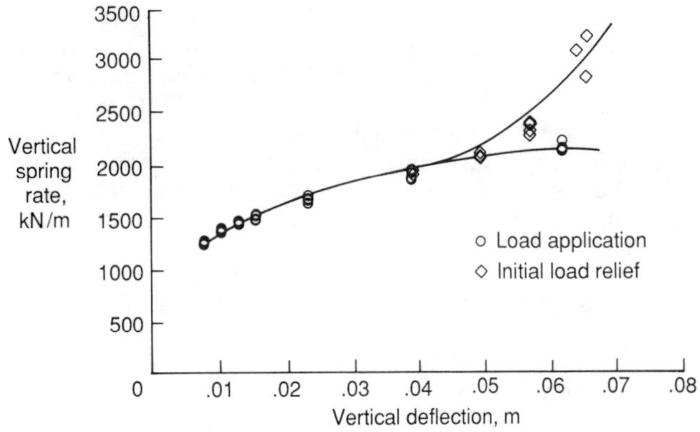

Figure 10 - Vertical spring rate of the radial tire at 310 psi.

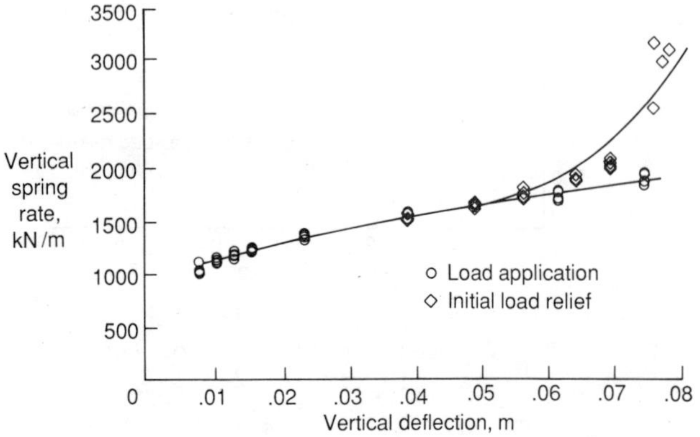

Figure 11 - Vertical spring rate of the radial tire at 245 psi.

Hysteresis Loss - From Figures 6, 7, and 8, it can be seen that the hysteresis loss (area inside the load-deflection curve) is less for the radial tire at 2.1 MPa (310 psi) than for the bias-ply tire, suggesting that there is less heat generated in the radial tire during the loading and unloading cycle. The hysteresis loss for the radial tire at 1.7 MPa (245 psi) is greater than for the bias-ply tire at the same inflation pressure. This was due to the higher tire deflection of 51% for the radial tire versus 35% for the bias-ply tire. The average vertical hysteresis loss for the bias-ply tire was 325 N-m (2,879 inch-lbs), for the radial tire inflated to 2.1 MPa (310 psi) the average loss was 283 N-m (2,506 inch-lbs), and for the radial tire inflated to 1.7 MPa (245 psi) the average loss was 329 N-m (2,913 inch-lbs).

STATIC LATERAL LOADING-Combined static vertical-lateral loading tests were conducted on two 30 X 11.5 - 14.5 bias-ply and radial-belted tires. Results of the combined static vertical-lateral load curves, lateral spring rate, and hysteresis loss of the tires are presented. Four lateral loading curves were generated for each tire.

LATERAL LOAD-DEFLECTION RELATIONSHIPS-A typical lateral load deflection curve for the bias-ply tire is shown in Figure 12. After the rated load of 111.2 kN (25,000 lbs) was applied vertically, the tire was loaded laterally beginning at zero lateral load and deflection. As the load was applied up to 17.8 kN (4,000 lbs), the slope continued to increase indicating a stiffening spring. At initial load relief, the slope is steeper than at load application. With a further load release, the spring softens and continues to soften as the load is returned to zero. The load is then applied in the opposite direction up to -17.8 kN (-4,000 lbs). The load is released where upon the slope is again steeper than at load application. The slope decreases and the spring softens as the load is returned to zero. This peak loading resulted in a lateral friction coefficient for the bias-ply tire of 0.16.

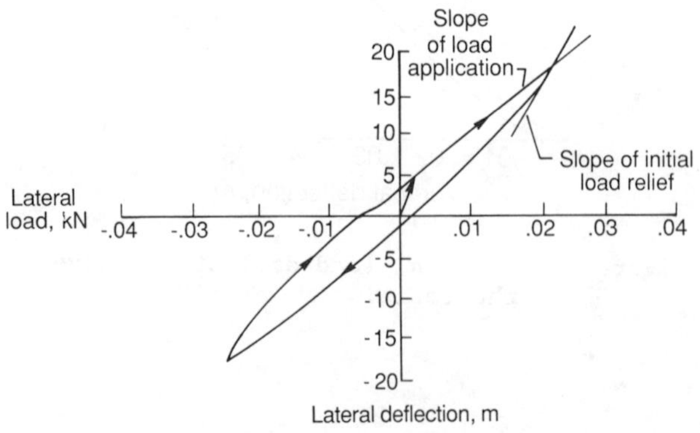

Figure 12 - Lateral load-deflection curve of the bias-ply tire at 245 psi.

Figures 13 and 14 are plots of the lateral load deflection curve of the radial tire at 2.1 MPa and 1.7 MPa (310 psi and 245 psi) respectively. The load was applied in a similar manner as for the bias-ply tire. However, the maximum lateral load was ±13.3kN (±3,000 lbs) because of tire slippage on the frictionless table that prevented testing to higher loads. This resulted in a maximum lateral friction coefficient of 0.12 for the radial tire. Similar spring stiffness relationships were observed for the radial tire at each inflation pressure.

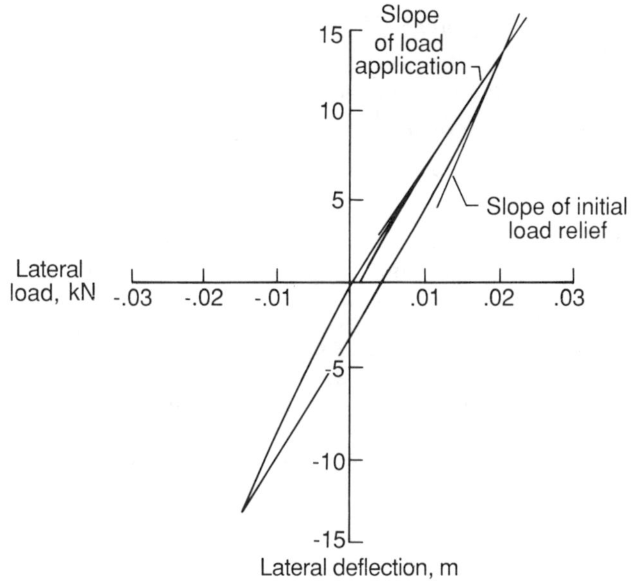

Figure 13 - Lateral load-deflection curve ot the radial tire at 310 psi.

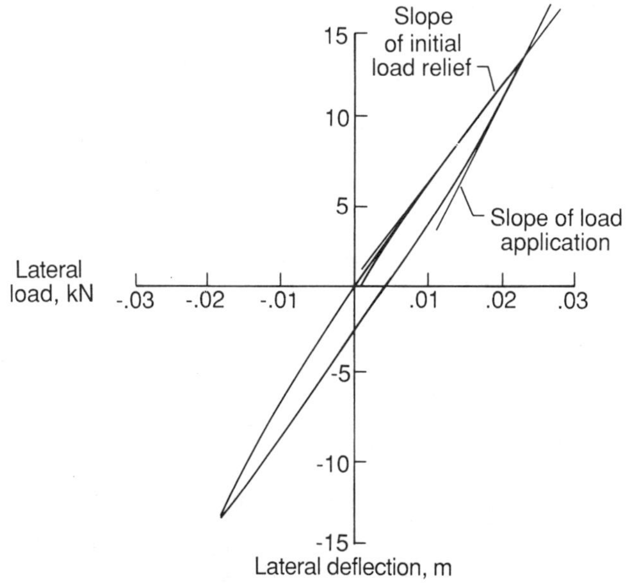

Figure 14 - Lateral load-deflection curve of the radial tire at 245 psi.

Spring Rate - The lateral spring rates for each tire were determined from the instantaneous slope of the combined vertical and lateral-load deflection curves. In Figure 15, the lateral spring rate for the bias-ply tire is plotted as a function of lateral deflection. The spring rate is nearly constant at 656 kN/m (3,750 lbs/inch) during load application. A maximum spring rate of 1,531 kN/m (8,750 lbs/inch) was observed at initial load relief and as the load is relieved further, the spring softens and the spring rate decreases to 787 kN/m (4,500 lbs/inch).

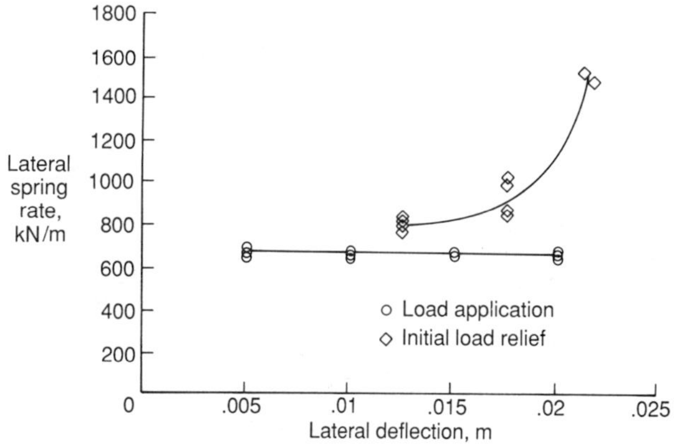

Figure 15 - Lateral spring rate of the bias-ply tire at 245 psi.

The lateral spring rate for the radial tire is plotted as a function of lateral deflection in Figures 16 and 17. As with the bias-ply tire, the slope of the lateral load deflection curve for the radial tire at 2.1 MPa (310 psi) is nearly linear during load application which results in a nearly constant spring rate of 744 kN/m (4,250 lbs/inch). The maximum spring rate of 1,225 kN/m (7,000 lbs/inch) was again obtained at initial load relief. The spring continues to soften as the load is continued to be relieved to a spring rate of 787 kN/m (4,500 lbs/inch). The radial tire at 1.7 MPa (245 psi) had similar characteristics as at its rated inflation pressure of 2.1 MPa (310 psi). The nearly linear load application curve results in a nearly constant spring rate of 569 kN/m (3,250 lbs/inch). The maximum spring rate at initial load relief was again at 1,225 kN/m (7,000 lbs/inch). The softening spring as the load is relieved results in a spring rate of 656 kN/m (3,750 lbs/inch). A comparison of the three spring rate plots indicates that the lateral stiffness of the radial tire inflated to 2.1 MPa (310 psi) is nearly equivalent to the bias-ply tire inflated to 1.7 MPa (245 psi). If the radial tire is operated at 1.7 MPa (245 psi), then its lateral stiffness values are 13 percent to 20 percent less than the bias-ply tire at 1.7 MPa (245 psi).

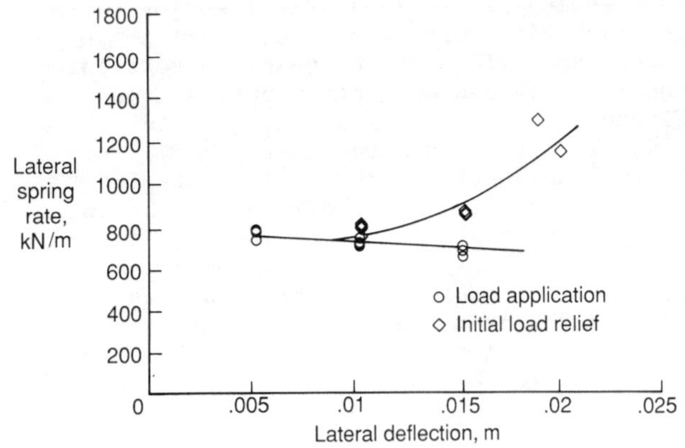

Figure 16 - Lateral spring rate of the radial tire at 310 psi.

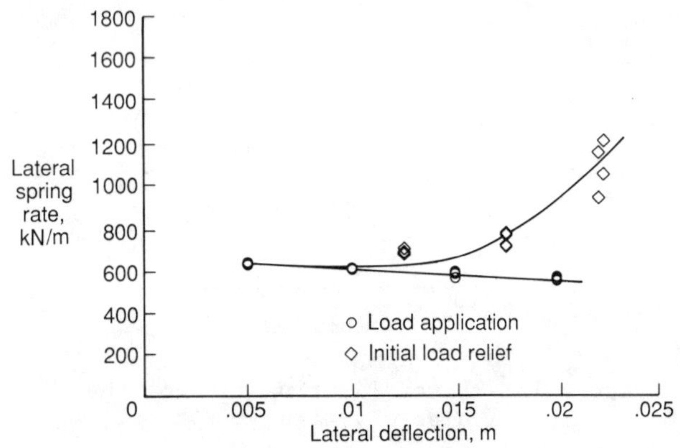

Figure 17 - Lateral spring rate of the radial tire at 245 psi.

<u>Hysteresis Loss</u> - An analysis of the lateral-load deflection curves, indicates that the radial tire exhibits hysteresis losses which are about 55 percent of the bias-ply tire. Thus, the radial tire generates less heat during the loading cycles than the bias-ply tire. The hysteresis loss for the bias-ply tire was 185 N-m (1,639 inch-lbs) and for the radial tire 83 N-m (732 inch-lbs) at 2.1 MPa (310 psi) and 85 N-m (751 inch-lbs) at 1.7 MPa (245 psi). The radial tire expended more energy at the lower inflation pressure than at the rated pressure. It should be noticed, however, that the maximum lateral load was 4.4 kN (1,000 lbs) less for the radial tire than for the bias-ply tire due to slippage problems with the radial tire.

MASS MOMENT OF INERTIA-The mass moment of inertia was calculated for each tire and was first calculated for the plates alone as 0.06 kg-m-sec^2 (5.18 inch-lbs-sec^2) using the following equation:

$$J_{plates} = \frac{\tau^2 W R^2}{4 \pi^2 L} \quad (1)$$

The mass moment of inertia of the tire and rotating parts was calculated using the following equation:

$$J_{tire} = \frac{\tau^2 W R^2}{4 \pi^2 L} - J_{plates} \quad (2)$$

where J_{plates} was 0.06 kg-m-sec^2 (5.18 inch-lbs-sec^2).

The respective moment of inertia values were 0.35 kg-m-sec^2 (30.7 inch-lbs-sec^2) for the bias-ply tire and 0.314 kg-m-sec^2 (27.2 inch-lbs-sec^2) and 0.311 kg-m-sec^2 (27.0 inch-lbs-sec^2) for the radial tire at 2.1 MPa and 1.7 MPa (310 psi and 245 psi) respectively. Thus the moment of inertia of the radial tire is 10% less than that of the bias ply tire. This suggests that the radial tire would require less energy dissipation to spin up during the landing touchdown than the bias-ply tire. This in turn could mean reduced tread wear for the radial tire for high speed landings.

CONCLUSIONS

An investigation has been conducted to determine and compare the mechanical characteristics of 30 x 11.5 - 14.5 bias-ply and radial-belted aircraft tires. Stiffness and damping characteristics were obtained from load-deflection curves under various static loading conditions. Inertia properties of the test articles were obtained using a torsional pendulum.

The results obtained show the following:

1. Vertical stiffness characteristics for the radial and bias-ply tires are nearly equal under the conditions tested.
2. Lateral stiffness of the radial tire inflated to 2.1 MPa (310 psi) is nearly equivalent to that of the bias-ply tire inflated to 1.7 MPa (245 psi). At 1.7 MPa (245 psi), the lateral stiffness of the radial tire is from 13 percent to 20 percent less than the bias-ply tire at its rated inflation pressure.
3. Hysteresis loss of the radial tire at 2.1 MPa (310 psi) is significantly less than the bias-ply tire under vertical or lateral loading conditions. This result indicates that the radial tire will dissipate less heat during normal operation and therefore, will run cooler.

4. The mass moment of inertia of the radial tire is about 10% less than the bias-ply tire, indicating that the radial tire will exhibit less spin-up wear than the bias-ply tire under the same touchdown conditions.

NOMENCLATURE

τ	average period of system oscillation, seconds
J_G	mass moment of inertia about the center of gravity, kg-m-sec^2 (inch-lbs-sec^2)
L	length of support cables, m (inches)
R	radial distance from the center of the plate to the support cables, m (inches)
W	weight of object being measured, kg (lbs)

REFERENCES

1. Cesar, J. P.; Musy, J.; Olds, R.: Development of Radial Aircraft Tires. Presented at the 38th International Air Safety Seminar, Boston, Massachusetts, November 4-7, 1985.
2. Tanner, J. A.: Fore-and-Aft Elastic Response Characteristics of 34 x 9.9, Type VII, 14 Ply-Rating Aircraft Tire of Bias-Ply, Bais-Belted, and Radial-Belted Design, NASA TN D-7449, 1974.
3. Tanner, J. A.; Stubbs, S. M.; and McCarty, J. L.: Static and Yawed-Rolling Mechanical Properties of Two Type VII Aircraft Tires, NASA TP 1963.
4. Tanner, J. A.; Stubbs, S. M.; Dreher, R. C.; and Smith, E. G.: Dynamics of Aircraft Antiskid Braking Systems, NASA TP 1959, February 1982.
5. Volterra, E.; Zachmanoglou, E. C.: Dynamics of Vibrations, Merrill Books, Inc., 1965, pp. 58-60.

Fore-and-Aft Stiffness and Damping Characteristics of 30 x 11.5-14.5, Type VIII, Bias-Ply and Radial-Belted Aircraft Tires

Mercedes C. Lopez and Pamela A. Davis
NASA Langley Research Center
William A. Vogler
PRC Systems Services
A Division of Planning Research Corp.
Robert B. Yeaton
NASA Langley Research Center

* Paper 881357 presented at the Aerospace Technology Conference and Exposition, Anaheim, California, October, 1988.

ABSTRACT

An investigation was conducted to determine the fore and aft elastic response characteristics and footprint geometrical properties of 30 x 11.5 - 14.5, Type VIII, bias-ply and radial-belted aircraft tires. Stiffness and damping characteristics of each tire were obtained from load-deflection curves generated from static tests. Tire footprints were obtained for various vertical loads, and geometrical measurements were obtained from the resulting silhouettes. Results of this investigation show considerable differences in stiffness and damping characteristics between the bias-ply and radial-belted tire designs. Footprint geometrical data indicate that footprint aspect ratio effects may interfere with improved hydroplaning potential associated with the radial-belted tire operating at higher inflation pressures. Tire-wheel slippage problems encountered when testing the radial-belted tire design required special attention.

RADIAL-BELTED AIRCRAFT TIRES have been developed and are being tested in Europe and in the United States. So far, results show that the radial-belted tire offers several advantages over the bias-ply tire such as reduced weight, lower operating temperatures and increased tread life; thus resulting in less cost per landing. Studies on tire performance are progressing because of the need for a database of the mechanical properties of radial-belted tires. NASA Langley Research Center in a joint effort with the Air Force's Wright Aeronautical Laboratories is developing a database on the mechanical properties of bias-ply and radial-belted 30 X 11.5 - 14.5, Type VIII, aircraft tires. A recent publication (1) provides data on vertical and lateral stiffness and damping characteristics of both tire designs.

To extend these studies, an investigation was conducted to measure footprint geometrical properties and fore and aft stiffness and damping characteristics of the 30 x 11.5 -14.5 bias-ply and radial-belted aircraft tires. The bias-ply tire was tested at an inflation pressure of 245 psi (1.7 MPa) and the radial-belted tire was tested at inflation pressures of 310 psi (2.1 MPa) and 245 psi (1.7 MPa). Maximum loads applied were 25,000 lb (111 kN) vertical load and 3000 lb (13.3 kN) fore-and-aft load. Tire footprints were obtained at various vertical loading conditions. The tires and wheel assemblies used during this investigation were supplied by the U. S. Air Force.

TEST APPARATUS AND PROCEDURES

Fore-and-aft static tests and footprint geometrical measurements were conducted on bias-ply and radial-belted 30 x 11.5 - 14.5, Type VIII, aircraft tires. Tire specifications are shown in Table I and details of tire construction are illustrated in figure 1. The bias-ply tire is constructed with carcass plies arranged on a bias to form an angle between the reinforcing cords of alternating plies.
The carcass is then capped with the tire tread (2).

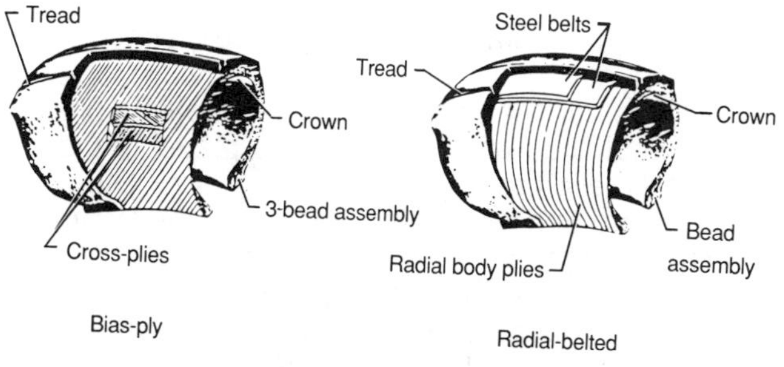

Figure 1 - Bias-ply and radial-belted tire construction.

opez

Table 1:Characteristics of Bias-ply and Radial-belted Tires

Parameter	Bias-ply	Radial-belted
Size	30 x 11.5 - 14.5	30 x 11.5 - 14.5
Ply rating	24	26
Weight,kg (lbm)	31.25 (68.75)	25.2 (55.5)
Rated vertical load, kN (lbf)	111 (25,000)	111 (25,000)
Rated inflation pressure at 35 percent load deflection,MPa (psi)	1.7 (245)	2.1 (310)
Outside diameter of unloaded tire, m (inches)	0.76 (30)	0.76 (30)
Maximum carcass width of unloaded tire, m (inches)	0.2 (8)	0.2 (8)
Tread grooves	3	4

The radial-belted tire is constructed with the reinforcing cords of the carcass oriented radially about the tire. The carcass is reinforced with circumferential belts and capped with the tire tread (2). During this investigation, the bias-ply tire was tested at its rated inflation pressure of 245 psi (1.7 MPa) and the radial-belted tire was tested at its rated inflation pressure of 310 psi (2.1 MPa) and also at an inflation pressure of 245 psi (1.7 MPa).

The test set-up is shown in figure 2. The main structure consists of two three-bay portal frames joined overhead by four beams and along the floor by a thick plate (3). Adapter plates fixed to vertical beams suspended from the upper part of the structure support the wheel and tire assembly and prevent any axial rotation of the wheel. A steel platen suspended from four vertical cables serves as the loading medium. Each cable is suspended from a load cell connected to a screwjack. The screwjacks, mechanically driven by an electric motor, move simultaneously displacing the platen in the vertical direction to load or unload the tire. During the test, fore and aft loading of the tire was attained by displacing the platen in the fore or aft directions by means of a hydraulic cylinder (figure 3). Braking forces were measured by a load cell connected in series to the hydraulic cylinder. Vertical and fore-and-aft displacements of the platen were measured by displacement transducers. Platen displacements were considered to be equal to the

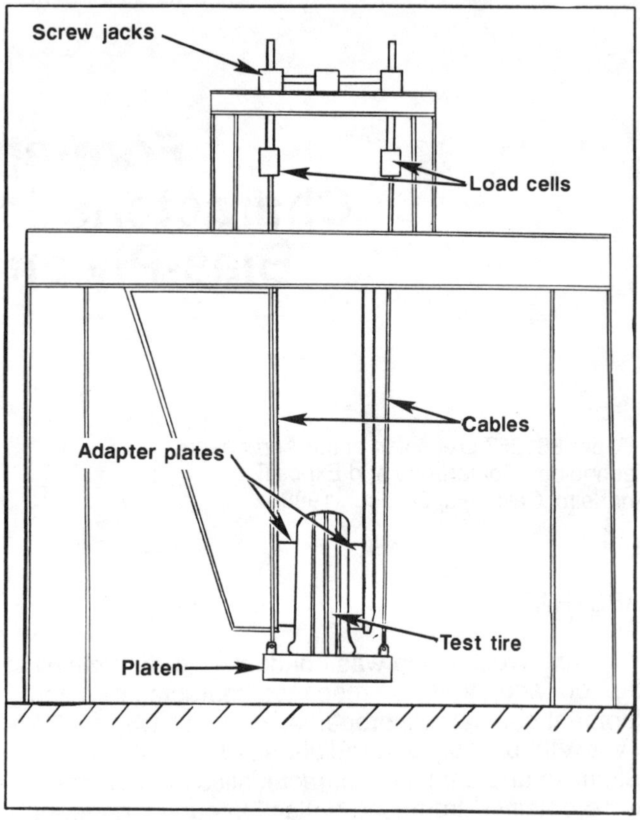

Figure 2 - Static test apparatus.

tire footprint displacements. Axial rotational movements due to excessive tolerance between wheel and adapter plates were measured with a DCDT mounted as shown in figure 3.
An extensiometer, a 3 inch (7.62 cm) copper berylium strain gage arch capable of measuring small displacements, was used to measure tire-wheel slippage when testing the radial-belted tire (figure 4). The data acquisition system was manually triggered at specific load intervals. Data from the various sensors was fed into a computer and afterwards saved for further reduction and analysis.

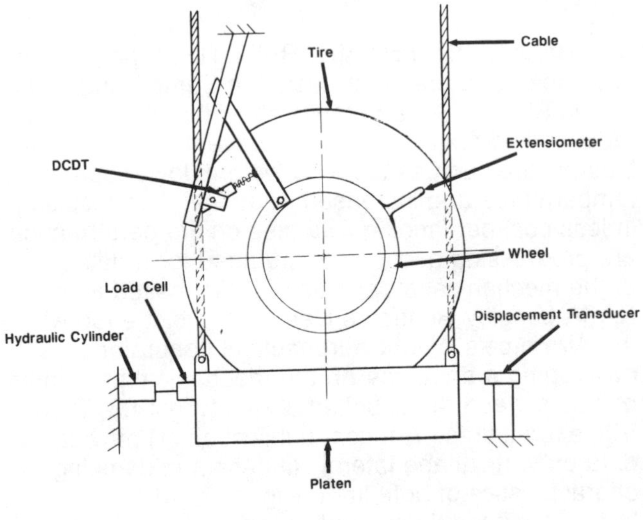

Figure 3 - Side view of tire test apparatus.

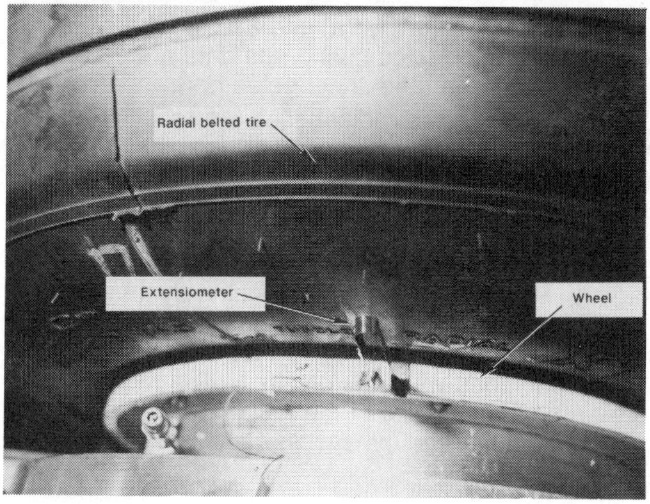

Figure 4 - Tire-wheel displacement monitor.

STATIC FORE-AND-AFT TEST - Static fore-and-aft tests were conducted to determine the stiffness and damping characteristics of the two different tire designs. The test involved applying maximum vertical rated load of 25,000 lb (111 kN) through one complete cycle for the bias-ply tire and through three complete cycles for the radial-belted tire. Force and displacement data were recorded at 250 lb (1.11 kN) load increments. Stiffness and damping characteristics were then obtained from the resulting hysteresis loops.

FOOTPRINT MEASUREMENTS - Tire footprints were obtained for vertical loads ranging from 0 to 25,000 lb (111 kN) in 5,000 lb (22.2 kN) increments. The process to obtain the footprints involved coating the tire tread with ink or chalk and applying the desired vertical load to the tire on a cardboard sheet located between the tire and the platen. Geometrical characteristics were obtained from the resulting silhouette (figure 5) with a computerized planimeter.

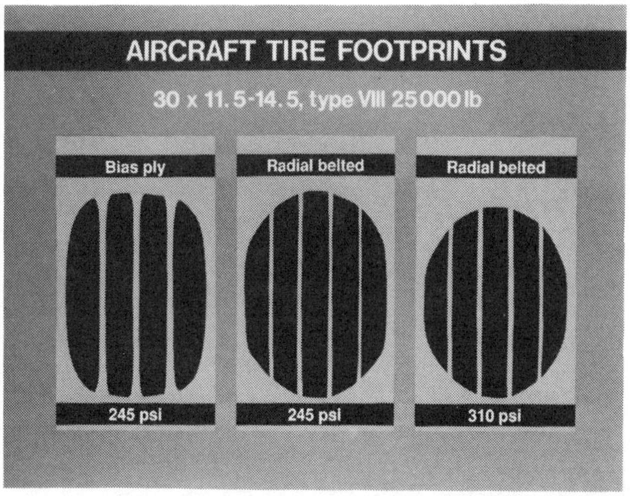

Figure 5 - Bias-ply and radial-belted tire footprints.

RESULTS

Results from fore-and-aft tests for two 30 x 11.5 - 14.5 bias-ply and radial-belted tires are presented in this section. Stiffness and damping characteristics of each tire were obtained from static load deflection curves.

LOAD DEFLECTION - Four hysteresis loops, each one corresponding to a different location around the periphery of the tire, were obtained for each tire. One of the resulting hysteresis loops for the bias-ply tire is shown in Figure 6. Parameters in the abscissa correspond to tire footprint displacements values corrected for excessive tolerance between wheel and adapter plates.

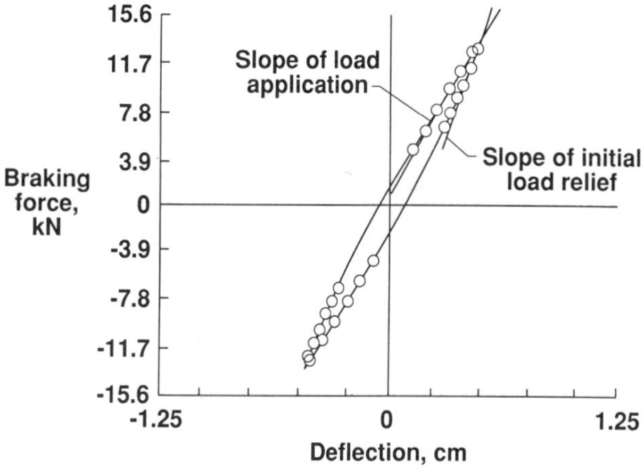

Figure 6 - Load deflection curve for bias-ply tire at 245 psi (1.7 MPa).

Parameters in the ordinate corresponded to braking force values. The loop is composed of four individual curves. Based on observed nonlinear trends, a second degree curve fit was chosen to fit the data points obtained from load application in the fore or aft directions and a third degree curve fit was selected for load relief data. Results shown in Figure 6 represent nonlinear load deflection characteristics for the bias-ply tire and demonstrate the hysteretic nature of the loading-unloading cycle. A maximum deflection of 0.2 inch (0.46 cm) was obtained for this tire under 3,000 lb (13.3 kN) braking force. Similar load-deflection characteristics were obtained for the radial-belted tire at inflation pressures of 245 psi (1.7 MPa) and 310 psi (2.1 MPa) respectively. Figures 7 and 8 show load deflection data for the radial-belted tire at the different inflation pressures where displacement values were not corrected for tire-wheel slippage. Load-deflection curves of corrected data shown in Figures 9 and 10 indicate steeper slopes and less hysteresis when compared with the uncorrected data in Figures 7 and 8. Maximum displacement values obtained from uncorrected data were 0.32 inch (0.79 cm) under the same conditions as those for the bias-ply tire. The maximum displacements were 0.27 inch (0.69 cm) for the corrected data.

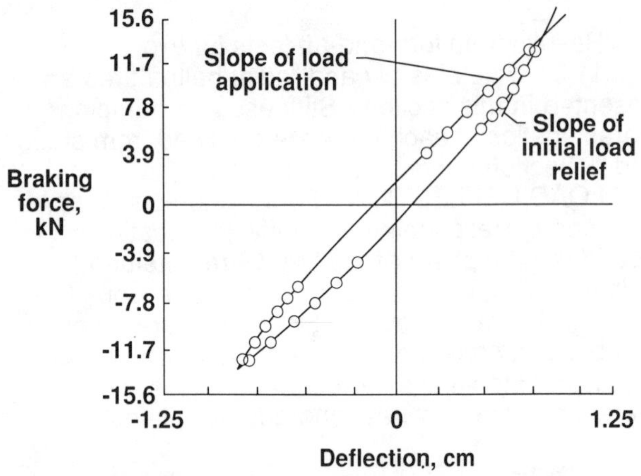

Figure 7 - Uncorrected load-deflection curve of radial-belted tire at 245 psi (1.7 MPa).

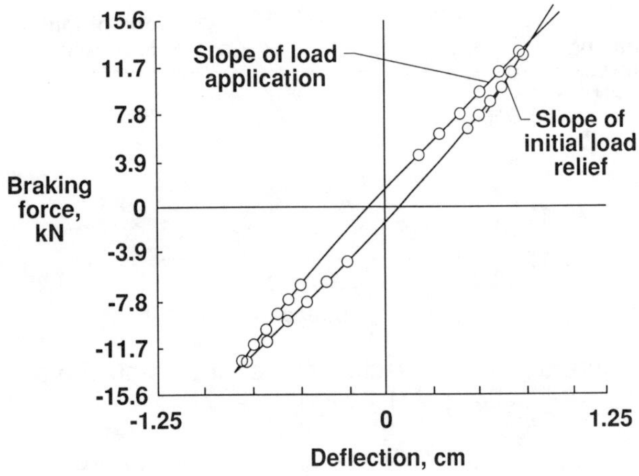

Figure 8 - Uncorrected load-deflection curve of radial-belted tire at 310 psi (2.1 MPa).

SPRING RATE - Fore-and-aft spring rates were obtained evaluating the first derivative of the functions defined for each loading or unloading interval at specific points along the curves. Fore-and-aft spring rates are plotted as a function of fore and aft displacements for the bias-ply tire in figure 11. Spring rate values linearly decreased from 15,277 lb/in (2675 kN/m) to 13,087 lb/in (2297 kN/m) during the load application process. During load relief, spring rate values decreased from a maximum value of 25,878 lb/in (4532 kN/m) to a minimum value of 18,160 lb/in (3180 kN/m). The fore-and-aft stiffness of the radial-belted tire was relatively insensitive to variations in inflation pressure over the range of pressures tested as indicated by comparing the uncorrected data in figures 12 and 13 and by comparing the corrected data in figures 14 and 15. The maximum spring rate values obtained from uncorrected data were 8508 lb/in (1490 kN/m) and 8841 lb/in (1548 kN/m) for load application and 15291 lb/in (2678 kN/m) and 13298 lb/in (2329 kN/m) for load relief at inflation pressures 245 psi (1.7 MPa) and 310 psi (2.1 MPa) respectively. Corrected spring rate data are 12 to 27 percent higher than the uncorrected data.

Comparing corrected radial-belted tire results with those obtained from the bias-ply tire, maximum stiffness values for the radial-belted tire were 26 to 37 percent less than the values obtained for the bias-ply tire. The differences in fore-and-aft stiffness characteristics between the bias-ply tire and the radial-belted tire imply that the landing dynamic characteristics of an aircraft equipped with radial tires would be considerably different than those of an aircraft equipped with bias-ply tires. The reduced fore-and-aft spring rate of the radial-belted tire could be detrimental to the dynamics of antiskid braking systems.

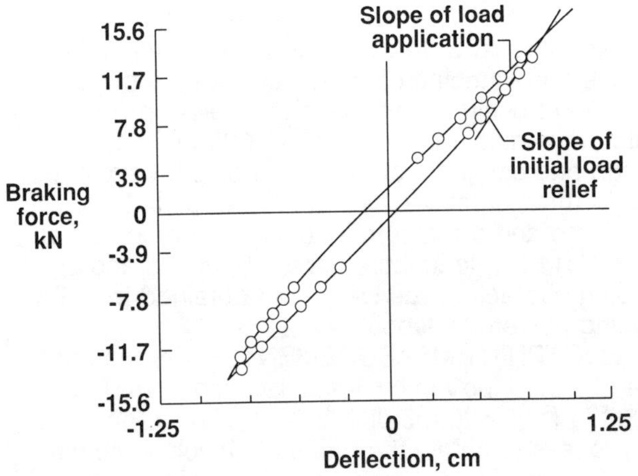

Figure 9 - Corrected load-deflection curve of radial-belted tire at 245 psi (1.7 MPa).

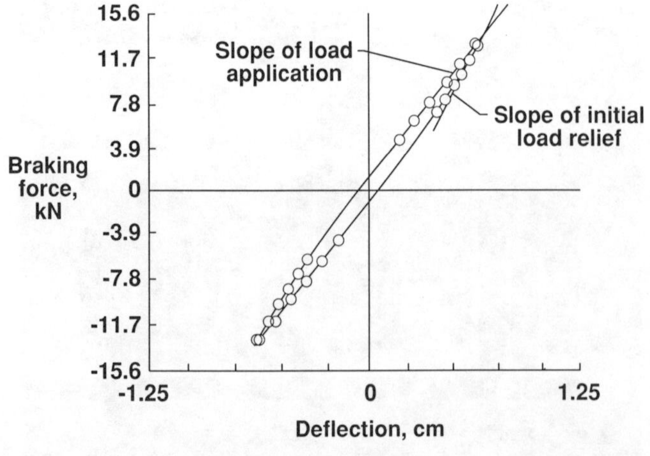

Figure 10 - Corrected load-deflection curve of radial-belted tire at 310 psi (2.1 MPa).

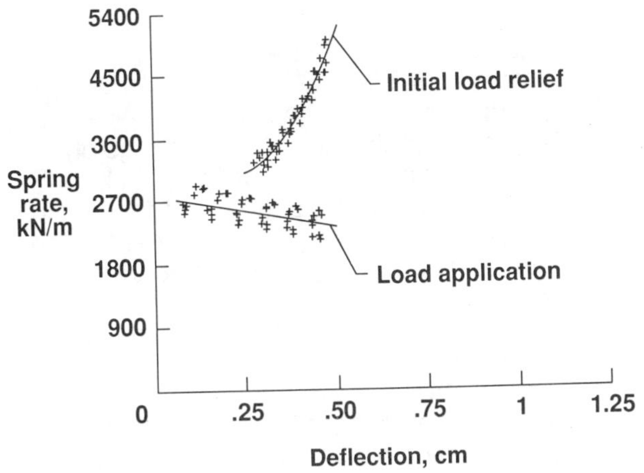

Figure 11 - Spring rate distribution curve of bias-ply tire at 245 psi (1.7 MPa).

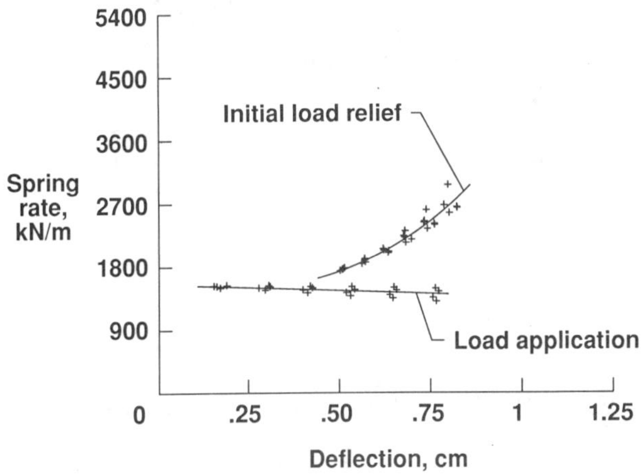

Figure 12 - Uncorrected spring rate distribution curve of radial-belted at 245 psi (1.7 MPa).

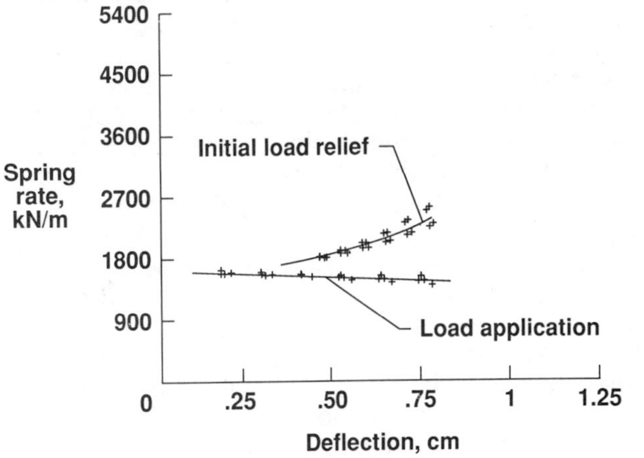

Figure 13 - Uncorrected spring rate distribution curve of radial-belted tire at 310 psi (2.1 MPa).

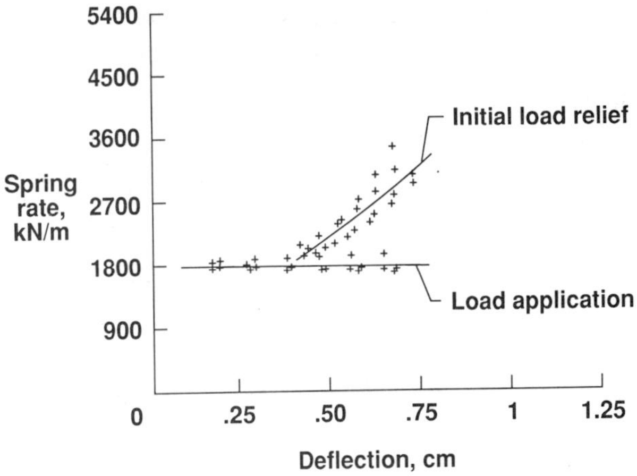

Figure 14 - Corrected spring rate distribution curve of radial-belted tire at 245 psi (1.7 MPa).

HYSTERETIC LOSS - Energy loss during the loading and unloading cycle is represented by the area of the hysteresis loops shown in figures 6 to 10. The hysteresis loss for the bias-ply tire was 230 in-lb (26 kN m). Data from uncorrected hysteresis loops for the radial-belted tire at 245 psi (1.7 MPa) and 310 psi (2.1 MPa) showed hysteretic losses that were 40 to 16 percent higher respectively than the hysteretic loss for the bias-ply tire. Corrected radial-belted tire data showed a hysteretic loss of 283 in-lb (32 kN m) at 245 psi (1.7 MPa) which was 19 percent higher than the bias-ply tire hysteresis. At 310 psi (2.1 MPa), corrected radial-belted data showed a hysteretic loss of 159 in-lb (18 kN m) which was 31 percent lower than the bias-ply hysteresis. Variations in hysteresis values for corrected and uncorrected data for both inflation pressures ranges from 25 to 42 percent. These results indicate that the radial-belted tire will run cooler than the bias-ply tire when both are operated at rated inflation pressures.

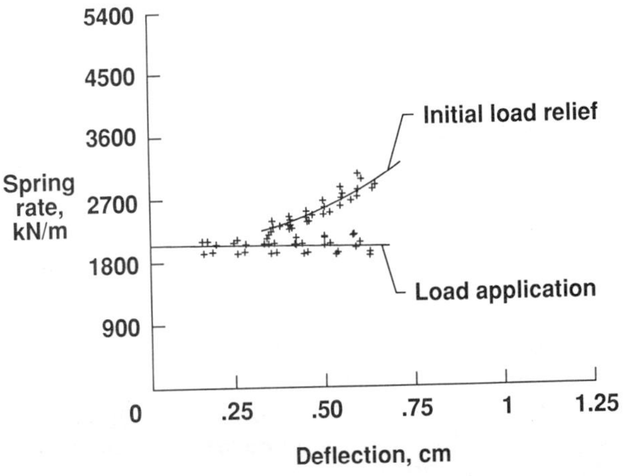

Figure 15 - Corrected spring rate distribution curve of radial-belted tire at 310 psi (2.1 MPa).

FOOTPRINT GEOMETRICAL MEASUREMENTS

Tire footprints were obtained for each tire under various vertical loading conditions. Geometrical parameters such as area, length, and width were measured from the resulting silhouettes. A plot of net footprint area as a function of vertical load is shown in figure 16. Vertical stiffness characteristics for the radial-belted tire and the bias-ply tire are nearly equal at inflation pressures of 245 psi (1.7 MPa). At rated inflation pressure of 310 psi (2.1 MPa) the radial-belted tire was shown to be stiffer vertically than the bias-ply tire. Figure 5 shows tire footprints for the bias-ply and radial-belted tires at maximum rated load. At 245 psi (1.7 MPa) the predicted hydroplaning speed of the bias-ply and radial-belted tire is 141 kts. At 310 psi (2.1 MPa), the

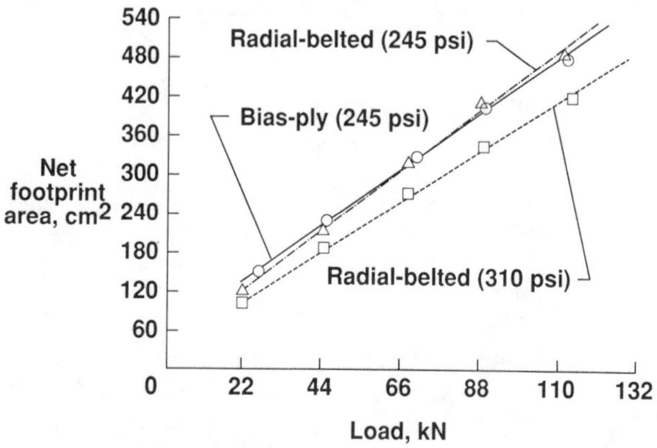

Figure 16 - Net footprint area as a function of load for 30 x 11.5 - 14.5 tire designs.

predicted hydroplaning speed of the radial-belted tire is 158 kts (4). However, as pointed out in reference 5, the increased hydroplaning speed for the radial-belted tire at the higher inflation pressure may not be realized because of adverse effects associated with the aspect ratio of the nearly circular footprint of the radial-belted tire.

CONCLUSIONS

Fore-and-aft static tests and footprint geometrical measurements were made on bias-ply and radial-belted 30 x 11.5 -14.5 aircraft tires in an effort to define tire mechanical properties. Stiffness and damping characteristics where obtained from load deflection curves. Tire-wheel slippage problems were encountered while testing the radial-belted tire. Data showing both corrected and uncorrected footprint displacement values for the radial-belted tire were presented for comparison. Footprint geometrical properties were obtained for various vertical loading conditions.

The results of this investigation indicate the following:

1. The radial-belted tire at inflation pressures of 245 psi (1.7 MPa) and 310 psi (2.1 MPa) is about 50 percent less stiff in the fore-and-aft directions than the bias-ply tire at an inflation pressure of 245 psi (1.7 MPa) when tire-wheel slippage is not taken into account. Correcting for tire-wheel slippage, the radial-belted tire is about 36 percent less stiff than the bias-ply tire. The reduced stiffness of the radial-belted tire may be detrimental to the performance of aircraft antiskid braking systems.
2. Fore-and-aft stiffness characteristics for the radial-belted tire are insensitive to inflation pressure variations over the range of pressures tested.
3. Hysteretic loss of the radial-belted tire at an inflation pressure of 310 psi (2.1 MPa) is significantly less than the hysteretic loss of the bias-ply tire at 245 psi (1.7 MPa). Hysteretic loss for the radial-belted tire is inversely proportional to inflation pressure variations.
4. The nearly circular shape of the radial footprint at an inflation pressure of 310 psi (2.1 MPa) may interfere with the improved hydroplaning potential associated with higher inflation pressures.

REFERENCES

1. Davis, Pamela A.; and Lopez, Mercedes C.: Static Mechanical Properties at 30 x 11.5 - 14.5, Type VIII, Aircraft Tires at Bias-Ply and Radial-Belted Design. NASA TP 2810, 1988.

2. Tanner, John A.: Fore and Aft Elastic Response Characteristics of 34 x 9.9, Type VII, 14 Ply-Rating Aircraft Tires of Bias-Ply, Bias-Belted, and Radial-Belted Design. NASA TN D-7449, 1974.

3. Sleeper, Robert K.; and Dreher, Robert C.: Tire Stiffness and Damping Determined from Static and Free-Vibration Tests. NASA TP 1671, 1980.

4. Horne, Walter B.; and Dreher, Robert C.: Phenomena of Tire Hydroplaning. NASA TN-D 2056, 1963.

5. Horne, Walter B.; Yager, T. J., and Ivey, D. L.: Recent Studies to Investigate Effects of Tire Footprint Aspect Ratio on Dynamic Hydroplaning Speed. The Tire Pavement Interface, ASTM STP 929, M. G. Pottinger and T. J. Yager, Eds., American Society for Testing and Materials, Philadelphia, 1986

Spin-Up Studies of the Space Shuttle Orbiter Main Gear Tire

Robert H. Daugherty and Sandy M. Stubbs
NASA Langley Research Center

* Paper 881360 presented at the Aerospace Technology Conference and Exposition, Anaheim, California, October, 1988.

ABSTRACT

One of the factors needed to describe the wear behavior of the Space Shuttle Orbiter main gear tires is their behavior during the spin-up process. An experimental investigation of tire spin-up processes was conducted at the NASA Langley Research Center's Aircraft Landing Dynamics Facility (ALDF). During the investigation, the influence of various parameters such as forward speed and sink speed on tire spin-up forces were evaluated. A mathematical model was developed to estimate drag forces and spin-up times and is presented. The effect of prerotation was explored and is discussed. Also included is a means of determining the sink speed of the orbiter at touchdown based upon the appearance of the rubber deposits left on the runway during spinup.

THE SPACE SHUTTLE ORBITER is a unique spacecraft designed to land on conventional landing gear and runways. Excessive wear encountered during landings at the Kennedy Space Center Shuttle Landing Facility has shown that a detailed understanding of the tire spin-up process is needed since it is the catalyst for further tire wear during the rollout process. Not only does tire spinup cause wear, but the associated drag loads can excite fore-and-aft dynamic response of the landing gear. Thus a method to predict those loads is needed.

The purpose of this paper is to present the results of tests conducted at the NASA Langley Research Center to investigate the spin-up behavior of the Space Shuttle Orbiter main gear tire. Results of spin-up tests are plotted as time histories of vertical and drag loads as well as wheel velocity and angular acceleration. Simple equations of motion are used to describe the spin-up process and to estimate drag loads obtained during spinup. The effects of surface characteristics on spin-up time and loads are discussed. The effect of prerotation on spin-up wear is shown, as well as a method of determining orbiter sink speed at touchdown based upon the appearance of the spin-up rubber deposits on the runway.

APPARATUS

This investigation was conducted at the NASA Langley Research Center Aircraft Landing Dynamics Facility (ALDF). The facility consists of a set of rails 850 m long on which a 49,000 kg. carriage travels. The facility is shown in figure 1.

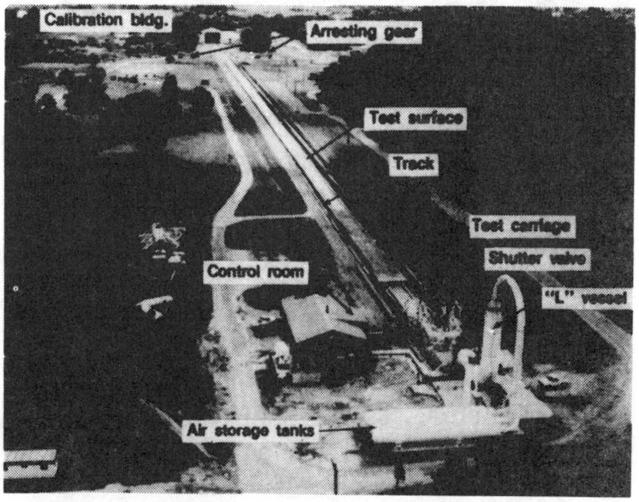

Figure 1. Aircraft Landing Dynamics Facility.

The carriage is propelled at speeds up to 220 kts. using a high pressure water jet directed at a turning bucket mounted on the carriage. Arrestment is achieved using a set of water turbines connected by nylon tapes and steel cables that are engaged by a nose block on the front of the carriage. A drop carriage mounted on vertical rails in the center of the main carriage allows the test fixture to be lowered to the simulated runway under a variety of sink speed and load combinations. A more detailed description of this facility is given in reference 1.

The tires used in this study were 44.5 x 16.0 - 21 bias-ply aircraft tires with a 34-ply rating. A photograph of a new tire is shown in figure 2. The tires have a 5-groove tread pattern made of natural rubber. The tread grooves are 2.5 mm deep and there is an additional 2.5 mm of rubber between the bottom of the grooves and the first carcass ply. The tires have 16 carcass plies, their rated load is 271 kN, and their rated pressure is 2.17 MPa. The tires were mounted on orbiter main wheels and masses were added to simulate the brake rotor inertia so that the inertia of the entire wheel/mass/tire combination was 19.06 kg m^2. This inertia was measured using a tri-filar pendulum method.

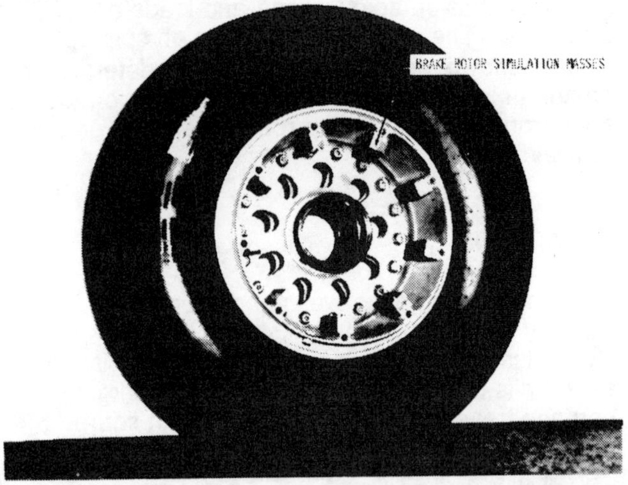

Figure 2. Space Shuttle Orbiter main tire.

Measurements obtained during a test run were digitally telemetered to a receiving station where they were converted into engineering units by a desktop computer and stored. The force measurement dynamometer sketched in figure 3 allowed all forces and moments generated at the tire/runway interface to be measured.
A device was built to prerotate the orbiter tire before the carriage was launched during a test run. After achieving slightly more than the desired tire prerotation test speed, the carriage was launched and the spin-up device was ratcheted and locked away from the orbiter tire which was then ready for the landing impact.

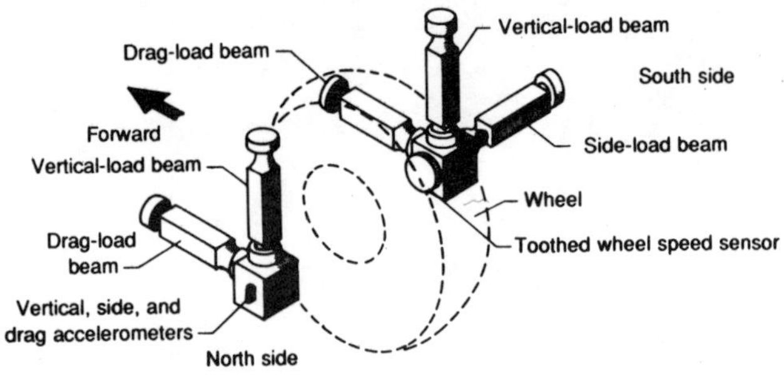

Figure 3. Force measurement dynamometer.

Most tests were conducted on a 550 m long concrete runway designed to simulate the rough Kennedy Space Center (KSC) Shuttle Landing Facility runway. The KSC runway shown in figure 4 has an extremely rough longitudinally-brushed texture combined with transverse grooves 6.4 mm square in cross section with a 29 mm spacing. This rough runway contributes greatly to tire wear during spinup.

Figure 4. Rough Kennedy Space Center runway.

A spin-up test normally consisted of setting the dynamometer to the desired yaw angle, setting hydraulic orifices to allow the desired sink speed when commanded, and accelerating the carriage to the desired test speed. At a preselected location, a limit switch commanded downward motion of the drop carriage with hydraulic load cylinders applying vertical force downwards as well. After tire spinup, another limit switch commanded upward motion of the drop carriage and the test was concluded. As mentioned, data were obtained using the force measurement dynamometer, but in addition, wheel speed data were obtained using a magnetic transducer observing a toothed plate mounted on the orbiter wheel. Wheel angular acceleration data were obtained by mathematically differentiating the wheel speed data using a desktop computer.

RESULTS AND DISCUSSION

In this investigation, the behavior of the orbiter main tire during spinup was determined. Effects of forward and vertical speed of the tire at spinup were observed, as well as the effect of runway surface type on tire behavior. Prerotation tests were conducted to observe possible decreases in tire wear during spinup. Knowledge gained during this program was used to develop a method of determining aircraft sink speed at touchdown based upon the rubber deposits left on the runway.

SPIN-UP DYNAMICS - Because of the size, inertia, and landing speed of the Space Shuttle Orbiter main gear tire, more energy is expended spinning it up to landing speed than perhaps any other tire in existence. The extreme texture of the KSC runway produces tire spin-up wear patches which are deeper than the first carcass cord level. As vertical load increases on the tire during spinup the drag forces produced by the tire increase while the drag force friction coefficient varies only slightly.

The vertical load time history from a typical spinup on the simulated KSC surface at the ALDF is shown in figure 5. The load rate, the slope of the initial load trace, has been shown to be quite similar to that actually experienced during orbiter landings. The sink rate for the test shown was about 0.8 m/sec. The horizontal velocity was 190 knots.

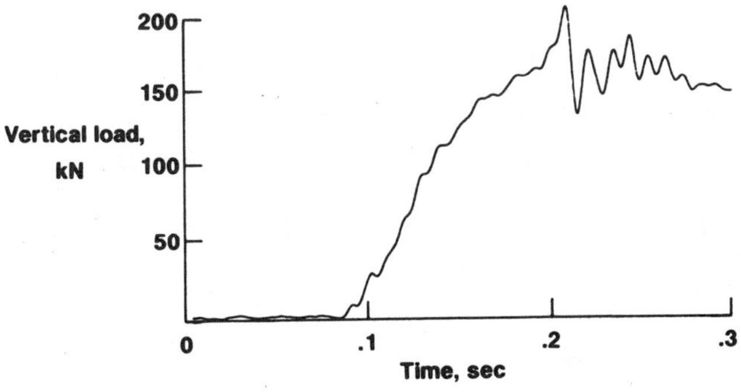

Figure 5. Vertical load time history during spinup.

Figure 6 shows the drag load time history during the spinup. Note that the drag load ceases at about 0.2 sec. The wheel speed time history shown in figure 7 indicates that at 0.2 sec the wheel speed was approximately 66 percent of its ground-synchronous value when the drag load abrubtly ceased. If the time scale in figure 6 is multiplied by forward speed the drag force can then be plotted as

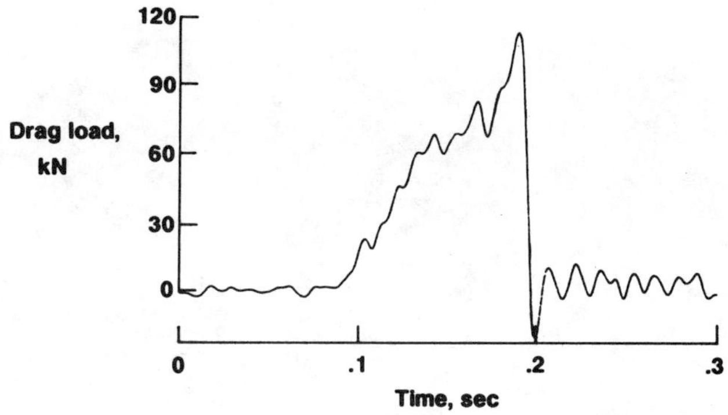

Figure 6. Drag load time history during spinup.

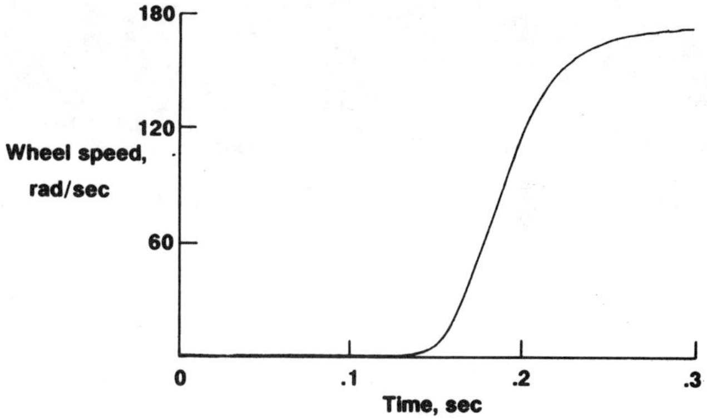

Figure 7. Wheel speed time history during spinup.

a function of distance and the area under the drag force curve is a measure of energy expended in the tire spin-up process. For the data shown in figure 6 this energy was about 0.57 MJ. After spinup is complete, calculation of the kinetic energy of the rotating tire based on its angular speed and inertia shows it to be about 0.28 MJ. The difference between the kinetic energy of the rotating tire and the energy expended at the ground/tire interface during spinup is a measure of energy dissipated by the tire as heat and tire wear. Since the spin-up drag load occurred over a 0.1 second time period, the average power required to spinup the tire was about 5.7 MW. A photograph of the tire after this spinup is shown in figure 8. Figure 9 shows the angular acceleration time history of the wheel along with a plot of the drag load from figure 6. One can see that a significant amount of energy is expended prior to any wheel acceleration and also that there is non-zero wheel acceleration after the time at which drag load ceases. This is because the tire behaves as a torsional spring and "winds up" as drag load, or torque, is applied. At some point, the time at which drag load ceases, the tire delivers its stored energy back into the surface and spins the wheel up to its full synchronous velocity.

Figure 8. Tire wear after a single KSC runway spinup.

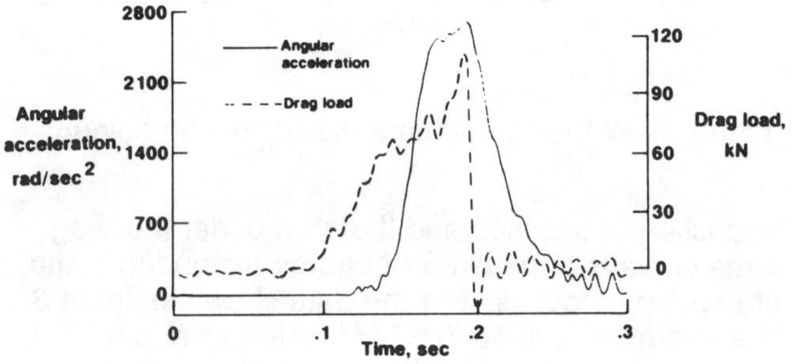

Figure 9. Angular acceleration and drag load time histories.

A method to predict the drag loads during spinup was developed. It begins with the simple equation of motion:

$$T = I\ddot{\theta}$$

where
T = torque on the system
I = inertia of the system
$\ddot{\theta}$ = angular acceleration

The torque on the tire is the drag load multiplied by the instantaneous loaded radius. One can set a boundary condition that states that the angular speed, $\dot{\theta}$, equals about 75 percent of the synchronous speed at the time, t_o, at which drag load ceases. Integration of the torque equation and substitution of the boundary condition yields the following equation.

$$t_o = \left(\frac{V}{K\,SR\mu}\right)^{1/2}$$

where V = ground speed at touchdown in knots
SR = sink rate in m/sec
μ = average friction coefficient during spinup
K = a constant that reflects tire static load-deflection characteristics and unit conversions

Figure 10 shows the typical drag force friction coefficient during spinup on the simulated KSC surface. The plot shows that the average coefficient during spinup is about 0.6. Knowledge of the tire vertical spring rate combined with sink rate yields a tire vertical loading rate and allows the drag force to be calculated during spinup as:

$$F = K_1 t\,SR\mu$$

where F = drag force
t = time
K_1 = tire spring rate in N/m in region of 5 cm tire deflection

And from the shape of typical drag force curves, a further assumption of this analysis is that the maximum drag force occurs at a time equal to about 0.9 t_o and thereafter drag force linearly approaches zero at time t_o. This model is strongly affected by the linearity of the vertical load time history so it is used only as a rough predictor of drag loads, but the model works well in rollout simulators to allow pilots to "feel" the drag effects associated with spinup.

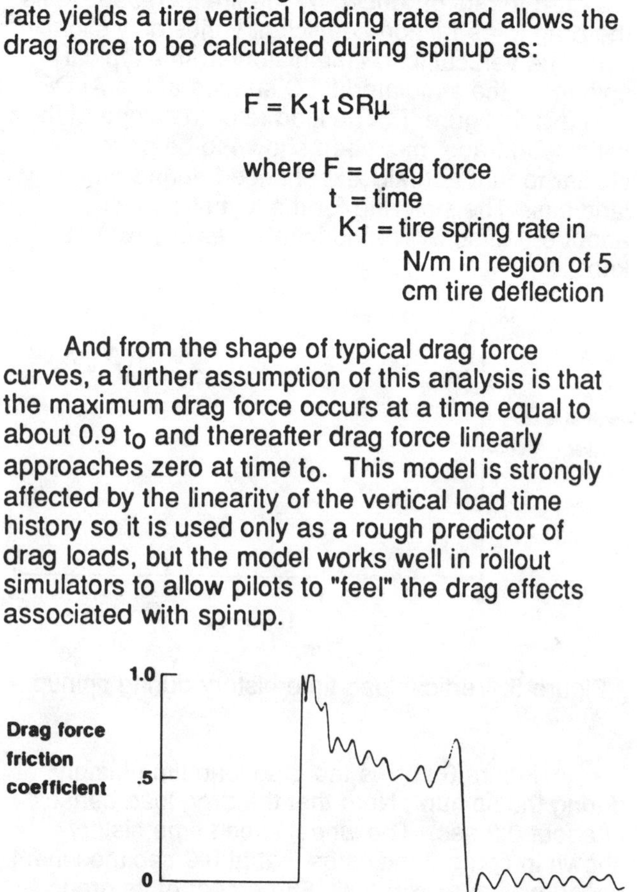

Figure 10. Drag force coefficient time history during spinup.

SURFACE EFFECTS - The texture of the surface on which spinup occurs appears to have a significant effect on both the spin-up loads and tire wear. Tests discussed in reference 2 show that friction differences of almost a factor of 2 exist between a concrete runway and an aircraft carrier non-skid deck during spinup. All of the tire response during spinup can be linked to the surface characteristics. For example, as shown on the KSC runway, typical spin-up drag loads last for about 0.1 seconds and the maximum drag load is about 111 kN. Figure 11 shows the drag load and drag force coefficient time histories for a spinup conducted on a smooth, ungrooved concrete runway at approximately the same sink rate as for the KSC spinup. The horizontal velocity for this test was 209 knots. The characteristics of the surface are such that the average drag force friction coefficient during the spinup is about 0.3 rather than 0.6.

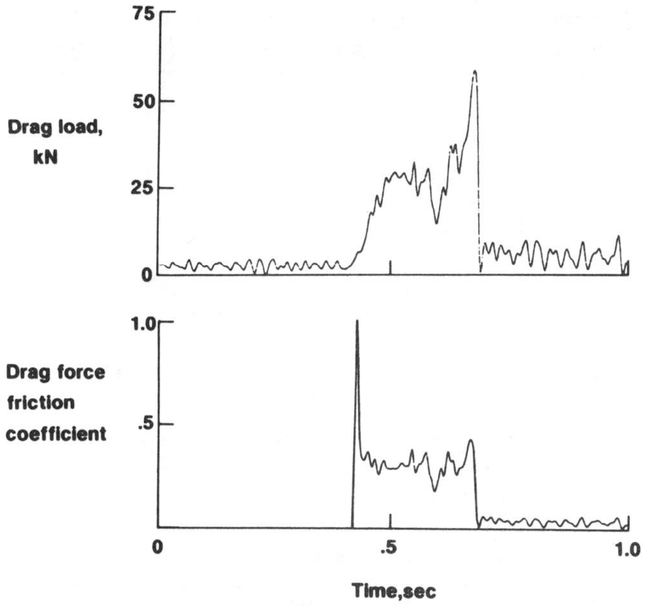

Figure 11. Drag load and drag force coefficient time histories for spinup on smooth, ungrooved concrete.

This causes the spin-up time to be longer (as shown by the t_o equation) and it is in fact about 0.25 seconds as opposed to 0.1 seconds for the KSC spinup. Consequently, the drag loads do not rise to as high a level. The energy expended during this spinup was approximately 0.72 MJ, while the kinetic energy of the wheel at 209 knots was 0.36 MJ. Again, the difference between the two energy levels is the heat and tire wear losses that occur during spinup. The depth of tire wear, though, for spinup on the smooth, ungrooved concrete runway is only about 1.6 mm - not even through the tread ribs. The reason the tire does not wear as deeply on the smooth surface as it does for the KSC spinup is again related to the surface friction characteristics. The lower friction surface and corresponding longer spin-up time allows the tire to turn through a larger angle during spinup, thus distributing the heat and wear energy losses over a larger portion of the circumference. Consequently, the depth of wear for the smooth runway spinup is not as great as for the KSC spinup.

SINK RATE - A test was conducted to determine if tire wear during spinup was dependent on sink rate. The equation given for spin-up time, t_o, shows that increasing the sink rate decreases the spin-up time because of the higher tire loading rate which causes drag forces to build up more quickly. One would also be led to believe that since a large amount of energy must now be put into the tire in a shorter period of time and therefore into a smaller portion of the circumference, that tire wear depth would increase. Figure 12 shows the rubber deposit left by the tire after a 3.4 m/sec sink rate test at 207 kts. One can see the rapid buildup of vertical load as a widening in the rubber deposit.

Figure 12. High sink rate test rubber deposit on runway.

Figure 13 shows the tire after the test and it can be seen that tire wear depth is no worse than for a normal landing. The reason for this is that although more energy is input into the tire per unit circumference, the tire footprint width grows very quickly because of the high sink rate, exposing more tire area to the wear mechanism. The end result is that the amount of energy per unit area is probably only slightly affected by sink rate, therefore, tire wear depth is more or less independent of sink rate.

Figure 13. Tire wear caused by high sink rate test.

Figure 15. Tire wear for 27 knot prerotation test.

PREROTATION - Several tests were run to evaluate the effect of prerotating the tire prior to touchdown on tire wear. Three prerotation speeds, 7, 15, and 27 knots, were evaluated using landing speeds of approximately 215 kts. and sink rates of 0.6 to 0.9 m/sec. Previous tests described in reference 3 showed substantial tire wear benefits with pre-spin velocities of about 11 percent of landing speed. The results of the three prerotation tests are presented in figure 14 where spin-up wear is plotted as a function of prerotation speed. Figure 15 shows a photograph of the tire after a 27 kt. prerotation speed test. Comparing it with figure 8 shows the benefit in spin-up wear attained.
The wear depth of the tire shown in figure 8 was about 5.6 mm whereas the wear depth for the 27 kt. prerotation test was only about 2.5 mm.

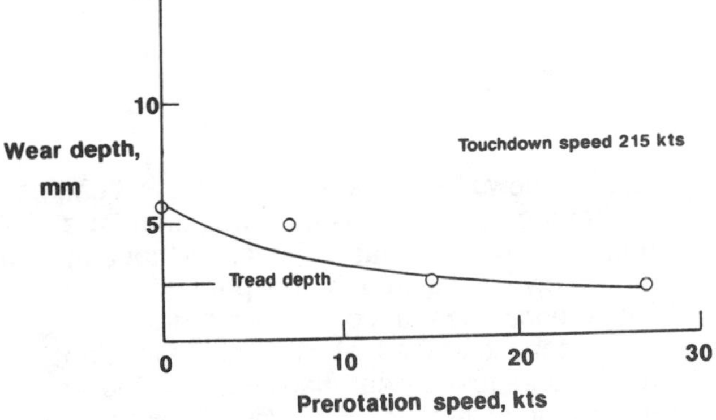

Figure 14. Spin-up wear depth as a function of prerotation speed.

MEASURING SINK RATE - A method to measure the sink rate of an aircraft based upon the appearance of the spin-up rubber deposits was developed. Several assumptions must be made to use the procedure. The first assumption is that wheel angular velocity is nearly zero for the first several feet of the spinup. This assumption is easily made for the orbiter tire because of its high inertia and figure 7 supports the assumption. Another assumption is that in the first two centimeters or so of tire vertical deflection during spinup, the aircraft sink speed is constant.
The larger the tire, the more accurate the assumption is. Static tests were conducted to determine the effect of vertical load on both tire footprint width and tire vertical deflection.

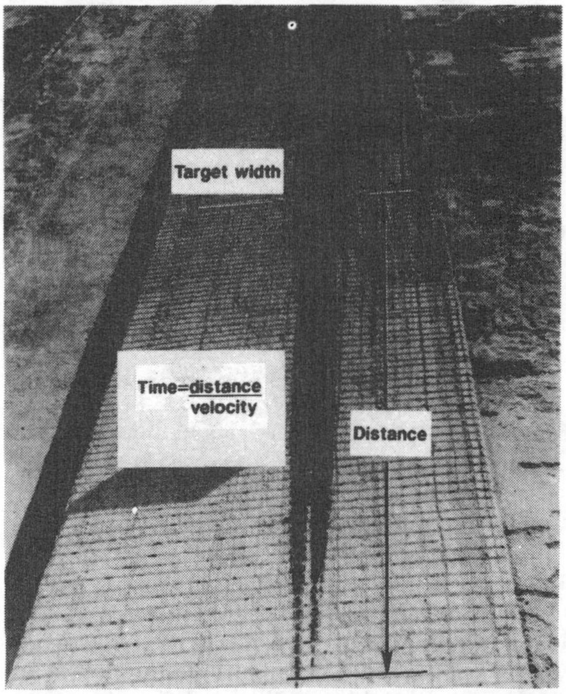

Figure 16. Rubber deposit on runway from typical spinup.

By knowing approximately what load it takes to get about two centimeters of tire deflection, for example, the corresponding footprint width is also known. Then the rubber deposit on the runway is measured as shown in figure 16. The distance it takes to reach the target footprint width can be converted into time knowing the landing speed. Then the target deflection (two centimeters or whatever is convenient) can be divided by the time it took to reach that deflection to yield the sink rate that the tire or strut had at touchdown. Measurements of the tire sink rate made at the ALDF using this method have been in extremely good agreement with actual measurements of sink rate using instrumentation onboard the carriage.

CONCLUDING REMARKS

An experimental investigation was conducted at the NASA Langley Research Center's Aircraft Landing Dynamics Facility to gain insight into the spin-up dynamics of the orbiter main gear tire. Tests showed that severe wear occurs during spinup on the extremely rough KSC runway surface. Energy levels near 0.7 MJ are expended during the spin-up process, with about 50 percent of the energy becoming tire rotational kinetic energy and the other half dissipated as heat and tire wear. It was learned that torsional spring characteristics of the tire cause the wheel angular acceleration history to lag the drag force time history. The torsional energy is released later in the spin-up process when the wheel speed approaches 70 to 75 percent of synchronous velocity. This causes the drag loads to drop to zero which causes the cessation of runway rubber deposits. The remainder of the spinup occurs with no externally applied forces.

Friction coefficients provided by the KSC runway during spinup average about 0.6, and the time during which drag loads are produced is about 0.1 sec. for nominal spinups. Spinup on the smooth, ungrooved concrete runway produces average friction coefficients of about 0.3, with an increase of the nominal drag load production time to about 0.25 sec.

High sink rates apparently do no increase tire wear depth during spinup because energy is distributed over a wider footprint area due to the increased vertical loading rate.

Prerotation of the main tire was found to significantly reduce the spin-up wear. Prerotation velocities of about 10 percent of synchronous velocity reduced spin-up wear by over 50 percent.

A method to calculate sink rate based on measurements of the runway spin-up rubber patch was developed and close correlation to actual sink rate measurements was obtained.

REFERENCES

1. Davis, Pamela A.; Stubbs, Sandy M.; Tanner, John A.: Aircraft Landing Dynamics Facility, A Unique Facility with New Capabilities. SAE 851938 Aerospace Technology Conference and Exposition, Long Beach, CA October 1985

2. Horne, Walter B.: Experimental Investigation of Spin-up Friction Coefficients on Concrete and Nonskid Carrier Deck Surfaces. NASA TN D-214 , April 1960

3. Byrdsong, Thomas A.; McCarty, John Locke; Yager, Thomas J.: Investigation of Aircraft Tire Damage Resulting From Touchdown on Grooved Runway Surfaces. NASA TN D-6690, March 1972

Cornering and Wear Characteristics of the Space Shuttle Orbiter Nose-Gear Tire

Pamela A. Davis and Sandy M. Stubbs
NASA Langley Research Center
William A. Vogler
Planning Research Corp.

* Paper 892347 presented at the Aerospace Technology Conference and Exposition, Anaheim, California, September, 1989.

ABSTRACT

Tests of the Space Shuttle Orbiter nose-gear tire have been completed at NASA Langley's Aircraft Landing Dynamics Facility. The purpose of these tests was to determine the cornering and wear characteristics of the Space Shuttle Orbiter nose-gear tire under realistic operating conditions. The tire was tested on a simulated Kennedy Space Center runway surface at speeds from 100 to 180 kts. The results of these tests defined the cornering characteristics which included side forces and associated side force friction coefficient over a range of yaw angles from 0° to 12°. Wear characteristics were defined by tire tread and cord wear over a yaw angle range of 0° to 4° under dry and wet runway conditions. Wear characteristics were also defined for a 15 kt crosswind landing with two blown right main-gear tires and nose-gear steering engaged.

THE SPACE SHUTTLE ORBITER is a unique space vehicle designed to land like a conventional airplane. As conventional airplanes are subject to crosswind landings, so too is the Space Shuttle Orbiter. During the later part of the rollout, the pilot must use either differential braking or nose-gear steering in order to maintain the vehicle on the proper runway heading during a crosswind landing. In the late 1970's, there was a desire to define the response of the Space Shuttle Orbiter to nose-gear steering input which was satisfied by measuring the cornering characteristics of the nose-gear tire under realistic operating conditions. These tests were conducted on a dry light broom finish runway surface in a speed range from 50 to 100 kts with a yaw angle range of 0° to 12°. More information concerning these earlier tests is available in reference 1. Since these tests were conducted, there has been interest in obtaining cornering as well as wear data on the Space Shuttle Orbiter nose-gear tire at its higher landing speeds and on a test surface that simulates the Kennedy Space Center runway, which is very rough and caused more tire wear than expected during Space Shuttle Orbiter landings. The cornering tests were conducted on a dry KSC surface with a yaw angle range of 0° to 10° at speeds of 180 knots. Dry and wet surface tests were completed to measure tire wear with a yaw angle range of 0° to 4° at similar speeds. A worst condition wear test was also conducted which simulated a 15 knot crosswind with two right main-gear tires blown and nose-gear steering engaged.

The purpose of this paper is to present the data obtained from these two test programs conducted with the Space Shuttle Orbiter nose-gear tire and to analyze these data with respect to tire cornering and wear characteristics.

APPARATUS AND TEST PROCEDURES

TEST FACILITY-Tests of the Space Shuttle Orbiter nose-gear tire were performed at NASA Langley Research Center's Aircraft Landing Dynamics Facility (ALDF) as shown in figure 1. This

Figure 1 - Aircraft Landing Dynamics Facility.

is a unique facility that is designed for testing aircraft tires, landing gear, and landing gear systems under simulated takeoff and landing conditions on actual runway surfaces. The test tire was mounted on the 48,100 kg (106,000 lb_m) test carriage which reached speeds of 180 kts and was propelled down the 853 m (2800 ft) track by a high-pressure water jet system. An arrestment system consisting of five sets of water turbines connected by nylon tapes and steel cabling at the end of the 549 m (1800 ft) test section was used to stop the carriage. A more detailed description of the facility is available in reference 2.

TEST SURFACE-The Space Shuttle Orbiter nose-gear tire was tested on a simulated KSC runway surface. The KSC surface shown in figure 2 consisted of a longitudinal brushed texture in combination with transverse grooves 6.4 mm (0.25 in.) wide, 6.4 mm (0.25 in.) deep and spaced 29 mm (1.15 in.) apart. The surface is extremely rough and provides good traction during a wet landing (Ref. 3). The surface was level with no crown. Dry surface cornering tests and both dry and wet surface tire wear tests were conducted.

Figure 2 - Kennedy Space Center Runway.

TEST TIRES-The tires used in these tests were 32 X 8.8, type VII, bias-ply aircraft tires with a 20-ply rating. Figure 3 shows a new Space Shuttle Orbiter nose-gear tire. The tire has a three groove tread pattern with the grooves spaced 30 mm (1.2 in.) apart and has 10 actual carcass plies with a single under tread ply. The original groove depth was 2.4 mm (0.094 in.). A total of five tires were used and tested up to a vertical load of 133.4 kN (30,000 lb) and at the rated pressure of 2.07 MPa (300 psi).

INSTRUMENTATION-Tire friction characteristics were measured using a dynamometer as shown in the photograph and schematically in figure 4. The dynamometer is instrumented with strain gage load beams to measure axle loads in the vertical, lateral and fore-and-aft directions. In addition, three strain gage type accelerometers mounted on the test wheel axle were used to measure axle acceleration along the three axes so that inertial corrections to the load data could be made.

Figure 3 - New Space Shuttle Orbiter nose-gear tire.

The carriage on board instrumentation system uses telemetry to transmit the data to a ground station where the digital data is stored on a 26-megabyte computer for further analysis.

TEST PROCEDURES-The test procedures for the cornering tests consisted of rotating the dynamometer and wheel assembly to the specified yaw angle, accelerating the carriage to the desired speed and then lowering the tire on to the test surface. Once the maximum vertical load was applied, the tire was raised off the runway. The yaw angle was held constant during these tests and ranged from 0° to 10°. All cornering tests were conducted on a dry surface at approximately 180 kts with peak vertical loads of approximately 133.4 kN (30,000 lb).

Tire wear test procedures consisted of setting the tire wheel assembly at the specified yaw angle, accelerating the carriage to the desired speed and rolling the tire on to the test surface. The tire was rolled for 488 m (1600 ft) and then raised off the runway. The yaw angle was again held constant during these tests and ranged from 0° to 4°. Dry and wet runway surface tire wear tests were conducted over a speed range of 100 to 180 kts with vertical loads of approximately 80 kN (18,000 lb).

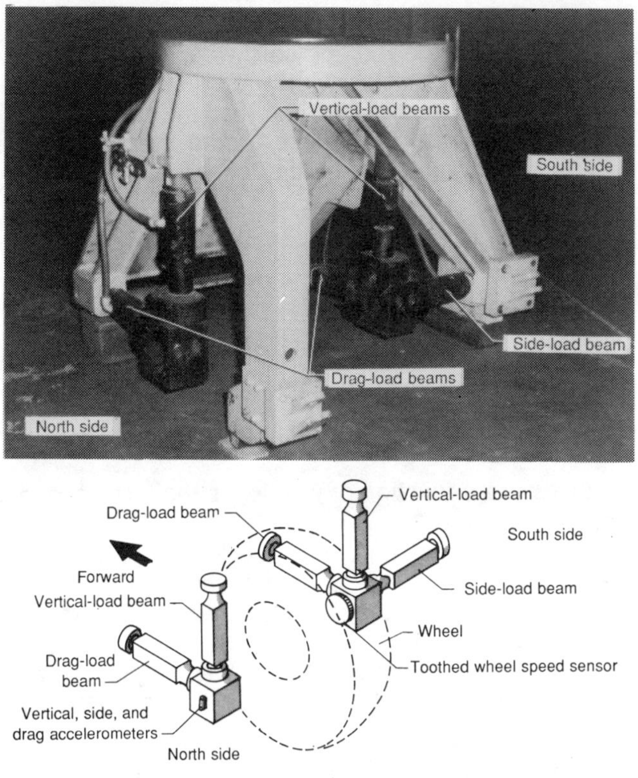

Figure 4 - Photograph and sketch of dynamometer.

Table 1 - Space Shuttle Nose-Gear Tire Test Matrix

Run Number	Speed (kts)	Yaw Angle (Degrees)	Vertical Load (kN/klb)	Test Surface Condition
Test Type: Cornering Characteristics				
1	180	0	133/30	Dry
2	↓	1	↓	↓
3		2		
4		4		
5		7		
6		7		
7		10		
Test Type: Tire Wear Characteristics				
8	180	0	80/18	Dry
9	180	2	↓	↓
10	160	2		
11	150	2		
12	130	2		
13	100	2		
14	↓	2		
15		2		
16		4		
17	150	1		Wet
18	160	2		↓
19	160	4		
20	100	1		
21	↓	2		
22		4		
Test Type: 15 knot Crosswind Simulation				
23	165	1	93/21	Dry
24	150	3	↓	↓
25	130	4		
26	100	2.5		
27	100	2.5		

The final test series conducted was to determine tire wear characteristics under conditions that would simulate a 15 kt crosswind landing with two right main-gear tires blown and nose-gear steering engaged. The tests were conducted on a dry surface with a yaw angle range of 1° to 4° for the first three test runs and 2.5° for the last two test runs. The speeds ranged from 165 to 100 kts with a vertical load of approximately 93.3 kN (21,000 lb). Table 1 gives a complete test matrix for the above tests.

RESULTS AND DISCUSSION

CORNERING CHARACTERISTICS-Tests were conducted at the ALDF in the late 1970's to measure the cornering characteristics of the Space Shuttle Orbiter nose-gear tire. These cornering tests analyzed the effects of tire vertical load, yaw angle and ground speed on side force and its associated friction coefficient. These tests were conducted at yaw angles of 0° to 12° over a speed range of 50 to 100 kts. The data from these tests were published in reference 1. In the mid 1980's, there was a desire to verify that the results from the above tests were valid at the higher landing speed of 180 knots. The latter tests were conducted at similar yaw angles and vertical loads. The cornering data presented is the combined data from both test programs. This data is presented as carpet plots in order to illustrate the relationship between the cornering characteristics and the test parameters. These characteristics are presented as a function of both yaw angle and vertical load. The speed at which the tests were conducted is represented by the test-point symbols identified in the plot legend. A least-squares bi-cubic curve fit to the data was used.

Side Force-The side force was measured normal to the wheel plane and is presented in figure 5 for the various yaw angles and vertical loads tested. As the yaw angle increases, the effect on side force due to changes in the vertical force is more predominant. The trends in the data show that side force increases with increasing yaw angle with a constant vertical load. Side force also increases with increasing vertical load at a constant yaw angle. At the lower vertical loads, the side force appears to reach a maximum at the maximum yaw angle tested. There is no evidence that ground speed had an effect on side force.

Side force friction coefficient, μ_s, is a ratio of the side force to the vertical force and is plotted as a function of vertical load and yaw angle in figure 6. For the tests conducted, the data show that μ_s decreases with increasing vertical load at a constant yaw angle. The coefficient values increased with increasing yaw angle at a constant vertical force. The yaw angle at which maximum μ_s occurred increases as the vertical force increases. The coefficient values varied between 0 and 0.5. There is no evidence that ground speed had an effect on side force friction coefficient.

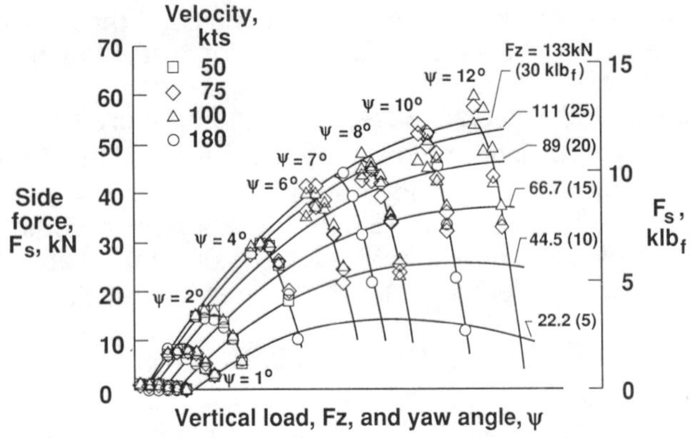

Figure 5 - Variation of side force with vertical load and yaw angle over a range of ground speed.

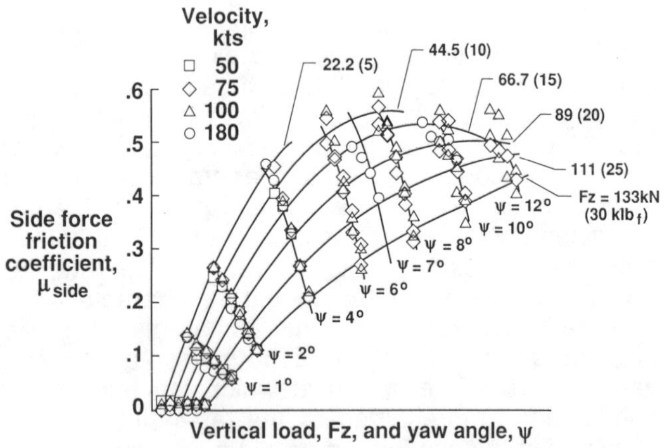

Figure 6 - Variation of side-force friction coefficient with vertical load and yaw angle over a range of ground speed.

WEAR CHARACTERISTICS-When more than expected wear appeared on the Space Shuttle Orbiter main-gear tires during KSC landings, tests were conducted with a main-gear tire at the ALDF to determine its wear characteristics and possible solutions to the wear problem (3). As a result of this program, wear tests were conducted on the Space Shuttle Orbiter nose-gear tire at the ALDF in order to determine its wear characteristics. A total of 15 tests were completed on dry and wet KSC surfaces. In order to simulate wear on landing rollout, the yaw angle ranged from 0° to 4° with a decreasing speed of 180 to 100 kts at a vertical load of 80 kN (18,000 lb). A final wear scenario that was conducted was considered to be a worse case situation; a 15 kt crosswind landing with two right main-gear tires blown and nose-gear steering engaged. These tests were conducted on a dry surface with a yaw angle range of 1° to 4° over a speed range of 160 to 100 kts with a vertical load of 93.3 kN (21,000 lb).

The tire wear data for all these tests is presented as tire tread and cord wear as a function of side energy as well as with photographs of the tire. Side energy is defined as the amount of work produced laterally by the tire during cornering. It is defined mathematically by the following:

$$\text{Side Energy} = \int_0^S F \sin\psi \, dx$$

where: S = total rollout distance
F = side force
ψ = yaw angle

F and ψ are a function of x which is the instantaneous rollout distance (3).

Dry Surface-Nine tests were conducted to look at tire wear on a dry surface. The first test was a spin-up test to simulate spin-up wear at nose gear pitch over. The spin-up patch was long and narrow due to a slow sink rate but showed negligible wear as can be seen in figure 7. The other eight tests

Figure 7 - Spin-up wear on a dry surface.

were to simulate tire wear during rollout with a 2° yaw angle and a 4° yaw angle on the last test. There was little wear during the 2° yaw angle tests with only the first cord exposed in the spin-up patch area and around the right shoulder. With the increase in yaw angle to 4° for the last test, there was significantly more wear. As a result of this test, the fifth cord was broken in the spin-up patch area and is shown in figure 8.

Wet Surface-Six tests were conducted to study the effects of a wet KSC surface on nose-gear tire wear. The initial test was, again, to determine spin-up wear on the tire at a 1° yaw angle. The spin-up wear was negligible and the tire remained in a nearly new condition as shown in figure 9. The yaw angle was then increased up to 4°, reduced to 1° and

Figure 8 - Roll-out wear on a dry surface.

Figure 9 - Spin-up wear on a wet surface.

Figure 10 - Roll-out wear on a wet surface.

Figure 11 - Blown tire.

increased again to 4° as the speed decreased to study the roll-out wear under wet surface conditions. After the first three runs, the first cord was exposed circumferentially and broken in a few places. After the fifth test, the second cord was exposed circumferentially and broken in many areas with the third cord exposed in some areas as shown in figure 10. The final test was at a 4° yaw angle and at 100 kts. This test resulted in a blown tire at approximately 305 m (1000 ft) into the test run. The tire blew when cord wear was between the sixth and seventh cord and is shown in figure 11.

Tire wear is plotted as a function of side energy in figure 12. This plot shows that tire wear realtive to side energy on the KSC runway is not sensitive to surface wetness. The wet tests indicated less spin-up wear than during a dry test. The tire blow out under wet conditions occurred because enough side energy was generated to wear the tire to such an extremely thin state. It is apparent from figure 12, that further testing under dry conditions would have resulted in a similar situation.

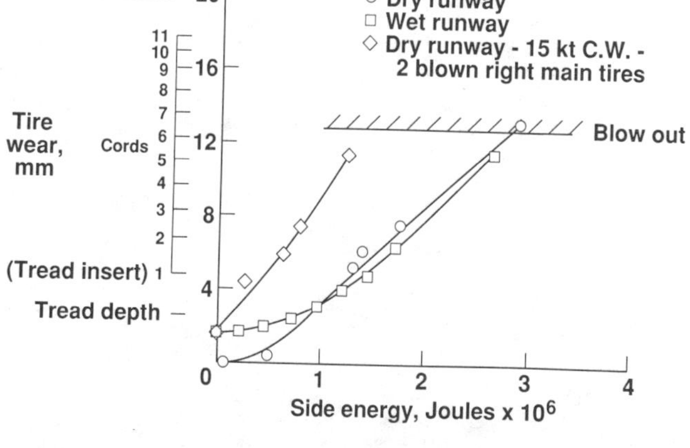

Figure 12 - Space Shuttle Orbiter nose-gear tire wear as a function of side energy.

Worst Case Simulation-Five tests were completed on a dry KSC surface to simulate a 15 kt crosswind landing. The tire was first spun-up for approximately 16.2 m (250 ft) at 160 kts with a 1° yaw angle and a vertical load of 93.3 kN (21,000 lb). The spin-up wear was negligible with some scuffing occurring on both shoulders as shown in figure 13.

Figure 13 - Spin-up wear on a dry surface with a 15 kt crosswind simulation.

The second and third tests were at 3° and 4° yaw angles respectively, at the same vertical load and at speeds of 150 and 130 kts respectively. At a 4° yaw angle, the first cord was exposed in the spin-up patch area and at various other places on the tire. The final two tests were at a 2.5° yaw angle, similar vertical load and at 100 kts. The final test resulted in the fifth cord being broken in the spin-up patch area as shown in figure 14.

Figure 14 - Rollout wear on a dry surface with a 15 kt crosswind simulation.

From figure 12, the tire wear at spin-up was the same as for the dry runway tests as expected. However, for this worst case situation, the wear increased more quickly as a function of side energy than for the other two tires tested. It is important to note that although the tire did not produce as much side energy as the other two tires, total wear depth was comparable. The reason for the increased wear rate is not fully understood.

CONCLUSIONS

Cornering and wear tests were conducted on the Space Shuttle Orbiter nose-gear tire at NASA Langley's Aircraft Landing Dynamics Facility. The results of the tests indicate the following conclusions:

(1) Cornering characteristics of the Space Shuttle Orbiter nose-gear tire are insensitive to variations in ground speed over the speed range of 50 to 180 kts.

(2) Side forces generated during cornering tests were significant and the effect of vertical force on side load became more predominant as the yaw angle increased .

(3) The side force friction coefficient values varied between 0 and 0.5 and increased with increasing yaw angle at a constant vertical force.

(4) Tire wear as a function of tire side energy produced was similar under dry and wet surface conditions.

(5) The reason for the increase in wear rate for the tire used in the crosswind simulation tests compared with the other two tires is not known.

REFERENCES

1. Vogler, William A.; Tanner, John A.: Cornering Characteristics of the Nose-Gear Tire of the Space Shuttle Orbiter. NASA TP 1917, October, 1981.
2. Davis, Pamela A., Stubbs, Sandy M; Tanner, John A.: Langley Aircraft Landing Dynamics Facility. NASA RP 1189, October, 1987.
3. Daugherty, Robert H.; Stubbs, Sandy M.: Cornering and Wear Behavior of the Space Shuttle Orbiter Main Gear Tire. Presented at the 1987 SAE Aerospace Technology Conference and Exposition, Long Beach, California, October, 1987.

BIBLIOGRAPHY

APPENDIX 1

This Appendix references additional literature available on aircraft landing gear systems. Due to space constraints, the valuable technical papers included in the bibliography were unable to be reprinted in their entirety in the book. Instead, abstracts of each paper have been included. The papers in this Appendix have been arranged in alphabetical order by the primary author's last name.

All of these papers have been published by SAE and are available in original or photocopy form. For ordering information, contact the Customer Service Department, SAE, 400 Commonwealth Drive, Warrendale, Pennsylvania, 15096-0001, USA, 412/776-4970.

Black, Raymond J., "Realistic Evaluation of Landing Gear Shimmy Stabilization by Test and Analysis," SAE Technical Paper 760496

An experimental and analytical program for prediction of airplane landing gear shimmy stability is outlined. The method makes use of laboratory shimmy tests on a flywheel which simulates the runway and a landing gear mounting structure which simulates the fuselage. Differences between the laboratory tests and airplane tests are detailed. Because of the latter differences, the prediction of airplane results is carried out by an experimentally verified analysis rather than a direct application of the laboratory test results. The analytical model is outlined including the tire mechanics. Samples of correlation between analytical results and experimental results (laboratory and airplane) are given.

Bobo, Stephen N.; Johnson, Richard A.; Durup, Paul C., "Comparative Tests of Aircraft Radial and Bias Ply Tires," SAE Technical Paper 881359

Laboratory dynamometer tests are being conducted to assess the difference in performance between radial and bias ply tires, both new and retreaded, of various sizes and manufacturers. Tire properties that affect the operation and safety of landing gear systems, such as temperature performance, cornering power, dynamic loaded radius and wheel stresses are being compared. The tests are described along with some initial findings.

Elsaie, A. M.; Santillan, R. Jr., " Structural Optimization of Landing Gears Using STARSTRUC," SAE Technical Paper 871047

The impact of structural optimization is growing in many industries due to economic pressures demanding efficiency in the design process. This efficiency implies developing products which are cost effective and ahead of the competition at the same time. The motivation of the present work is to provide the structural design engineer with tools of optimization techniques and practices that have been applied successfully to landing gears.

Modern landing gears have to meet a multitude of landing and ground handling design loads whose magnitudes are several times the gross weight of the aircraft. All the design loads have to be investigated and their effect on each component must be evaluated. Furthermore, the response of the landing gears to all the design loads must be constrained to satisfy the design requirements while minimizing its structural weight. The weight of the landing gear is becoming an ever more important factor, as inefficient design can add unnecessary weight to the aircraft and, consequently, decrease the payload or useful load.

Typical design examples of components of landing gears are presented that demonstrate the performance of STARSTRUC as an effective weight optimization design tool. The minimum weight design is achieved when the landing gear is subjected to behavior constraints on stresses, deflections, buckling, and frequencies of vibration.

Gehrett, Larry J., "Proper Aircraft Tire Size Selection-Optimum Performance with Minimum Maintenance," SAE Technical Paper 790598

High speeds with heavy loads represent the type of operating condition to which the aircraft tire is subjected during its utilization on business aircraft. This type of operation produces severe dynamic forces that challenge the tire design engineer. This challenge is met by selecting the proper tire size, incorporating the necessary design features into the tire construction, and then thoroughly testing the tire to ensure compliance with the performance requirements established by the airframe engineer. The result of selecting the proper tire size is optimum tire and aircraft performance achieved with minimum cost for both airframe manufacturer and end user — the aircraft customer.

Grossman, Daniel T., "F-15 Nose Landing Gear Shimmy, Taxi Test and Correlative Analyses," SAE Technical Paper 801239

F-15 taxi tests and analyses were performed to evaluate the effects of aircraft design changes on nose landing gear shimmy. Preliminary analytical studies indicated that these changes would have an adverse effect on shimmy speed. This was of particular concern because limit cycle shimmy had been experienced on the baseline gear for cases with out-of-tolerance strut torsional freeplay. The trade-offs considered in the choice of a taxi test over a laboratory dynamometer test are presented. Operational aspects of the taxi test are discussed. Several instances of limit cycle shimmy were encountered during testing and results indicate that shimmy speed is a function of strut torsional freeplay. A description of the math model used in the non-linear analyses is provided. Analytical results are presented in terms of shimmy speed versus strut torsional freeplay. These results confirm the limit cycle nature of the shimmy phenomenon and correlate well with the taxi test results. Additional analyses are presented indicating the sensitivity of the shimmy to changes in tire parameter values and strut frictional coefficients. Assumptions used in the development of an equivalent linear math model are given.

RELATED READING

APPENDIX 2

This appendix is a collection of papers suggested for related reading by the individuals who assisted with the development of this publication. Due to space constraints, the abstracts of these papers were not included. These papers, nevertheless, are valuable contributions to aircraft landing gear systems technology. The titles in this Appendix have been arranged in alphabetical order by the primary author's last name, followed by the publisher. Most of these references have not been published by SAE, therefore, copies should be obtained directly from the publisher listed.

"Aircraft Tire Engineering Data," B.F. Goodrich Company (1983).

Batterson, Sidney A., "A Study of the Dynamics of Airplane Braking Systems as Affected by Tire Elasticity and Brake Response," NASA TN D-3081, 1965.

Berry, R. W., and Tsien, V. C., "Mathematical Analysis of Corotating Nose Gear Shimmy Phenomena," Journal of the Aeronautical Science, December 1962, pp. 1462 - 1470.

Brewer, H. K., "Stresses and Deformations in Multi-ply Aircraft Tires Subject to Inflation Pressure Loadings," Wright Patterson Air Force Base, Ohio. Techn. Rep. AFFDL-TR-70-62, June 1970.

Cesar, J. P., Musy, J., and Olds, R., "Development of Radial Aircraft Tires," Michelin paper presented to 38th International Air Safety Seminar (Boston, Massachusetts), Nov. 4-7, 1985.

Clark, Samuel K., and Dodge, Richard N., "Heat Generation in Aircraft Tires Under Free Rolling Conditions," NASA CR-3629, 1982.

Clark, S. K., Dodge, R. N., Lackey, J. I., and Nybakken, G. H., "Structural Modeling of Aircraft Tires," NASA CR-2220, 1973.

Clark, S. K., "Theory of the Elastic Net Applied to Cord-Rubber Composites," Rubber Chemical Technol. 56, 372-389 (May - June 1983).

Clark, S. K., Dodge, R. N., and Nybakken, G. H., "Dynamic Properties of Aircraft Tires," AIAA Journal of Aircraft, March 1974, pp. 166-172.

Clark, S. K., Dodge, R. N., Lackey, J. I., and Nybakken, G. H., "Structional Modeling of Aircraft Tires," AIAA Journal of Aircraft, February 1972, pp 162-167.

Clark, Samuel K., "Mechanics of Pneumatic Tires," U.S. Department of Transportation, 1981.

Collins, R. L., and Black, R. J., "Tire Parameter for Landing Gear Shimmy Studies," AIAA Journal of Aircraft, May-June 1969, pp 252-258.

Conant, F. S., "Tire Temperatures," Rubber Chem. Technol., vol. 44, no. 2, 1971, pp. 397-439.

Daugherty, Robert H., and Stubbs, Sandy M., "A Study of the Cornering Forces Generated by Aircraft Tires on a Tilted, Free-Swiveling Nose Gear," NASA TP 2481, October 1985.

Daugherty, Robert H., Stubbs, Sandy M., and Robinson, Martha P., "Cornering Characteristics of the Main-Gear Tire of the Space Shuttle Orbiter," NASA TP 2790, March 1988.

Davis, Pamela A., and Lopez Mercedes C., "Static Mechanical Properties of 30x 11.5-14.5, Type VIII Aircraft Tires of Bias-Ply and Radial-Belted Design," NASA TP 2810, May 1988.

DeCarbon, Christian Bourcier, "Analytical Study of Shimmy of Airplane Wheels," NACA TM 1337, September 1952. (Translation of paper published in 1948.)

DeEskinazi, J., Werner, S., and Yang, T.Y., "Contact of an Inflated Toroidal Membrane With a Flat Surface as an Approach to the Tire Deflection Program," Tire Sci. Technol. 3. 43-61 (1975).

Dengler, M., Goland, M., Hermman, G., "A Bibliographic Survey of Automobile and Aircraft Shimmy," WADC Technical Report 52-141, December 1951, p 41.

Dodge, R. N., Larson, R. B., Clark, S. K., and Nybakken, G. H., "Testing Techniques for Determining Static Mechanical Properties of Pneumatic Tires," NASA CR-2412, 1974.

Dreher, Robert C., and Tanner, John A., "Experimental Investigation of the Cornering Characteristics of a C40 x 14-21 Cantilever Aircraft Tire," NASA TN D-7203, 1973.

Dreher, Robert C., and Yager, Thomas J., "Friction Characteristics of 20 x 4.4, Type VII, Aircraft Tires Constructed With Different Tread Rubber Compounds," NASA TN D-8252, 1976.

Dreher, Robert C., and Tanner, John A., "Experimental Investigation of the Cornering Characteristics of 18 x 5.5, Type VII, Aircraft Tires with Different Tread Patterns," NASA TN D-7815, 1974.

Durup, Paul C., "Improvement of Overload Capability of Air Carrier Aircraft Tires," Rep. No. FAA-RD-78-133, Oct. 1978.

Edman, J. L., "Experimental Study of Moreland's Theory of Shimmy," WADC Technical Report 56-197, July 1956.

Ellis, J.R., "Vehicle Dynamics," London Books Ltd., London 1969, page 61 and page 95.

Feng, W. W., Tielking, J.T., and Huang, P., "The Inflation and Contact Constraint of a Rectangular Mooney Membrane," J. Appl. Mech. 73. 979-983 (1974).

Frank, F., and Hofferberth, W., "Mechanics of the Pneumatic Tire," Rubber Chemical Technol. 40, 271-322 (1967).

Fromm, H., Von Schlippe, B., and Dietrich, R., "Papers on Shimmy and Rolling Behavior of Landing Gear," Presented at Stuttgart Conference, NACA TM 1365, August 1954. (Translation of paper published in 1941.)

Hadekel, R., "The Mechanical Characteristics of Pneumatic Tyres - A Digest of Present Knowledge," S&T Memo, No. 10/52, TPA 3/TIB, British Min. Supply, Nov. 1952. (Supercedes S&T Memo. No. 5/50.)

Hample, W. G., "Friction Study of Aircraft Tire Material on Concrete," NACA TN 3294, 1955.

Hirzel, E. A., "Antiskid and Modern Aircraft," SAE Technical Paper 720868.

Ho, F. H. and Lai, J. L., "Parametric Shimmy of a Nose Gear," AIAA Journal of Aircraft, July-August 1970, pp. 373-375.

Horne, Walter B., and Dreher, Robert C., "Phenomena of Pneumatic Tire Hydroplaning," NASA TN-D-2056, 1963.

Horne, W. B., "Experimental Investigation of Spin-Up Friction Coefficients on Concrete and Nonskid Carrier - Deck Surfaces," NASA TN D-214, April 1960.

Horne, Walter B., Stephenson, Bertrand H., and Smiley, Robert F., "Low-Speed Yawed-Rolling and Some Other Elastic Characteristics of Two 56-Inch-Diameter, 24-Ply-Rating Aircraft Tires," NACA TN 3235, 1954.

Horne, Walter B., Smiley, Robert F., and Stephenson, Bertrand H., "Low-Speed Yawed-Rolling Characteristics and Other Elastic Properties of a Pair of 26-Inch-Diameter, 12-Ply-Rating, Type VII Aircraft Tires," NACA TN 3604, 1956.

Horne, Walter B., "Static Force-Deflection Characteristics of Six Aircraft Tires Under Combined Loading," NACA TN 2926, 1953.

Horne, Walter B., and Smiley, Robert F., "Low Speed Yawed-Rolling Characteristics and Other Elastic Properties of a Pair of 40-Inch-Diameter, 14-Ply-Rating, Type VII Aircraft Tires," NACA TN 4109, 1958.

Joyner, Upshur T., Horne, Walter B., and Leland, Trafford J. W., "Investigations on the Ground Performance of Aircraft Relating to Wet Runway Braking and Slush Drag," AGARD Rep. 429, Jan. 1963.

Kaga, H., Okamota, K., and Toxawa, Y., "Internal Stress Analysis of Tire Under Vertical Loads Using Finite Element Method," Tire Sci. Technol. 5, 102-118 (1977).

Kainradl, P., and Kaufmann, G., "Heat Generation in Pneumatic Tires," Rubber Chem. & Technol., vol. 49, no. 3, July-Aug. 1976, pp. 823-861.

Kalnins, A., "Analysis of Shells of Revolution Subjected to Symmetrical and Non-Symmetrical Loads," J. Appl. Mech. 31.223 (1964).

Kim, Kyun O., Noor, Ahmed K., and Tanner, John A., "Modeling and Analysis of the Space Shuttle Nose-Gear Tire with Semianalytic Finite Elements," NASA TP 2977 April 1990.

Leve, H. L., "Designing Stable Dual Wheel Gears," AIAA Paper No. 69-769, July 1969.

Lou, A.Y.C., "Modification of Linear Membrane Theory for Predicting Inflated Tire Shapes," Developments in Mechanics, Vo. 7. Proc. 13th Midwestern Mechanics Conf., University of Pittsburgh, Pittsburgh, Pennsylvania Aug 13-15, 637-656 (1973).

McCarty, John L., and Tanner, John A., "Temperature Distribution in an Aircraft Tire at Low Ground Speeds," NASA TP-2195, August 1983.

McCarty, J. L., "Results From Recent NASA Tire Thermal Studies," pp. 211-222, NASA CP-2264 (1983).

McCarty, John Locke, "Wear and Related Characteristics of an Aircraft Tire During Braking," NASA TN D-6963, 1972.

McCarty, John L., Yager, Thomas J., and Riccitiello, S. R., "Wear, Friction, and Temperature Characteristics of an Aircraft Tire Undergoing Braking and Cornering," NASA TP-1569, 1979.

Moreland, W. J., "The Story of Shimmy," Journal of the Aeronautical Sciences, December 1954, pp. 793-808.

Noor, A. K., Andersen, C. M., and Tanner, J. A., "Mixed Models and Reduction Techniques for Large-Rotation Nonlinear Analysis of Shells of Revolution with Application to Tires," NASA TP-2343 (1984).

Noor, Ahmed K., and Tanner, John A., "Advances in Contact Algorithms and Their Application to Tires," NASA TP-2781, 1988.

Noor, Ahmed K., Andersen, Carl M., and Tanner, John A., "Exploiting Symmetries in the Modeling and Analysis of Tires," NASA TP-2649, 1987.

Nybakken, G. H., Dodge, R. N., and Clark, S. K., "A Study of Dynamic Tire Properties Over a Range of Tire Constructions," NASA CR-2219, 1973.

Pacejka, H. B., "The Wheel Shimmy Phenomenon," Doctorial Thesis, Delft Technical Institute, December 1966.

"Recommended Practice for Measurement of Static Mechanical Stiffness Properties of Aircraft Tires," AIR 1380, SAE, August, 1975.

Ridha, R. A., "Analysis for Tire Mold Design," Tire Sci. Technol. 1, 195-210 (1974).

Ridha, R. A., "Computation of Stresses, Strains and Deformation of Tires," Rubber Chemical Technol. 53, (4), 849-902 (Sept. - Oct., 1980).

Rogers, L. C. and Brewer, H. K., "Synthesis of Tire Equations for Shimmy and Other Dynamic Studies," AIAA Journal of Aircraft, September 1971, pp 689-697.

Rogers, L. C., "Theoretical Tire Equations for Shimmy and Other Dynamic Studies," AIAA Journal of Aircraft, August 1972, pp 585-589.

Rogers, L. C., "Bi-Normal Coordinates in Descrete Systems with Application to an Aircraft Shimmy Problem," AFFDL TR-72-79, June 1972.

Sleeper, Robert K., and Dreher, Robert C., "Tire Stiffness and Damping Determined From Static and Free-Vibration Tests," NASA TP-1671, 1980.

Smiley, R. F., "Correlation: Evaluation, and Extension of Linearized Theories for Tire Motion and Wheel Shimmy," NACA TM 3632, June 1956.

Smiley, Robert F., and Horne, Walter B., "Mechanical Properties of Pneumatic Tires With Special Reference to Modern Aircraft Tires," NASA TR R-64, 1960. (Supercedes NACA TN 4110.)

Smiley, R. F., and Horne, W. B., "Mechanical Properties of Pneumatic Tires with Special Reference to Modern Aircraft Tires," NASA TR R-64, 1960.

Stevens, J. E., "Shimmy of a Nose Gear with Dual Corotating Wheels," Journal of the Aeronautical Sciences, August 1961, pp. 622-630.

Stevens, J. E., "Relaxation Characteristics of Pneumatic Tires," Journal of Aero/Space Sciences, June 1959, pp 343-350.

Straub, H. H., Yurczyk, R. F., and Attri, N. S., "Development of a Pneumatic-Fluidic Antiskid System," AFFDL-TR-74-117, U.S. Air Force, October 1974. (Available from DTIC as AD A009 170.)

Straub, H. H., Attri, N. S., and Yurczyk, R. F., "Test and Performance Criteria for Airplane Antiskid Systems," AFFDL-TR-74-118, U.S. Air Force, Oct. 1974. (Available from DTIC as AD A008 536.)

Stubbs, Sandy M., Tanner, John A., and Smith, Eunice G., "Behavior of Aircraft Antiskid Braking Systems on Dry and Wet Runway Surfaces - A Slip-Velocity-Controlled, Pressure-Bias-Modulated System," NASA TPO-1051, 1979.

Stubbs, Sandy M., "Landing Characteristics of a Dynamic Model of the HL-10 Manned Lifting Entry Vehicle," NASA TN D-3570, 1966.

Stubbs, Sandy M., and Tanner, John A., "Behavior of Aircraft Antiskid Braking Systems on Dry and Wet Runway Surfaces - A Velocity-Rate-Controlled, Bias-Modulated System," NASA TN D-8332, 1976.

Tanner, John A., Stubbs, Sandy M., and McCarty, John L., "Static and Yawed-rolling Mechanical Properties of Two Type VII Aircraft Tires," NASA TP-1863, May 1981.

Tanner, John A., Stubbs, Sandy M., Dreher, Robert C., and Smith Eunice G., "Dynamics of Aircraft Antiskid Braking Systems," NASA TP-1959, February 1982.

Tanner, John A., Dreher, Robert C., Stubbs, Sandy M., and Smith Eunice, G., "Tire Tread Temperatures During Antiskid Braking and Cornering on a Dry Runway," NASA TP-2009, May 1982.

Tanner, John A., and Dreher, Robert C., "Cornering Characteristics of a 40 x 14-16 Type VII Aircraft Tire and a Comparison with Characteristics of a C40 x 14-21 Cantilever Aircraft Tire," NASA TN D-7351, October 1973.

Tanner, John A., McCarty, John L., and Batterson, Sidney A., "The Elastic Response of Bias-Ply Aircraft Tires to Braking Forces," NASA TN D-6246, 1971.

Tanner, John A., and Stubbs, Sandy M., "Behavior of Aircraft Antiskid Braking Systems on Dry and Wet Runway Surfaces - A Slip-Ratio-Controlled System With Ground Speed Reference From Unbraked Nose Wheel," NASA TN D-8455, 1977.

Tanner, John A., "Fore-and-Aft Elastic Response Characteristics of 34 x 9.9, Type VII, 14 Ply-Rating Aircraft Tires of Bias-Ply, Bias-Belted, and Radial-Belted Design," NASA TN D-7449, 1974.

Tanner, John A., McCarty, John L., and Clark, S. K., "Current Research in Aircraft Tire Design and Performance," 1980 Aircraft Safety and Operating Problems, Joseph W. Stickle, compiler, NASA CP-2170, Part 2, 1981, pp. 543-553.

Tanner, John A., Stubbs, Sandy M., and Smith, Eunice G., "Behavior of Aircraft Antiskid Braking Systems on Dry and Wet Runway Surfaces Hydromechanically Controlled System," NASA TP-1877, 1981.

Thompson, Wilbur E., and Horne, Walter B., "Low-Speed Yawed-Rolling Characteristics of a Pair of 56-Inch-Diameter, 32-Ply-Rating Type VII Aircraft Tires," NASA MEMO 2-7-59L, 1959.

Tielking, J. T., and Feng, W. W., "The Application of the Maximum Potential Energy Principle to Nonlinear Axisymmetric Membrane Problems," J. Appl. Mech. 73. 491-496 (1974).

Vahldiek, A. M., "Stability Studies for the Twin Wheel Landing Gear," ARL TR 60-282, December 1960.

Walter, J. D., and Patel, H. P., "Approximate Expressions for the Elastic Constants of Cord-Rubber Laminates," Rubber Chemical Technol. 52, (4), 710-724 (Sept. - Oct. 1979).

Walter, J. D., "Cord-Rubber Tire Composites: Theory and Applications," Rubber Chemical Technol. 52, (4), 710-724 (Sept-Oct, 1979).

Williams, D., "The Theory and Prevention of Aeroplane Nose-Wheel Shimmy," Royal Aircraft Establishment Report - Structures 125, August 1952.

Yager, Thomas J., Phillips, W. Pelham, Horne, Walter B., and Sparks, Howard C., (Appendix D by R. W. Sugg), "A Comparison of Aircraft and Ground Vehicle Stopping Performance on Dry, Wet, Flooded, Slush-, Snow-, and Ice-Covered Runways," NASA TN D-6098, 1970.

INDEX

A

B

C

D

E

F

G

H

I

J

K

L

M

N

O

P

R

S

T

U

V

W

Y

CURRENT TITLES IN THE SAE PROGRESS IN TECHNOLOGY SERIES

AUTOMOTIVE FUEL ECONOMY
PT-15
315 pp., 22 Papers, ISBN 0-89883-103-2
Casebound 1976

THE MEASUREMENT AND CONTROL OF DIESEL PARTICULATE EMISSIONS
PT-17
388 pp., 16 Papers, ISBN 0-89883-105-9
Casebound 1979

AUTOMOTIVE FUEL ECONOMY
PT-18
325 pp., 17 Papers, ISBN 0-89883-106-7
Casebound 1979

ALCOHOLS AS MOTOR FUELS
PT-19
342 pp., 20 Papers, ISBN 0-89883-107-5
Casebound 1980

GAS TURBINES FOR AUTOS AND TRUCKS
PT-20
420 pp., 17 Papers, ISBN 0-89883-108-3
Casebound 1981

ELECTRIC AND HYBRID VEHICLES
PT-21
344 pp., 25 Papers, ISBN 0-89883-109-1
Casebound 1981

SYNTHETIC AUTOMOTIVE ENGINE OILS
PT-22
356 pp., 26 Papers, ISBN 0-89883-110-5
Casebound 1981

TURBOCHARGED DIESEL AND SPARK IGNITION ENGINES
PT-23
440 pp., 23 Papers, ISBN 0-89883-111-3
Casebound 1981

PASSENGER CAR DIESELS
PT-24
376 pp., 16 Papers, ISBN 0-89883-112-1
Casebound 1981

THE MEASUREMENT AND CONTROL OF DIESEL PARTICULATE EMISSIONS (PART 2)
PT-25
412 pp., 23 Papers, ISBN 0-89883-113-X
Casebound 1982

TWO-STROKE CYCLE SPARK-IGNITION ENGINES
PT-26
468 pp., 30 Papers, ISBN 0-89883-114-8
Casebound 1982

ENGINE-OIL EFFECTS ON VEHICLE FUEL ECONOMY
PT-27
404 pp., 24 Papers, ISBN 0-89883-115-6
Casebound 1982

THE ADIABATIC ENGINE: PAST, PRESENT, AND FUTURE DEVELOPMENTS
PT-28
464 pp., 36 Papers, ISBN 0-89883-116-4
Casebound 1984

ANTI-LOCK BRAKING SYSTEMS FOR PASSENGER CARS AND LIGHT TRUCKS: A REVIEW
PT-29
408 pp., 34 Papers, ISBN 0-89883-117-2
Casebound 1987

CONTINUOSLY VARIABLE TRANSMISSIONS FOR PASSENGER CARS
PT-30
232 pp., 19 Papers, ISBN 0-89883-118-0
Casebound 1987

PASSENGER CAR INFLATABLE RESTRAINT SYSTEMS: A COMPENDIUM OF PUBLISHED SAFETY RESEARCH
PT-31
412 pp., 33 Papers, ISBN 0-89883-119-9
Casebound 1987

PLASTICS IN AUTOMOTIVE APPLICATIONS: AN OVERVIEW
PT-32
300 pp., 33 Papers, ISBN 0-89883-121-0
Casebound 1989

ADVANCES IN TWO STROKE CYCLE ENGINE TECHNOLOGY
PT-33
496 pp., 36 Papers, ISBN 0-89883-120-2
Casebound 1989

RECONSTRUCTION OF MOTOR VEHICLE ACCIDENTS: A TECHNICAL COMPENDIUM
PT-34
504 pp., 33 Papers, ISBN 0-89883-122-9
Casebound 1989

ACCIDENT RECONSTRUCTION TECHNOLOGIES: PEDESTRIANS AND MOTORCYCLES IN AUTOMOTIVE COLLISIONS
PT-35
600 pp., 42 Papers, ISBN 1-56091-010-0
Casebound 1990

FUEL METHANOL - - A DECADE OF PROGRESS
PT-36
424 pp., 30 Papers, ISBN 1-56091-011-9
Casebound 1990

For further information or to order the above titles, contact:

Society of Automotive Engineers, Inc.
400 Commonwealth Drive
Warrendale, PA 15096 U.S.A.

Telephone: (412) 776-4970
FAX: (412 776-5760
Telex: 866-355